A PRIMER FOR
FINITE ELEMENTS
IN ELASTIC STRUCTURES

A PRIMER FOR FINITE ELEMENTS IN ELASTIC STRUCTURES

W. F. CARROLL
Professor Emeritus
Department of Civil and Environmental Engineering
University of Central Florida

JOHN WILEY & SONS, INC.
New York / Chichester / Weinheim / Brisbane / Singapore / Toronto

MATLAB is a registered trademark of
The Math Works, Inc.
24 Prime Park Way
Natick, MA 01760-1500
phone: (508) 647-7000
fax: (508) 647-7001
e-mail: info@mathworks.com
WWW: http://www.mathworks.com

This book is printed on acid-free paper.

This publication is designed to provide accurate and authoritative information in regard to the subject matter covered. It is sold with the understanding that the publisher is not engaged in rendering professional services. If professional advice or other expert assistance is required, the services of a competent professional person should be sought.

Library of Congress Cataloging-in-Publication Data:

Carroll, W. F. (William F.)
     A primer for finite elements in elastic structures / W. F. Carroll.
         p. cm.
     Includes bibliographical references and index.
     ISBN 0-471-28345-2 (cloth : alk. paper)
     1. Elastic analysis (Engineering) 2. Finite element method.
     I. Title.
     TA 653.C37 1998
     620.1'1232--dc21   98-13435

Printed in the United States of America.
10 9 8 7 6 5 4 3 2 1

For my wife, Kitty

CONTENTS

PREFACE

This book was written in response to an observed need for the clear presentation of fundamentals of the finite element method as it is applied to elastic structures. Most individual parts of the method are not difficult to grasp, but because there are many parts to it, all of which must be understood clearly and implemented correctly, its full understanding may be elusive. The method does not work if its execution is mostly correct; execution must be completely correct.

What students need is basic, informative, and coherent complementary course material to solidify their grasp of the method. Considerable effort is devoted to developing, explaining, and illustrating the mathematical and structural bases. The emphasis is on completeness and clarity of coverage as these topics apply. Though the theory is covered fully, some of its mathematical rigor is left to others in the belief that first-time students or unassisted readers ought not to clutter their basic understanding attempting to grasp the rigor of these matters. Instead, they need to know what the theory is about, how it applies, and that it is valid.

The first chapter is devoted to the matrix algebra and algebraic equation systems used in the finite element method. Though most of today's students enter a first course on the finite element method with some exposure to linear algebra, most need to review what they know. Moreover, most will also need to add to their knowledge some aspects of the subject which they have missed and which they will find important for coping with the development of procedures, methods, and equations pertinent to the subject. It is necessary for understanding that students possess a good measure of comprehension of and manipulative skill with the matrix algebra and equation systems of the finite element method. This chapter serves as a source reference for this mathematics.

Except for the most trivial structural problems, students must use matrix manipulation software to perform the addition, transposing, multiplication, inversion, partitioning, rearrangement, and assembly of matrices that the finite element method requires. There are many matrix manipulation computer codes, such as spreadsheets and mathematics codes, which may be used effectively to this end. The availability of the code, its ability to do what its users require, and their familiarity and facility with it should govern their choice of which code to use.

Chapters 2 through 5 present the basics of the finite element method in terms of one-dimensional (1D) truss and framework elements. The mathematics from Chapter 1 are employed with applicable concepts from mechanics and structural theory. These chapters should be studied carefully and mastered fully. Their focus is on fundamentals. The illustrations used are primarily 1D structural elements to facilitate understanding the fundamentals. The chapters following Chapter 5 treat 2D structural elements in depth. The geometry, transformations, and mapping employed with multidimensional elements can be grasped with less difficulty if the fundamentals of the finite element method are already in hand.

The book is intended for students at the senior or first-year graduate student level, taking essentially a first course in the finite element method. Prior study of linear algebra is most helpful. Prior study of basic structural analysis is vital. Some lack of basic knowledge of these subjects can be overcome, however, by mastering the fundamentals presented in the first five chapters reinforced by a good treatment of structural analysis from an appropriate text. Though the book is intended to complement a student's study in a formal course on the finite element method, the presentation is also suitable for diligent self-study. Readers who undertake to master the finite element method with this book through self-study should take the chapters in sequence.

Listed at the end of each chapter is a bibliography which readers may wish to consult. These lists are not intended to be comprehensive. Instead, they identify other treatments of the chapter's subjects presented at a comparable level of sophistication and rigor.

Also at each chapter's end are problems for readers to test their understanding of the text material. Some of these problems are similar to or extensions of the examples treated in the text. Others go beyond these examples. Some are theoretical; others have a practical focus. The latter should be accomplished with matrix manipulation software and a computer. Answers to most of the problems are consolidated at the end of the text.

For the convenience of readers, the appendix describes the use of the interactive matrix computation code, MATLAB®, which can implement efficiently the matrix operations of the finite element method. MATLAB is a registered trademark of The Math Works, Inc., 24 Prime Park Way, Natick, MA 01760-1500. MATLAB employs a step-by-step manual entry of data and commands, displaying results as the user operates the code. It also allows the development of programmed routines called m-files, written by the user, which accept simplified data entry, perform the required analyses automatically, and display results in a variety of formats. The appendix illustrates both modes of operation. The programmed routines, or m-files, illustrated for plane trusses, plane frames, plane stress thin plates, and thin flexure plates were written by the author. To download these m-files, see Section A.3 of the appendix.

A number of persons have contributed to this text. The most significant have been my students in the two courses on finite elements in structures I taught often over a period of eleven years. It has been my interaction with them that caused the development of lecture notes, course materials, and finally this manuscript. I also wish to acknowledge and express appreciation to my colleagues at the University of Central Florida for the time and effort they spent reviewing and commenting on this manuscript.

W. F. CARROLL

CHAPTER 1

FINITE ELEMENT METHOD PREREQUISITES

1.1. INTRODUCTION

The finite element method is a numerical procedure which can model the behavior of a structure with great accuracy. It can deal with complicated one-, two- and three-dimensional (1D, 2D, 3D) geometry, variations in material properties, and various structural restraints. It does so by reducing the governing partial differential equations for the structure to systems of linear algebraic equations. An alternative to dealing directly with the governing partial differential equations to obtain the algebraic equation systems is to treat the structure in terms of its energy and the work done on it. Either approach can produce the same systems of equations. These equations, however produced, are then solved using well-established procedures in linear algebra.

The understanding and application of the finite element method to structural systems requires the analyst to have a grasp of a number of topics in mathematics and structural mechanics. None require a comprehensive in depth mastery, but all require recognition of their limitations and a good level of manipulative skill in using the theorems and procedures as they apply to the finite element method.

Many of these supporting topics are most effectively explained in conjunction with the steps in the finite element method to which they apply. In most cases, that is what has been done in this book. An understanding of two topics is necessary to even begin to discuss the finite element method. These are selected subjects in matrix algebra and in linear algebraic equation systems. Both are presented briefly in this chapter.

1.2. MATRIX OPERATIONS FOR THE FINITE ELEMENT METHOD

Matrix algebra is a discipline in mathematics whose importance to the finite element method rests with the fact that the equations of the finite element method are systems of linear algebraic equations and that such systems are conveniently represented as matrix equations. Though one might deal directly with the algebraic equations, the shorthand representation by a single matrix equation for hundreds of algebraic equations saves an enormous amount of space, spares the reader long lists of algebraic equations, and facilitates understanding the role of the equation systems in the method. Moreover, important steps in the finite element method involve manipulating the equation systems: to develop other equation systems, to solve some equation systems for unknown quantities, and to use some of the other equation systems for the calculation of other quantities sought for the structure. The use of the theorems and rules of matrix algebra greatly facilitates explaining and performing these manipulations.

One should bear in mind that the operations performed on a matrix are based directly on the operations that are performed on an algebraic equation system. To this end, if the reader has a question regarding the validity of a particular matrix operation, or precisely what the operation should entail, the question should be resolved by stepping back to the relevant algebraic equation system and relating to the operations on that system that must be performed.

1.2.1. Basic Matrix Definitions

What follows are definitions and explanations of selected matrix terms and entities which are pertinent to the finite element method as treated in this book.

A *matrix* is a rectangular array of numbers or variables. Its number of rows is an integer m; its number of columns is an integer n. The size of a matrix is stated as its number of rows times its number of columns, or $m \times n$. A specific row in a matrix is often designated by an integer i; a specific column by an integer j.

The symbol for a matrix is a boldface letter. The array of entries, or the numbers or variables, in a matrix is enclosed in square brackets. Often the symbol for the matrix may also be enclosed in square brackets, for example, $[\mathbf{A}]$. As an illustration,

$$[\mathbf{A}] = \begin{bmatrix} 5 & 1 & 9 \\ 4 & 7 & 2 \\ 3 & 6 & 8 \\ 1 & 7 & 3 \end{bmatrix}.$$

Matrix **A** is a 4×3 array of numbers ($m = 4$, $n = 3$).

A *row matrix*, also called a *row vector*, is a matrix containing one row. A *column matrix*, also called a *column vector* or simply a *vector*, is a matrix containing one column. The symbol for a row matrix is the lower half of the square brackets enclosing the matrix symbol, for example $\lfloor \mathbf{A} \rfloor$. The symbol for a column matrix is braces enclosing the matrix symbol, for example $\{\mathbf{A}\}$.

A *square matrix* is a matrix with the same number of rows and columns. Square matrices are especially important in the finite element method, and there are a number of special square matrices that are employed in it.

A *diagonal matrix* is a square matrix whose entries are zeros, except along its major diagonal, where some or all of the entries are nonzero. The major diagonal entries are those that run from the top left of the array to the bottom right.

A *unit matrix* is a diagonal matrix whose major diagonal entries are all ones. The unit matrix plays an important role in matrix inversion operations and in the solution of algebraic equation systems. Note that the unit matrix may be any size, as long as it is square, with ones as entries on its diagonal and zeros elsewhere. The unit matrix has its own special symbol, **I**.

An *upper triangular matrix* is a square matrix with all zero entries below its major diagonal. There are no restrictions on the values of the entries on or above the diagonal. A *lower triangular matrix* is similar, except the zero entries are above the major diagonal. A *unit triangular matrix* is a triangular matrix whose diagonal entries are all ones. Triangular matrices are employed in the solution of algebraic equation systems and in finding the inverse of square matrices. The following is a unit upper triangular matrix:

$$[\mathbf{A}] = \begin{bmatrix} 1 & 2 & 9 \\ 0 & 1 & 3 \\ 0 & 0 & 1 \end{bmatrix}.$$

A *symmetric matrix* is a square matrix in which each entry in row i and column j above the major diagonal is equal to its corresponding entry in row j and column i below the major diagonal. In other words, entry a_{ij} in a symmetric matrix **A** is equal to entry a_{ji}. The entries along the major diagonal have no restrictions on their values. The following is a symmetric matrix:

$$[\mathbf{A}] = \begin{bmatrix} 5 & 1 & 9 & 6 \\ 1 & 7 & 2 & 7 \\ 9 & 2 & 8 & 3 \\ 6 & 7 & 3 & 2 \end{bmatrix}.$$

1.2.2. Basic Matrix Operations

Though the theory of matrix algebra is an important discipline and the subject of many definitive texts (see the bibliography at the end of the chapter), only some matrix operations are used extensively in the finite element method. Moreover, the mathematical rigor of their complete development is less important to the structural analyst than a clear understanding of their limitations and how they are applied. To this end, what follows are the definitions and descriptions of the matrix operations employed in this book. The rigor of their complete proof and development is left to others.

Matrix addition is defined as the sum of two matrices of identical size. It is accomplished by adding algebraically the corresponding entries of the two matrices. Corresponding entries are entries in each matrix which occupy the same position in their respective arrays. The result of matrix addition is a third matrix the same size as the original two, but of course different from either of those two. Matrix addition is not defined for two matrices which are of different size.

Matrix subtraction is defined similarly to matrix addition, except that the corresponding entries are subtracted algebraically. In order to obtain the correct algebraic sign for the resulting matrix in matrix subtraction, attention must be paid to which matrix is subtracted from which.

Multiplication of a matrix by a scalar is accomplished by multiplying each entry of the matrix by the same scalar constant. It is a simple concept, but a useful one for scaling the magnitudes of the entries of a matrix when matrices are compared to one another or assembled into a single matrix.

Matrix multiplication is the multiplication of two matrices by one another. Symbolically,

$$[\mathbf{A}][\mathbf{B}] = [\mathbf{C}]. \tag{1.1}$$

Matrix **A** is the premultiplier, matrix **B** is the postmultiplier, and matrix **C** is the product. To obtain the entry c_{ij} at row i and column j in the product matrix **C**, one must compute the sum of the products of the entries in row i of matrix **A** with the corresponding entries in column j of matrix **B**. That is,

$$c_{ij} = a_{i1}b_{1j} + a_{i2}b_{2j} + \cdots + a_{in}b_{nj}. \tag{1.2}$$

Matrix multiplication is defined when the premultiplier **A** has its number of columns equal to the number of rows in the postmultiplier **B**. If the premultiplier **A** is $m \times n$, then the postmultiplier **B** must be $n \times p$. There is no restriction on the number of rows m in the premultiplier **A** or the number of columns p in the postmultiplier **B**.

To develop the complete matrix **C**, equation 1.2 is executed $m \times p$ times. The resulting product matrix **C** will have m rows and p columns. For example,

$$\begin{bmatrix} 2 & 3 & 1 \\ 4 & 6 & 5 \end{bmatrix} \begin{bmatrix} 1 & 4 & 2 & 1 \\ 5 & 2 & 1 & 3 \\ 8 & 3 & 4 & 2 \end{bmatrix} = \begin{bmatrix} 25 & 17 & 11 & 13 \\ 74 & 43 & 34 & 32 \end{bmatrix}.$$

The premultiplier is 2×3 in size; the postmultiplier is 3×4. Thus the multiplication is defined with three columns in the premultiplier and three rows in the postmultiplier. The size of the product is 2×4. The first entry in row 1 of the product is obtained by multiplying row 1 of the premultiplier by column 1 of the postmultiplier. The expression, row times column, is meant to describe the sum of the products of the corresponding entries in the row and column. Row 1 of the premultiplier times column 1 of the postmultiplier is $2 \times 1 + 3 \times 5 + 1 \times 8 = 25$. The first entry in row 2 of the product is row 2 of the premultiplier times column 1 of the postmultiplier or, $4 \times 1 + 6 \times 5 + 5 \times 8 = 74$; and so forth.

The order of the multiplication cannot be reversed. In the example above, the multiplication would not be defined if the premultiplier was the 3×4 matrix and the postmultiplier was the 2×3 matrix. The number of columns in the premultiplier would not be the same as the number of rows in the postmultiplier in this instance. Even in those instances when the multiplication is defined for the reversed order of the two matrices, the resulting product matrix is not the same. The mathematical expression for this situation is that matrix multiplication is not commutative.

Matrix multiplication is distributive, however. For example,

$$[A][B] + [A][C] = [A]([B] + [C]). \tag{1.3}$$

Equation 1.3 indicates that the two matrix products may be formed and then added, or that the matrices **B** and **C** may be added and then the resulting sum premultiplied by matrix **A**. The result is the same in either sequence of operations as long as the order of the matrix multiplication involved is unchanged.

The associative property of matrix multiplication indicates that

$$([A][B])[C] = [A]([B][C]). \tag{1.4}$$

The product of matrices **A** and **B** may be formed and then matrix **C** may be pre-multiplied by this result, or the product of matrices **B** and **C** may be formed and then the result premultiplied by matrix **A**. Either sequence produces the same end result as long as the order of individual matrix multiplication is not reversed.

It is the definition of matrix multiplication that allows a system of algebraic equations to be represented by a single matrix equation. Equations 1.5 illustrate this:

$$a_{11}x_1 + a_{12}x_2 + a_{13}x_3 = b_1.$$

$$a_{21}x_1 + a_{22}x_2 + a_{23}x_3 = b_2. \tag{1.5a}$$

$$a_{31}x_1 + a_{32}x_2 + a_{33}x_3 = b_3;$$

$$\begin{bmatrix} a_{11} & a_{12} & a_{13} \\ a_{21} & a_{22} & a_{23} \\ a_{31} & a_{32} & a_{33} \end{bmatrix} \begin{Bmatrix} x_1 \\ x_2 \\ x_3 \end{Bmatrix} = \begin{Bmatrix} b_1 \\ b_2 \\ b_3 \end{Bmatrix}; \tag{1.5b}$$

$$[A]\{X\} = [B]. \tag{1.5c}$$

Equations 1.5a, 1.5b, and 1.5c each represent the same system of equations. Equations 1.5b and 1.5c afford the opportunity to deal with this system using matrix algebra. Note that matrix **A** must be the premultiplier of the vector **X**. Note also that a matrix equation is a statement of the equality of two matrices, which is true if the corresponding entries of each matrix are equal.

When the unit matrix is one of the multipliers in a matrix multiplication, the resulting product is the other multiplier. That is, $[A][I] = [A]$ and $[I][A] = [A]$. Hence the name unit matrix.

Transposing a matrix is an operation whereby each of the corresponding rows and columns of a matrix are interchanged. That is, row 1 becomes column 1, row 2 becomes column 2, and so on. If a matrix **A** has m rows and n columns, then its transpose $\mathbf{A}^T$ has n rows and m columns. The symbol for the transpose of a matrix is the symbol of the original matrix with the superscript T.

If a_{ij} is an entry in matrix $\mathbf{A}$ at the position corresponding to row i and column j, then it is also an entry in matrix $\mathbf{A}^T$ but will be in the position corresponding to row j and column i of matrix $\mathbf{A}^T$. An illustration of a matrix and its transpose is

$$[\mathbf{A}] = \begin{bmatrix} 3 & 6 \\ 2 & 1 \\ 5 & 4 \end{bmatrix}, \quad [\mathbf{A}^T] = \begin{bmatrix} 3 & 2 & 5 \\ 6 & 1 & 4 \end{bmatrix}.$$

The following transpose operations on matrices are useful:

$$\lfloor \mathbf{A}^T \rfloor = \{\mathbf{A}\}, \tag{1.6a}$$

$$\{\mathbf{A}^T\} = \lfloor \mathbf{A} \rfloor, \tag{1.6b}$$

$$[\mathbf{A}^T]^T = \mathbf{A}, \tag{1.6c}$$

$$[\mathbf{A}^T] + [\mathbf{B}^T] = [\mathbf{A} + \mathbf{B}]^T, \tag{1.6d}$$

$$[\mathbf{AB}]^T = [\mathbf{B}^T][\mathbf{A}^T]. \tag{1.6e}$$

The validity of equations 1.6a–c is clear from the definition of the transpose. Equation 1.6d should be clear from the definitions of the matrix addition and transpose operations.

Equation 1.6e may be rationalized by forming the matrix products

$$[\mathbf{C}]_{m \times n} = [\mathbf{A}]_{m \times o}[\mathbf{B}]_{o \times n}, \tag{1.7a}$$

$$[\mathbf{D}]_{n \times m} = [\mathbf{B}^T]_{n \times o}[\mathbf{A}^T]_{o \times m}. \tag{1.7b}$$

The matrix sizes are shown; the matrix multiplications are defined; matrix $\mathbf{A}^T$ is the transpose of matrix $\mathbf{A}$; and matrix $\mathbf{B}^T$ is the transpose of matrix $\mathbf{B}$.

An entry c_{ij} from matrix $\mathbf{C}$ in equation 1.7a may be regarded as row i of matrix $\mathbf{A}$ times column j of matrix $\mathbf{B}$. An entry d_{ji} from matrix $\mathbf{D}$ in equation 1.7b may be regarded as row j of matrix $\mathbf{B}^T$ times column i of matrix $\mathbf{A}^T$. However, because of the properties of the transpose, d_{ji} may also be regarded as column j of matrix $\mathbf{B}$ times row i of matrix $\mathbf{A}$. But this latter expression is also the expression for c_{ij}. Thus c_{ij} must equal d_{ji}. This is the necessary and sufficient condition for $\mathbf{C}^T = \mathbf{D}$, which establishes the validity of equation 1.6e.

Equation 1.6e is an exception to the prohibition on reversing the order of matrix multiplication. To form the transpose of a matrix product of two matrices in conformity with equation 1.6e, you must form the transpose of each multiplier, reverse the order of the matrices in the multiplication sequence, and then implement the matrix multiplication.

Matrix partitioning is the subdivision of a matrix into several smaller arrays, called submatrices. The usual partitioning of a matrix, in the finite element method, is into two or four submatrices. One usually partitions so that the number of columns to the left of the vertical partition in the coefficient matrix equals the number of rows above the horizontal partition. The advantage of partitioning this way is that the resulting submatrices may be manipulated with matrix operations in the same manner that the entries of a matrix are manipulated. For example, the equation system represented by equations 1.5 might be partitioned and represented in terms of eight submatrices and two submatrix equations as

$$\left[\begin{array}{cc|c} a_{11} & a_{12} & a_{13} \\ a_{21} & a_{22} & a_{23} \\ \hline a_{31} & a_{32} & a_{33} \end{array}\right] \left\{\begin{array}{c} x_1 \\ x_2 \\ \hline x_3 \end{array}\right\} = \left\{\begin{array}{c} b_1 \\ b_2 \\ \hline b_3 \end{array}\right\}, \tag{1.8a}$$

$$\left[\begin{array}{c|c} \mathbf{A}_{11} & \mathbf{A}_{12} \\ \hline \mathbf{A}_{21} & \mathbf{A}_{22} \end{array}\right] \left\{\begin{array}{c} \mathbf{X}_1 \\ \hline \mathbf{X}_2 \end{array}\right\} = \left\{\begin{array}{c} \mathbf{B}_1 \\ \hline \mathbf{B}_2 \end{array}\right\}, \tag{1.8b}$$

$$[\mathbf{A}_{11}]\{\mathbf{X}_1\} + [\mathbf{A}_{12}]\{\mathbf{X}_2\} = [\mathbf{B}_1], \tag{1.8c}$$

$$[\mathbf{A}_{21}]\{\mathbf{X}_1\} + [\mathbf{A}_{22}]\{\mathbf{X}_2\} = [\mathbf{B}_2]. \tag{1.8d}$$

The submatrices in equations 1.8b–d correspond to the partitioned submatrices in equation 1.8a. For example,

$$[\mathbf{A}_{11}] = \left[\begin{array}{cc} a_{11} & a_{12} \\ a_{21} & a_{22} \end{array}\right], \qquad [\mathbf{A}_{21}] = \lfloor a_{31} \quad a_{32} \rfloor, \qquad \text{etc.}$$

The ability to deal with a system of equations, such as equations 1.5a, either as one matrix equation (equation 1.5c) or two matrix equations (equations 1.8c and 1.8d) provides the analyst with a convenient means for solving an equation system when some of the unknowns are in the load vector **B** and the remainder are in the displacement vector **X**. As we shall see, this is frequently the case when applying the finite element method to structures. Though not especially useful when the system comprises only a few algebraic equations, when it has tens or hundreds of such equations, its treatment by matrix and submatrix equations is advantageous.

The *matrix inverse* is a square matrix derived from another square matrix of the same size. The symbol for the inverse of matrix **A** is $\mathbf{A}^{-1}$. The inverse is defined by the expression $[\mathbf{A}][\mathbf{B}] = [\mathbf{I}]$. That is, if the product of two square matrixes is the unit matrix, then one matrix is the inverse of the other. Thus,

$$[\mathbf{A}][\mathbf{A}^{-1}] = [\mathbf{B}^{-1}][\mathbf{B}] = [\mathbf{I}]. \tag{1.9a}$$

If equation 1.5c is premultiplied by matrix $\mathbf{A}^{-1}$, then

$$[\mathbf{A}^{-1}][\mathbf{A}]\{\mathbf{X}\} = [\mathbf{A}^{-1}][\mathbf{B}]. \tag{1.9b}$$

Substituting equation 1.9a into 1.9b yields

$$[\mathbf{I}]\{\mathbf{X}\} = [\mathbf{A}^{-1}][\mathbf{B}]. \qquad (1.9c)$$

Since a matrix multiplied by the unit matrix is not changed,

$$\{\mathbf{X}\} = [\mathbf{A}^{-1}][\mathbf{B}]. \qquad (1.10)$$

If the unknowns in the equation system of equation 1.5 are represented by x_i and the vector $\mathbf{X}$, then equation 1.10 represents the solution for the unknowns. One must obtain the inverse of the known matrix $\mathbf{A}$ and premultiply the known matrix $\mathbf{B}$ by this inverse. The solution of an algebraic equation system using matrix algebra then becomes focused on obtaining the inverse of the coefficient matrix of the matrix equation which represents the algebraic equation system.

Not all square matrices have an inverse. If the solution to a specific equation system as represented by equations 1.5c does not exist, then the specific solution represented by equation 1.10 cannot be valid. Since the specific matrices $\mathbf{A}$ and $\mathbf{B}$ would appear to be valid in equation 1.5c and matrix $\mathbf{B}$ in equation 1.10, the logical conclusion to draw is that there is no matrix $\mathbf{A}^{-1}$ for the specific system which satisfies equation 1.10. A square matrix for which there is no inverse is called *singular*.

Some methods for obtaining the inverse of a square matrix, or determining if the matrix is singular, are discussed later.

An *orthogonal matrix* is a square matrix whose inverse is its transpose, $\mathbf{A}^{-1} = \mathbf{A}^{T}$, or

$$[\mathbf{A}][\mathbf{A}^{T}] = [\mathbf{I}] \quad \text{or} \quad [\mathbf{A}^{T}][\mathbf{A}] = [\mathbf{I}]. \qquad (1.11)$$

Most square matrices whose inverses are needed are not orthogonal, but one important one employed in the finite element method is. It relates to coordinate transformation and is discussed later.

Clearly if one can recognize that a matrix is orthogonal, its inverse, which is its transpose, is easily obtained. The tests to determine if a square matrix is orthogonal may be illustrated by expanding equation 1.11. Though the matrices used in the illustration below are 3×3, they may be any size:

$$\begin{bmatrix} a_{11} & a_{12} & a_{13} \\ a_{21} & a_{22} & a_{23} \\ a_{31} & a_{32} & a_{33} \end{bmatrix} \begin{bmatrix} a_{11} & a_{21} & a_{31} \\ a_{12} & a_{22} & a_{32} \\ a_{13} & a_{23} & a_{33} \end{bmatrix} = \begin{bmatrix} 1 & 0 & 0 \\ 0 & 1 & 0 \\ 0 & 0 & 1 \end{bmatrix}. \qquad (1.12)$$

The products obtained from the multiplication of any row of matrix $\mathbf{A}$ in equation 1.12 with the column in matrix $\mathbf{A}^{T}$ whose number is the same as the row number from matrix $\mathbf{A}$ indicate that the sum of the squares of any row or any column in an orthogonal matrix must equal 1. The products obtained from the multiplication of any row in matrix $\mathbf{A}$ with a column in matrix $\mathbf{A}^{T}$ whose number is different from the row number from matrix $\mathbf{A}$ indicate that the products of any two different rows or any two different columns in an orthogonal matrix must be zero. A square matrix which satisfies these two conditions is orthogonal and its inverse is its transpose.

1.2.3. Determinants

The determinant of a square matrix is a single number, a scalar quantity. Its symbol is either det[**A**] or |**A**|. Determinants of 2×2 or 3×3 matrices are usually calculated by procedures which employ summing the products of diagonals and taking the differences. For example,

$$\det[\mathbf{A}] = \begin{vmatrix} a_{11} & a_{12} \\ a_{21} & a_{22} \end{vmatrix} = (a_{11})(a_{22}) - (a_{12})(a_{21}), \tag{1.13a}$$

$$\det[\mathbf{A}] = \begin{vmatrix} a_{11} & a_{12} & a_{13} \\ a_{21} & a_{22} & a_{23} \\ a_{31} & a_{32} & a_{33} \end{vmatrix}$$

$$= (a_{11})(a_{22})(a_{33}) + (a_{12})(a_{23})(a_{31}) + (a_{13})(a_{21})(a_{32})$$
$$- (a_{13})(a_{22})(a_{31}) - (a_{11})(a_{23})(a_{32}) - (a_{12})(a_{21})(a_{33}). \tag{1.13b}$$

Equations 1.13 are valid only for 2×2 and 3×3 determinants; they are not valid for higher order determinants. For higher order determinants, and if desired also for 2×2 and 3×3 determinants, the Laplace expansion formula, described below in terms of cofactors, must be employed.

A *minor* of a square matrix, $\bar{M}_{ij}$, is the determinant calculated from the matrix with row i and column j deleted. The symbol for a minor of a matrix is $\bar{M}$, with subscripts which indicate the rows and columns deleted from the original matrix when computing the necessary determinant. If matrix **A** is $m \times m$, then there are m^2 minors $\bar{M}_{ij}$, one for each entry a_{ij} of matrix **A**.

A *cofactor* of a square matrix, $\bar{A}_{ij}$, is the corresponding minor with its sign modified as follows:

$$\bar{A}_{ij} = (-1)^{(i+j)} \bar{M}_{ij}. \tag{1.14}$$

The symbol for a cofactor is the letter symbol of its matrix with an overbar, with the subscripts of the corresponding minor which indicate the row and column deleted from the original matrix when computing the necessary determinant. There is of course one cofactor for each entry of a matrix.

The *cofactor matrix* $\bar{\mathbf{A}}$ of matrix **A** is the matrix of cofactors calculated in accordance with equation 1.14. The subscripts of the cofactor entries $\bar{A}_{ij}$ also indicate the position of each cofactor in the cofactor matrix $\bar{\mathbf{A}}$. If matrix **A** is $m \times m$, then its cofactor matrix $\bar{\mathbf{A}}$ is $m \times m$, with each of its entries a determinant of order $(m-1) \times (m-1)$. An illustration of a 3×3 matrix and its 3×3 cofactor matrix calculated using equation 1.14 is

$$[\mathbf{A}] = \begin{bmatrix} 1 & 2 & 3 \\ 5 & 2 & 1 \\ 6 & 1 & 3 \end{bmatrix}, \qquad [\bar{\mathbf{A}}] = \begin{bmatrix} 5 & -9 & -7 \\ -3 & -15 & 11 \\ -4 & 14 & -8 \end{bmatrix}.$$

The *adjoint of matrix* **A** is the transpose of the cofactor matrix of matrix **A**. Its symbol is adj **A** or $\bar{\mathbf{A}}^T$. The adjoint is an important player in determining the inverse of a matrix.

The *Laplace expansion formula* for the calculation of the determinant of a square matrix is described in terms of the entries of the matrix and the corresponding entries of its cofactor matrix. The Laplace expansion formula is a general formula which may be applied to any size square matrix. To apply it for the determinant of matrix **A**, any row or column of matrix **A** and the corresponding row or column of its cofactor matrix $\bar{\mathbf{A}}$ are multiplied. Thus,

$$[\mathbf{A}] = \begin{bmatrix} a_{11} & a_{12} & a_{13} \\ a_{21} & a_{22} & a_{23} \\ a_{31} & a_{32} & a_{33} \end{bmatrix}, \quad [\bar{\mathbf{A}}] = \begin{bmatrix} \bar{A}_{11} & \bar{A}_{12} & \bar{A}_{13} \\ \bar{A}_{21} & \bar{A}_{22} & \bar{A}_{23} \\ \bar{A}_{31} & \bar{A}_{32} & \bar{A}_{33} \end{bmatrix};$$

$$\det[\mathbf{A}] = a_{11}\bar{A}_{11} + a_{12}\bar{A}_{12} + a_{13}\bar{A}_{13}, \quad \text{or}$$

$$= a_{21}\bar{A}_{21} + a_{22}\bar{A}_{22} + a_{23}\bar{A}_{23}, \quad \text{or}$$

$$= a_{11}\bar{A}_{11} + a_{21}\bar{A}_{21} + a_{31}\bar{A}_{31}, \quad \text{etc.}$$

The formula applied to the first row of the 3×3 matrix **A** of numbers, illustrated above, and the first row of its cofactor matrix $\bar{\mathbf{A}}$ results in $(1)(5) + (2)(-9) + (3)(-7) = -34$. Applying it to any other corresponding rows or columns produces the same value for its determinant.

Some important properties of determinants are as follows:

- $\det[\mathbf{A}] = \det[\mathbf{A}^T]$.
- If all the entries in any row or any column of matrix [**A**] are 0's, $\det[\mathbf{A}] = 0$.
- If any two rows or any two columns in matrix [**A**] are interchanged, the determinant of the resulting modified matrix is $-\det[\mathbf{A}]$.
- If any two rows or any two columns of matrix [**A**] are identical, then $\det[\mathbf{A}] = 0$.
- If a multiple of any row(or column) in matrix [**A**] is added to a multiple of any other row (or column) in the matrix, the value of $\det[\mathbf{A}]$ is unchanged.
- $\det[[\mathbf{A}][\mathbf{B}]] = (\det[\mathbf{A}])(\det[\mathbf{B}])$.
- $\det[\mathbf{A}^{-1}] = (\det[\mathbf{A}])^{-1}$.

The first three properties listed are easily deduced from the characteristics of the matrices described and the application of the Laplace expansion formula to them. The fourth property is based on the third and the application of the Laplace expansion formula. Proof of the last three properties is more complicated and will not be repeated here. Each of the properties listed is easily demonstrated for a square matrix of numerical values using a computer and software which calculates the determinant.

1.2.4. Adjoint Method of Matrix Inversion

The basic method for determining the inverse of a matrix **A** involves $\det[\mathbf{A}]$ and adj[**A**]. This method will be referred as the *adjoint* method. Other methods which are computationally more efficient than the adjoint method and thus used extensively in the finite element method are the *Gauss–Jordan* elimination and back-substitution method and the *Choleski* substitution method.

The *adjoint method* is important to the understanding of matrix inverses and is useful when the entries of the inverse involve variables or functions rather than numbers. To develop the adjoint method, first form the equation system

$$[\mathbf{P}] = [\mathbf{A}] \, \text{adj}[\mathbf{A}] = [\mathbf{A}][\bar{\mathbf{A}}^T]. \tag{1.15a}$$

$$
\begin{bmatrix}
p_{11} & p_{12} & p_{13} \\
p_{21} & p_{22} & p_{23} \\
p_{31} & p_{32} & p_{33}
\end{bmatrix}
=
\begin{bmatrix}
a_{11} & a_{12} & a_{13} \\
a_{21} & a_{22} & a_{23} \\
a_{31} & a_{32} & a_{33}
\end{bmatrix}
\begin{bmatrix}
\bar{A}_{11} & \bar{A}_{21} & \bar{A}_{31} \\
\bar{A}_{12} & \bar{A}_{22} & \bar{A}_{32} \\
\bar{A}_{13} & \bar{A}_{23} & \bar{A}_{33}
\end{bmatrix}
\tag{1.15b}
$$

The entry p_{11} equals row 1 of matrix $\mathbf{A}$ times column 1 of adj[$\mathbf{A}$] or

$$p_{11} = a_{11}\bar{A}_{11} + a_{12}\bar{A}_{12} + a_{13}\bar{A}_{13}. \tag{1.15c}$$

Equation 1.15c is an application of the Laplace expansion formula on matrix $\mathbf{A}$ for det[$\mathbf{A}$]. Thus $p_{11} = \det[\mathbf{A}]$. Similarly, p_{22} and p_{33} also equal det[$\mathbf{A}$]. Indeed the diagonal entries of matrix $\mathbf{P}$, whatever its size, are equal to det[$\mathbf{A}$].

The off-diagonal entries of matrix $\mathbf{P}$ may be determined by considering the entry p_{12} of equation 1.15b as follows:

$$p_{12} = a_{11}\bar{A}_{21} + a_{12}\bar{A}_{22} + a_{13}\bar{A}_{23}. \tag{1.15d}$$

If matrix $\mathbf{A}$ is now modified so that row 2 is identical to row 1, the cofactors in column 2 of adj[$\mathbf{A}$] are unchanged since they are determined by deleting row 2 of matrix $\mathbf{A}$, the row that was modified. The other columns of adj[$\mathbf{A}$], of course, will change. Equation 1.15b so modified is

$$
\begin{bmatrix}
-- & p_{12} & -- \\
-- & p_{mm} & -- \\
-- & -- & --
\end{bmatrix}
=
\begin{bmatrix}
a_{11} & a_{12} & a_{13} \\
a_{11} & a_{12} & a_{13} \\
a_{31} & a_{32} & a_{33}
\end{bmatrix}
\begin{bmatrix}
-- & \bar{A}_{21} & -- \\
-- & \bar{A}_{22} & -- \\
-- & \bar{A}_{23} & --
\end{bmatrix}.
\tag{1.15e}
$$

The entries p_{12}, those for rows 1 and 3 of matrix $\mathbf{A}$, and those for column 2 of the adj[$\mathbf{A}$] are shown, since they are unchanged. Notice that in equation 1.15e

$$p_{mm} = a_{11}\bar{A}_{21} + a_{12}\bar{A}_{22} + a_{13}\bar{A}_{23}. \tag{1.15f}$$

The right-hand side of equation 1.15f involves terms whose values have not been affected by the modifications to matrix $\mathbf{A}$. Thus p_{mm} from equations 1.15e and 1.15f must be equal to the original value of p_{12} as given in equation 1.15d. The right-hand side of equation 1.15f is also an application of the Laplace expansion formula for the determinant of the modified matrix $\mathbf{A}$ of equation 1.15e. But since this modified matrix **A** has two identical rows (rows 1 and 2), its determinant is equal to zero. Therefore p_{mm} and the original p_{12} must equal zero.

Similar reasoning can be undertaken for each off-diagonal entry of matrix **P** in equation 1.15b to show that each must be zero.

Matrix **P** is thus a diagonal matrix whose diagonal entries are each det[**A**]. Referring to the definition of multiplication of a matrix by a scalar, matrix **P** may also be regarded as the unit matrix **I** multiplied by a scalar which is equal to det[**A**], or

$$[\mathbf{P}] = \begin{bmatrix} \det[\mathbf{A}] & 0 & 0 \\ 0 & \det[\mathbf{A}] & 0 \\ 0 & 0 & \det[\mathbf{A}] \end{bmatrix} = \det[\mathbf{A}] \begin{bmatrix} 1 & 0 & 0 \\ 0 & 1 & 0 \\ 0 & 0 & 1 \end{bmatrix}$$

$$= \det[\mathbf{A}][\mathbf{I}]. \tag{1.15g}$$

Substituting equation 1.15g into equation 1.15a and comparing the result to equation 1.9 give

$$\det[\mathbf{A}][\mathbf{I}] = [\mathbf{A}] \, \text{adj}[\mathbf{A}],$$

$$\frac{[\mathbf{A}] \, \text{adj}[\mathbf{A}]}{\det[\mathbf{A}]} = [\mathbf{I}].$$

Then

$$[\mathbf{A}^{-1}] = \frac{\text{adj}[\mathbf{A}]}{\det[\mathbf{A}]}. \tag{1.15h}$$

Equation 1.15h is a basic expression for calculating the inverse of a square matrix. It also indicates the test for the existence of the inverse. The determinant of the matrix cannot be zero. If it is, equation 1.15h cannot be executed. If the determinant is zero, the inverse does not exist and the matrix is called singular.

As an illustration of the use of the matrix inverse, consider the following equation system and solve for the values of the unknowns in vector {**X**}:

$$\begin{array}{rcrcrcr} 2x_1 & + & x_2 & - & 2x_3 & = & 24, \\ -2x_1 & + & 3x_2 & + & 3x_3 & = & -8, \\ 4x_1 & + & 2x_2 & - & x_3 & = & -12; \end{array} \tag{1.16a}$$

$$\begin{bmatrix} 2 & 1 & -2 \\ -2 & 3 & 3 \\ 4 & 2 & -1 \end{bmatrix} \begin{Bmatrix} x_1 \\ x_2 \\ x_3 \end{Bmatrix} = \begin{Bmatrix} 24 \\ -8 \\ -12 \end{Bmatrix}; \tag{1.16b}$$

$$[\mathbf{A}]\{\mathbf{X}\} = \{\mathbf{B}\}. \tag{1.16c}$$

Applying equation 1.14 to matrix $\mathbf{A}$, the cofactor matrix of $\mathbf{A}$ is given as

$$[\bar{\mathbf{A}}] = \begin{bmatrix} -9 & 10 & -16 \\ -3 & 6 & 0 \\ 9 & -2 & 8 \end{bmatrix}. \tag{1.16d}$$

The $\det[\mathbf{A}]$ is obtained using the Laplace expansion formula as

$$\det[\mathbf{A}] = (-2)(-3) + (3)(6) + (3)(0) = 24.$$

Substituting into equation 1.15h yields

$$[\mathbf{A}^{-1}] = \frac{1}{24} \begin{bmatrix} -9 & -3 & 9 \\ 10 & 6 & -2 \\ -16 & 0 & 8 \end{bmatrix} = \begin{bmatrix} -\frac{3}{8} & -\frac{1}{8} & \frac{3}{8} \\ \frac{5}{12} & \frac{1}{4} & -\frac{1}{12} \\ -\frac{2}{3} & 0 & \frac{1}{3} \end{bmatrix}. \tag{1.16e}$$

The solution of equations 1.16b is given by

$$\{\mathbf{X}\} = [\mathbf{A}^{-1}]\{\mathbf{B}\},$$

or

$$\begin{Bmatrix} x_1 \\ x_2 \\ x_3 \end{Bmatrix} = \begin{bmatrix} -\frac{3}{8} & -\frac{1}{8} & \frac{3}{8} \\ \frac{5}{12} & \frac{1}{4} & -\frac{1}{12} \\ -\frac{2}{3} & 0 & \frac{1}{3} \end{bmatrix} \begin{Bmatrix} 24 \\ -8 \\ -12 \end{Bmatrix} = \begin{Bmatrix} -12.5 \\ 9 \\ -20 \end{Bmatrix}. \tag{1.16f}$$

Thus $x_1 = -12.5$, $x_2 = 9$, and $x_3 = -20$.

It is interesting to consider the number of arithmetic operations that are required to invert a square matrix of size $n \times n$ using the adjoint method. Considering equation 1.15h, one must undertake the operations to calculate the n calculations to implement the Laplace expansion rule for $\det[\mathbf{A}]$, the n^2 determinants, each $(n-1) \times (n-1)$, to obtain adj$[\mathbf{A}]$, and then the n^2 divisions when $\det[\mathbf{A}]$ is divided into each entry of adj$[\mathbf{A}]$.

In accordance with the Laplace expansion formula, the number of operations D_n to calculate a determinant of order n is the sum of the following:

n	Multiplications of a_{ij} and $\bar{A}_{ij}$ from a row of matrix $\mathbf{A}$ and the corresponding row of $\bar{\mathbf{A}}$
$n-1$	Additions of these products
n	Times the number of operations to calculate the n $\bar{A}_{ij}$ terms, each a determinant of order $(n-1) \times (n-1)$

Thus,

$$D_n = n + (n-1) + nD_{n-1} = (2n-1) + nD_{n-1}. \tag{1.17a}$$

The number of operations I_n to invert an $n \times n$ square matrix then is the sum of the following:

D_n	Number of operations for one $n \times n$ determinant
n^2	Times D_{n-1} to compute the terms in adj[**A**] less the nD_{n-1} operations done to determine D_n
n^2	Divisions of det[**A**] into each entry of adj[**A**]

Using equation 1.17a,

$$I_n = (2n - 1) + nD_{n-1} + n^2 D_{n-1} - nD_{n-1} + n^2$$
$$= (2n - 1) + n^2(D_{n-1} + 1). \tag{1.17b}$$

Equation 1.17b will understate the number of operations for the inversion slightly, since no account has been taken of the sign changes needed to form adj[**A**] (see equation 1.14) and none has been taken for the effort to arrange adj[**A**] as the transpose of cofactor matrix $\bar{\mathbf{A}}$.

Using the fact that $D_n = 3$ for a 2×2 determinant, $(a_{11})(a_{22}) - (a_{12})(a_{21})$, a table of the number of operations for matrix inversion which employs the adjoint method can be established using equations 1.17, as shown below:

n	D_n	I_n
2	3	
3	14	41
4	63	247
5	324	1,609
6	1,955	11,711
7	13,698	95,857
8	105,599	876,751
9	986,408	8,877,617
10	9,864,099	98,640,919

Obviously the effort involved in inverting a matrix with the adjoint method increases sharply as the size of the matrix increases. A 10×10 square matrix would represent an almost trivial model of a realistic structural system, yet the number of arithmetic operations to be performed just for a matrix inversion is significant. A system of equations involving 100 or more equations, representing a more realistic structural system, demands a more efficient solution algorithm if it is to be feasible.

1.2.5. Gauss–Jordan Method of Matrix Inversion

The Gauss–Jordan elimination back-substitution method is an established and efficient algorithm for the inversion of a matrix **A**. The concept may be understood from the matrix equation

$$\mathbf{AX} = \mathbf{I},\tag{1.18a}$$

$$\begin{bmatrix} a_{11} & a_{12} & a_{13} \\ a_{21} & a_{22} & a_{23} \\ a_{31} & a_{32} & a_{33} \end{bmatrix} \begin{bmatrix} x_{11} & x_{12} & x_{13} \\ x_{21} & x_{22} & x_{23} \\ x_{31} & x_{32} & x_{33} \end{bmatrix} = \begin{bmatrix} 1 & 0 & 0 \\ 0 & 1 & 0 \\ 0 & 0 & 1 \end{bmatrix}.\tag{1.18b}$$

Matrix $\mathbf{X}$ is the unknown inverse of matrix $\mathbf{A}$. Equations 1.18 represent a system of nine linear algebraic equations for which there are nine unknown quantities x_{ij}. Algebraic operations will be performed on these equations to solve for these unknowns.

In the first phase of the algorithm, the operations transform matrix $\mathbf{A}$ to a unit upper triangular matrix. To retain the validity of the equation system, each algebraic operation performed on the rows of matrix $\mathbf{A}$ must also be performed on the corresponding rows of matrix $\mathbf{I}$. These steps involve eliminating, in a systematic manner, the unknown x_{21} from the second algebraic equation of the system, the unknown x_{31} and x_{32} from the third equation of the system, the unknowns x_{41}, x_{42} and x_{43} from the fourth equation for a larger system of equations than illustrated, and so on. The diagonal terms x_{ii} of the coefficient matrix are made unity by dividing each of the corresponding equations by its diagonal term. The resulting equation system is

$$\mathbf{UX} = \mathbf{B},\tag{1.18c}$$

$$\begin{bmatrix} 1 & u_{12} & u_{13} \\ 0 & 1 & u_{23} \\ 0 & 0 & 1 \end{bmatrix} \begin{bmatrix} x_{11} & x_{12} & x_{13} \\ x_{21} & x_{22} & x_{23} \\ x_{31} & x_{32} & x_{33} \end{bmatrix} = \begin{bmatrix} b_{11} & b_{12} & b_{13} \\ b_{21} & b_{22} & b_{23} \\ b_{31} & b_{32} & b_{33} \end{bmatrix}.\tag{1.18d}$$

The second phase of the algorithm includes performing algebraic operations on equation 1.18d until the coefficient matrix $\mathbf{U}$ is transformed to the unit matrix. These steps involve systematic back substitution of values of x_i into the equation system 1.18d as they become known. In matrix operations, the process is accomplished using the last row of equation 1.18d in algebraic operations on the next to the last row to transform the entry to the right of the diagonal of matrix $\mathbf{U}$ to zero. The operations performed on this next to the last row of matrix $\mathbf{U}$ must also be performed on the next to the last row of matrix $\mathbf{B}$ to retain the validity of the equation system. To eliminate the nonzero entries to the right of the diagonal in the next higher row of matrix $\mathbf{U}$, both the last row and the newly modified next to the last row of matrixes $\mathbf{U}$ and $\mathbf{B}$ must be employed on this next higher row. The process is continued for each succeeding higher row until matrix $\mathbf{U}$ has been transformed to the unit matrix and matrix $\mathbf{B}$ has become matrix $\mathbf{C}$. The resulting equation system is

$$\mathbf{IX} = \mathbf{C},\tag{1.18e}$$

$$\begin{bmatrix} 1 & 0 & 0 \\ 0 & 1 & 0 \\ 0 & 0 & 1 \end{bmatrix} \begin{bmatrix} x_{11} & x_{12} & x_{13} \\ x_{21} & x_{22} & x_{23} \\ x_{31} & x_{32} & x_{33} \end{bmatrix} = \begin{bmatrix} c_{11} & c_{12} & c_{13} \\ c_{21} & c_{22} & c_{23} \\ c_{31} & c_{32} & c_{33} \end{bmatrix}. \qquad (1.18f)$$

Observing equations 1.18a and 1.18e, matrix C equals matrix X and matrix X is matrix A^{-1}.

Matrix inversion by the Gauss–Jordan algorithm is illustrated in the following example (see also matrix A of equation 1.16b). Numerals shown in angle brackets identify the entries in the entire row of both matrices:

$$[A] = \begin{bmatrix} 2 & 1 & -2 \\ -2 & 3 & 3 \\ 4 & 2 & -1 \end{bmatrix}.$$

Row Number	Matrix A			Matrix I			Row Operations
$\langle 1 \rangle$	2	1	-2	1	0	0	
$\langle 2 \rangle$	-2	3	3	0	1	0	
$\langle 3 \rangle$	4	2	-1	0	0	1	
							Elimination
$\langle 4 \rangle$	1	$\frac{1}{2}$	1	$\frac{1}{2}$	0	0	$\langle 1 \rangle / 2$
$\langle 5 \rangle$	0	4	1	1	1	0	$\langle 2 \rangle + \langle 4 \rangle \times 2$
$\langle 6 \rangle$	0	0	3	-2	0	1	$\langle 3 \rangle - \langle 4 \rangle \times 4$
$\langle 7 \rangle$	0	1	$\frac{1}{4}$	$\frac{1}{4}$	$\frac{1}{4}$	0	$\langle 5 \rangle / 4$
$\langle 8 \rangle$	0	0	3	-2	0	1	$\langle 6 \rangle + \langle 7 \rangle \times 0$
$\langle 9 \rangle$	0	0	1	$-\frac{2}{3}$	0	$\frac{1}{3}$	$\langle 8 \rangle / 3$

	Matrix U			Matrix B			
$\langle 4 \rangle$	1	$\frac{1}{2}$	-1	$\frac{1}{2}$	0	0	
$\langle 7 \rangle$	0	1	$\frac{1}{4}$	$\frac{1}{4}$	$\frac{1}{4}$	0	
$\langle 9 \rangle$	0	0	1	$-\frac{2}{3}$	0	$\frac{1}{3}$	
							Back substitution
$\langle 10 \rangle$	0	1	0	$\frac{5}{12}$	$\frac{1}{4}$	$-\frac{1}{12}$	$\langle 7 \rangle - \langle 9 \rangle \times (\frac{1}{4})$
$\langle 11 \rangle$	1	0	-1	$\frac{7}{24}$	$-\frac{1}{8}$	$\frac{1}{24}$	$\langle 4 \rangle - \langle 10 \rangle \times (\frac{1}{2})$
$\langle 12 \rangle$	1	0	0	$-\frac{3}{8}$	$-\frac{1}{8}$	$\frac{3}{8}$	$\langle 11 \rangle + \langle 9 \rangle \times 1$

	Matrix I			Matrix $C = X = A^{-1}$			
$\langle 12 \rangle$	1	0	0	$-\frac{3}{8}$	$-\frac{1}{8}$	$\frac{3}{8}$	(see matrix A^{-1}
$\langle 10 \rangle$	0	1	0	$\frac{5}{12}$	$\frac{1}{4}$	$-\frac{1}{12}$	of equation 1.16d)
$\langle 9 \rangle$	0	0	1	$-\frac{2}{3}$	0	$\frac{1}{3}$	

The number of arithmetic operations required to invert a matrix of size $n \times n$ is different for the Gauss–Jordan method from what it is for the adjoint method. An estimate of that number can be made as follows. During the elimination phase on the 3×3 matrix in the example above the operations performed were:

Rows	Number of Operations
4, 7, 9	6
5, 6, 8	12

One can observe that if the matrix were $n \times n$ instead of 3×3, the operations performed would be as follows:

Rows	Number of Operations
First n	$2n$
Next $n - 1$	$4n$
Next $n - 2$	$4n$
$\vdots$	$\vdots$
Last 1	$4n$

The sum of the number of row operations from the last to the second is $1 + 2 + \cdots + (n - 2) + (n - 1)$, which may be regarded as the sum of an arithmetic progression. Its total is equal to its average value times the number of terms, or

$$S_E = (n - 1)\{\tfrac{1}{2}[1 + (n - 1)]\} = \tfrac{1}{2}n(n - 1). \tag{1.19a}$$

The number of operations during the elimination phase then is

$$E_n = (2n)(n) + (4n)(S_E),$$

or

$$E_n = 2n^2 + (4n)[\tfrac{1}{2}n(n - 1)] = 2n^3. \tag{1.19b}$$

During the back-substitution phase, the number of operations are as follows

Rows	Number of Operations
1	$4n$
2	$4n$
$\vdots$	$\vdots$
$n - 2$	$4n$
$n - 1$	$4n$

The sum of the number of row operations is given by equation 1.19a; thus the number of operations during back substitution is

$$B_n = (4n)[\tfrac{1}{2}n(n - 1)] = 2n^3 - 2n^2. \tag{1.19c}$$

The total number of arithmetic operations for the Gauss–Jordan algorithm in the inversion of an $n \times n$ matrix is estimated by

$$GJ_n = E_n + B_n = 4n^3 - 2n^2. \tag{1.19d}$$

The table for the number of operations for matrix inversion developed earlier for the adjoint method is reproduced below with numbers of operations for the Gauss–Jordan algorithm added:

n	D_n	I_n	GJ_n	GJ_n as percent of I_n
2	3	—	24	
3	14	41	90	225
4	63	247	224	90.7
5	324	1,609	450	28.0
6	1,955	11,711	792	6.76
7	13,698	95,857	1274	1.33
8	105,599	876,751	1920	0.219
9	986,408	8,877,617	2754	0.031
10	9,864,099	98,640,919	3800	0.004

Clearly the Gauss–Jordan algorithm for matrix inversion enjoys an enormous advantage over the adjoint method as matrices exceed 4 × 4 in size. For very large systems of equations it is a feasible method whereas the adjoint method is not.

The Gauss–Jordan method, however, cannot distinguish between singular and nonsingular matrices. It requires each row of matrix entries to be divided by a diagonal value of the matrix **A** or its subsequently modified diagonal entries. If the diagonal value happens to be zero, the algorithm fails unless special actions are taken. When coding the algorithm for use on a computer, these actions must be incorporated or the algorithm may not produce the inverse, even though the inverse exists. Indeed the user may not know whether the matrix is singular or if the algorithm being used cannot deal with an unusual circumstance. The adjoint method, on the other hand, will always produce the inverse if it exists.

1.2.6. Choleski's Method for Matrix Inversion

Choleski's method is also an efficient algorithm for matrix inversion. It is a more recent approach than the Gauss–Jordan algorithm, but it employs a comparable number of arithmetic operations. Choleski's method involves substitution operations only, and there are several variations of it. One variation begins by decomposing the matrix **A** into a product of a lower triangular matrix **L** and a unit upper triangular matrix **U** in accordance with

$$[\mathbf{L}][\mathbf{U}] = [\mathbf{A}]. \tag{1.20a}$$

Illustrating equation 1.20a with 3 × 3 square matrices yields

$$\begin{bmatrix} l_{11} & 0 & 0 \\ l_{21} & l_{22} & 0 \\ l_{31} & l_{32} & l_{33} \end{bmatrix} \begin{bmatrix} 1 & u_{12} & u_{13} \\ 0 & 1 & u_{23} \\ 0 & 0 & 1 \end{bmatrix} = \begin{bmatrix} a_{11} & a_{12} & a_{13} \\ a_{21} & a_{22} & a_{23} \\ a_{31} & a_{32} & a_{33} \end{bmatrix}. \tag{1.20b}$$

Equation 1.20b represents nine algebraic equations with nine unknowns for a 3×3 matrix $\mathbf{A}$. If the matrix $\mathbf{A}$ were $n \times n$, there would be n equations with n unknowns. As with the Gauss–Jordan method during back substitution, a matrix equation in which one of the multipliers is a triangular matrix may be solved by substitution, beginning with the row or column at the apex of the triangle.

Going down the first column of matrix $\mathbf{A}$ in equation 1.20b

$$l_{11} = a_{11}, \tag{1.20c}$$

$$l_{21} = a_{21}, \tag{1.20d}$$

$$l_{31} = a_{31} \tag{1.20e}$$

Going down the second column of matrix $\mathbf{A}$,

$$(l_{11})(u_{12}) = a_{12} \quad \text{or} \quad u_{12} = \frac{a_{12}}{l_{11}}, \tag{1.20f}$$

$$(l_{21})(u_{12}) + (l_{22})(1) = a_{22} \quad \text{or} \quad l_{22} = a_{22} - (l_{21})(u_{12}), \tag{1.20g}$$

$$(l_{31})(u_{12}) + (l_{32})(1) = a_{32} \quad \text{or} \quad l_{32} = a_{32} - (l_{31})(u_{12}). \tag{1.20h}$$

Going down the third column of matrix $\mathbf{A}$,

$$(l_{11})(u_{13}) = a_{13} \quad \text{or} \quad u_{13} = \frac{a_{13}}{l_{11}}, \tag{1.20i}$$

$$(l_{21})(u_{13}) + (l_{22})(u_{23}) = a_{23} \quad \text{or} \quad u_{23} = \frac{a_{23} - (l_{21})(u_{13})}{l_{22}}, \tag{1.20j}$$

$$(l_{31})(u_{13}) + (l_{32}(u_{23}) + (l_{33})(1) = a_{33} \quad \text{or} \quad l_{33} = a_{33} - (l_{31})(u_{13}) - (l_{32})(u_{23}). \tag{1.20k}$$

These equations provide a means for the direct calculation of the nine unknown l_{ij} and u_{ij}. Thus matrices $\mathbf{L}$ and $\mathbf{U}$ are determined.

In the second phase of Choleski's method, the matrix $\mathbf{U}$ is inverted by substitution operations in accordance with

$$[\mathbf{U}^{-1}][\mathbf{U}] = [\mathbf{I}], \tag{1.21a}$$

$$\begin{bmatrix} t_{11} & t_{12} & t_{13} \\ t_{21} & t_{22} & t_{23} \\ t_{31} & t_{32} & t_{33} \end{bmatrix} \begin{bmatrix} 1 & u_{12} & u_{13} \\ 0 & 1 & u_{23} \\ 0 & 0 & 1 \end{bmatrix} = \begin{bmatrix} 1 & 0 & 0 \\ 0 & 1 & 0 \\ 0 & 0 & 1 \end{bmatrix}. \tag{1.21b}$$

Finding the nine unknown t_{ij}, the entries of matrix $\mathbf{U}^{-1}$, follows the same process that was used to decompose matrix $\mathbf{A}$ into matrices $\mathbf{L}$ and $\mathbf{U}$. The equations are written by going down column 1 of matrix $\mathbf{I}$, then column 2, and then column 3. The process leads to

$$t_{21} = t_{31} = t_{32} = 0, \qquad (1.21c)$$

$$t_{11} = t_{22} = t_{33} = 1, \qquad (1.21d)$$

$$t_{12} = -u_{12}, \qquad (1.21e)$$

$$t_{13} = -u_{13} + (u_{12})(u_{23}), \qquad (1.21f)$$

$$t_{23} = -u_{23}. \qquad (1.21g)$$

These equations allow the direct calculation of nine unknown entries for matrix $\mathbf{U}^{-1}$, which is also a unit triangular matrix.

The last phase of the method is to find the inverse of matrix $\mathbf{A}$ by substitution methods. If equation 1.20a is premultiplied by matrix $\mathbf{A}^{-1}$ and postmultiplied by matrix $\mathbf{U}^{-1}$, the result is

$$[\mathbf{A}^{-1}][\mathbf{L}][\mathbf{U}][\mathbf{U}^{-1}] = [\mathbf{A}^{-1}][\mathbf{A}][\mathbf{U}^{-1}],$$

or

$$[\mathbf{A}^{-1}][\mathbf{L}] = [\mathbf{U}^{-1}], \qquad (1.22a)$$

$$
\begin{bmatrix} b_{11} & b_{12} & b_{13} \\ b_{21} & b_{22} & b_{23} \\ b_{31} & b_{32} & b_{33} \end{bmatrix}
\begin{bmatrix} l_{11} & 0 & 0 \\ l_{21} & l_{22} & 0 \\ l_{31} & l_{32} & l_{33} \end{bmatrix}
=
\begin{bmatrix} 1 & t_{12} & t_{13} \\ 0 & 1 & t_{23} \\ 0 & 0 & 1 \end{bmatrix}. \qquad (1.22b)
$$

The entries l_{ij} and u_{ij} of matrices $\mathbf{L}$ and $\mathbf{U}^{-1}$ are known from the first two phases of the algorithm. The entries b_{ij} of matrix $\mathbf{A}^{-1}$ are determined by a substitution process similar to the one used in the first two phases. The process here, however, starts by going down column 3 of matrix $\mathbf{U}^{-1}$, then down column 2, and finally down column 1. The result is nine equations for the direct calculation of the entries for $\mathbf{A}^{-1}$:

$$b_{13} = \frac{t_{13}}{l_{33}}, \qquad (1.22c)$$

$$b_{23} = \frac{t_{23}}{l_{33}}, \qquad (1.22d)$$

$$b_{33} = \frac{1}{l_{33}}, \qquad (1.22e)$$

$$b_{12} = \frac{t_{12} - (b_{13})(l_{32})}{l_{22}}, \qquad (1.22f)$$

$$b_{22} = \frac{1 - (b_{23})(l_{32})}{l_{22}}, \tag{1.22g}$$

$$b_{32} = -\frac{(b_{33})(l_{32})}{l_{22}}, \tag{1.22h}$$

$$b_{11} = \frac{1 - (b_{12})(l_{21}) - (b_{13})(l_{31})}{l_{11}}, \tag{1.22i}$$

$$b_{21} = -\frac{(b_{22})(l_{21}) + (b_{23})(l_{31})}{l_{11}}, \tag{1.22j}$$

$$b_{31} = -\frac{(b_{32})(l_{21}) + (b_{33})(l_{31})}{l_{11}}. \tag{1.22k}$$

Thus matrix $\mathbf{A}^{-1}$ is determined. It is left to the reader to use Choleski's method to verify these equations and to obtain the numerical inverse of the matrix $\mathbf{A}$ given earlier in equation 1.16b.

The process applies for $n \times n$ matrices, though clearly equations 1.20–1.22 must be extended for matrices larger than 3×3. The Choleski algorithm is usually implemented by programming the sequence of steps just illustrated for a computer, obtaining and using numerical values for each quantity at each step along the way in the sequence.

1.3. ALGEBRAIC EQUATION SYSTEMS

A linear algebraic equation system in the finite element method is represented by the matrix equation

$$[\mathbf{K}]\{\mathbf{q}\} = \{\mathbf{F}\}. \tag{1.23}$$

Matrix $\mathbf{K}$ is the stiffness matrix; usually its entries are determined before solution of the equation system is attempted. Matrix $\mathbf{q}$ is the displacement vector; a few of its entries will be known and the rest will not when solution of the equation system is attempted. Matrix $\mathbf{F}$ is the load vector; usually many of its entries, but not all, are known before solution of the equation system. Often many of the known entries in the load vector are zeros. The total number of unknown entries $(q_i + f_i)$ must equal the number of equations in the system for a solution to be obtained.

In the execution of the finite element method a major amount of computational effort must be devoted to the manipulation and solution of the algebraic equation systems developed, so that efficient solution procedures are needed.

1.3.1. Reduction Solution Procedure

An equation system in which all of the unknowns appear in the load vector may be solved by inverting the stiffness matrix and then premultiplying the load vector by the inverse of the stiffness matrix. For a large equation system whose load vector contains a significant number of zeros, a procedure exists which takes advantage of the presence of these zeros to partition the equation system and deal with submatrices which are significantly smaller than the stiffness matrix of the original system. Though two of these submatrices will have to be inverted instead of the single original stiffness matrix and others will have to be multiplied and added, the total number of arithmetic operations needed to solve the equation system can be reduced by orders of magnitude. The procedure is called a reduction procedure.

The following is an equation system with all its unknowns in the displacement vector and some zero entries in its load vector:

$$
\begin{bmatrix}
k_{11} & k_{12} & k_{13} & k_{14} & k_{15} \\
k_{21} & k_{22} & k_{23} & k_{24} & k_{25} \\
k_{31} & k_{32} & k_{33} & k_{34} & k_{35} \\
k_{41} & k_{42} & k_{43} & k_{44} & k_{45} \\
k_{51} & k_{52} & k_{53} & k_{54} & k_{55}
\end{bmatrix}
\begin{Bmatrix}
q_1 \\ q_2 \\ q_3 \\ q_4 \\ q_5
\end{Bmatrix}
=
\begin{Bmatrix}
f_1 \\ 0 \\ f_3 \\ 0 \\ f_5
\end{Bmatrix}.
\tag{1.24a}
$$

Equation 1.24a will be rearranged so that the zeros appear last in vector **F**. A simple rearrangement is to list the first, third, and fifth equations of the system first and then list the second and fourth equations. The matrix representation of the rearranged system is

$$
\left[
\begin{array}{ccc|cc}
k_{11} & k_{12} & k_{13} & k_{14} & k_{15} \\
k_{31} & k_{32} & k_{33} & k_{34} & k_{35} \\
k_{51} & k_{52} & k_{53} & k_{54} & k_{55} \\
\hline
k_{21} & k_{22} & k_{23} & k_{24} & k_{25} \\
k_{41} & k_{42} & k_{43} & k_{44} & k_{45}
\end{array}
\right]
\begin{Bmatrix}
q_1 \\ q_2 \\ q_3 \\ q_4 \\ q_5
\end{Bmatrix}
=
\begin{Bmatrix}
f_1 \\ f_3 \\ f_5 \\ 0 \\ 0
\end{Bmatrix}.
\tag{1.24b}
$$

Partitioning matrix **K** of equation 1.24b is done horizontally on the zeros and vertically in the corresponding position so that the number of columns to the left of the vertical partition equals the number of rows above the horizontal partition. Partitioning in this manner allows the equation system to be treated in terms of submatrices as

$$
\left[
\begin{array}{c|c}
\mathbf{K}_{ff} & \mathbf{K}_{fs} \\
\hline
\mathbf{K}_{sf} & \mathbf{K}_{ss}
\end{array}
\right]
\begin{Bmatrix}
\mathbf{q}_f \\ \mathbf{q}_s
\end{Bmatrix}
=
\begin{Bmatrix}
\mathbf{F}_f \\ \mathbf{0}
\end{Bmatrix}.
\tag{1.24c}
$$

Submatrix equations may be written as

$$
[\mathbf{K}_{ff}]\{\mathbf{q}_f\} + [\mathbf{K}_{fs}]\{\mathbf{q}_s\} = \{\mathbf{F}_f\},
\tag{1.24d}
$$

$$
[\mathbf{K}_{sf}]\{\mathbf{q}_f\} + [\mathbf{K}_{ss}]\{\mathbf{q}_s\} = \{\mathbf{0}\}.
\tag{1.24e}
$$

Rearranging equation 1.24e yields

$$
\{\mathbf{q}_s\} = -[\mathbf{K}_{ss}^{-1}][\mathbf{K}_{sf}]\{\mathbf{q}_f\}.
\tag{1.24f}
$$

Substituting equation 1.24f into equation 1.24d yields

$$[\mathbf{K}_{ff}]\{\mathbf{q}_f\} - [\mathbf{K}_{fs}][\mathbf{K}_{ss}^{-1}][\mathbf{K}_{sf}]\{\mathbf{q}_f\} = \{\mathbf{F}_f\},$$

or

$$([\mathbf{K}_{ff}] - [\mathbf{K}_{fs}][\mathbf{K}_{ss}^{-1}][\mathbf{K}_{sf}])\{\mathbf{q}_f\} = \{\mathbf{F}_f\},$$

or

$$[\mathbf{K}_r]\{\mathbf{q}_f\} = \{\mathbf{F}_f\}, \tag{1.24g}$$

where

$$[\mathbf{K}_r] = [\mathbf{K}_{ff}] - [\mathbf{K}_{fs}][\mathbf{K}_{ss}^{-1}][\mathbf{K}_{sf}]. \tag{1.24h}$$

Finally, from equation 1.24g

$$\{\mathbf{q}_f\} = [\mathbf{K}_r^{-1}]\{\mathbf{F}_f\}. \tag{1.24i}$$

Thus the equation system is solved. Vector $\mathbf{q}_s$ may be evaluated using equation 1.24f. The entries of vectors $\mathbf{q}_f$ and $\mathbf{q}_s$ are the partitioned entries of the original vector $\mathbf{q}$.

The effort involved in solving equation 1.24a for the five unknown q_i, either by the direct inversion procedure or by employing the reduction procedure just described, would require the following number of arithmetic operations:

	Number of Operations	
	Adjoint Method	Gauss–Jordan
Direct inversion		
Matrix inversions (1 at 5 × 5)	1609	450
Matrix multiplications	45	45
total	1654	495
Reduction procedure		
Matrix inversions (2 at 2 × 2, 3 × 3)	48	114
Matrix multiplications	70	70
Matrix additions	9	9
total	127	193

The savings in computation effort is substantial when the reduction procedure is employed, even on the small equation system used in the illustration. The savings increase dramatically as the size of the equation system increases. Clearly, if one had to solve by hand a system of five equations where two of the entries in the load vector were zero, use of the reduction procedure and the adjoint method for matrix inversion would be the most efficient method. If none of the entries in the load vector were zero, then this reduction procedure would not apply.

A more *general reduction procedure* not requiring the presence of zeros in the load vector can be developed similarly. The equation system need not be rearranged. It is partitioned to make the submatrices $\mathbf{K}_{ff}$ and $\mathbf{K}_{ss}$ square and roughly equal in size. In symbols,

$$\left[\begin{array}{c|c} \mathbf{K}_{ff} & \mathbf{K}_{fs} \\ \hline \mathbf{K}_{sf} & \mathbf{K}_{ss} \end{array}\right] \left\{\begin{array}{c} \mathbf{q}_f \\ \mathbf{q}_s \end{array}\right\} = \left\{\begin{array}{c} \mathbf{F}_f \\ \mathbf{F}_s \end{array}\right\}. \tag{1.25a}$$

Submatrix equations are written as

$$[\mathbf{K}_{ff}]\{\mathbf{q}_f\} + [\mathbf{K}_{fs}]\{\mathbf{q}_s\} = \{\mathbf{F}_f\}, \tag{1.25b}$$

$$[\mathbf{K}_{sf}]\{\mathbf{q}_f\} + [\mathbf{K}_{ss}]\{\mathbf{q}_s\} = \{\mathbf{F}_s\}. \tag{1.25c}$$

Rearranging equation 1.25c yields

$$\{\mathbf{q}_s\} = [\mathbf{K}_{ss}^{-1}](\{\mathbf{F}_s\} - [\mathbf{K}_{sf}]\{\mathbf{q}_f\}). \tag{1.25d}$$

Substituting equation 1.25d into equation 1.25b gives

$$[\mathbf{K}_{ff}]\{\mathbf{q}_f\} + [\mathbf{K}_{fs}][\mathbf{K}_{ss}^{-1}](\{\mathbf{F}_s\} - [\mathbf{K}_{sf}]\{\mathbf{q}_f\}) = \{\mathbf{F}_f\},$$

or

$$([\mathbf{K}_{ff}] - [\mathbf{K}_{fs}][\mathbf{K}_{ss}^{-1}][\mathbf{K}_{sf}])\{\mathbf{q}_f\} = \{\mathbf{F}_f\} - [\mathbf{K}_{fs}][\mathbf{K}_{ss}^{-1}]\{\mathbf{F}_s\},$$

or

$$[\mathbf{K}_r]\{\mathbf{q}_f\} = \{\mathbf{F}_f\} - [\mathbf{K}_{fs}][\mathbf{K}_{ss}^{-1}]\{\mathbf{F}_s\}, \tag{1.25e}$$

where the matrix $\mathbf{K}_r$ is as shown in equations above and defined in equation 1.24h. Thus from equations 1.25e,

$$\{\mathbf{q}_f\} = [\mathbf{K}_r^{-1}](\{\mathbf{F}_f\} - [\mathbf{K}_{fs}][\mathbf{K}_{ss}^{-1}]\{\mathbf{F}_s\}). \tag{1.25f}$$

The vector $\mathbf{q}_s$ is calculated using equation 1.25d, once the vector $\mathbf{q}_f$ is determined from equation 1.25f.

If the equation system is partitioned roughly in half, two matrix inversions will be done on submatrices about one-half the size of the original coefficient matrix. Reference to the numbers of arithmetic operations tabulated earlier suggests that considerable computation effort will be saved.

1.3.2. Submatrix Solution Procedure

When the finite element method is applied to structural systems, usually some of the entries of the displacement vector are unknown and the remaining unknowns are in the load vector. The equation system might look like

$$\begin{bmatrix} k_{11} & k_{12} & k_{13} & k_{14} & k_{15} \\ k_{21} & k_{22} & k_{23} & k_{24} & k_{25} \\ k_{31} & k_{32} & k_{33} & k_{34} & k_{35} \\ k_{41} & k_{42} & k_{43} & k_{44} & k_{45} \\ k_{51} & k_{52} & k_{53} & k_{54} & k_{55} \end{bmatrix} \begin{Bmatrix} q_1 \\ q_{o2} \\ q_3 \\ q_{o4} \\ q_5 \end{Bmatrix} = \begin{Bmatrix} f_{o1} \\ f_2 \\ f_{o3} \\ f_4 \\ f_{o5} \end{Bmatrix}. \qquad (1.26a)$$

The subscripts in vectors $\mathbf{q}$ and $\mathbf{F}$ preceded by o indicate known quantities. Equations 1.26a cannot be solved by the direct inversion procedure. The unknowns in the load vector make the multiplication step impossible.

A procedure which deals with this circumstance is similar to the reduction procedures discussed above. The equation system 1.26a is rearranged so that the known entries in vector $\mathbf{q}$ appear last and the known entries in vector $\mathbf{F}$ appear first.

This rearrangement requires that both the sequence of the equations in the system and the sequence of appearance of the entries q_i in the algebraic equations be changed. The former requires that the rows of matrix $\mathbf{K}$ and vector $\mathbf{F}$ change. The latter requires that the columns of matrix $\mathbf{K}$ and the rows of vector $\mathbf{q}$ change for the matrix representation of the equation system to be consistent with the dictates of matrix multiplication. Thus matrix $\mathbf{K}$ must undergo a reordering by both rows and columns, a process easily done in two steps. First reorder its rows; then reorder the columns of the newly reordered matrix $\mathbf{K}$. Equation 1.26a so reordered is

$$\left[\begin{array}{ccc|cc} k_{11} & k_{13} & k_{15} & k_{12} & k_{14} \\ k_{31} & k_{33} & k_{35} & k_{32} & k_{34} \\ k_{51} & k_{53} & k_{55} & k_{52} & k_{54} \\ \hline k_{21} & k_{23} & k_{25} & k_{22} & k_{24} \\ k_{41} & k_{43} & k_{45} & k_{42} & k_{44} \end{array}\right] \begin{Bmatrix} q_1 \\ q_3 \\ q_5 \\ q_{o2} \\ q_{o4} \end{Bmatrix} = \begin{Bmatrix} f_{o1} \\ f_{o3} \\ f_{o5} \\ f_2 \\ f_4 \end{Bmatrix}. \qquad (1.26b)$$

Partitioning equations 1.26b on the demarcation between known and unknown entries in the displacement and load vectors is done as described in the reduction procedure above. This makes submatrices $\mathbf{K}_{ff}$ and $\mathbf{K}_{ss}$ square and allows the equation system to be treated in terms of submatrices as

$$\left[\begin{array}{c|c} \mathbf{K}_{ff} & \mathbf{K}_{fs} \\ \hline \mathbf{K}_{sf} & \mathbf{K}_{ss} \end{array}\right] \begin{Bmatrix} \mathbf{q}_f \\ \mathbf{q}_o \end{Bmatrix} = \begin{Bmatrix} \mathbf{F}_o \\ \mathbf{F}_s \end{Bmatrix}. \qquad (1.26c)$$

Submatrix equations may be written as

$$[\mathbf{K}_{ff}]\{\mathbf{q}_f\} + [\mathbf{K}_{fs}]\{\mathbf{q}_o\} = \{\mathbf{F}_o\}, \qquad (1.26d)$$

$$[\mathbf{K}_{sf}]\{\mathbf{q}_f\} + [\mathbf{K}_{ss}]\{\mathbf{q}_o\} = \{\mathbf{F}_s\}. \qquad (1.26e)$$

Equations 1.26d and 1.26e may be rewritten as

$$\{\mathbf{q}_f\} = [\mathbf{K}_{ff}^{-1}](\{\mathbf{F}_o\} - [\mathbf{K}_{fs}]\{\mathbf{q}_o\}), \tag{1.26f}$$

$$\{\mathbf{F}_s\} = [\mathbf{K}_{sf}]\{\mathbf{q}_f\} + [\mathbf{K}_{ss}]\{\mathbf{q}_o\}. \tag{1.26g}$$

Equations 1.26f and 1.26g provide the means to calculate directly the unknowns appearing in the global stiffness equations.

1.3.3. Solution without Rearrangement

There is another solution method for an algebraic equation system whose unknowns appear in both the load and displacement vectors. Though the submatrix procedure described above is an efficient algorithm for the solution of equation systems, it is cumbersome to program its requirement to rearrange the global stiffness equations. The solution-without-rearrangement method requires no rearrangement. It does possess the disadvantage that the entire global stiffness matrix must be inverted, rather than two smaller submatrices.

The method entails modifying the global stiffness equations so that the trivial equations specifying nodal displacements are embedded directly into the equation system. In the equation system 1.26a described earlier, the trivial equations are

$$q_2 = q_{o2} \quad \text{and} \quad q_4 = q_{o4}.$$

If these are inserted into the equation system, the second of equations 1.26a becomes

$$(0)q_1 + (1)q_2 + (0)q_3 + (0)q_4 + (0)q_5 = q_{o2}$$

and the fourth of equations 1.26a becomes

$$(0)q_1 + (0)q_2 + (0)q_3 + (1)q_4 + (0)q_5 = q_{o4}.$$

To retain the correct displacement vector, the first of equations 1.26a must be modified as

$$k_{11}q_1 + (0)q_2 + k_{13}q_3 + (0)q_4 + k_{15}q_5 = f_{o1} - k_{12}q_{o2} - k_{14}q_{o4}.$$

Since q_{o2} and q_{o4} are known values, they are placed on the right side of the equation. The third and fifth equations of 1.26a are modified similarly. The resulting modified equation system in matrix form is

$$\begin{bmatrix} k_{11} & 0 & k_{13} & 0 & k_{15} \\ 0 & 1 & 0 & 0 & 0 \\ k_{31} & 0 & k_{33} & 0 & k_{35} \\ 0 & 0 & 0 & 1 & 0 \\ k_{51} & 0 & k_{53} & 0 & k_{55} \end{bmatrix} \begin{Bmatrix} q_1 \\ q_2 \\ q_3 \\ q_4 \\ q_5 \end{Bmatrix} = \begin{Bmatrix} f_{o1} - k_{12}q_{o2} - k_{14}q_{o4} \\ q_{o2} \\ f_{o3} - k_{32}q_{o2} - k_{34}q_{o4} \\ q_{o4} \\ f_{o5} - k_{52}q_{o2} - k_{54}q_{o4} \end{Bmatrix}. \tag{1.27}$$

Equation 1.27 may now be solved by the direct inversion procedure, since all of the unknowns are in the displacement vector and all the entries in the load vector are known. Invert the modified stiffness matrix and then premultiply the modified load vector by this inverse. In so doing, one obtains values for all entries of the displacement vector, including the ones already known. The values for the unknown entries in the original load vector may be obtained by premultiplying the newly obtained vector $\mathbf{q}$ by the original unmodified stiffness matrix.

1.3.4. Penalty Solution Procedure

Another procedure which is easier to program than either method described above is the penalty procedure. It also does not require rearranging the global stiffness equations, and it reduces considerably the modifications that must be made to the global stiffness matrix and load vector to render the global stiffness equations ready for solution.

In the penalty procedure one forces the algebraic equations in the equation system, by numerical means, to produce the known values which appear in vector $\mathbf{q}$. For example, letting $q_2 = q_{o2}$, the second equation of equation system 1.26a may be written as

$$k_{21}q_1 + k_{22}q_2 + k_{23}q_3 + k_{24}q_4 + k_{25}q_5 = f_2. \tag{1.28a}$$

A quantity called the penalty P is calculated as

$$P_{22} = (\text{big nr})(k_{22}). \tag{1.28b}$$

where "big nr" is a very large number, perhaps 10 orders of magnitude larger than any term in equation 1.28a. To form the penalty term P_{22}, "big nr" is multiplied by the diagonal entry k_{22} in the stiffness matrix, which corresponds to the value of q_{o2} in vector $\mathbf{q}$.

Since $q_2 = q_{o2}$,

$$(P_{22})(q_2) = (P_{22})(q_{o2}). \tag{1.28c}$$

Equations 1.28a and 1.28c are added to produce

$$k_{21}q_1 + (P_{22} + k_{22})q_2 + k_{23}q_3 + k_{24}q_4 + k_{25}q_5 = P_{22}q_{o2} + f_2. \tag{1.28d}$$

If P_{22} is large enough, say 10 times as large as any other term in the equation, then equation 1.28d will be closely approximated by

$$k_{21}q_1 + P_{22}q_2 + k_{23}q_3 + k_{24}q_4 + k_{25}q_5 = P_{22}q_{o2}. \tag{1.28e}$$

The fourth equation of equation system 1.26a may be similarly modified as

$$k_{41}q_1 + k_{42}q_2 + k_{43}q_3 + P_{44}q_4 + k_{45}q_5 = P_{44}q_{o4}, \tag{1.28f}$$

where $P_{44} = (\text{big nr})(k_{44})$.

Replacing the second and fourth equations of the equation system 1.26a by their modified versions (equations 1.28e and 1.28f) yields

$$
\begin{bmatrix}
k_{11} & k_{12} & k_{13} & k_{14} & k_{15} \\
k_{21} & P_{22} & k_{23} & k_{24} & k_{25} \\
k_{31} & k_{32} & k_{33} & k_{34} & k_{35} \\
k_{41} & k_{42} & k_{43} & P_{44} & k_{45} \\
k_{51} & k_{52} & k_{53} & k_{54} & k_{55}
\end{bmatrix}
\begin{Bmatrix}
q_1 \\ q_2 \\ q_3 \\ q_4 \\ q_5
\end{Bmatrix}
=
\begin{Bmatrix}
f_{o1} \\ P_{22}q_{o2} \\ f_{o3} \\ P_{44}q_{o4} \\ f_{o5}
\end{Bmatrix}.
\tag{1.28g}
$$

Equation 1.28g may be solved by the direct inversion procedure since all the unknowns are in the displacement vector and all the entries in the load vector are known. The procedure from this point on is identical to the procedure employed in the solution–without-rearrangement method described above.

If "big nr" is 10 orders of magnitude larger than the other entries of the modified global stiffness equations 1.28g, the values obtained in the solution will be accurate to about nine significant figures.

1.4. CLOSING REMARKS

The definitions and descriptions of matrices and matrix operations discussed in this chapter, and those for algebraic equation systems, are what is needed to begin to treat and understand the concepts employed in the finite element method. They are the language of the method. There are other topics in mathematics, and some in structural mechanics, that must also be mastered. These will be treated as they are needed in the chapters that follow.

BIBLIOGRAPHY

Bathe, K. J., *Finite Element Procedures in Engineering Analysis* (Chapter 1), Prentice-Hall, Englewood Cliffs, NJ, 1982.

Cook, R. D., Malkus, D. S., and Plesha, M. E., *Concepts and Applications of Finite Element Analysis* (Chapter 2), Wiley, New York, 1989.

Froberg, C. E., *Introduction to Numerical Analysis*, Addison-Wesley, Reading, MA, 1969.

Hibbeler, R. C., *Structural Analysis* (Chapter 13), 3rd ed., Prentice-Hall, Englewood Cliffs, NJ, 1994.

Huebner, K. H., *The Finite Element Method for Engineers* (Appendix A), Wiley, New York, 1975.

McGuire, W., and Gallagher, R. H., *Matrix Structural Analysis* (Chapter 11), Wiley, New York, 1979.

Noble, B., *Applied Linear Algebra*, Prentice-Hall, Englewood Cliffs, NJ, 1969.

Yang, T. Y., *Finite Element Structural Analysis* (Chapter 2), Prentice-Hall, Englewood Cliffs, NJ, 1986.

PROBLEMS

1.1. Given:

$$[A] = \begin{bmatrix} 1 & 3 & 5 & 2 \\ 4 & 1 & 8 & 9 \\ 3 & 3 & 9 & 7 \\ 9 & 3 & 2 & 2 \end{bmatrix} \quad \text{and} \quad [B] = \begin{bmatrix} 3 & 4 & 1 & 5 \\ 4 & 1 & 1 & 9 \\ 5 & 2 & 7 & 7 \\ 4 & 9 & 5 & 2 \end{bmatrix}.$$

Using computer software which manipulates individual matrices, calculate:
a. Transpose of matrix **A**
b. Inverse of matrix **A**
c. Product matrix **AB**
d. Determinant of matrix **A**
e. Determinant of matrix **B**
f. Determinant of matrix $\mathbf{A}^T$
g. Determinant of matrix $\mathbf{A}^{-1}$
h. Reciprocal of det $\mathbf{A}^{-1}$
i. Determinant of product matrix **AB**
j. Product of det **A** and det **B**
k. Determinant of matrix $\mathbf{A}^{\text{mod}\,1}$, where

$$\mathbf{A}^{\text{mod}\,1} = \mathbf{A} \quad \text{with 3 times row 2 added to row 3}$$

l. Determinant of matrix $\mathbf{A}^{\text{mod}\,2}$, where

$$\mathbf{A}^{\text{mod}\,2} = \mathbf{A} \quad \text{with columns 2 and 4 interchanged}$$

What observations can you make regarding possible relationships among the results of these calculations?

1.2. For the matrix

$$[A] = \begin{bmatrix} 1 & 3 & 4 \\ 2 & 8 & 12 \\ 3 & 3 & 4 \end{bmatrix}$$

a. Decompose matrix **A** into a unit upper triangular matrix **U** and a lower triangular matrix **L**.
b. Calculate the determinants of matrices **U** and **L**.
c. Calculate the determinant of matrix **A** using your calculated values for det **U** and det **L**.

1.3. Solve the equation system shown, by hand but with a calculator for simple arithmetic calculations:

$$\begin{bmatrix} 1 & 2 & 6 \\ 2 & 5 & 15 \\ 6 & 15 & 46 \end{bmatrix} \begin{Bmatrix} x_1 \\ x_2 \\ x_3 \end{Bmatrix} = \begin{Bmatrix} 2 \\ -1 \\ -2 \end{Bmatrix}.$$

a. Use the adjoint method for direct matrix inversion.
b. Use the Gauss–Jordan elimination method.
c. Use the Choleski substitution method.

1.4. Repeat Problem 1.3 with the following equation system:

$$
\begin{aligned}
x + 3y + z &= 17, \\
3x + 2y + 3z &= 23, \\
x + 3y + 2z &= 20.
\end{aligned}
$$

1.5. Repeat Problem 1.3 with the following equation system:

$$
\begin{aligned}
x_1 + 3x_2 + x_3 &= 16, \\
3x_1 + 2x_2 + 3x_3 &= 27, \\
x_1 + 3x_2 + 2x_3 &= 21.
\end{aligned}
$$

1.6. Repeat Problem 1.3 with the following equation system:

$$\begin{bmatrix} 1 & 3 & 1 \\ 3 & 2 & 3 \\ 1 & 3 & 2 \end{bmatrix} \begin{bmatrix} x_{11} & x_{12} & x_{13} \\ x_{21} & x_{22} & x_{23} \\ x_{31} & x_{32} & x_{33} \end{bmatrix} = \begin{bmatrix} 17 & 16 & 15 \\ 23 & 27 & 17 \\ 20 & 21 & 17 \end{bmatrix}.$$

1.7. With appropriate matrix symbols and matrix operations prove that:

$$\text{If} \quad [C] = [A][B], \quad \text{then} \quad [C^{-1}] = [B^{-1}][A^{-1}].$$

1.8. Using the two matrices shown below and computer software which manipulates individual matrices, demonstrate the relationship among inverses you proved in Problem 1.7:

$$[A] = \begin{bmatrix} 2 & 5 & -4 & 6 \\ 3 & -3 & 9 & 7 \\ 5 & 1 & 8 & 2 \\ 4 & 6 & 3 & -7 \end{bmatrix}, \quad [B] = \begin{bmatrix} 2 & 2 & -3 & 9 \\ 3 & -5 & 6 & 6 \\ 8 & 1 & 4 & -9 \\ 7 & 9 & 4 & 2 \end{bmatrix}$$

1.9. Solve the equation system shown using computer software which manipulates individual matrices:

$$
\begin{array}{rcrcrcrcrcl}
6x_1 & + & 2x_2 & + & x_3 & + & x_4 & & & = & 14, \\
5x_1 & + & 4x_2 & + & x_3 & + & x_4 & - & 3x_5 & = & 0, \\
2x_1 & + & 2x_2 & + & 3x_3 & + & 2x_4 & - & x_5 & = & 8, \\
x_1 & + & x_2 & + & x_3 & + & 3x_4 & + & 2x_5 & = & 5, \\
3x_1 & - & x_2 & + & 5x_3 & - & x_4 & + & 5x_5 & = & 0.
\end{array}
$$

   a. Use the direct matrix inversion method of solution.

   b. Use a reduction method, taking advantage of the zeros to the right of the equal signs.

   c. Use the general reduction method applied on the equations in the order shown.

1.10. Solve the system of equations shown by using a reduction method so that only 2×2 and 2×1 matrices are manipulated:

$$
\begin{bmatrix}
1 & -1 & 1 & -1 \\
4 & -3 & 0 & -2 \\
3 & 1 & -1 & -2 \\
-3 & 2 & 2 & -1
\end{bmatrix}
\begin{Bmatrix}
x_1 \\ x_2 \\ x_3 \\ x_4
\end{Bmatrix}
=
\begin{Bmatrix}
8 \\ 4 \\ 0 \\ 0
\end{Bmatrix}.
$$

1.11. Repeat Problem 1.10 with the following equation system but using the general reduction method:

$$
\begin{array}{rcrcrcrcl}
x_1 & - & x_2 & + & x_3 & - & x_4 & = & 8 \\
-3x_1 & + & 2x_2 & + & 2x_3 & - & x_4 & = & 0 \\
4x_1 & - & 3x_2 & & & - & 2x_4 & = & 4 \\
3x_1 & + & x_2 & - & x_3 & - & 2x_4 & = & 0
\end{array}
$$

1.12. Repeat Problem 1.10 with the following equation system:

$$
\begin{bmatrix}
1 & -1 & 1 & -1 \\
-3 & 2 & 2 & -4 \\
4 & -3 & -6 & -2 \\
3 & 5 & -1 & -2
\end{bmatrix}
\begin{Bmatrix}
x_1 \\ x_2 \\ x_3 \\ x_4
\end{Bmatrix}
=
\begin{Bmatrix}
10 \\ 12 \\ -7 \\ 3
\end{Bmatrix}.
$$

1.13. Solve the equation system shown using computer software which manipulates individual matrices:

$$
\begin{bmatrix}
1 & 5 & -2 & 4 & 8 \\
2 & -7 & 6 & 3 & -1 \\
1 & 1 & 3 & -2 & -4 \\
8 & 6 & -2 & 5 & 1 \\
-3 & -2 & 1 & 1 & 4
\end{bmatrix}
\begin{Bmatrix}
x_1 \\
x_2 \\
x_3 \\
6 \\
-8
\end{Bmatrix}
=
\begin{Bmatrix}
-20 \\
21 \\
40 \\
b_4 \\
b_5
\end{Bmatrix}
$$

a. Use the submatrix solution procedure.
b. Use the solution-without-rearrangement procedure.
c. Use the penalty solution procedure.

1.14. Repeat Problem 1.13 with the following equation system:

$$
\begin{bmatrix}
3 & -1 & -2 & 1 & 2 \\
1 & 4 & -1 & 2 & 5 \\
-2 & 2 & 3 & 1 & 4 \\
2 & -1 & 4 & 1 & -5 \\
1 & 3 & -2 & -1 & -1
\end{bmatrix}
\begin{Bmatrix}
x_1 \\
3 \\
x_3 \\
-4 \\
x_5
\end{Bmatrix}
=
\begin{Bmatrix}
41 \\
b_2 \\
26 \\
b_4 \\
-21
\end{Bmatrix}
$$

1.15. Solve the equation system shown using computer software which manipulates individual matrices:

$$
\begin{bmatrix}
1 & 5 & -2 & 4 & 8 & 3 \\
2 & -7 & 6 & 3 & -1 & 5 \\
1 & 1 & 3 & -2 & -4 & -1 \\
8 & 6 & -2 & 5 & 1 & -4 \\
-3 & -2 & 1 & 1 & 4 & 5 \\
2 & -1 & 4 & 6 & -3 & 1
\end{bmatrix}
\begin{Bmatrix}
x_1 \\
x_2 \\
x_3 \\
x_4 \\
12 \\
-15
\end{Bmatrix}
=
\begin{Bmatrix}
-12 \\
-79 \\
-10 \\
-64 \\
b_5 \\
b_6
\end{Bmatrix}
$$

a. Use the submatrix solution procedure, but within this procedure apply the general reduction method so that the largest matrix inverted is 2×2.
b. Use the penalty solution procedure.

1.16. Shown below is a homogeneous algebraic equation system. The condition for which a solution to these equations exists is that the determinant of the coefficient matrix must be zero. Determine the values of the scalar λ which will cause this condition to occur:

$$\left[\begin{bmatrix} 10 & 2 & 1 \\ 2 & 10 & 1 \\ 2 & 1 & 10 \end{bmatrix} - \lambda \begin{bmatrix} 1 & 0 & 0 \\ 0 & 1 & 0 \\ 0 & 0 & 1 \end{bmatrix} \right] \begin{Bmatrix} x_1 \\ x_2 \\ x_3 \end{Bmatrix} = \begin{Bmatrix} 0 \\ 0 \\ 0 \end{Bmatrix},$$

or

$$[[A] - \lambda[I]]\{X\} = \{0\}.$$

CHAPTER 2

THE FINITE ELEMENT METHOD

2.1. INTRODUCTION

A finite element of a structural system is a discrete piece of that system. Some structural systems, such as trusses and frames, are assemblages of inherently discrete pieces so that their finite elements are physically apparent. Other structural systems, such as beams and plates, are continuous systems. They may be regarded as assemblages of discrete pieces, but the choice of pieces is less obvious and may even be arbitrary.

2.2. NODES, EQUILIBRIUM, AND CONTINUITY

Whether the structural system is regarded as an assemblage of physically apparent finite elements or arbitrary ones, the finite elements for each possess similar characteristics. The finite elements join one another to form the assemblage at key points called nodes. In general, both the equilibrium of forces and moments and the continuity of displacements are enforced at nodes. The continuity of both translation and rotation displacements is enforced at nodes when both forces and moments are in equilibrium there.

Nodal equilibrium and continuity requirements are easy to visualize and enforce for 1D finite elements, those requiring one geometric coordinate to describe them, such as truss, beam, and frame elements. On the other hand, when the structural system is continuous in two or more dimensions, such as in a plate, equilibrium and continuity enforced at the nodes may not be sufficient. There may not be displacement continuity along the edges of the elements between the nodes. If continuity is achieved along the edges of 2D or 3D finite elements as well as at nodes, the elements are called conforming. Otherwise they are nonconforming. Both conforming and some nonconforming elements are used to good advantage in finite element analyses.

Figure 2.1 illustrates a plane truss represented as an assemblage of finite elements. The finite elements of the truss are its members, and its nodes are the joints.

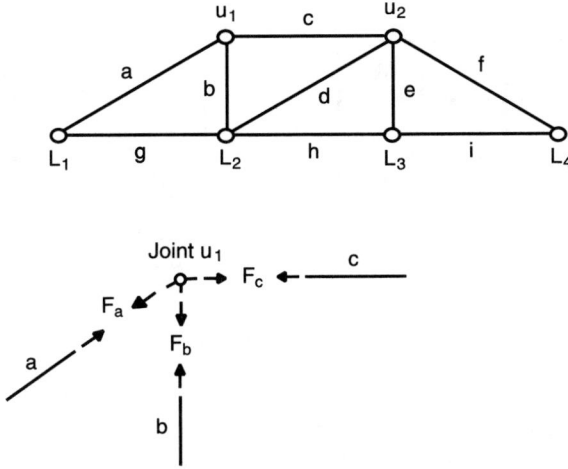

Figure 2.1 Truss

Equilibrium of forces and continuity of displacement translations are enforced at the nodes. Moments are not shown because moments are assumed to be nonexistent in a truss. Rotation displacements are not addressed either.

Figure 2.2. illustrates a plane frame as an assemblage of finite elements. The frame element are its members and the joints are its nodes. Equilibrium of forces and moments and continuity of displacement translations and rotations are enforced at nodes.

Figure 2.3 illustrates a beam as an assemblage of finite elements, and Figure 2.4 illustrates a plate. With beams and plates, the finite elements may be arbitrarily

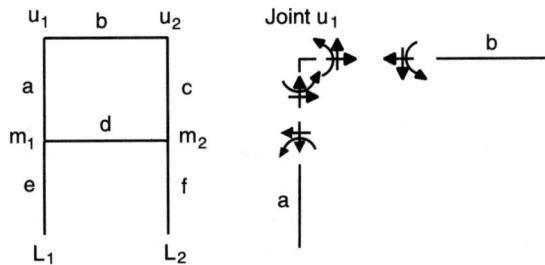

Figure 2.2 Frame

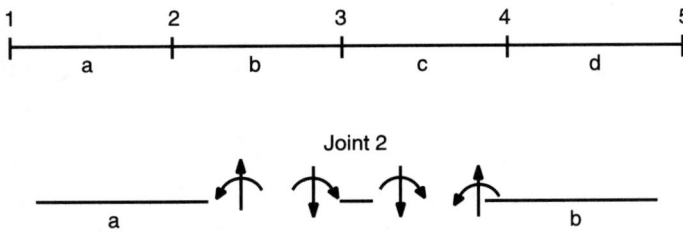

Figure 2.3 Beam

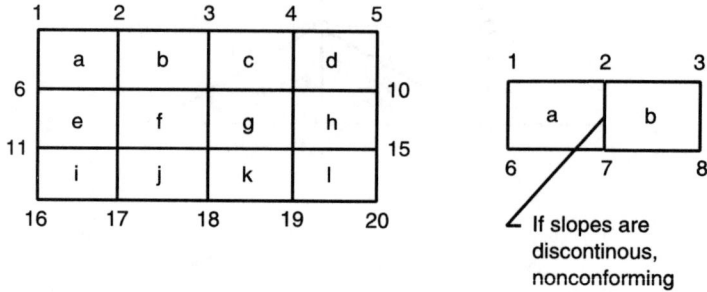

Figure 2.4 Plate

selected. At the nodes in both beams and plates, force and moment are in equilibrium and displacement translation and rotation are continuous. In the plates, slopes may be discontinuous along edges in between nodes, resulting in nonconforming elements. They may also be made continuous to produce conforming elements.

2.3. DEGREES OF FREEDOM

In addition to nodes, the points in the structure where the finite elements join, degrees of freedom (dof) at nodes are vital entities in a finite element analysis. The simplest and most easily visualized dof are the nodal displacements. These dof must be sufficient in number and of the correct character to completely determine the deformed equilibrium position of the structural system and its finite elements under the action of the applied loads. The deformed equilibrium position of the structure is called its "displaced state." For example, if the horizontal and vertical components of each of the joint displacements of the truss shown in Figure 2.1 are known, then the displaced states of the truss and each of its finite elements are determined.

Since displacement continuity is enforced at the nodes in a structural system, the nodal displacements and dof of the structural system are the same both for the structural system as a whole and for its finite elements. The dof in the structural system are referenced to its finite elements at times and to the structural system as a whole at other times. The former are called element dof (edof) and the latter are called global dof (gdof). Though it is convenient to have different numbering schemes for each, there is a one-to-one correspondence between edof and gdof.

2.4. STIFFNESS EQUATIONS

Equilibrium conditions and the force-deformation characteristics of structural members dictate that there be deterministic relationships among the forces, moments, displacement translations, and displacement rotations acting on the members. The relationships are usually developed among the nodal forces and moments and their dof, and they are linear, algebraic equations for linear, elastic structural systems. Depending on their form, they are called either stiffness equations or flexibility equations. In stiffness equations, forces and moments are expressed in terms of the displacements, whereas in flexibility equations it is the reverse. For most finite element analyses, element stiffness equations are developed for the finite elements, and these are used to develop global stiffness equations for the structural system as a whole.

2.4.1. Element Stiffness Equations

The 2-node truss element shown in Figure 2.5 has nodal forces F_1, F_2, F_3, and F_4 acting along the lines of action of the four edof q_1, q_2, q_3, and q_4. The lines of action of the edof are shown as horizontal and vertical, even though the element is at an inclination. These horizontal and vertical lines of action usually refer to the reference axes for the structure as a whole, called the global reference axes. The nodal forces and displacements in matrix form are

$$\{\mathbf{F}\} = \begin{Bmatrix} F_1 \\ F_2 \\ F_3 \\ F_4 \end{Bmatrix}, \qquad \{\mathbf{q}\} = \begin{Bmatrix} q_1 \\ q_2 \\ q_3 \\ q_4 \end{Bmatrix}.$$

The relationships among these eight quantities will be written as

$$\begin{Bmatrix} F_1 \\ F_2 \\ F_3 \\ F_4 \end{Bmatrix} = \begin{bmatrix} k_{11} & k_{12} & k_{13} & k_{14} \\ k_{21} & k_{22} & k_{23} & k_{24} \\ k_{31} & k_{32} & k_{33} & k_{34} \\ k_{41} & k_{42} & k_{43} & k_{44} \end{bmatrix} \begin{Bmatrix} q_1 \\ q_2 \\ q_3 \\ q_4 \end{Bmatrix}. \tag{2.1a}$$

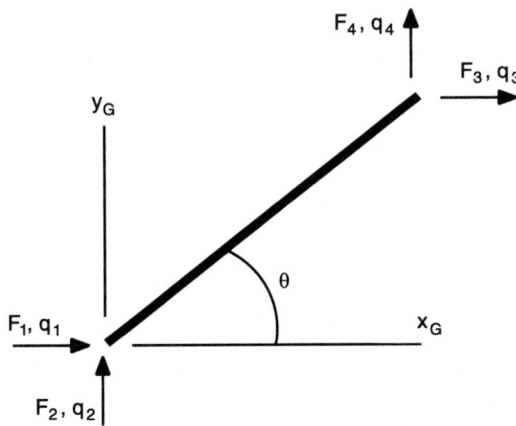

Figure 2.5 Truss element

or

$${F} = [K]{q}. \tag{2.1b}$$

The vector **F** is the element force vector; the matrix **K** is the element stiffness matrix; and the vector **q** is the element displacement vector. Each of these is shown referred to the global reference axes.

Equations 2.1a and 2.1b represent four linear algebraic equations and, expressed in the form shown, are element stiffness equations. The four individual algebraic equations are

$$F_1 = k_{11}q_1 + k_{12}q_2 + k_{13}q_3 + k_{14}q_4, \tag{2.1c}$$

$$F_2 = k_{21}q_1 + k_{22}q_2 + k_{23}q_3 + k_{24}q_4, \tag{2.1d}$$

$$F_3 = k_{31}q_1 + k_{32}q_2 + k_{33}q_3 + k_{34}q_4, \tag{2.1e}$$

$$F_4 = k_{41}q_1 + k_{42}q_2 + k_{43}q_3 + k_{44}q_4. \tag{2.1f}$$

If the nodal displacements q_i are expressed in terms of the nodal forces F_i instead of the reverse, as shown above, then the equations are flexibility equations. Most modern finite element analyses are stiffness formulations. Consequently, this book is restricted to the stiffness approaches used in the finite element method.

The stiffness equations for other finite elements can be represented also by equations 2.1. However, the number of equations in the system would equal the number of edof for the finite element in question. For example, a plane frame element with a node on each end and three edof at each node has six dof. At each node there are the two components of the nodal translation along the lines of action of the gdof and the rotation or slope at the node. The element stiffness equations for a plane frame finite element then are a system of six linear algebraic equations.

2.4.2. Global Stiffness Equations

In a manner like the finite elements of a structural system, structural systems as a whole also possess relationships among the forces and moments applied at their nodes and the displacement translations and rotations of nodes. The displacements are the gdof of the system and the reference axes they coincide with are the global reference axes. The applied forces and moments must be resolved into components whose lines of action coincide with these reference axes and gdof. The equilibrium of the system as a whole and the force-translational and moment-rotational deformation characteristics of the structural system determine the relationships. Symbolically they are represented as

$${F_G} = [K_G]{q_G}. \tag{2.2}$$

Though equation 2.2 is similar in appearance to equation 2.1b, it is not the same. The global stiffness equations of a structural system, however, are related to the element stiffness equations of its finite elements. They may be obtained from a superposition of all of its element stiffness equations if the element stiffness equations are referred to the same reference axes as the global stiffness equations. The process of doing so employs and enforces the equilibrium of forces and moments and the continuity of displacements at nodes. Details of the process, called the assembly of the global stiffness matrix and global force vector, are presented in Chapter 4.

The forces and moments in the global force vector $\mathbf{F}_G$ must be at the nodes and along the lines of action of the gdof. They should include all external forces and moments applied at nodes, including any unknown support reactions. The nodal displacements in the global displacement vector $\mathbf{q}_G$ include the unknown translations and rotations at each node and the displacements at supports which correspond to the support reactions of the structure. These support displacements must be known if the support reactions are unknown, or at least the relationship between the support reaction and its displacement must be known.

The number of linear algebraic equations in a structure's global stiffness equations equals its number of gdof. Consequently if a structural system has n gdof, its global force vector $\mathbf{F}_G$ and global displacement vector $\mathbf{q}_G$ each has n rows, and its global stiffness matrix $\mathbf{K}_G$ will have n rows and n columns. One should note there is only one set of global stiffness equations for a particular structure subjected to a particular set of loads and supports. On the other hand, the structure will have as many sets of element stiffness equations as it has finite elements.

One should also note that the total number of unknown quantities in vectors $\mathbf{F}_G$ and $\mathbf{q}_G$ must equal the number of equations in the global stiffness equation system in order to be able to solve these equations for the unknowns. If there are m unknown support reactions in vector $\mathbf{F}_G$, there will be m corresponding known support displacements in vector $\mathbf{q}_G$. The remaining number of forces and moments in vector $\mathbf{F}_G$ will be known quantities. The remaining number of displacements in vector $\mathbf{q}_G$ will be unknown quantities.

2.4.3. Global and Local Reference Axes

Global reference axes are a single set of rectangular coordinate axes to which the structure as a whole is referred. They often are aligned with the structure's supports. Local reference axes are identified with each finite element; there is one set for each element in the structure. Local axes are usually aligned with the geometry of the finite element and relate directly to its structural behavior.

Referral of element stiffness equations to global reference axes is necessary to obtain the structure's global stiffness equations. However, the element stiffness equations of all but the simplest finite elements cannot be obtained directly in a form referred to the global reference axes. Since element stiffness equations result from the force-deformation and moment-rotation behavior of the element, relationships among these quantities will be more easily obtained if they are referred to local axes which relate directly to this structural behavior. Having developed such element stiffness equations, it is straightforward mathematical procedure for 1D elements to transform them from reference to local axes to element stiffness equations referred to global axes. The transformation to global axes for 2D elements is also necessary, but it is a process different from what it is for 1D elements. The processes are discussed in Chapter 3 for 1D elements in 2D and 3D space and in Chapters 9–11 for 2D elements in 2D and 3D space.

There is another reason for dealing with both local and global axes. To obtain shears and axial force at element nodes, it is efficient to use element stiffness equations which are referred to local axes of the finite element, axes related to the structural behavior of the element. For example, the local axes of a plane frame finite element are along the axis of the element, perpendicular to this axis and in the plane of the member, and perpendicular to the plane of the member. The first provides the axial displacement and force, the second the transverse displacement and shear, and the third the rotation and moment. The solution of the global stiffness equations will provide the nodal displacements referred to the global axes. These are easily transformed to the local axes so that the internal axial force, shear, and moment for the element may be computed directly from the element stiffness equations.

2.5. SYSTEM RESTRAINTS: BOUNDARY CONDITIONS

Structural restraints may take the form of specified displacements at specific gdof at the nodes, or along the edges or on the surfaces of 2D or 3D structures. Another form of restraint to which structures are subjected are forces, moments, and stresses which may be specified at nodes or along the edges and surfaces of 1D, 2D, and 3D structures. A mixed restraint takes the form of a known relationship between a force or moment and its displacement and may be at a node, along an edge, or on a surface of a structure.

2.5.1. Essential Boundary Condition

At pinned and roller supports, the translations are specified. For the pinned support, translations are specified along the lines of action of each translation gdof. For the roller support, a translation is specified along the line perpendicular to the roller surface. At a fixed support, both the translation and rotation gdof are specified. When the restraints specified are displacements which have been defined as the gdof and contained in the global displacement vector $\mathbf{q}_G$, they are referred to as essential boundary conditions. Their imposition on the global stiffness equations is accomplished by simply assigning these known displacement values to the corresponding entries of vector $\mathbf{q}_G$ (equation 2.2) prior to solution of the equations for the other entries of vector $\mathbf{q}_G$. Note that the entries of vector $\mathbf{F}_G$ corresponding to these known entries of vector $\mathbf{q}_G$ must be unknown restraint forces and moments—the support reactions.

When the restraints specified are known displacements along an edge of a 2D or 3D structure, they must be specified as displacement translations and rotations at the appropriate nodes of the structure. These also are essential boundary conditions and, as nodal displacements, are treated as described above.

2.5.2. Natural Boundary Conditions

When known forces, moments, and stresses are specified at nodes or along edges or surfaces of 1D, 2D, or 3D structures, they pose a structural restraint. These are quantities which may be expressed in terms of the higher order derivatives of the displacements defined as the dof. For example if an axial stress is specified at a node of an axial structure, it may be regarded as proportional to the axial strain or first derivative of the axial displacement, where axial displacements are the dof defined in axial structures.

When the restraints on the structure are of the form of higher order derivatives of the displacements defined as the dof, they are referred to as natural boundary conditions. Their imposition on the global stiffness equations is accomplished by adding the known force or moment values at the nodes in question to the corresponding entries of the global load vector $\mathbf{F}_G$ before solving the global stiffness equations for the displacements at gdof. They will affect the solution of the global stiffness equations, but they impose no requirement on the corresponding entries of the global displacement vector $\mathbf{q}_G$ prior to solution.

When forces and moments are specified along edges of 2D or on edges and surfaces of 3D structures, they must be reduced to equivalent forces and moments at the nodes on the edges or surfaces where they are applied. They are natural boundary conditions and are treated as described above.

When stresses are specified along an edge or surface of a structure, they too are natural boundary conditions. Stresses are proportional to strains and the strains are first derivatives of the displacements defined as dof. As a practical matter, to impose them on the global stiffness equations, they must be reduced to equivalent forces and moments at the nodes on the edges or surfaces where they are applied and then entered in the global load vector as described above.

2.5.3. Mixed Boundary Conditions

Another restraint condition may have important application for structural systems. If a support displaces an unknown amount in proportion to its unknown reaction force or moment, it behaves like an elastic spring as

$$R_s = -k_s q_s, \tag{2.3a}$$

where R_s is the unknown support reaction, k_s is the spring constant, and q_s is the unknown support displacement. The negative sign is required since the reaction must be in the opposite direction of the corresponding displacement.

Such a restraint is one form of a mixed boundary condition. It may be introduced into the global stiffness equations by substituting for R_s in vector $\mathbf{F}_G$ of the global stiffness equations the expression contained above in equation 2.3a. Thus the stiffness equation, from among those of the global stiffness equations, which relates to the support reaction would be

$$R_s = k_{s1}q_1 + k_{s2}q_2 + \cdots + k_{ss}q_s + \cdots + k_{sn}q_n, \tag{2.3b}$$

$$-k_s q_s = k_{s1}q_1 + k_{s2}q_2 + \cdots + k_{ss}q_s + \cdots + k_{sn}q_n. \tag{2.3c}$$

Transferring the unknown displacement term $-k_s q_s$ to the right-hand side of equation 2.3b results in

$$0 = k_{s1}q_1 + k_{s2}q_2 + \cdots + (k_{ss} + k_s)q_s + \cdots + k_{sn}q_n. \tag{2.3d}$$

In short, this mixed boundary condition is imposed on the global stiffness equations by adding k_s to the corresponding diagonal entry in matrix $\mathbf{K}_G$ and assigning a zero for the corresponding entry in vector $\mathbf{F}_G$, prior to solving for the unknown displacements in vector $\mathbf{q}_G$. In this situation, the corresponding entry q_s in vector $\mathbf{q}_G$ remains unknown until after solving the global stiffness equations. The unknown reaction R_s may be determined from equation 2.3a or 2.3b after the global stiffness equations are solved and the nodal displacements are known.

If the relationship between the unknown force on the structure and its corresponding displacement occurs along the length of a beam, as with a beam on an elastic foundation, or along the surface of a plate, as with a plate on an elastic foundation, the result is also a mixed boundary condition. To treat this condition, the proportionality constant between the unknown reaction forces and displacements must be reduced to equivalent constants for each element. These equivalent constants are then applied to the element stiffness matrices. The result is reflected in the global stiffness matrix when the element stiffness matrices are superimposed to produce it. The beam on an elastic foundation is discussed in detail in Chapter 5 of this book, and the thin plate on an elastic foundation is discussed in Chapter 10.

2.5.4. Restraint Sufficiency

In stiffness formulations of structural problems, essential, natural, and mixed boundary conditions must be observed. However, the natural boundary conditions are regarded as known applied loads. Essential and mixed boundary conditions, on the other hand, are reactive in nature. They will develop whatever force or moment is necessary to ensure that the specified essential or mixed restraint condition is satisfied. If there are enough of these essential or mixed conditions imposed on the structure, then the structure will be unable to undergo rigid-body displacement. Its displacement field will be the result of structural deformations alone.

There must be enough restraints on the structure so that its displacements are the result of structural deformation without the effects of rigid-body motions. Otherwise, the solution of equation 2.2 for the displacements at gdof will not be possible; there will not be a unique set of displacements for the solution. Mathematically, this means that essential or mixed boundary conditions must be imposed on the global stiffness equations prior to attempting their solution. That is, either vector $\mathbf{q}_G$ or matrix $\mathbf{K}_G$ or both must modified prior to attempting to solve equation 2.2. The minimum number of essential and mixed restraints that are required are what results in the structure being stable and statically determinate with respect to its supports. The maximum is 1 less than the number of gdof employed in the analysis.

2.6. STEPS IN THE FINITE ELEMENT PROCESS

The development of a structure's global stiffness equations with the system's restraints imposed is a primary goal of the finite element analysis of the structure. It is these equations, when solved simultaneously, which yield the structure's nodal displacements. These global stiffness equations reflect the interaction of the gdof, applied forces and moments, support reactions, and support displacements on one another.

Once the global displacements of vector $\mathbf{q}_G$ are known, the support reactions, as part of vector $\mathbf{F}_G$, may be calculated. Moreover, the one-to-one correspondence of gdof and edof may be called on to identify the values of the nodal displacements for each finite element of the structural system. Then the element stiffness equations for each finite element may be employed to calculate the internal forces and moments at the nodes of the elements. These will be the element's axial force, shear, and moment at the nodes.

In finite element analyses of 2D or 3D continuous structural systems (plates, solid structures, etc.), the global or element displacements may be used further to calculate strains in the structure. The strains may then be substituted into the constitutive relationships for the material of the structure to compute its stresses.

The essential steps in the entire process may be summarized as follows:

- Identify the finite elements and nodes which will be employed to model the structural system. This step is called discretizing the structural system.
- Define the global reference axes for the structural system, its gdof, and the edof for each finite element of the system. This step includes establishing the numbering schemes which will be employed for finite elements, edof, and gdof.
- Write the element stiffness equations for each finite element of the structural system. In this step, it will be convenient to retain the element stiffness equations in two forms for 1D elements. One will be the initial form, referred to the element's own local reference axes. This is the form most easily obtained. The other will be the form referred to the structural system's global reference axes and is usually developed from mathematical transformations of the initial form. For 2D or 3D elements the transformation to global references axes is interwoven with the determination of element stiffness matrices and vectors of equivalent nodal loads.
- Assemble the global stiffness matrix from the superposition of each of the system's element stiffness matrices using the form of the element stiffness matrices referred to the structure's global axes. Write the global load vector and impose the restraints existing on the structural system in the global stiffness equations.
- Solve the global stiffness equations simultaneously for the system's unrestrained nodal displacements.
- Using the values of the unrestrained global nodal displacements, calculate the support reactions.
- Using the one-to-one correspondence of gdof and edof, identify the element displacements at nodes. Substitute these displacements into the element stiffness equations and calculate the element forces and moments at their nodes.

The chapters that follow will present detailed discussions and procedures for each of these steps.

BIBLIOGRAPHY

Becker, E. B., Cary, G. F., and Odom, J. T., *Finite Elements, An Introduction* (Chapter 1), Vol. 1, Prentice-Hall, Englewood Cliffs, NJ, 1976.

Cook, R. D., Malkus, D. S., and Plesha, M. E., *Concepts and Applications of Finite Element Analysis* (Chapter 1), Wiley, New York, 1989.

Desai, C. S., and Abel, J. F., *Introduction to the Finite Element Method* (Chapter 1), Van Nostrand Reinhold, New York, 1977.

Huebner, K. H., *The Finite Element Method for Engineers* (Chapter 1), Wiley, New York, 1975.

Logan, D. L., *A First Course in the Finite Element Method* (Chapter 1), PWS-Kent, Boston, 1992.

McGuire, W., and Gallagher, R. H., *Matrix Structural Analysis* (Chapter 1), Wiley, New York, 1979.

Reddy, J. N., *An Introduction to the Finite Element Method* (Chapter 1), 2nd ed., McGraw-Hill, New York, 1993.

Yang, T. Y., *Finite Element Structural Analysis* (Chapter 1), Prentice-Hall, Englewood Cliffs, NJ, 1986.

CHAPTER 3

ELEMENT STIFFNESS EQUATIONS BY DIRECT METHODS

3.1. DIRECT DETERMINATION OF ELEMENT STIFFNESS EQUATIONS

The direct determination of element stiffness equations employs the application of fundamental equilibrium concepts to the finite element and analysis of its force-deformation and moment-deformation behavior. It provides clear insights into the structural behavior of the finite element and thus into the finite element analysis of the structural system and perhaps an improved understanding of modeling structures using the finite element method. However, direct determination is practical only for simple 1D elements, such as prismatic truss, beam, and frame elements. For more complicated structural elements, other ways must be used to determine their element stiffness equations.

3.1.1. Element Stiffness Matrix: Global Axes

The entries of matrix $\mathbf{K}$ of equations 2.1 and 2.2 are called stiffness coefficients k_{ij}. They may be determined directly for a truss finite element from its force-deformation behavior and its equilibrium.

Observing the relationships expressed in equation 2.1b, if the nodal displacements of a 2-node truss element are specified as follows:

$$\text{If} \quad \{\mathbf{q}\} = \left\{ \begin{array}{c} 0 \\ 0 \\ 1 \\ 0 \end{array} \right\}, \quad \text{then} \quad \{\mathbf{F}\} = \left\{ \begin{array}{c} k_{13} \\ k_{23} \\ k_{33} \\ k_{43} \end{array} \right\}.$$

Figure 3.1 illustrates the truss element in this specified displaced state, showing its nodal forces and specified nodal displacements. Note the forces and displacements

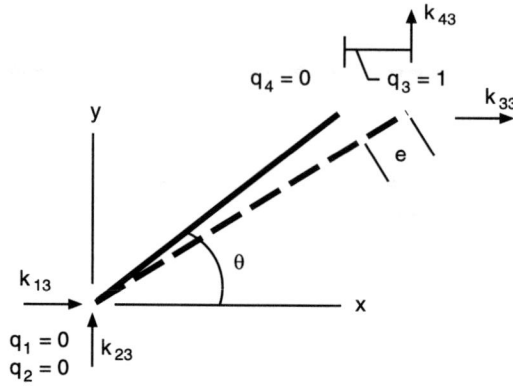

Figure 3.1 Truss element specified state

are aligned with axes different from the element itself. These axes will be referred to as global axes. The axial force P in the element is given as

$$P = \frac{k_{33}}{\cos \theta} \quad \text{or} \quad P = \frac{k_{43}}{\sin \theta}.$$

Geometrically, the elongation e is given by

$$e = q_3 \cos \theta = \cos \theta.$$

From basic mechanics

$$e = \frac{PL}{AE},$$

where L is the length of the element, A the cross-sectional area of the element, and E Young's modulus for the element. Substituting for the axial force P and the elongation e in this expression and rearranging yield

$$k_{33} = \frac{AE}{L} \cos^2 \theta \quad \text{or} \quad k_{43} = \frac{AE}{L} \sin \theta \cos \theta.$$

Equilibrium of the element requires that

$$k_{13} = -\frac{AE}{L} \cos^2 \theta \quad \text{and} \quad k_{23} = -\frac{AE}{L} \sin \theta \cos \theta.$$

Thus column 3 of the plane truss element stiffness matrix was obtained directly from the force-deformation characteristics of the truss element and its equilibrium by setting the displacement at edof 3 equal to one and the displacements at the other edof equal to zero. Columns 1, 2, and 4 of the element stiffness matrix may be obtained in a similar manner by setting displacements at edof 1, 2, and 4 equal to one respectively in three separate actions and the corresponding displacements at the other edof equal to zero.

With the angle θ measured from the positive global x axis to the axis of the element and letting $s = \sin\theta$ and $c = \cos\theta$, the complete plane truss element stiffness matrix referred to global reference axes is

$$[\mathbf{K}] = \frac{AE}{L}\begin{bmatrix} c^2 & sc & -c^2 & -sc \\ sc & s^2 & -sc & -s^2 \\ -c^2 & -sc & c^2 & sc \\ -sc & -s^2 & sc & s^2 \end{bmatrix}. \tag{3.1}$$

It is possible also to determine the element stiffness matrix of a beam element for a straight prismatic beam directly. The procedure is similar to the one described for the truss element. Observing the relationships expressed in equation 2.1b, if the displaced state of a 2-node beam element is specified with nodal displacements

$$\{\mathbf{q}\} = \begin{Bmatrix} 1 \\ 0 \\ 0 \\ 0 \end{Bmatrix},$$

then

$$\{\mathbf{F}\} = \begin{Bmatrix} k_{11} \\ k_{21} \\ k_{31} \\ k_{41} \end{Bmatrix}.$$

Figure 3.2 illustrates the beam element in this specified displaced state. Note that with a beam element edof 1 and 3 are transverse translations and 2 and 4 are slopes

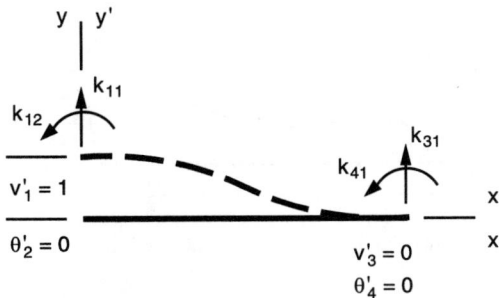

Figure 3.2 Beam element specified state

or rotations, and the global x–y and local x'–y' reference axes coincide. The corresponding forces and moments are end shears k_{11} and K_{31} and end moments k_{21} and k_{41}. A way to obtain the relationships among these quantities for a prismatic beam element is to use the slope deflection equations.

These may be developed from a beam element subjected to end moments, shears, rotations, and a relative end translation, as shown in Figure 3.3. The equation of the deformed element, shows as the dashed line, is $y = y(x)$. Its bending moment may be written as

$$M(x) = \frac{M_{AB} + M_{BA}}{L} x - M_{AB} = \frac{M_{AB}}{L}(x - L) + \frac{M_{BA}}{L} x. \tag{3.2}$$

From basic mechanics, $M(x) = EIy''(x)$, where E is the element's modulus of elasticity and I is its centroidal moment of inertia. Substituting and integrating,

$$EIy''(x) = \frac{M_{AB}}{L}(x - L) + \frac{M_{BA}}{L} x, \tag{3.3a}$$

$$EIy'(x) = \frac{M_{AB}}{L}\left(\frac{x^2}{2} - Lx\right) + \frac{M_{BA}}{L}\frac{x^2}{2} + C_1, \tag{3.3b}$$

$$EIy(x) = \frac{M_{AB}}{L}\left(\frac{x^3}{6} - \frac{Lx^2}{2}\right) + \frac{M_{BA}}{L}\frac{x^3}{6} + C_1 x + C_2. \tag{3.3c}$$

At point A on the element's left end,

$$y(0) = 0 \quad \text{and} \quad y'(0) = \theta_{AB}.$$

Applying these boundary conditions to equations 3.3b and 3.3c,

$$C_1 = EI\theta_{AB} \quad \text{and} \quad C_2 = 0.$$

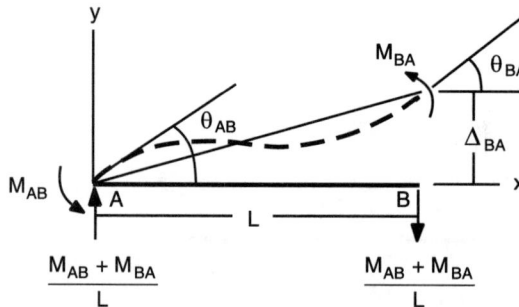

Figure 3.3 Beam element displaced state

Rewriting equations 3.3b and 3.3c,

$$EIy'(x) = \frac{M_{AB}}{L}\left(\frac{x^2}{2} - Lx\right) + \frac{M_{BA}}{2L}x^2 + EI\theta_{AB}, \tag{3.3d}$$

$$EIy(x) = \frac{M_{AB}}{L}\left(\frac{x^3}{6} - \frac{Lx^2}{2}\right) + \frac{M_{BA}}{6L}x^3 + EI\theta_{AB}x. \tag{3.3e}$$

At point B on the element's right end,

$$y(L) = \Delta_{BA} \quad \text{and} \quad y'(L) = \theta_{BA}.$$

Substituting into equations 3.3d and 3.3e and rearranging,

$$M_{AB} - M_{BA} = \frac{2EI}{L}\theta_{AB} - \frac{2EI}{L}\theta_{BA}, \tag{3.3f}$$

$$2M_{AB} - M_{BA} = \frac{6EI}{L}\theta_{AB} - \frac{6EI}{L^2}\Delta_{BA}. \tag{3.3g}$$

Solving equations 3.3f and 3.3g for M_{AB} and M_{BA},

$$M_{AB} = \frac{EI}{L}\left(4\theta_{AB} + 2\theta_{BA} - \frac{6\Delta_{BA}}{L}\right), \tag{3.4a}$$

$$M_{BA} = \frac{EI}{L}\left(2\theta_{AB} + 4\theta_{BA} - \frac{6\Delta_{BA}}{L}\right). \tag{3.4b}$$

Equations 3.4 are a form of the classical slope-deflection equations. The end moments k_{21} and k_{41} shown in Figure 3.2 may be written directly by applying these slope deflection equations,

$$k_{21} = k_{41} = \left(\frac{EI}{L}\right)\left(-\frac{6(-v_1)}{L}\right) = \frac{6EI}{L^2}.$$

Note that the relative displacement of the right end of the element with respect to its left end, Δ_{BA}, is negative in Figure 3.2.

Summing moments about the right end of the element in Figure 3.2 and using the expressions for k_{21} and k_{41},

$$k_{21} + k_{41} - K_{11}L = 0,$$

$$k_{11} = \frac{12EI}{L^3}.$$

Summing forces in the vertical direction,

$$k_{31} = -\frac{12EI}{L^3}.$$

Thus column 1 of the element stiffness matrix was obtained directly from the moment-deformation characteristics and equilibrium of the beam element by setting the displacement at edof 1 equal to one and the displacements at the other edof equal to zero. Columns 2, 3, and 4 of the element stiffness matrix may be obtained separately in a similar manner by setting the displacements at the other edof equal to one respectively in three separate actions and the corresponding displacements at the other edof equal to zero. The complete beam element stiffness matrix is

$$[\mathbf{K}] = \frac{EI}{L} \begin{bmatrix} \frac{12}{L^2} & \frac{6}{L} & -\frac{12}{L^2} & \frac{6}{L} \\ \frac{6}{L} & 4 & -\frac{6}{L} & 2 \\ -\frac{12}{L^2} & -\frac{6}{L} & \frac{12}{L^2} & -\frac{6}{L} \\ \frac{6}{L} & 2 & -\frac{6}{L} & 4 \end{bmatrix}. \tag{3.5}$$

The structural behavior of the beam element is apparent in this development, due in large measure to the local reference axes. One axis is the axis of the beam and one is transverse to the beam and in the plane of bending. They relate directly to transverse displacement and shear. The third is perpendicular to the plane of bending and relates directly to slope, rotation, and moment. Equation 3.5 may also be regarded as referred to the beam's global reference axes, since a straight beam's global and local element reference axes may coincide.

3.1.2. Element Stiffness Matrix: Local Axes

A 6-dof plane frame element inclined from the frame's global reference axes is illustrated in Figure 3.4. Both global and local edof are shown. The angle θ is measured from the positive global x axis to the positive local x' axis. It would be difficult to obtain its stiffness matrix directly when referencing the frame's global axes. However, as we shall see, its stiffness matrix when referred to its own local axes

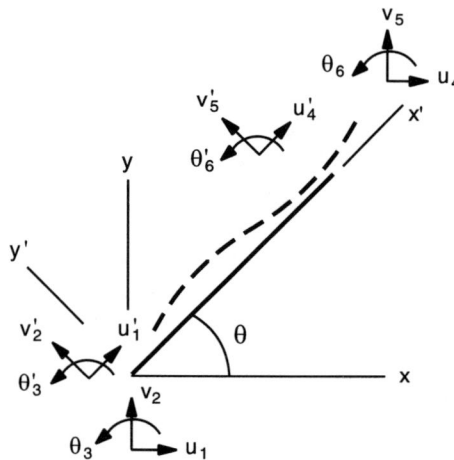

Figure 3.4 Plane frame element displaced state

can be obtained from a superposition of axial and beam structural behavior when both are referred to the same local axes. The resulting stiffness matrix can be transformed later to refer to the frame's global axes.

Since axial and flexural behavior of structural finite elements are unaffected by one another for small deformations, the stiffness equations for one behavior will be uncoupled from the other. Thus they may be developed separately, referred to the same coordinate axes, and grouped together and written as one matrix equation:

$$\{\mathbf{F}'\} = [\mathbf{K}']\{\mathbf{q}'\}. \tag{3.6}$$

Equation 3.6 is identical to equation 2.1b, except for the "primes" on each matrix symbol, which indicate reference to local coordinate axes. The force vector $\mathbf{F}'$ contains the six forces and moments along the lines of its local reference axes. They will be axial force, shear, and moment at the ends of the element. The displacement vector $\mathbf{q}'$ contains the six translations and rotations which are the edof. They will be axial deformation, transverse translation, and slope at the ends of the element. The stiffness matrix $\mathbf{K}'$ contains six rows and six columns of stiffness coefficients; 16 of them are the stiffness coefficients shown in equation 3.5.

Since equation 3.1 is an axial stiffness matrix referred to global reference axes, it cannot be combined directly with equation 3.5 to produce a plane frame element stiffness. Instead the axial stiffness matrix referred to local axes must be developed.

Figure 3.5 illustrates a 2-node axial finite element employing local reference axes. Note that since the element's axis is the reference axis, there can be only two edof. Consequently, the element stiffness equations must be of the form

$$\left\{ \begin{array}{c} F_1' \\ F_2' \end{array} \right\} = \left[\begin{array}{cc} k_{11}' & k_{12}' \\ k_{21}' & k_{22}' \end{array} \right] \left\{ \begin{array}{c} u_1' \\ u_2' \end{array} \right\}. \tag{3.7}$$

The stiffness coefficients in equation 3.7 are referred to local axes since they relate forces and displacements which are referred to the local axes. As we shall see, these stiffness coefficients are different from those referred to global axes.

If the displaced state of the truss element is specified with nodal displacements

$$\{\mathbf{q}'\} = \left\{ \begin{array}{c} 1 \\ 0 \end{array} \right\},$$

then

$$\{\mathbf{F}'\} = \left\{ \begin{array}{c} k_{11}' \\ k_{21}' \end{array} \right\}.$$

Figure 3.5 Truss element displaced state

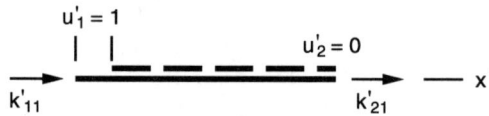

Figure 3.6 Truss element specified state

Figure 3.6 illustrates the truss element in this specified displaced state, referred to local axes. The element's axial force is K'_{11} and its displacement u'_1 equals its deformation. From basic mechanics, the deformation of the element is

$$u'_1 = 1 = \frac{k'_{11}L}{AE},$$

or

$$k'_{11} = \frac{AE}{L}.$$

Equilibrium of the element indicates that

$$k'_{21} = -\frac{AE}{L}.$$

Thus column 1 of the element stiffness matrix was obtained directly from its force-deformation behavior and its equilibrium by setting the displacement at edof 1 equal to one and the displacement at the other edof to zero. Column 2 may be obtained in a similar manner as

$$k'_{12} = -\frac{AE}{L} \quad \text{and} \quad k'_{22} = \frac{AE}{L}.$$

The complete 2-node truss element stiffness matrix when referred to its local axes is

$$[\mathbf{K}'] = \frac{AE}{L} \begin{bmatrix} 1 & -1 \\ -1 & 1 \end{bmatrix}. \tag{3.8}$$

Equation 3.8 could also have been obtained from equation 3.1 by setting the angle θ equal to zero, so that the local and global axes coincide. If rows 2 and 4 and columns 2 and 4, all zeros, are then deleted, the result is equation 3.8.

The relationships among the axial force, shear, and moment and the axial, transverse, and rotation displacements for the plane frame element shown in Figure 3.4 can be written as six stiffness equations, each referred to the element's local coordinate axes; in matrix form, they are equations 3.9. The axial stiffness equations, by convention, are assigned as algebraic equations 1 and 4; the flexural stiffness equations are assigned as algebraic equations 2, 3, 5, and 6. The zeros in rows 1 and 4 and columns 1 and 4 reflect the fact that the axial and flexural stiffness equations are uncoupled from one another:

$$\begin{Bmatrix} F_1' \\ F_2' \\ F_3' \\ F_4' \\ F_5' \\ F_6' \end{Bmatrix} = \frac{E}{L} \begin{bmatrix} A & 0 & 0 & -A & 0 & 0 \\ 0 & \frac{12I}{L^2} & \frac{6I}{L} & 0 & -\frac{12I}{L^2} & \frac{6I}{L} \\ 0 & \frac{6I}{L} & 4I & 0 & -\frac{6I}{L} & 2I \\ -A & 0 & 0 & A & 0 & 0 \\ 0 & -\frac{12I}{L^2} & -\frac{6I}{L} & 0 & \frac{12I}{L^2} & -\frac{6I}{L} \\ 0 & \frac{6I}{L} & 2I & 0 & -\frac{6I}{L} & 4I \end{bmatrix} \begin{Bmatrix} u_1' \\ v_2' \\ \theta_3' \\ u_4' \\ v_5' \\ \theta_6' \end{Bmatrix} \qquad (3.9)$$

or

$$\{\mathbf{F}'\} = [\mathbf{K}']\{\mathbf{q}'\}. \qquad (3.6)$$

3.1.3. Element Load Vector

If loads are applied only at nodes of a structure, the element load vector $\mathbf{F}'$ is an array of unknown internal axial forces, shears, and moments on the element at its nodes. The applied nodal forces and moments, however, must be reduced to components whose lines of action are along the global axes and added algebraically to the global load vector at the appropriate gdof positions. Clearly applied nodal forces and moments affect the global stiffness equations and its solution for the nodal displacements.

Applied loads are always at nodes for trusses since the joints are always taken as the nodes. Concentrated forces and couples can be applied only at nodes of a beam, frame, or other continuous structure if the analyst ensures that nodes are located where the forces and couples occur.

It may not always be convenient or efficient to place nodes where concentrated forces or couples are applied. It is not possible if the applied load is distributed over the length of a 1D element or over the area or volume of a 2D or 3D element. It is possible in these circumstances, however, to reduce the loads between nodes to equivalent concentrated forces and moments applied at the element nodes which produce the same nodal displacements that the actual applied loads between the nodes produce. These equivalent nodal loads must be regarded as applied external nodal nodes which are added algebraically to both the element and global load vectors.

The equivalent forces and couples are determined as an element force vector $\mathbf{F}_E'$, whose entries are force and moment components with lines of action along the element's local axes. The vector $\mathbf{F}_E'$ must be transformed later to a vector $\mathbf{F}_E$ whose components are along the global coordinate axes so they may be added algebraically to the global load vector at the appropriate gdof positions. Clearly these equivalent loads affect the global load vector and thus its solution for the nodal displacements of the structure. Indeed they should; they are determined so that they produce the same effect on the nodal displacements that the applied loads between nodes do.

The direct determination of the applied equivalent nodal loads for an element with applied loads between its nodes can be obtained for a 1D element through the superposition of two configurations of the element. In the first configuration, the actual restraints on all of the edof are increased to full restraints so that the nodal displacements are all zero. The restraint reactions resulting from the applied load between the nodes and the fully restrained edof are then determined. These restraint reactions are referred to as the element's fixed end moment, shear, and axial load vector $\mathbf{F}'_{\text{FEMVA}}$.

The second configuration is derived from the first by removing the applied load between the nodes and replacing it with the negatives of the entries of the element's nodal fixed end shear, moment, and axial load vector. In addition, "negative" restraints are applied to the edof so that the restraints are decreased from full restraint to the original actual restraints. The element in this configuration will have nodal displacements resulting from the applied load vector $-\mathbf{F}_{\text{FEMVA}}$ and the original restraints.

If the applied loads and solutions for nodal displacements of these two configurations are superimposed (added algebraically), what obtains is an element with the following:

- Its nodal restraints are the actual nodal restraints since the restraints applied in the second configuration are the negatives of those applied in the first.
- Its actual applied load is just the original applied load between nodes, since the applied nodal loads in the second configuration vector $-\mathbf{F}'_{\text{FEMVA}}$ are the negatives of the nodal restraint reactions occurring in the first.
- Its nodal displacements are only obtained from the second configuration, since those in the first configuration were zero.

The superimposed nodal displacements must be the correct nodal displacements for the element acting under its applied load between the nodes, because the element is subjected only to the applied load between its nodes and its actual restraints. Moreover, these nodal displacements were produced by the negative entries of the element's fixed moment, shear, and axial force vector. Consequently the element's equivalent applied nodal loads are

$$\{\mathbf{F}'_E\} = -\{\mathbf{F}'_{\text{FEMVA}}\}. \tag{3.10}$$

Fixed end moments and shears have been determined for a variety of transverse loads on a beam with fixed supports. The negatives of these are a beam element's equivalent applied nodal loads. Figure 3.7 displays them for a beam element with four common loads between nodes. These equivalent applied nodal moments and shears may be superimposed on one another to determine the equivalent applied nodal moments and shears for elements with several loads between nodes.

When the nodal displacements obtained from solving the global stiffness equations are used with the element stiffness equations to calculate the element's internal force vector, what is obtained is the algebraic sum of the element's internal forces plus any equivalent applied nodal loads. Consequently, the equivalent applied load vector $\mathbf{F}'_E$ must be subtracted from the result of this calculation to obtain the correct nodal internal forces, or

$$\{\mathbf{F}'\} = [\mathbf{K}']\{\mathbf{q}'\} - \{\mathbf{F}'_E\}. \tag{3.11}$$

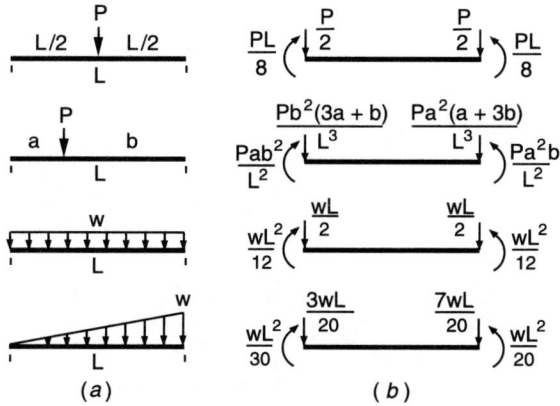

Figure 3.7 Equivalent applied nodal loads

3.2. TRANSFORMATION OF ELEMENT STIFFNESS EQUATIONS: 1D ELEMENTS IN 2D AND 3D SPACE

The assembly of global stiffness equations of a structure from the superposition of its element stiffness equations, to be discussed in Chapter 4, requires that the global stiffness equations and each of the element stiffness equations be referred to a common set of coordinate axes, the global axes. The transformation of each set of element stiffness equations from reference to its local axes to the global axes in a plane structural system involves a rotation through the angle θ of matrices $\mathbf{F}'$ to $\mathbf{F}$, $\mathbf{K}'$ to $\mathbf{K}$, and $\mathbf{q}'$ to $\mathbf{q}$. The transformation is the same in concept for each element of a structure, though each element transformation differs because each will have a different angle θ measured from the global x axis of the structure. It is similar for different types of structural systems, though clearly the transformations of truss or frame elements will differ in their details.

3.2.1. Displacement Transformation

The vector $\mathbf{q}$ (Figure 3.8) has components q_1 and q_2 along the global x–y axes; its components along the local x'–y' axes are q_1' and q_2'. The primes on coordinates, displacements, or matrices indicate reference to local axes; the absence of the prime indicates reference to global axes.

From Figure 3.8,

$$q_1' = a + b, \qquad a = (\cos \theta)q_1, \qquad b = (\sin \theta)q_1,$$

so that

$$q_1' = (\cos \theta)q_1 + (\sin \theta)q_2.$$

Also, from Figure 3.8,

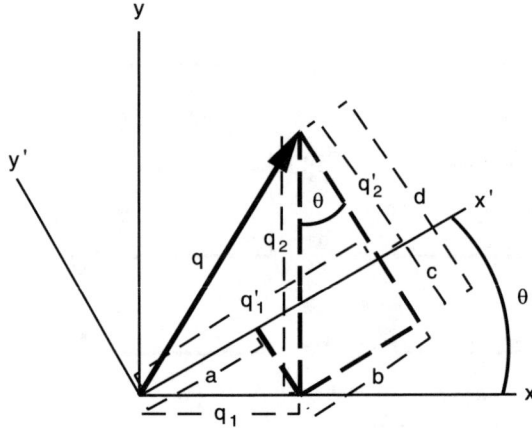

Figure 3.8 Transformation of vector components

$$q_2' = -c + d, \qquad c = (\sin \theta)q_1, \qquad d = (\cos \theta)q_2,$$

so that

$$q_2' = -(\sin \theta)q_1 + (\cos \theta)q_2.$$

In matrix form with $s = \sin \theta$ and $c = \cos \theta$,

$$\begin{Bmatrix} q_1' \\ q_2' \end{Bmatrix} = \begin{bmatrix} c & s \\ -s & c \end{bmatrix} \begin{Bmatrix} q_1 \\ q_2 \end{Bmatrix}, \tag{3.12}$$

or

$$\{\mathbf{q}'\} = [\mathbf{T}]\{\mathbf{q}\}. \tag{3.13}$$

The transformation matrix $\mathbf{T}$ in equations 3.12 and 3.13 transforms the global components of the vector $\mathbf{q}$ to local components.

Though the entries of matrix $\mathbf{T}$ are sines or cosines of the angle θ, a more general interpretation of these entries is that they are the direction cosines of the local axes with respect to the global axes. The first and second entries in row 1 are the direction cosines of the x' axis with respect to the x and y axes, respectively; the first and second entries in row 2 are the direction cosines of the y' axis with respect to the x and y axes.

By extension, it should be clear that if the vector $\mathbf{q}$ were in 3D space, the entries of its transformation matrix would be the direction cosines of the x'–y'–z' axes with respect to the x–y–z axes. The first-row entries would be the direction cosines of the x' axis with respect to the x, y, and z axes, respectively; the second-row entries would be the direction cosines of the y' axis with respect to the x, y, and z axes; and the third-row entries would be the direction cosines of the z' axis with respect to the x, y, and z axes. Equation 3.13 would apply; however, vector $\mathbf{q}'$ would contain the three local vector components q_1', q_2' and q_3'; vector $\mathbf{q}$ would contain the three global vector components q_1, q_2, and q_3; and matrix **T** would contain the nine direction cosines described. Equation 3.14 shows the 3D vector component transformation:

$$
\left\{ \begin{array}{c} q_1' \\ q_2' \\ q_3' \end{array} \right\} = \left[\begin{array}{ccc} l_1 & m_1 & n_1 \\ l_2 & m_2 & n_2 \\ l_3 & m_3 & n_3 \end{array} \right] \left\{ \begin{array}{c} q_1 \\ q_2 \\ q_3 \end{array} \right\},
\tag{3.14}
$$

or

$$
[\mathbf{q}'] = [\mathbf{T}]\{\mathbf{q}\}.
\tag{3.13}
$$

The plane truss elements illustrated in Figures 2.5, 3.1, 3.5, and 3.6 have a set of translation displacement components at each node, represented by the vector **q**. One could write the transformation of the components at both nodes as

$$
\left\{ \begin{array}{c} q_1' \\ q_2' \\ q_3' \\ q_4' \end{array} \right\} = \left[\begin{array}{cccc} c & s & 0 & 0 \\ -s & c & 0 & 0 \\ 0 & 0 & c & s \\ 0 & 0 & -s & c \end{array} \right] \left\{ \begin{array}{c} q_1 \\ q_2 \\ q_3 \\ q_4 \end{array} \right\}.
\tag{3.15a}
$$

It is clear from Figures 3.5 and 3.6 that the local axes used allow only 2 edof; q_2' and q_4' are zero. However, there are four edof when the nodal displacements are referred to global axes. The transformation of plane truss edof from a global reference to a local reference is more simply represented as

$$
\left\{ \begin{array}{c} q_1' \\ q_2' \end{array} \right\} = \left[\begin{array}{cccc} c & s & 0 & 0 \\ 0 & 0 & c & s \end{array} \right] \left\{ \begin{array}{c} q_1 \\ q_2 \\ q_3 \\ q_4 \end{array} \right\},
\tag{3.15b}
$$

or

$$
\{\mathbf{q}'\} = [\mathbf{T}]\{\mathbf{q}\}.
\tag{3.13}
$$

Equation 3.15b may be obtained from equation 3.15a by simply deleting the second and fourth equations from the system of equations 3.15a and redesignating q_3' as q_2'.

If the truss element were in 3D space, equation 3.13 would be

$$
\left\{ \begin{array}{c} q_1' \\ q_2' \end{array} \right\} = \left[\begin{array}{cccccc} l_1 & m_1 & n_1 & 0 & 0 & 0 \\ 0 & 0 & 0 & l_1 & m_1 & n_1 \end{array} \right] \left\{ \begin{array}{c} q_1 \\ q_2 \\ q_3 \\ q_4 \\ q_5 \\ q_6 \end{array} \right\}. \tag{3.15c}
$$

The plane frame edof shown in Figure 3.4 may be regarded as two sets of three vector components, one set at each node. The translation vector components at each node are transformed with equations 3.15a. There will be four translation components in vector $\mathbf{q}'$ since a plane frame element has both axial and transverse translation displacements at each node. It also has a rotation displacement at each node aligned with the axis perpendicular to the plane of the element. The rotation is the same in both the local and global reference frame. The plane frame element's displacement component transformation equation may be written as

$$
\left\{ \begin{array}{c} u_1' \\ v_2' \\ \theta_3' \\ u_4' \\ v_5' \\ \theta_6' \end{array} \right\} = \left[\begin{array}{cccccc} c & s & 0 & 0 & 0 & 0 \\ -s & c & 0 & 0 & 0 & 0 \\ 0 & 0 & 1 & 0 & 0 & 0 \\ 0 & 0 & 0 & c & s & 0 \\ 0 & 0 & 0 & -s & c & 0 \\ 0 & 0 & 0 & 0 & 0 & 1 \end{array} \right] \left\{ \begin{array}{c} u_1 \\ v_2 \\ \theta_3 \\ u_4 \\ v_5 \\ \theta_6 \end{array} \right\}, \tag{3.16}
$$

or

$$
\{\mathbf{q}'\} = [\mathbf{T}]\{\mathbf{q}\}. \tag{3.13}
$$

The form of equations 3.13, transforming displacements referred to global axes to local axes, is the required form for displacement transformation in the finite element method. As we shall see, it is the form needed to develop the transformation of the element stiffness matrix referred to local axes, from local to global axes. We saw earlier (Section 2.6) that it is the form needed to transform the nodal displacements calculated from the solution of the global stiffness equations which are identified as element displacements to displacements referred to local axes. These are then used to obtain the element forces and moments referred to local axes using the element stiffness equations referred to local axes.

3.2.2. Force and Moment Transformation

Forces and moments must also be transformed between local and global reference axes. The transformations developed for displacements were for any vector, so they apply also for forces and moments. One has only to replace vectors $\mathbf{q}'$ with $\mathbf{F}'$ and $\mathbf{q}$

with $\mathbf{F}$. However, the form of these transformations is not what is needed for the finite element method. What is required is the inverse of this form, or

$$\{\mathbf{F}\} = [\mathbf{T}_f]\{\mathbf{F}'\}. \tag{3.17}$$

Matrix $\mathbf{T}_f$ may not be the inverse of matrix $\mathbf{T}$. If matrix $\mathbf{T}$ is not square, there is no inverse defined for it. One notes that the matrices $\mathbf{T}$ in equations 3.15b and 3.15c are not square. When it is square (equations 3.12, 3.14, 3.15a, and 3.16), its inverse exists, and it equals its transpose. When matrix $\mathbf{T}$ is square, it is orthogonal.

It is possible to define dummy equations and edof when the number of edof in one reference frame differs from the number in the other. In this manner the matrix $\mathbf{T}$ can be made square. Its inverse is then obtained by transposing it. For example, equation 3.15b is not changed in substance if dummy equations $q_d' = 0$ are included. In matrix notation, equation 3.15b becomes

$$\begin{Bmatrix} q_1' \\ q_d' \\ q_3' \\ q_d' \end{Bmatrix} = \begin{bmatrix} c & s & 0 & 0 \\ 0 & 0 & 0 & 0 \\ 0 & 0 & c & s \\ 0 & 0 & 0 & 0 \end{bmatrix} \begin{Bmatrix} q_1 \\ q_2 \\ q_3 \\ q_4 \end{Bmatrix}. \tag{3.18}$$

It is an inefficient process, however, requiring the storage, accounting for, and handling of many nonessential zeros.

A general way to develop a more efficient form of the needed force and moment transformations is to use the displacement transformations and energy conservation in the structural system. Most structural systems may be regarded as conservative. That is, the work done on them, by their applied forces and moments as they move through their corresponding displacements, is transformed into elastic potential energy which is completely recoverable when the forces and moments are removed.

The work done by these forces and moments, referred to global axes, as they move through the displacements they cause is given as

$$W = \tfrac{1}{2}(q_1 F_1 + q_2 F_2 + q_3 F_3 + \cdots + q_n F_n). \tag{3.19a}$$

The factor $\tfrac{1}{2}$ is included because the displacements are zero before the forces and moments are applied, and the load-deformation behavior is assumed to be linear. In matrix form

$$W = \tfrac{1}{2}\lfloor q_1 \quad q_2 \quad q_3 \quad \cdots \quad q_n \rfloor \begin{Bmatrix} F_1 \\ F_2 \\ F_3 \\ \vdots \\ F_n \end{Bmatrix}, \tag{3.19b}$$

$$W = \tfrac{1}{2}\lfloor \mathbf{q} \rfloor \{\mathbf{F}\}. \tag{3.19c}$$

If the forces, moments, and displacements are referred to local axes, the total amount of work done is unchanged, since work is a scalar quantity and the structural system is conservative. The form of their equations will be the same as equations 3.19. The difference is that the forces, moments, and displacements will have primes. Thus,

$$W = \tfrac{1}{2}\lfloor \mathbf{q}' \rfloor \{\mathbf{F}'\}. \tag{3.20}$$

Equating equations 3.20 and 3.19c results in

$$\lfloor \mathbf{q} \rfloor \{\mathbf{F}\} = \lfloor \mathbf{q}' \rfloor \{\mathbf{F}'\}. \tag{3.21}$$

The displacement transformation is

$$\{\mathbf{q}'\} = [\mathbf{T}]\{\mathbf{q}\}. \tag{3.13}$$

Its transpose is

$$\lfloor \mathbf{q}' \rfloor = \lfloor \mathbf{q} \rfloor [\mathbf{T}^{\mathsf{T}}]. \tag{3.22}$$

Substituting equation 3.22 into equation 3.21,

$$\lfloor \mathbf{q} \rfloor \{\mathbf{F}\} = \lfloor \mathbf{q} \rfloor [\mathbf{T}^{\mathsf{T}}]\{\mathbf{F}'\}. \tag{3.23}$$

Since equation 3.23 is a valid matrix equation, so also is

$$\{\mathbf{F}\} = [\mathbf{T}^{\mathsf{T}}]\{\mathbf{F}'\}. \tag{3.24}$$

This is the desired force and moment transformation for the finite element method. It is called a contragredient transformation. In a conservative structural system, if displacements transform as in equation 3.13, then forces and moments must transform as in equation 3.24. The transformation matrix in the force–moment transformation is the transpose of the transformation matrix in the displacement transformation.

As an example, the contragredient counterpart to equation 3.15b is

$$\begin{Bmatrix} F_1 \\ F_2 \\ F_3 \\ F_4 \end{Bmatrix} = \begin{bmatrix} c & 0 \\ s & 0 \\ 0 & c \\ 0 & s \end{bmatrix} \begin{Bmatrix} F_1' \\ F_2' \end{Bmatrix}. \tag{3.25}$$

Equation 3.24 is applied in the finite element method to transform from the local axes of the element to the global axes of the structure, the vector of equivalent applied axial forces, shears, and moments $\mathbf{F}_E'$ obtained for elements with loads between nodes:

$$\{\mathbf{F}_E\} = [\mathbf{T}^{\mathsf{T}}]\{\mathbf{F}_E'\}. \tag{3.26}$$

Note that with equations 3.13 and 3.24 the analyst has two convenient ways to obtain the transformation matrix **T**. The displacement components of the element can be analyzed to obtain matrix **T** for equation 3.13 and then its transpose $\mathbf{T}^{\mathrm{T}}$ written for equation 3.24. Alternatively, the force and moment components of the element can be analyzed to obtain $\mathbf{T}^{\mathrm{T}}$ for equation 3.24 and then its transpose **T** written for use with equation 3.13.

Both matrices **T** and $\mathbf{T}^{\mathrm{T}}$ are required to transform displacements and forces and moments effectively. As we shall see, both are also required to transform the element stiffness matrices , **K**′ to **K**.

3.2.3. Element Stiffness Matrix Transformation

To complete the development of the transformation of element stiffness equations from local to global axes, an effective procedure to transform the element stiffness matrices is needed. The development is based on the conservative properties of structural systems which led to the contragredient transformations described above for the force and displacement transformations.

The element stiffness equations with respect to local axes are

$$\{\mathbf{F}'\} = [\mathbf{K}']\{\mathbf{q}'\}. \tag{3.6}$$

If both sides of equations 3.6 are premultiplied by matrix $\mathbf{T}^{\mathrm{T}}$, the result is

$$[\mathbf{T}^{\mathrm{T}}][\mathbf{F}'] = [\mathbf{T}^{\mathrm{T}}][\mathbf{K}']\{\mathbf{q}'\}. \tag{3.27}$$

Substituting for the matrix product $[\mathbf{T}^{\mathrm{T}}]\{\mathbf{F}'\}$ from equation 3.24 and for vector $\mathbf{q}'$ from equation 3.13 results in

$$\{\mathbf{F}\} = [\mathbf{T}^{\mathrm{T}}][\mathbf{K}'][\mathbf{T}]\{\mathbf{q}\}. \tag{3.28}$$

The element stiffness equations referred to global axes are

$$\{\mathbf{F}\} = [\mathbf{K}]\{\mathbf{q}\}. \tag{2.2}$$

Comparing equations 2.2 and 3.28, it is clear that

$$[\mathbf{K}] = [\mathbf{T}^{\mathrm{T}}][\mathbf{K}'][\mathbf{T}]. \tag{3.29}$$

Equation 3.29 is the desired element stiffness matrix transformation for the finite element method. One develops the element stiffness matrix **K**′ referred to its local coordinate axes, develops the element's transformation matrix **T** from either an analysis of the element's displacements or its forces and moments, and performs the matrix operations indicated by equation 3.29 to put the element stiffness matrix into a form which refers it to the structure's global axes so that it may be assembled into the global stiffness matrix.

3.3. CLOSING REMARKS

It is worth noting that the development of element stiffness equations so far has been limited to the direct approach, one that has been suitable for simple 1D finite elements. If more sophisticated 1D elements and useful 2D and 3D elements are to be studied, other approaches will be needed. These will developed beginning with Chapter 5 and through the remainder of this book.

At this point, however, to obtain a more complete grasp of the essentials of the finite element method in its entirety, it is desirable to study next the global stiffness equations: their formation, solution, and the results from their solution.

BIBLIOGRAPHY

Beer, F. P., and Johnston, E. R. Jr., *Mechanics of Materials*, 2nd ed., McGraw-Hill, New York, 1992.

Cook, R. D., Malkus, D. S., and Plesha, M. E., *Concepts and Applications of Finite Element Analysis* (Chapters 2, 7), Wiley, New York, 1989.

Hibbeler, R. C., *Structural Analysis* (Chapters 10, 14, 15), 3rd ed., Prentice-Hall, Englewood Cliffs, NJ, 1994.

Logan, D. L., *A First Course in the Finite Element Method* (Chapter 3), PWS-Kent, Boston, 1992.

McGuire, W., and Gallagher, R. H., *Matrix Structural Analysis* (Chapters 2, 5), Wiley, New York, 1979.

Weaver, W. Jr., and Johnston, P. R., *Finite Elements for Structural Analysis* (Chapter 1), Prentice-Hall, Englewood Cliffs, NJ, 1984.

Weaver, W. Jr., and Gere, J. M., *Matrix Analysis of Framed Structures* (Chapters 1, 3), 3rd ed., Van Nostrand Reinhold, New York, 1990.

Yang, T. Y., *Finite Element Structural Analysis* (Chapter 4, 5), Prentice-Hall, Englewood Cliffs, NJ, 1986.

PROBLEMS

3.1. Given the two-member plane truss shown for which joint B displaces in-plane right 0.006 in. and downward 0.0002 in.

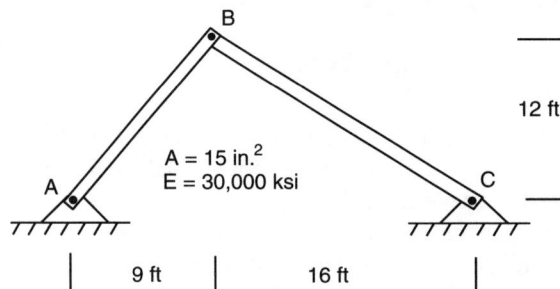

Problem 3.1

a. Write the element stiffness matrix $\mathbf{K}'$ for each member with respect to local element axes.

b. Write each member's transformation matrix $\mathbf{T}$ using x–y axes through points A and C as the structure's global axes.

c. Determine each member's change in length using results from part b and software which manipulates individual matrices.

d. Determine the forces in each member using results from parts a and c and software which manipulates individual matrices.

e. What force must be applied at joint B to cause it to displace as described?

3.2. Repeat Problem 3.1 for the three-member plane truss shown.

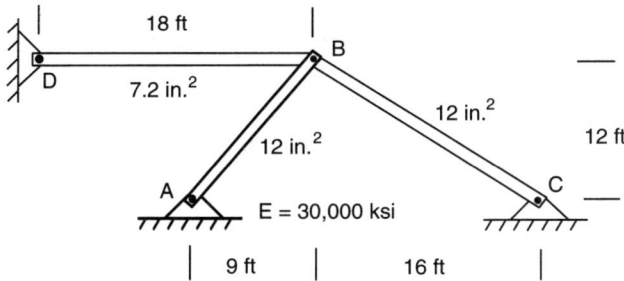

Problem 3.2

3.3. Repeat Problem 3.1 for the three-member plane truss shown for which point B displaces in-plane 6 mm to the right and 3 mm downward. Numbers shown near each member are its cross-sectional area in square millimeters $\times 10^3$.

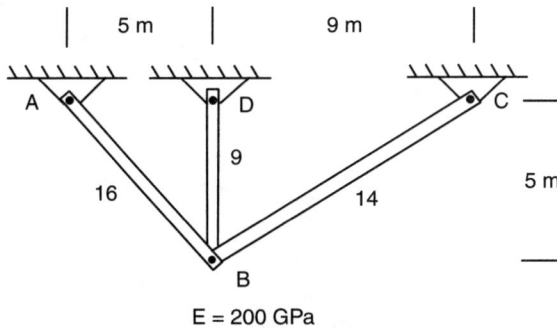

Problem 3.3

3.4. For the truss shown:

a. Write the element stiffness matrix $\mathbf{K}'$ for each member with respect to local element axes.

b. Write each member's transformation matrix $\mathbf{T}$ using x–y axes through points A and C as the structure's global axes.

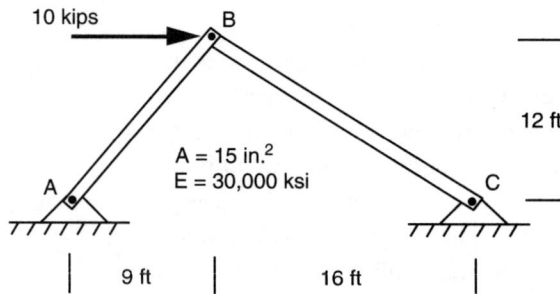

10 kips

B

12 ft

A = 15 in.²
E = 30,000 ksi

A

C

9 ft | 16 ft

Problem 3.4

c. Using results from parts *a* and *b* and software which manipulates individual matrices, write each element's stiffness equations with respect to the global axes.

d. Using appropriate individual algebraic element stiffness equations and software which manipulates individual matrices and considering the equilibrium of forces and the compatibility of displacements at joint *B*, determine the displacement of joint *B*.

e. Determine each member's change in length using results from parts *b* and *d* and software which manipulates individual matrices.

f. Determine the forces in each member using results from parts *a* and *e* and software which manipulates individual matrices.

3.5. The beam shown has the following slopes at each support:

$$\theta_A = 0.00045 \text{ rad clockwise (CW)},$$

$$\theta_B = 0.00030 \text{ rad counterclockwise (CCW)},$$

$$\theta_c = 0.00025 \text{ rad CW}.$$

a. Write the stiffness matrices for members *AB* and *BC*.

b. Using software which manipulates individual matrices, calculate the shears and moments in members *AB* and *BC*.

c. What are the applied forces and moments and support reactions needed to cause these slopes, shears, and moments?

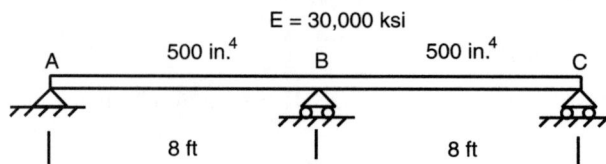

E = 30,000 ksi

A   500 in.⁴   B   500 in.⁴   C

8 ft | 8 ft

Problem 3.5

3.6. Repeat Problem 3.5 for the beam shown.

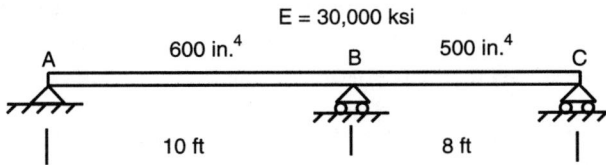

E = 30,000 ksi

A   600 in.^{4} B   500 in.^{4} C

10 ft   8 ft

Problem 3.6

3.7. Repeat Problem 3.5 for the beam shown, which has the following slopes at supports B and C:

$$\theta_B = 0.00025 \text{ rad CW},$$
$$\theta_C = 0.00040 \text{ rad CCW},$$

and whose deflection at C is

$$\delta_c = 0.050 \text{ in.} \downarrow$$

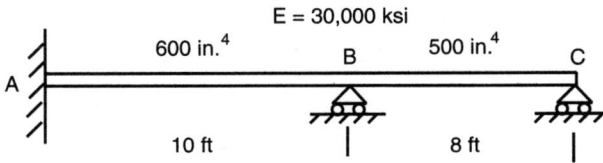

E = 30,000 ksi

A   600 in.^{4} B   500 in.^{4} C

10 ft   8 ft

Problem 3.7

3.8. For the beam shown:

a. Write the element stiffness matrices for members AB and BC.

b. Determine the vector of equivalent forces and moments for the 10-kip concentrated load on member AB.

c. Using appropriate individual element stiffness equations and software which manipulates individual matrices and considering the equilibrium of

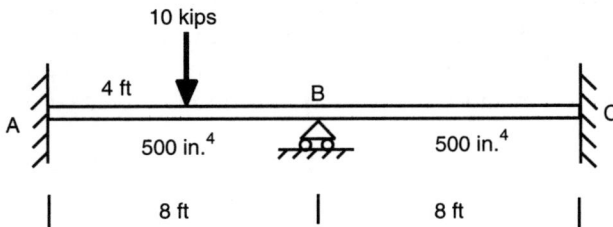

10 kips

4 ft   B

A   500 in.^{4} 500 in.^{4} C

8 ft   8 ft

E = 30,000 ksi

Problem 3.8

forces and moments and the compatibility of displacements at joint *B*, determine the slope of the beam at support *B*.

  d. Calculate the shears and moments in members *AB* and *BC* using results from parts a, b, and c and software which manipulates individual matrices.

  e. Determine the beam's support reactions.

3.9. Write the vector of equivalent forces and moments for each of the beam elements shown.

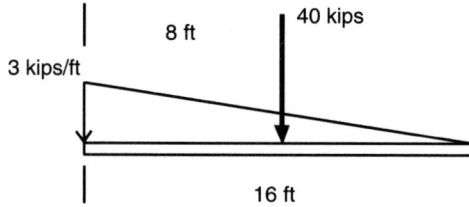

Problem 3.9a

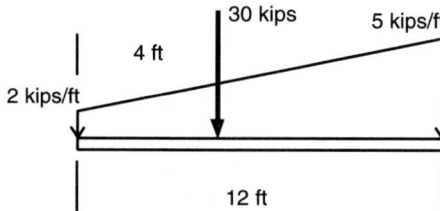

Problem 3.9b

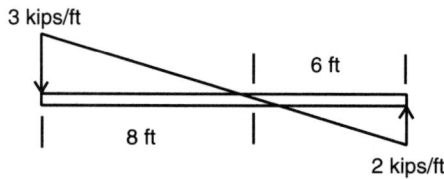

Problem 3.9c

3.10. Write the vector of equivalent forces and moments for elements *AB* and *BC* of the plane frames shown, first with respect to each element's local axes and then with respect to the structure's global *x*–*y* axes. Note that the distributed load shown on member *AB* of each structure is expressed per unit of length along *AB*.

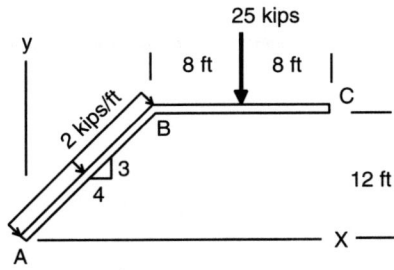

Problem 3.10a

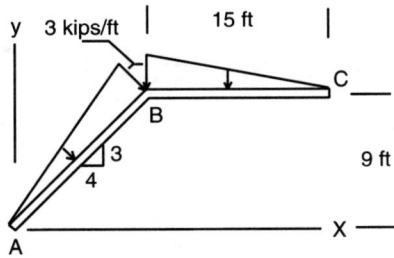

Problem 3.10b

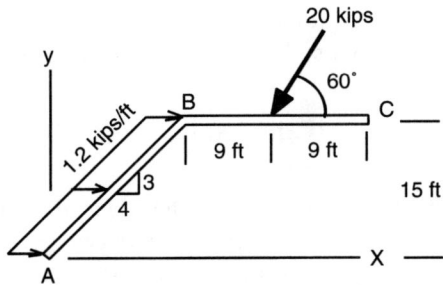

Problem 3.10c

CHAPTER 4

GLOBAL STIFFNESS EQUATIONS

4.1. INTRODUCTION

The notion of global stiffness equations was introduced earlier in Chapter 2. These equations represent the relationships among the external forces and moments applied at the structure's nodes and the translations and rotations which occur at the nodes. Each of these must be referred to common reference axes, the structure's global reference axes. The global stiffness equations of a structure are given as

$$\{\mathbf{F}_G\} = [\mathbf{K}_G]\{\mathbf{q}_G\}. \tag{2.2}$$

The entries in the global load vector $\mathbf{F}_G$ are the externally applied nodal forces and moments on the structure. Any loads applied between nodes must be reduced to equivalent nodal forces and moments for inclusion into vector $\mathbf{F}_G$. In their entirety, these entries comprise the known nodal and equivalent nodal forces and moments, many of which may be zeros, and the unknown support reactions for the structure which must be at nodes.

The entries in the global displacement vector $\mathbf{q}_G$ are the gdof, or nodal translations and rotations referred to the global reference axes. Though a few of these entries will be known and specified at the structure's supports, most will be unknown. Ultimately it is these gdof which must be solved for from the global stiffness equations along with the structure's reactions. The gdof values are also needed to calculate element internal shears, moments, and stresses from the element stiffness and constitutive equations.

The entries in the global stiffness matrix $\mathbf{K}_G$ are the global stiffness coefficients. They must be obtained independently of equation 2.2 and then used in equation 2.2 to solve for the gdof.

4.2. ASSEMBLY OF GLOBAL STIFFNESS MATRIX

The stiffness coefficients in the global stiffness matrix are directly related to the stiffness coefficients of the element stiffness matrices. Each element stiffness coefficient of each element stiffness matrix will contribute to the global stiffness matrix. As a first step in developing the global stiffness matrix, all of the element stiffness matrices of the structure must be determined and referred to the global reference axes.

Equations 3.5, 3.8, and 3.9 indicate how to obtain the element stiffness coefficients for some specific types of finite elements, each referred to the elements' local reference axes. Equation 3.29 provides a procedure for transforming stiffness matrices of 1D finite elements for reference from local to global axes.

4.2.1. Equilibrium and Compatibility at Nodes

The assembly of a structure's finite elements into an aggregate which models the structure's behavior satisfactorily is based on requiring the points where the elements join (nodes) to be in equilibrium. In addition, at each node the displacements defined as dof must be identical for each element joining at the node and for the structure at that node. This latter requirement is called compatibility of nodal displacements. Figures 2.1–2.4 illustrate these requirements for four different structural finite elements.

The enforcement of equilibrium and compatibility at nodes provides the means to assemble the element stiffness coefficients as the global stiffness matrix. Figure 4.1 is a very simple structural system, a three-member plane truss, which will be used to illustrate the procedure. The structure should first be considered in its entirety, without supports, restraints, or applied loads. All of its members and gdof must be identified. For the structure shown, there are global reference axes $(x–y)$, three nodes (n_1, n_2, n_3), three members (boxed letters), and six gdof (circled numbers). The order of the gdof chosen does not affect the validity of the procedure, but it will affect the computation effort necessary to solve the global stiffness equations, as will be discussed later. Once chosen, the gdof order establishes the order within the global stiffness matrix. The global stiffness equations for this structure are

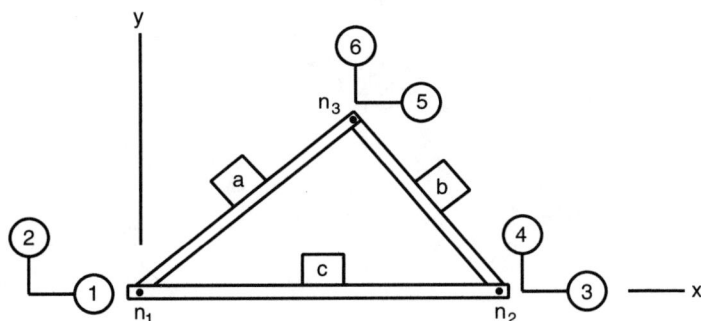

Figure 4.1 Truss structure

$$
\begin{Bmatrix} P_1 \\ P_2 \\ P_3 \\ P_4 \\ P_5 \\ P_6 \end{Bmatrix}
=
\begin{bmatrix}
k_{G11} & k_{G12} & k_{G13} & k_{G14} & k_{G15} & k_{G16} \\
k_{G21} & k_{G22} & k_{G23} & k_{G24} & k_{G25} & k_{G26} \\
k_{G31} & k_{G32} & k_{G33} & k_{G34} & k_{G35} & k_{G36} \\
k_{G41} & k_{G42} & k_{G43} & k_{G44} & k_{G45} & k_{G46} \\
k_{G51} & k_{G52} & k_{G53} & k_{G54} & k_{G55} & k_{G56} \\
k_{G61} & k_{G62} & k_{G63} & k_{G64} & k_{G65} & k_{G66}
\end{bmatrix}
\begin{Bmatrix} q_{G1} \\ q_{G2} \\ q_{G3} \\ q_{G4} \\ q_{G5} \\ q_{G6} \end{Bmatrix}.
\tag{4.1}
$$

Figure 4.2 contains a free body diagram (FBD) of node 3, showing gdof 5 and 6, and members a and b as they join at node 3. An externally applied load P at node 3 is shown as components along the lines of gdof 5 and 6. The components of the unknown internal axial forces of members a and b and their edof are shown, all with respect to the global reference axes. The edof for members a and b are designated by the numerals followed by the member letter. Not shown are the other two pairs of edof at the other end of each member. These are $1a$ and $2a$ for member a and $3b$ and $4b$ for member b.

Imposing the conditions for the equilibrium of node 3 requires that the forces acting along gdof 5 and 6 add to zero independently. Thus,

$$
P_5 = F_{3a} + F_{1b}, \tag{4.2a}
$$

$$
P_6 = F_{4a} + F_{2b}. \tag{4.2b}
$$

Expressions for the components of the internal member forces may be written from the truss element stiffness equations referred to the global reference axes, developed earlier as equation 2.1,

$$
F_{3a} = k_{a31}q_{1a} + k_{a32}q_{2a} + k_{a33}q_{3a} + k_{a34}q_{4a}, \tag{4.3a}
$$

$$
F_{1b} = k_{b11}q_{1b} + k_{b12}q_{2b} + k_{b13}q_{3b} + k_{b14}q_{4b}, \tag{4.3b}
$$

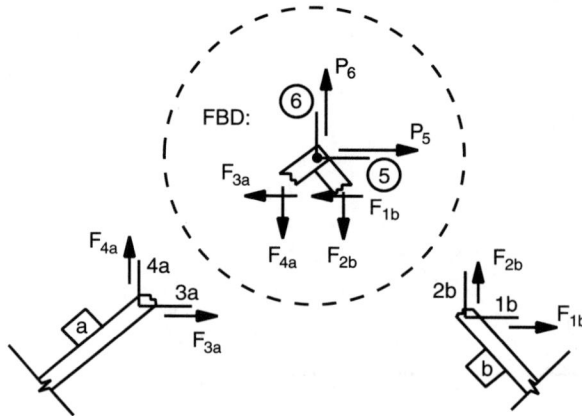

Figure 4.2 Node equilibrium and compatibility

$$F_{4a} = k_{a41}q_{1a} + k_{a42}q_{2a} + k_{a43}q_{3a} + k_{a44}q_{4a}, \tag{4.3c}$$

$$F_{2b} = k_{b21}q_{1b} + k_{b22}q_{2b} + k_{b23}q_{3b} + k_{b24}q_{4b}. \tag{4.3d}$$

Enforcing the compatibility requirement at node n_3 means that

$$q_{G5} = q_{3a} = q_{1b}, \tag{4.4a}$$

$$q_{G6} = q_{4a} = q_{2b}. \tag{4.4b}$$

Note also that

$$q_{G1} = q_{1a}, \qquad q_{G2} = q_{2a}, \qquad q_{G3} = q_{3b}, \qquad q_{G4} = q_{4b}. \tag{4.4c}$$

Substituting equations 4.3 and 4.4 into equations 4.2 results in

$$P_5 = k_{a31}q_{G1} + k_{a32}q_{G2} + k_{b13}q_{G3} + k_{b14}q_{G4} + (k_{a33} + k_{b11})q_{G5} + (k_{a34} + k_{b12})q_{G6}, \tag{4.5a}$$

$$P_6 = k_{a41}q_{G1} + k_{a42}q_{G2} + k_{b23}q_{G3} + k_{b24}q_{G4} + (k_{a43} + k_{b21})q_{G5} + (k_{a44} + k_{b22})q_{G6}. \tag{4.5b}$$

Equations 4.5 show the relationships which exist among the externally applied loads at gdof 5 and 6 and the displacements of the structure at each gdof. Thus they must be the fifth and sixth global stiffness equations. The fifth and sixth global stiffness equations written from matrix equation 4.1 are

$$P_5 = k_{G51}q_{G1} + k_{G52}q_{G2} + k_{G53}q_{G3} + k_{G54}q_{G4} + k_{G55}q_{G5} + k_{G56}q_{G6}, \tag{4.5c}$$

$$P_6 = k_{G61}q_{G1} + k_{G62}q_{G2} + k_{G63}q_{G3} + k_{G64}q_{G4} + k_{G65}q_{G5} + k_{G66}q_{G6}. \tag{4.5d}$$

Comparing equations 4.5a with 4.5c and 4.5b with 4.5d, the relationships among element and global stiffness coefficients may be seen. The element stiffness coefficients from each element stiffness equation which contributes to equilibrium along a gdof combine to produce all of the global stiffness coefficients for the global stiffness equation of that gdof. Where two or more elements join so that each has an edof coincident with a gdof, the element stiffness coefficients add algebraically to produce the corresponding global stiffness coefficients. For the illustration here,

$$k_{G51} = k_{a31}, \qquad k_{G52} = k_{a32}, \qquad k_{G53} = k_{b13}, \qquad k_{G54} = k_{b14},$$
$$k_{G55} = k_{a33} + k_{b11}, \qquad k_{G56} = k_{a34} + k_{b12}; \tag{4.6a}$$
$$k_{G61} = k_{a41}, \qquad k_{G62} = k_{a42}, \qquad k_{G63} = k_{b23}, \qquad k_{G64} = k_{b24},$$
$$k_{G65} = k_{a43} + k_{b21}, \qquad k_{G66} = k_{a44} + k_{b22}. \tag{4.6b}$$

These determinations are a consequence of the requirements for equilibrium at node 3 (equations 4.2), the element stiffness relationships for the elements joining at node

3 (equations 4.3), and the requirements for compatibility (equations 4.4) among the edof and gdof at node 3 and the other edof and gdof occurring in these relationships.

A similar analysis for the other four global stiffness equations can be made. The process of enforcing equilibrium and compatibility at nodes is repeated at each of the other nodes. At node 1, it will produce the stiffness coefficients for the first and second global stiffness equations, and at node 2 it will produce them for the third and fourth global stiffness equations. The process applies to other structures also, when the dof are translations, rotations, or both.

By obtaining global stiffness coefficients from element stiffness coefficients in this manner, the resulting solution of the global stiffness equations for nodal displacements and the calculation of element internal nodal shears and moments from these nodal displacements must result in the equilibrium of all forces and moments at nodes. This in return results in equilibrium of each element of the structure and of the structure as a whole. Moreover, it must also result in the continuity at the nodes of the element displacements defined as edof and nodal displacements defined as gdof.

4.2.2. A Process for the Assembly

A process for assembling the global stiffness matrix from the element stiffness coefficients is illustrated in Figures 4.3 and 4.4. The key step is to identify the gdof and

$$\mathbf{K}_a = \begin{array}{cc|c|c|c|c|} & & \textcircled{1} & \textcircled{2} & \textcircled{5} & \textcircled{6} \\ & & 1 & 2 & 3 & 4 \\ \hline \textcircled{1} & 1 & k_{a11} & k_{a12} & k_{a13} & k_{a14} \\ \hline \textcircled{2} & 2 & k_{a21} & k_{a22} & k_{a23} & k_{a24} \\ \hline \textcircled{5} & 3 & k_{a31} & k_{a32} & k_{a33} & k_{a34} \\ \hline \textcircled{6} & 4 & k_{a41} & k_{a42} & k_{a43} & k_{a44} \\ \hline \end{array}$$

$$\mathbf{K}_b = \begin{array}{cc|c|c|c|c|} & & \textcircled{5} & \textcircled{6} & \textcircled{3} & \textcircled{4} \\ & & 1 & 2 & 3 & 4 \\ \hline \textcircled{5} & 1 & k_{b11} & k_{b12} & k_{b13} & k_{b14} \\ \hline \textcircled{6} & 2 & k_{b21} & k_{b22} & k_{b23} & k_{b24} \\ \hline \textcircled{3} & 3 & k_{b31} & k_{b32} & k_{b33} & k_{b34} \\ \hline \textcircled{4} & 4 & k_{b41} & k_{b42} & k_{b43} & k_{b44} \\ \hline \end{array}$$

$$\mathbf{K}_c = \begin{array}{cc|c|c|c|c|} & & \textcircled{1} & \textcircled{2} & \textcircled{3} & \textcircled{4} \\ & & 1 & 2 & 3 & 4 \\ \hline \textcircled{1} & 1 & k_{c11} & k_{c12} & k_{c13} & k_{c14} \\ \hline \textcircled{2} & 2 & k_{c21} & k_{c22} & k_{c23} & k_{c24} \\ \hline \textcircled{3} & 3 & k_{c31} & k_{c32} & k_{c33} & k_{c34} \\ \hline \textcircled{4} & 4 & k_{c41} & k_{c42} & k_{c43} & k_{c44} \\ \hline \end{array}$$

Figure 4.3 Element stiffness matrices

$$\mathbf{K}_G = \begin{bmatrix}
K_{a11}+K_{c11} & K_{a12}+K_{c12} & K_{c13} & K_{c14} & K_{a13} & K_{a14} \\
K_{a21}+K_{c21} & K_{a22}+K_{c22} & K_{c23} & K_{c24} & K_{a23} & K_{a24} \\
K_{c31} & K_{c32} & K_{b33}+K_{c33} & K_{b34}+K_{c34} & K_{b31} & K_{b32} \\
K_{c41} & K_{c42} & K_{b43}+K_{c43} & K_{b44}+K_{c44} & K_{b41} & K_{b42} \\
K_{a31} & K_{a32} & K_{b13} & K_{b14} & K_{a33}+K_{b11} & K_{a34}+K_{b12} \\
K_{a41} & K_{a42} & K_{b23} & K_{b24} & K_{a43}+K_{b21} & K_{a44}+K_{b22}
\end{bmatrix}$$

(gdof columns: ① ② ③ ④ ⑤ ⑥; gdof rows: ① ② ③ ④ ⑤ ⑥)

Figure 4.4 Global stiffness matrix

their coincident edof on the element stiffness matrices. Figure 4.3 shows the element stiffness matrices for the three finite elements of the structure shown in Figure 4.1. The edof numbers for each are shown as numerals, and the numbers of the coincident gdof are shown as circled numerals. The edof–gdof correspondence is taken directly from Figure 4.1, where the choices of gdof for the structure are shown and where the decision was made to let the start node for each element be its leftmost node. The order of these dof sequences does not affect the validity of the finite element method of analysis, but once it is established, it must be adhered to.

Figure 4.4 shows the global stiffness matrix for the plane truss assembled from its element stiffness matrices. Its gdof numbers are circled. The element stiffness coefficients shown in Figure 4.3 are mapped to the locations in the global stiffness matrix indicated by their coincident gdof numbers. For example, K_{a33} from element a and k_{b11} from element b are addressed to the position k_{G55}. No other element stiffness coefficients from any of the three element stiffness matrices is identified for this position; thus these two are added and entered there as k_{G55}. Each of the 48 element stiffness coefficients is similarly mapped to one of the 36 global stiffness coefficient locations.

The process illustrated in Figures 4.3 and 4.4 is general and easily implemented. What it requires is identifying the correspondence of the edof and gdof numbering schemes. It ensures equilibrium in the structure at its nodes and compatibility of displacements, defined as dof, at its nodes.

4.3. ASSEMBLY OF GLOBAL LOAD VECTOR

The entries in the global load vector must be concentrated forces or moments applied along corresponding gdof. They will reflect the applied loads on the structure, which were discussed in general terms as load restraints or natural boundary conditions in Section 2.5.2. The global load vector entries also will reflect support reactions which were discussed in general terms as displacement restraints or essential boundary conditions in Section 2.5.1 and as load-displacement restraints or mixed boundary conditions in Section 2.5.3. These support reactions are unknown quantities and must be treated as unknowns in the global load vector.

4.3.1. Applied Loads

For known concentrated forces and couples applied to the structure, the analyst usually ensures that a node is located at each of their points of application. The forces or couples then are reduced to components of force or moment acting at the node and along the lines of action of the gdof. Their values are inserted directly into the global load vector at the positions corresponding to their gdof.

If the analyst chooses not to locate a node at the point of application of a concentrated force or moment so that it acts on an element between nodes, the concentrated force or moment must be converted to equivalent forces and moments acting at the element's nodes along its edof.

If there is distributed load acting on the structure, the analyst should define that portion of the structure in terms of finite elements on which the distributed load acts. The distributed load on the elements then must be converted to equivalent forces and moments acting at the elements' nodes along their edof.

Section 3.1.3 described a direct method for the conversion process for either concentrated loads between nodes or distributed loads, and Figure 3.7 provided the results for four cases for prismatic beam or frame elements. The resulting equivalent forces and moments are referred to local element reference axes, and they must be transformed and referred to the structure's global reference axes before they can be included in the global load vector. Section 3.2.2 described a procedure for the transformation of forces and moments of 1D frame elements.

Once the values for the equivalent forces and moments are obtained with respect to the global reference axes, the one-to-one correspondence of the edof and gdof may be invoked to identify their gdof. The values are then inserted into the global load vector at the positions corresponding to their gdof. If a value for an applied concentrated force or moment at a node and an equivalent force or moment at a node are both identified with the same gdof, the values are added algebraically and inserted into the position in the global load vector corresponding to the gdof. Similarly, if two adjacent elements have loads between nodes or distributed loads, the equivalent forces and moments at the gdof and node where they join will add and be inserted into the global load vector at the positions corresponding to the gdof.

4.3.2. Support Reactions

In addition to the known force and moment entries in the global load vector discussed above, the global load vector also contains the structure's unknown support reactions. When a gdof is specified as a known value, often zero, this displacement restraint causes a reactive force or moment at that gdof which will be whatever it has to be to cause the specified displacement restraint to occur. The reactive force or moment is an unknown support reaction and occurs as such in the global load vector at the position of its corresponding gdof. It is added to any other entry occurring at that gdof. The global stiffness equations account for its presence when the value of the specified displacement is entered in the global displacement vector at the position of its corresponding gdof.

A restraint on the structure may be a known linear relationship between the unknown displacement at a gdof and the unknown force or moment there, such as discussed in Section 2.5.3. The unknown force or moment is an unknown support reaction at the gdof. It occurs as an entry in the global load vector. The unknown displacement is an unknown quantity in the global displacement vector. The global stiffness equations account for this load-displacement restraint by adding the proportionality constant of the linear relationship to the stiffness coefficient of the global stiffness matrix in the diagonal position corresponding to the gdof. A zero must be assigned to the corresponding position in the global load vector.

4.4. GLOBAL DISPLACEMENT VECTOR

The global displacement vector contains mostly the unknown displacements at the nodes. Its only known entries will be the displacements specified at the gdof at supports where a displacement restraint (essential boundary condition) is imposed. Load restraints (natural boundary conditions) and load-displacement restraints (mixed boundary conditions) do not alter the structure of the global displacement vector.

4.5. SOLUTION OF GLOBAL STIFFNESS EQUATIONS

The solution of the global stiffness equations—a linear algebraic equation system— for the structure's unknown nodal displacements and support reactions may be undertaken when the following have been accomplished:

- The global stiffness matrix has been assembled from the structure's element stiffness coefficients.
- The global load vector has been assembled from the concentrated forces and moments applied at nodes and from the equivalent concentrated forces and moments due to loads between nodes.
- The structure's displacement restraints have been specified in the global displacement vector and the corresponding unknown support reactions have been added as entries to the global load vector.
- The structure's load-displacement restraints, if any, have been imposed by modifying the global stiffness matrix and corresponding entries in the global load vector.

There will be as many unknowns to solve for as there are gdof. Some will be unknown gdof in the global displacement vector, and the rest will be unknown support reactions in the global load vector. The submatrix, the solution without rearrangement, or the penalty methods (discussed in Section 1.3) may be applied. The method employed should be dictated by the advantages and disadvantages of each as they apply to the global stiffness equations to be solved. For example, if the equation system is large, routines which implement the submatrix method will be most efficient since smaller matrices will be manipulated. On the other hand, the submatrix method is a complicated procedure to implement and program. If the size of the equation system is not so important but simplicity and programming ease are, then either the penalty or solution-without-rearrangement procedure should be used.

4.5.1. Plane Truss Example

As an illustration, the three-member plane truss used above is shown in Figure 4.5 with supports and an applied load. Discretized, it is shown in Figure 4.6.

The element properties may be tabulated as follows:

Member	A (in.^{2})	L (in.)	AE/L (kips/in.)	$\cos \theta$	$\sin \theta$
a	12	240	1500.00	0.800	0.600
b	12	180	2000.00	0.600	−0.800
c	8	300	800.00	1.000	0.000

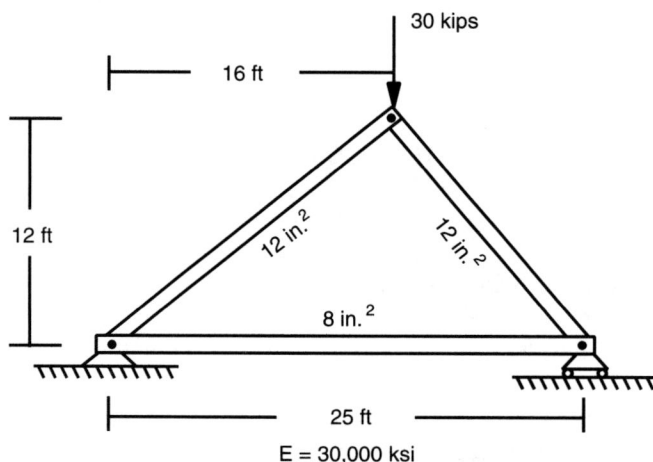

Figure 4.5 Truss

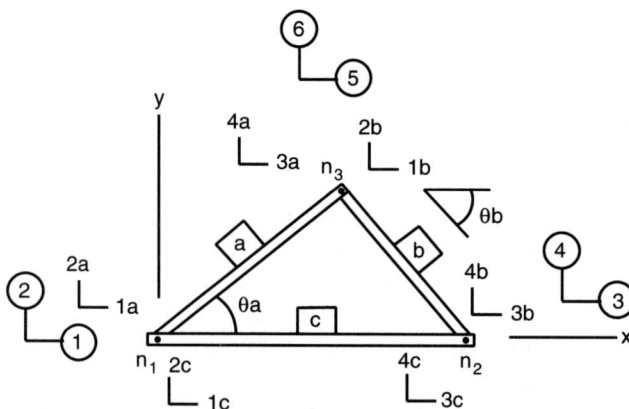

Figure 4.6 Discreticized truss

The element stiffness matrices with respect to global reference axes are (see equations 3.8, 3.15b, and 3.29)

$$[\mathbf{K}_a] = 1500.00 \begin{bmatrix} 0.800 & 0.000 \\ 0.600 & 0.000 \\ 0.000 & 0.800 \\ 0.000 & 0.600 \end{bmatrix} \begin{bmatrix} 1 & -1 \\ -1 & 1 \end{bmatrix} \begin{bmatrix} 0.800 & 0.600 & 0.000 & 0.000 \\ 0.000 & 0.000 & 0.800 & 0.600 \end{bmatrix}$$

$$= \begin{array}{c} ① \\ ② \\ ⑤ \\ ⑥ \end{array} \begin{array}{c} 1 \\ 2 \\ 3 \\ 4 \end{array} \begin{bmatrix} 960 & 720 & -960 & -720 \\ 720 & 540 & -720 & -540 \\ -960 & -720 & 960 & 720 \\ -720 & -540 & 720 & 540 \end{bmatrix} \text{kips/in.,}$$

$$[\mathbf{K}_b] = 2000.00 \begin{bmatrix} 0.600 & 0.000 \\ -0.800 & 0.000 \\ 0.000 & 0.600 \\ 0.000 & -0.800 \end{bmatrix} \begin{bmatrix} 1 & -1 \\ -1 & 1 \end{bmatrix} \begin{bmatrix} 0.600 & -0.800 & 0.000 & 0.000 \\ 0.000 & 0.000 & 0.600 & -0.800 \end{bmatrix}$$

$$= \begin{array}{c} ⑤ \\ ⑥ \\ ③ \\ ⑥ \end{array} \begin{array}{c} 1 \\ 2 \\ 3 \\ 4 \end{array} \begin{bmatrix} 720 & -960 & -720 & -960 \\ -960 & 1280 & 960 & -1280 \\ -720 & 960 & 720 & -960 \\ 960 & -1280 & -960 & 1280 \end{bmatrix} \text{kips/in.,}$$

$$[\mathbf{K}_c] = 800.00 \begin{bmatrix} 1.000 & 0.000 \\ 0.000 & 0.000 \\ 0.000 & 1.000 \\ 0.000 & 0.000 \end{bmatrix} \begin{bmatrix} 1 & -1 \\ -1 & 1 \end{bmatrix} \begin{bmatrix} 1.000 & 0.000 & 0.000 & 0.000 \\ 0.000 & 0.000 & 1.000 & 0.000 \end{bmatrix}$$

$$= \begin{array}{c} ① \\ ② \\ ③ \\ ④ \end{array} \begin{array}{c} 1 \\ 2 \\ 3 \\ 4 \end{array} \begin{bmatrix} 800 & 0 & -800 & 0 \\ 0 & 0 & 0 & 0 \\ -800 & 0 & 800 & 0 \\ 0 & 0 & 0 & 0 \end{bmatrix} \text{kips/in.,}$$

The global stiffness matrix, assembled using the process illustrated in Figures 4.3 and 4.4, is

$$[\mathbf{K}_G] = \begin{matrix} & ① & ② & ③ & ④ & ⑤ & ⑥ \\ ① & 1760 & 720 & -800 & 0 & -960 & -720 \\ ② & 720 & 540 & 0 & 0 & -720 & -540 \\ ③ & -800 & 0 & 1520 & -960 & -720 & 960 \\ ④ & 0 & 0 & -960 & 1280 & 960 & -1280 \\ ⑤ & -960 & -720 & -720 & 960 & 1680 & -240 \\ ⑥ & -720 & -540 & 960 & -1280 & -240 & 1820 \end{matrix} \quad \text{kips/in.},$$

The downward applied 30-kip load at the top of the truss is entered as −30 kips at the position corresponding to gdof 6 in the global load vector. The other unrestrained gdof 3 and 5 have no load applied so that zeros are entered at their corresponding positions in the global load vector.

Taking the restraints imposed at the supports as zero translation displacements, zeros are entered for the global displacements at the positions of gdof 1, 2, and 4 in the global load vector, and the unknown support reactions are entered at these positions in the global load vector.

The global stiffness equations, ready for solution, are

$$\begin{bmatrix} 1760 & 720 & -800 & 0 & -960 & -720 \\ 720 & 540 & 0 & 0 & -720 & -540 \\ -800 & 0 & 1520 & -960 & -720 & 960 \\ 0 & 0 & -960 & 1280 & 960 & -1280 \\ -960 & -720 & -720 & 960 & 1680 & -240 \\ -720 & -540 & 960 & -1280 & -240 & 1820 \end{bmatrix} \begin{Bmatrix} 0 \\ 0 \\ q_{G3} \\ 0 \\ q_{G5} \\ q_{G6} \end{Bmatrix} = \begin{Bmatrix} R_{G1} \\ R_{G2} \\ 0 \\ R_{G4} \\ 0 \\ -30 \end{Bmatrix}. \tag{4.7a}$$

Applying the penalty solution procedure to equation 4.7a, the "big nr" used will be 1×10^{10}. Stiffness coefficients k_{G11}, k_{G22}, and k_{G44} are each replaced by their current values multiplied by 1×10^{10}. Support reactions R_{G1}, R_{G2}, and R_{G4} are each replaced with the product of the specified value of their corresponding displacement, the value of their corresponding stiffness coefficient, and the value of "big nr." In this case each is replaced with a zero, since each of their specified displacement values was taken as zero.

The global stiffness equations, modified as described, are

$$
\begin{bmatrix}
\mathbf{1.76 \times 10^{13}} & 720 & -800 & 0 & -960 & -720 \\
720 & \mathbf{0.54 \times 10^{13}} & 0 & 0 & -720 & -540 \\
-800 & 0 & 1520 & -960 & -720 & 960 \\
0 & 0 & -960 & \mathbf{1.28 \times 10^{13}} & 960 & -1280 \\
-960 & -720 & -720 & 960 & 1680 & -240 \\
-720 & -540 & 960 & -1280 & -240 & 1820
\end{bmatrix}
\begin{Bmatrix}
q_{G1} \\ q_{G2} \\ q_{G3} \\ q_{G4} \\ q_{G5} \\ q_{G6}
\end{Bmatrix}
=
\begin{Bmatrix}
0 \\ 0 \\ 0 \\ 0 \\ 0 \\ -30
\end{Bmatrix}.
$$

(4.7b)

Equation 4.7b may be solved for the six displacements by inverting the modified global stiffness matrix and using it as a premultiplier on the modified global load vector. The result is

$$
\{\mathbf{q}_G\} =
\begin{Bmatrix}
0.000000 \\
0.000000 \\
0.018000 \\
0.000000 \\
0.004080 \\
-0.025440
\end{Bmatrix}
\quad \text{in.}
$$

(4.7c)

Substituting the values from equation 4.7c into equation 4.7a and performing the matrix multiplication indicated, the values for the complete global load vector are obtained:

$$
\{\mathbf{F}_G\} =
\begin{Bmatrix}
0.000 \\
10.800 \\
0.000 \\
19.200 \\
0.000 \\
-30.000
\end{Bmatrix}
\quad \text{kips.}
$$

(4.7d)

Values for R_{G1}, R_{G2}, and R_{G4} are at entries 1, 2, and 4.

The edof with respect to the global reference axes may be identified from equation 4.7c and premultiplied by their respective element stiffness matrices referenced to the global axes. The result will be the truss member force components along the global reference axes. A better approach is to transform the edof so that they refer to local axes (equation 3.15b) and premultiply them by their respective element stiffness matrices referenced to local axes (equation 3.7). This result gives the truss member forces referred to local axes or in terms of axial force. For this example,

$$\{\mathbf{F}_a\} = 1500 \begin{bmatrix} 1 & -1 \\ -1 & 1 \end{bmatrix} \begin{bmatrix} 0.8 & 0.6 & 0.0 & 0.0 \\ 0.0 & 0.0 & 0.8 & 0.6 \end{bmatrix} \begin{Bmatrix} 0.000000 \\ 0.000000 \\ 0.004080 \\ -0.025440 \end{Bmatrix} = \begin{Bmatrix} 18.00 \\ -18.00 \end{Bmatrix} \text{ kips,}$$

$$\{\mathbf{F}_b\} = 2000 \begin{bmatrix} 1 & -1 \\ -1 & 1 \end{bmatrix} \begin{bmatrix} 0.6 & -0.8 & 0.0 & 0.0 \\ 0.0 & 0.0 & 0.6 & -0.8 \end{bmatrix} \begin{Bmatrix} 0.004080 \\ -0.025440 \\ 0.018000 \\ 0.000000 \end{Bmatrix} = \begin{Bmatrix} 24.00 \\ -24.00 \end{Bmatrix} \text{ kips,}$$

$$\{\mathbf{F}_c\} = 800 \begin{bmatrix} 1 & -1 \\ -1 & 1 \end{bmatrix} \begin{bmatrix} 1.0 & 0.0 & 0.0 & 0.0 \\ 0.0 & 0.0 & 1.0 & 0.0 \end{bmatrix} \begin{Bmatrix} 0.000000 \\ 0.000000 \\ 0.018000 \\ 0.000000 \end{Bmatrix} = \begin{Bmatrix} -14.40 \\ 14.40 \end{Bmatrix} \text{ kips.}$$

Referring to the member local axes, F_a and F_b are compression and F_c is tension.

4.5.2. Plane Frame Example

The method is illustrated again with the three-member plane frame shown in Figure 4.7. The frame has a fixed support on its lower left corner, a pinned support on its lower right corner, and an elastic horizontal spring support on its upper right corner. The fixed support is fully restrained in all three gdof. The pinned support, however, settles 0.100 in. vertically downward but is fully restrained for horizontal translation.

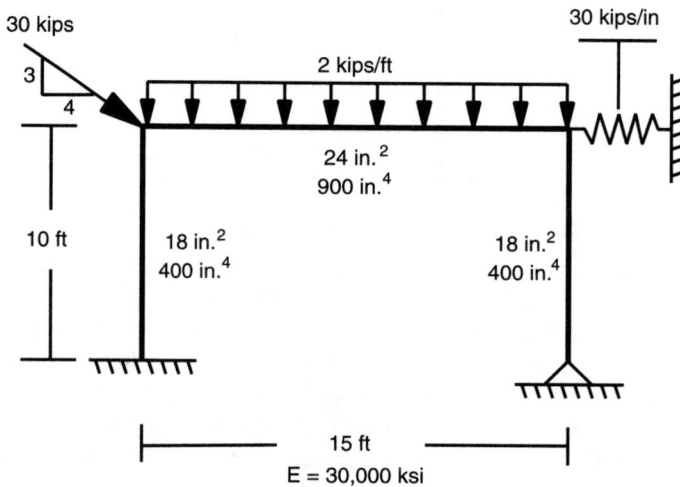

Figure 4.7 Frame

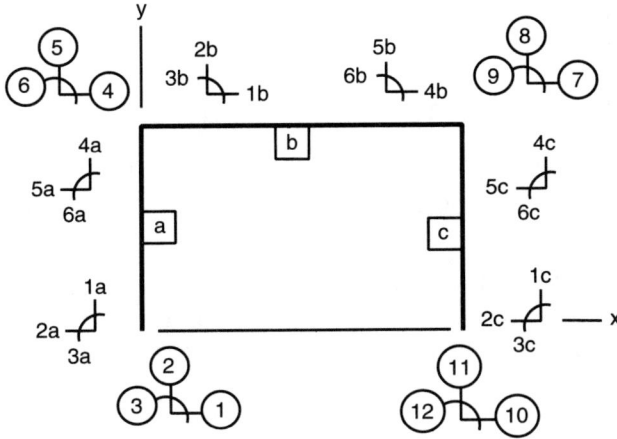

Figure 4.8 Discreticized frame

The spring support yields linearly at a rate of 30 kips/in. There is an inclined concentrated load at the upper left corner and a vertical, downward acting, uniformly distributed load across the top of the frame. The structure, discretized with reference axes, is shown in Figure 4.8.

The element properties may be tabulated as

Member	A (in.^{2})	I (in.^{4})	L (in.)	AE/L (kips/in.)	EI/L (kip-in.)	$\cos\theta$	$\sin\theta$
a	18	400	120	4500.00	100,000.00	0.000	1.000
b	24	900	180	4000.00	150,000.00	1.000	0.000
c	18	400	120	4500.00	100,000.00	0.000	1.000

The element stiffness matrices with respect to global reference axes are (see equations 3.9, 3.16, and 3.29)

$$[\mathbf{K}_a] = [\mathbf{T}_a]^T[\mathbf{K}_a'][\mathbf{T}_a],$$

$$[\mathbf{K}_b] = [\mathbf{T}_b]^T[\mathbf{K}_b'][\mathbf{T}_b],$$

$$[\mathbf{K}_c] = [\mathbf{T}_c]^T[\mathbf{K}_c'][[\mathbf{T}_c];$$

$$[\mathbf{K}_a] = [\mathbf{K}_c] = \begin{bmatrix} 83.33 & 0 & -5000 & -83.33 & 0 & -5000 \\ 0 & 4500 & 0 & 0 & -4500 & 0 \\ -5000 & 0 & 400,000 & 5000 & 0 & 200,000 \\ -83.33 & 0 & 5000 & 83.33 & 0 & 5000 \\ 0 & -4500 & 0 & 0 & 4500 & 0 \\ -5000 & 0 & 200,000 & 5000 & 0 & 400,000 \end{bmatrix};$$

$$[\mathbf{K}_b] = \begin{bmatrix} 4000 & 0 & 0 & -4000 & 0 & 0 \\ 0 & 55.56 & 5000 & 0 & -55.56 & 5000 \\ 0 & 5000 & 600,000 & 0 & -5000 & 300,000 \\ -4000 & 0 & 0 & 4000 & 0 & 0 \\ 0 & -55.56 & -5000 & 0 & 55.56 & -5000 \\ 0 & 5000 & 300,000 & 0 & -5000 & 600,000 \end{bmatrix}.$$

Units for the stiffness coefficients are kips per inch, kips, or kip-inches, depending on which edof they multiply and which corresponding internal force they produce.

The global stiffness matrix, assembled using the process illustrated in Figures 4.3 and 4.4, is

$$[\mathbf{K}_G] = \begin{bmatrix} 83.33 & 0 & -5000 & -83.33 & 0 & -5000 & 0 & 0 & 0 & 0 & 0 & 0 \\ 0 & 4500 & 0 & 0 & -4500 & 0 & 0 & 0 & 0 & 0 & 0 & 0 \\ -5000 & 0 & 400,000 & 5000 & 0 & 200,000 & 0 & 0 & 0 & 0 & 0 & 0 \\ -83.33 & 0 & 5000 & 4083.33 & 0 & 5000 & -4000 & 0 & 0 & 0 & 0 & 0 \\ 0 & -4500 & 0 & 0 & 4555.56 & 5000 & 0 & -55.56 & 5000 & 0 & 0 & 0 \\ -5000 & 0 & 200,000 & 5000 & 5000 & 1 \times 10^6 & 0 & -5000 & 300,000 & 0 & 0 & 0 \\ 0 & 0 & 0 & -4000 & 0 & 0 & 4083.33 & 0 & 5000 & -83.33 & 0 & 5000 \\ 0 & 0 & 0 & 0 & -55.56 & -5000 & 0 & 4555.56 & -5000 & 0 & -4500 & 0 \\ 0 & 0 & 0 & 0 & 5000 & 300,000 & 5000 & -5000 & 1 \times 10^6 & -5000 & 0 & 200,000 \\ 0 & 0 & 0 & 0 & 0 & 0 & -83.33 & 0 & -5000 & 83.33 & 0 & -5000 \\ 0 & 0 & 0 & 0 & 0 & 0 & 0 & -4500 & 0 & 0 & 4500 & 0 \\ 0 & 0 & 0 & 0 & 0 & 0 & 5000 & 0 & 200,000 & -5000 & 0 & 400,000 \end{bmatrix}.$$

$$(4.8a)$$

The 30-kip concentrated load at the upper left corner of the frame is resolved into a horizontal 24-kip component acting to the right along gdof 4 and an 18-kip vertical component acting downward along gdof 5. Thus 24 kips is entered in position 4 of the global load vector, and −18 kips is entered in position 5.

The 2-kip/ft uniformly distributed load along member b is replaced by equivalent forces and moments as illustrated in Figure 3.7. Thus in the global load vector, −15 kips is entered at positions 5 and 8, −450 kip-in. at position 6, and 450 kip-in. at position 9. At position 5, the −18-kip component from the concentrated load and the −15-kip equivalent load from the distributed load are added algebraically for a −33-kip final value at this position.

Displacement restraints occur at the fixed and pinned supports, gdof 1, 2, 3, 10, and 11. They are entered as specified values at these positions in the global displacement vector. The corresponding unknown support reactions are entered at the same positions in the global load vector.

The linear spring is a load-displacement restraint at gdof 7. It is imposed on the global stiffness equations by adding its proportionality constant (30 kips/in.) to the value of the corresponding global stiffness coefficient (k_{G77}) and by entering a zero at position 7 in the global load vector.

The global stiffness equations, ready for solution, are

$$
\begin{bmatrix}
83.33 & 0 & -5000 & -83.33 & 0 & -5000 & 0 & 0 & 0 & 0 & 0 & 0 \\
0 & 4500 & 0 & 0 & -4500 & 0 & 0 & 0 & 0 & 0 & 0 & 0 \\
-5000 & 0 & 400,000 & 5000 & 0 & 200,000 & 0 & 0 & 0 & 0 & 0 & 0 \\
-83.33 & 0 & 5000 & 4083.33 & 0 & 5000 & -4000 & 0 & 0 & 0 & 0 & 0 \\
0 & -4500 & 0 & 0 & 4555.56 & 5000 & 0 & -55.56 & 5000 & 0 & 0 & 0 \\
-5000 & 0 & 200,000 & 5000 & 5000 & 1\times10^6 & 0 & -5000 & 300,000 & 0 & 0 & 0 \\
0 & 0 & 0 & -4000 & 0 & 0 & 4113.33 & 0 & 5000 & -83.33 & 0 & 5000 \\
0 & 0 & 0 & 0 & -55.56 & -5000 & 0 & 4555.56 & -5000 & 0 & -4500 & 0 \\
0 & 0 & 0 & 0 & 5000 & 300,000 & 5000 & -5000 & 1\times10^6 & -5000 & 0 & 200,000 \\
0 & 0 & 0 & 0 & 0 & 0 & -83.33 & 0 & -5000 & 83.33 & 0 & -5000 \\
0 & 0 & 0 & 0 & 0 & 0 & 0 & -4500 & 0 & 0 & 4500 & 0 \\
0 & 0 & 0 & 0 & 0 & 0 & 5000 & 0 & 200,000 & -5000 & 0 & 400,000
\end{bmatrix}
\begin{Bmatrix}
0 \\ 0 \\ 0 \\ q_{G4} \\ q_{G5} \\ q_{G6} \\ q_{G7} \\ q_{G8} \\ q_{G9} \\ 0 \\ -0.100 \\ q_{G12}
\end{Bmatrix}
$$

$$
= \begin{Bmatrix}
R_{G1} \\ R_{G2} \\ R_{G3} \\ 24 \\ -33 \\ -450 \\ \mathbf{0} \\ -15 \\ 450 \\ R_{G10} \\ R_{G11} \\ 0
\end{Bmatrix} . \tag{4.8b}
$$

Applying the penalty solution procedure, "big nr" is again 1×10^{10}. Stiffness coefficients k_{G11}, k_{G22}, k_{G33}, $k_{G10,10}$, and $k_{G11,11}$ are each replaced by their current values multiplied by "big nr." Support reactions R_{G11}, R_{G22}, R_{G33}, $R_{G10,10}$, and $R_{G11,11}$ are each replaced by the product of the specified value of their displacement, the value of their corresponding stiffness coefficient, and the value of "big nr." In this case, each is zero, except for $R_{G11,11}$, since their specified displacements are zero. Reaction $R_{G11,11}$ is replaced with -0.45×10^{13}, the product of -0.100, 4500, and 1×10^{10}. The global stiffness equations, modified as described and ready for solution, are

$$
\begin{bmatrix}
0.833 \times 10^{12} & 0 & -5000 & -83.33 & 0 & -5000 & 0 & 0 & 0 & 0 & 0 & 0 \\
0 & 0.45 \times 10^{14} & 0 & 0 & -4500 & 0 & 0 & 0 & 0 & 0 & 0 & 0 \\
-5000 & 0 & 0.40 \times 10^{16} & 5000 & 0 & 200,000 & 0 & 0 & 0 & 0 & 0 & 0 \\
-83.33 & 0 & 5000 & 4083.33 & 0 & 5000 & -4000 & 0 & 0 & 0 & 0 & 0 \\
0 & -4500 & 0 & 0 & 4555.56 & 5000 & 0 & -55.56 & 5000 & 0 & 0 & 0 \\
-5000 & 0 & 200,000 & 5000 & 5000 & 1 \times 10^6 & 0 & -5000 & 300,000 & 0 & 0 & 0 \\
0 & 0 & 0 & -4000 & 0 & 0 & 4113.33 & 0 & 5000 & -83.33 & 0 & 5000 \\
0 & 0 & 0 & 0 & -55.56 & -5000 & 0 & 4555.56 & -5000 & 0 & -4500 & 0 \\
0 & 0 & 0 & 0 & 5000 & 300,000 & 5000 & -5000 & 1 \times 10^6 & -5000 & 0 & 200,000 \\
0 & 0 & 0 & 0 & 0 & 0 & -83.33 & 0 & -5000 & 0.833 \times 10^{12} & 0 & -5000 \\
0 & 0 & 0 & 0 & 0 & 0 & 0 & -4500 & 0 & 0 & 0.45 \times 10^{14} & 0 \\
0 & 0 & 0 & 0 & 0 & 0 & 5000 & 0 & 200,000 & -5000 & 0 & 400,000
\end{bmatrix}
\begin{Bmatrix}
0 \\ 0 \\ 0 \\ q_{G4} \\ q_{G5} \\ q_{G6} \\ q_{G7} \\ q_{G8} \\ q_{G9} \\ 0 \\ -0.100 \\ q_{G12}
\end{Bmatrix}
$$

$$
= \begin{Bmatrix}
0 \\ 0 \\ 0 \\ 24 \\ -33 \\ -450 \\ 0 \\ -15 \\ 450 \\ 0 \\ -0.45 \times 10^{13} \\ 0
\end{Bmatrix}. \tag{4.8c}
$$

The nodal displacements obtained from solving equation 4.8c are

$$
\{\mathbf{q}_G\} = \begin{Bmatrix}
0 \\
0 \\
0 \\
0.26478 \text{ in.} \\
-0.0060160 \text{ in.} \\
-0.0022609 \text{ rad} \\
0.26147 \text{ in.} \\
-0.10465 \text{ in.} \\
-0.000020645 \text{ rad} \\
0 \\
-0.10000 \text{ in.} \\
-0.0032580 \text{ rad}
\end{Bmatrix}. \tag{4.8d}
$$

The displacement of the spring at position 7 is $+0.26147$ in. Consequently, the force in the spring, or the spring reaction, may be calculated directly as

$$
R_{G7} = -30.000 \times 0.26147 = -7.8441 \text{ kips.}
$$

A global load vector may be calculated by premultiplying the global displacement vector (equation 4.8d) by the unmodified global stiffness matrix (equation 4.8a):

$$\{\mathbf{F}_G\}_{calc} = \begin{Bmatrix} -10.760 \text{ kips} \\ 27.072 \text{ kips} \\ 871.71 \text{ kip-in.} \\ 24.000 \text{ kips} \\ -33.000 \text{ kips} \\ -450.00 \text{ kip-in.} \\ -7.8440 \text{ kips} \\ -15.000 \text{ kips} \\ 450.00 \text{ kip-in.} \\ -5.3956 \text{ kips} \\ 20.928 \text{ kips} \\ 0 \end{Bmatrix}. \tag{4.8e}$$

Each of the support reactions acting on the frame are contained in the load vector of equation 4.8e at positions 1, 2, 3, 7, 10, and 11. Note the value of the spring reaction at position 7 is the same as calculated above.

Equation 4.8e includes nodal force and moment values derived from the equivalent nodal forces and moments of the distributed load on member b. These must be subtracted from the vector of equation 4.8e to obtain the global load vector reflecting the correct applied nodal forces and moments at all gdof:

$$\{\mathbf{F}_G\} = \begin{Bmatrix} -10.760 \\ 27.072 \\ 871.71 \\ 24.000 \\ -33.000 \\ -450.00 \\ -7.8440 \\ -15.000 \\ 450.00 \\ -5.3956 \\ 20.928 \\ 0 \end{Bmatrix} - \begin{Bmatrix} 0 \\ 0 \\ 0 \\ 0 \\ -15 \\ -450 \\ 0 \\ -15 \\ 450 \\ 0 \\ 0 \\ 0 \end{Bmatrix} = \begin{Bmatrix} -10.760 \text{ kips} \\ 27.072 \text{ kips} \\ 871.71 \text{ kip-in.} \\ 24.000 \text{ kips} \\ -18.000 \text{ kips} \\ 0 \\ -7.8440 \text{ kips} \\ 0 \\ 0 \\ -5.3956 \text{ kips} \\ 20.928 \text{ kips} \\ 0 \end{Bmatrix}. \tag{4.8f}$$

The displacements for each element with respect to global axes may be identified from equation 4.8d by involving the one-to-one correspondence of edof and gdof:

$$
\{q_a\} = \begin{Bmatrix} 0 \\ 0 \\ 0 \\ 0.26478 \\ -0.0060160 \\ -0.0022609 \end{Bmatrix}, \quad
\{q_b\} = \begin{Bmatrix} 0.26478 \\ -0.0060160 \\ -0.00226090 \\ 0.26147 \\ -0.10465 \\ 0.000026045 \end{Bmatrix}, \quad
\{q_c\} = \begin{Bmatrix} 0 \\ -0.1000 \\ -0.0032580 \\ 0.26147 \\ -0.10465 \\ 0.000026045 \end{Bmatrix}.
$$

Units for the element displacement vectors are inches for translation and radians for rotations.

These element displacement vectors must be transformed to reference to local element axes by premultiplying each by their respective transformation matrices. Then to obtain the internal forces of each element with respect to its local axes, the transformed element displacement vectors are substituted into the corresponding element stiffness equations, referred to local axes, and the matrix manipulations indicated are carried out. In symbols,

$$\{F'_a\} = [K'_a][T_a]\{q_a\},$$

$$\{F'_b\} = [K'_b][T_b]\{q_b\} = \{F_E\},$$

$$\{F'_c\} = [K'_c][T_c]\{q_c\}.$$

The results for the three elements of this example are

$$
\{F_a\} = \begin{Bmatrix} 27.072 \\ 10.761 \\ 871.72 \\ -27.072 \\ -10.761 \\ 419.55 \end{Bmatrix}, \quad
\{F_b\} = \begin{Bmatrix} 13.239 \\ 9.0722 \\ -419.55 \\ -13.239 \\ 20.928 \\ -647.48 \end{Bmatrix}, \quad
\{F_c\} = \begin{Bmatrix} 20.928 \\ 5.3956 \\ 0 \\ -20.928 \\ -5.3956 \\ 647.48 \end{Bmatrix}.
$$

Units for the element load vectors are kips for axial force and shear and kip-inches for moments. Internal axial forces are the first and fourth entries in each internal force vector, internal shears are the second and fifth entries, and internal moments are the third and sixth entries.

4.6. ELEMENT INTERNAL FORCES

As shown in the examples above, member internal forces at nodes are a simple calculation procedure once the global stiffness equations have been solved for the nodal displacements. The displacements thus obtained are initially referred to the global axes, and the element nodal displacements may be readily identified from among them. It is almost always simplest and most direct to transform element displacements to components along their local coordinate axes and then substitute them into the element stiffness equations referred to their local axes. The result then will be element internal forces as components along the local axes. For trusses, these are axial member forces. For beams and frames, they will be axial force, shear, and moment. For more complex structures, they will be comparable with relevant structural meanings.

4.7. BANDED GLOBAL STIFFNESS MATRIX

Many global stiffness matrices are sparsely populated, that is, a large proportion of their entries are zeros. The space occupied by the zeros is an unnecessary use of resources. Moreover, the computations done on the zeros in manipulating the global stiffness matrix are also an unnecessary expenditure of resources. In general, the larger the system of global stiffness equations, the higher will be the proportion of zeros in the global stiffness matrix. Very large global stiffness matrices may have in excess of 90% zeros as entries.

If the nonzero entries are left scattered throughout the positions of the global stiffness matrix, then little can be done to avoid wasting resources in storing and manipulating them. On the other hand, if the nonzero entries can be systematically placed in a specific region of the global stiffness matrix and that region's boundaries defined, then only the nonzero entries need be stored. Algorithms can be developed which identify them so that routine matrix operations can be undertaken with the global stiffness matrix and which identify when an entry in the original global stiffness matrix is zero so that no action needs to be taken.

For example, equations 4.9 illustrate two possible arrangements of a system of 10 global stiffness equations for a 4-element, 5-node, 10-dof beam. There are 52 nonzero entries in the global stiffness matrix and 48 zeros. Either arrangement is a valid form of the global stiffness equations for the beam. One may be obtained by rearranging the other:

$$
\begin{bmatrix}
k_{G11} & k_{G12} & k_{G13} & k_{G14} & 0 & 0 & 0 & 0 & 0 & 0 \\
k_{G21} & k_{G22} & k_{G23} & k_{G24} & 0 & 0 & 0 & 0 & 0 & 0 \\
k_{G31} & k_{G32} & k_{G33} & k_{G34} & k_{G35} & k_{G36} & 0 & 0 & 0 & 0 \\
k_{G41} & k_{G42} & k_{G43} & k_{G44} & k_{G45} & k_{G46} & 0 & 0 & 0 & 0 \\
0 & 0 & k_{G53} & k_{G54} & k_{G55} & k_{G56} & k_{G57} & k_{G58} & 0 & 0 \\
0 & 0 & k_{G63} & k_{G64} & k_{G65} & k_{G66} & k_{G67} & k_{G68} & 0 & 0 \\
0 & 0 & 0 & 0 & k_{G75} & k_{G76} & k_{G77} & k_{G78} & k_{G79} & k_{G7,10} \\
0 & 0 & 0 & 0 & k_{G85} & k_{G86} & k_{G87} & k_{G88} & k_{G89} & k_{G8,10} \\
0 & 0 & 0 & 0 & 0 & 0 & k_{G97} & k_{G98} & k_{G99} & k_{G9,10} \\
0 & 0 & 0 & 0 & 0 & 0 & k_{G10,7} & k_{G10,8} & k_{G10,9} & k_{G10,10}
\end{bmatrix}
\begin{Bmatrix}
q_1 \\ q_2 \\ q_3 \\ q_4 \\ q_5 \\ q_6 \\ q_7 \\ q_8 \\ q_9 \\ q_{10}
\end{Bmatrix}
=
\begin{Bmatrix}
F_1 \\ F_2 \\ F_3 \\ F_4 \\ F_5 \\ F_6 \\ F_7 \\ F_8 \\ F_9 \\ F_{10}
\end{Bmatrix},
$$

$$(4.9a)$$

$$
\begin{bmatrix}
k_{G11} & k_{G13} & 0 & 0 & 0 & 0 & 0 & 0 & k_{G14} & k_{G12} \\
k_{G31} & k_{G33} & k_{G35} & 0 & 0 & 0 & 0 & k_{G36} & k_{G34} & k_{G32} \\
0 & k_{G53} & k_{G35} & k_{G57} & 0 & 0 & k_{G58} & k_{G56} & k_{G54} & 0 \\
0 & 0 & k_{G75} & k_{G77} & k_{G79} & k_{G7,10} & k_{G78} & k_{G76} & 0 & 0 \\
0 & 0 & 0 & k_{G97} & k_{G99} & k_{G9,10} & k_{G98} & 0 & 0 & 0 \\
0 & 0 & 0 & k_{G10,7} & k_{G10,9} & k_{G10,10} & k_{G10,8} & 0 & 0 & 0 \\
0 & 0 & k_{G85} & k_{G87} & k_{G89} & k_{G8,10} & k_{G88} & k_{G86} & 0 & 0 \\
0 & k_{G63} & k_{G65} & k_{G67} & 0 & 0 & k_{G68} & k_{G66} & k_{G64} & 0 \\
k_{G41} & k_{G43} & k_{G45} & 0 & 0 & 0 & 0 & k_{G46} & k_{G44} & k_{G42} \\
k_{G21} & k_{G23} & 0 & 0 & 0 & 0 & 0 & 0 & k_{G24} & k_{G22}
\end{bmatrix}
\begin{Bmatrix}
q_1 \\ q_3 \\ q_5 \\ q_7 \\ q_9 \\ q_{10} \\ q_8 \\ q_6 \\ q_4 \\ q_2
\end{Bmatrix}
=
\begin{Bmatrix}
F_1 \\ F_3 \\ F_5 \\ F_7 \\ F_9 \\ F_{10} \\ F_8 \\ F_6 \\ F_4 \\ F_2
\end{Bmatrix}.
$$

$$(4.9b)$$

In the global stiffness matrix of equations 4.9b, the nonzero stiffness coefficients are spread across the range of row and column positions in a clustering about both diagonals of the matrix. In equations 4.9a, the nonzero coefficients are all clustered about the main diagonal, so that the greatest number of spaces in any row from the diagonal entry to the point where no more nonzero entries occur, in either direction, is 4. This number is called the half-bandwidth for the matrix. Clearly it will vary for different matrices. Indeed, for the global stiffness matrix of equation 4.9b it is 10.

Since structural stiffness matrices are symmetric, the nonzero coefficients in successive positions to the right of the diagonal entry in a given row are identical to the ones in successive positions in the column below the same diagonal entry. Thus there is no need to store both. In this example, only the diagonal entries and those for three positions to the right of the diagonal entry in each row need to be stored. This number of row spaces, four for equation 4.9a, again is the half-bandwidth.

A convenient array to store the nonzero entries of the global stiffness matrix of equation 4.9a would be 10 rows and 4 columns. The first column could contain the diagonal coefficients and the next 3 columns would contain the 3 entries to the right of each corresponding diagonal entry. If a particular row has fewer than 3 nonzero entries to the right of its diagonal entry, then zeros must be placed in the corresponding row positions. The storage requirement in this example for the banded region of the global stiffness matrix of equations 4.9a is 4×10, or 40, entries. It would be 10×10, or 100, entries for the full-size matrix. For matrices with greater sparseness, those of larger structural systems, the savings can be much greater. A 99-element beam with 100 nodes and 200 dof could have a half-bandwidth of 4, and its banded elements could be stored in a 200×4 matrix, requiring storage of 800 entries. The full-size global stiffness matrix for such a beam would be 200×200, requiring storage for 40,000 entries.

The solution of the global stiffness equations using just the nonzero global stiffness coefficients in the banded region of the global stiffness matrix requires that a precise transformation code be developed to identify them. Each stiffness coefficient must be applied to the correct global displacement as outlined in the global stiffness equations, notwithstanding the fact that they are stored in the reduced size, non-square matrix. Moreover, the zero coefficients of the full-size global stiffness matrix must be efficiently identified so that no action need be taken on them during the solution process.

Equation 4.9c illustrates the reduced matrix that could be used to store the banded nonzero coefficients of the global stiffness matrix of equation 4.9a:

$$
[\mathbf{K}_{banded}] = \begin{bmatrix}
k_{G11} & k_{G12} & k_{G13} & k_{G14} \\
k_{G22} & k_{G23} & k_{G24} & 0 \\
k_{G33} & k_{G34} & k_{G35} & k_{G36} \\
k_{G44} & k_{G45} & k_{G46} & 0 \\
k_{G55} & k_{G56} & k_{G57} & k_{G58} \\
k_{G66} & k_{G67} & k_{G68} & 0 \\
k_{G77} & k_{G78} & k_{G79} & k_{G7,10} \\
k_{G88} & k_{89} & k_{G8,10} & 0 \\
k_{G99} & k_{G9,10} & 0 & 0 \\
k_{G10,10} & 0 & 0 & 0
\end{bmatrix} .
\tag{4.9c}
$$

A transformation code can be developed as follows. The positions in the full-size global stiffness matrix of equation 4.9a are taken in row and column sequence from 1 to 10, so that the entries in the reduced matrix of equation 4.9c reflect these position indices. If i and j are these row and column position indices and m and n are the position indices for the reduced matrix, the transformation is (for $j \geq i$)

$$
i = m \quad \text{and} \quad j = m + n - 1
$$

or

$$
m = i \quad \text{and} \quad n = j - i + 1.
$$

Thus if the stiffness coefficient k_{Gij} is required, the indices m and n may be computed and the stiffness coefficient extracted from row m and column n of the reduced matrix. If $n >$ half-bandwidth,

$$
k_{Gij} = 0.
$$

Applying the symmetric property of the global stiffness matrix for the stiffness coefficients below the main diagonal of the matrix, for $j < i$,

$$
k_{Gij} = k_{Gji.}
$$

Thus, $m = j$ and $n = i - j + 1$. Again the stiffness coefficient would be extracted from row m and column n of the reduced banded matrix.

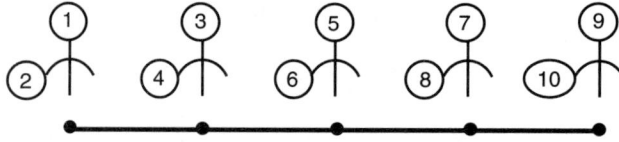

Figure 4.9 Discreticized beam

Though one might attempt to rearrange the global stiffness equations to achieve a minimum half-bandwidth along the main diagonal, it is usually achieved more readily through the numbering scheme for the gdof during the discretization of the structure. This numbering scheme controls the positions in the global stiffness matrix to which element stiffness coefficients are added during the assembly of the global stiffness matrix from the element stiffness matrices.

Since the diagonal entries of an element stiffness matrix occur also on the diagonal of the global stiffness matrix and its off-diagonal entries occur off it, the largest numerical difference between the corresponding gdof numbers on the element plus one for the diagonal position will be the half-bandwidth needed for that element. The largest half-bandwidth among all the elements will be the half-bandwidth for the global stiffness matrix. Thus to minimize the half-bandwidth for the global stiffness matrix, the analyst must use a gdof numbering scheme which has gdof numbers on each element whose differences are as small as possible.

For example, the discretized beam shown in Figure 4.9 has elements whose largest gdof number difference is 3. Thus its half-bandwidth is $3 + 1$, or 4. This is the minimum half-bandwidth that can be achieved with a beam. This numbering scheme is, of course, the one used for equation 4.9a.

For 2D structures it is less clear what numbering scheme will optimize the half-bandwidth. As a general rule, all gdof at a node should be numbered in sequence, and the sequence should proceed by node in the short direction of the structure. The rule is helpful for regular 2D structures. Figure 4.10 illustrates a rectangular plane stress plate discretized with nine elements. Each element has 4 nodes and there are 2 dof at each node. There are a total of 16 nodes and 32 gdof. For the numbering

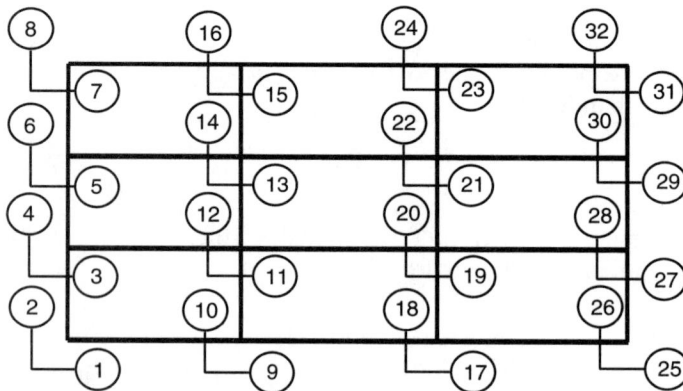

Figure 4.10 Discreticized plate

scheme shown, the largest difference in gdof numbering on any element is 11, so that the half-bandwidth of the plate's 32×32 global stiffness matrix is $11 + 1$, or12.

4.8. CLOSING REMARKS

Within the steps of the finite element method, choosing the type of finite elements employed and writing their element stiffness equations is the only step that differs significantly for different classes of structures. Discretization, assembly and solution of the global stiffness equations, and subsequent calculations using the gdof are not significantly different in concept for different classes of structures. The identification of the particular type of finite element which will be used and the development of its element stiffness equations are, however, the crux of obtaining a useful finite element model for the structure and are importantly different for different classes of structures.

The finite elements studied so far are 1D elements for which the element stiffness equations can be readily obtained by a direct approach. Other approaches will be needed to treat more sophisticated 1D elements and 2D and 3D elements. There are several approaches, each producing identical results. Those used most frequently employ the principle of virtual work, the principle of minimum potential energy, or Galerkin weighted residuals.

Many authors who deal with nonstructural applications use Galerkin weighted residuals, which adapt well to nonstructural applications. Many authors who apply the finite element method to structures use the principle of minimum potential energy, an efficient procedure for obtaining element stiffness matrices. It is this author's view that the principle of virtual work, which is equivalent to the principle of minimum potential energy, provides the best insights into the behavior of structural finite elements. Moreover, it seems to be easier to understand when students are first exposed to the development of element stiffness matrices, element equivalent load vectors, and element level boundary conditions. Consequently, in this book, the principle of virtual work will be used to develop the element stiffness equations of finite elements which cannot be readily obtained through the direct approach.

BIBLIOGRAPHY

Cook, R. D., Malkus, D. S., and Plesha, M. E., *Concepts and Applications of Finite Element Analysis* (Chapter 2), Wiley, New York, 1989.

Huebner, K. H., *The Finite Element Method for Engineers* (Chapter 2), Wiley, New York, 1975.

Logan, D. L., *A First Course in the Finite Element Method* (Chapter 2), PWS-Kent, Boston, 1992.

McGuire, W., and Gallagher, R. H., *Matrix Structural Analysis* (Chapter 3), Wiley, New York, 1979.

Weaver, W. Jr., and Gere, J. M., *Matrix Analysis of Framed Structures* (Chapter 3), 3rd ed., Van Nostrand Reinhold, New York, 1990.

Yang, T. Y., *Finite Element Structural Analysis* (Chapters 4, 5), Prentice-Hall, Englewood Cliffs, NJ, 1986.

PROBLEMS

4.1. For the plane truss shown:

a. Write the element stiffness equations for each member.

b. Using software which allows assembly of a global matrix from individual element matrices, write the global stiffness equations for the structure with all boundary conditions imposed.

c. Solve the global stiffness equations for joint displacements and support reactions using the submatrix solution procedure and software which manipulates individual matrices.

d. Solve the global stiffness equations for joint displacements and support reactions using the penalty solution procedure and software which manipulates individual matrices.

e. Use the member stiffness equations (part a), joint displacements (part c or d), and software which manipulates individual matrices to calculate the internal forces in each member.

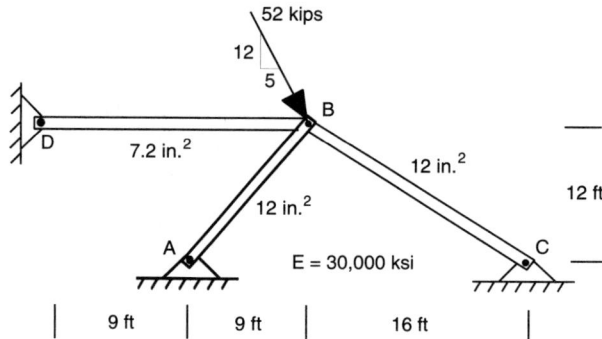

Problem 4.1

4.2. Repeat Problem 4.1 for the truss shown. Numbers shown near each member are its cross-sectional area in square millimeters $\times 10^3$.

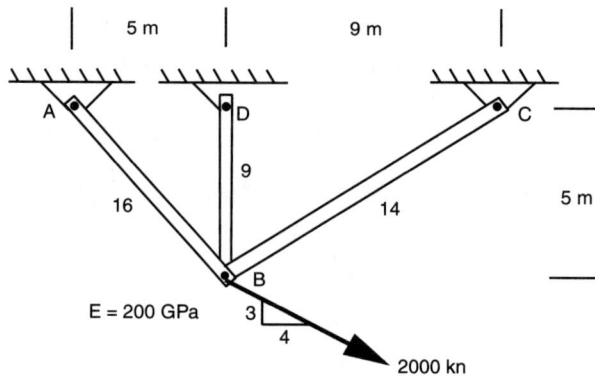

Problem 4.2

4.3. Using members AB and BC as beam elements and $E = 30,000$ ksi for the beam shown:

  a. Write its global stiffness equations with boundary conditions imposed using software which allows assembly of a global matrix from individual element matrices.

  b. Determine the beam's support reactions from its global stiffness equations and its shears and moments from its element stiffness equations. Use software which manipulates individual matrices.

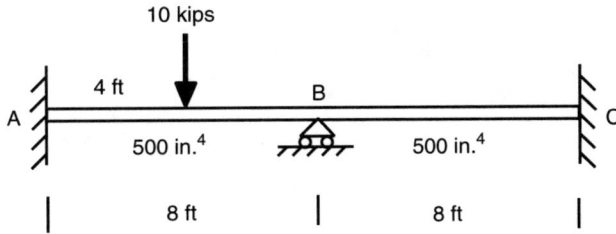

10 kips

4 ft

A

B

C

500 in.^{4}

500 in.^{4}

8 ft

8 ft

Problem 4.3

4.4. For the beam shown repeat Problem 4.3 for $E = 30,000$ ksi.

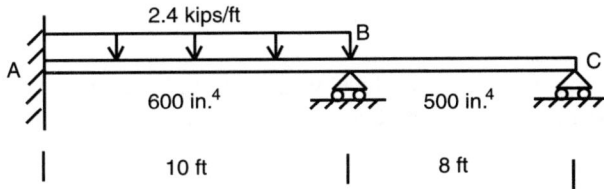

2.4 kips/ft

B

A

C

600 in.^{4}

500 in.^{4}

10 ft

8 ft

Problem 4.4

4.5. For the beam shown repeat Problem 4.3 for $E = 30,000$ ksi.

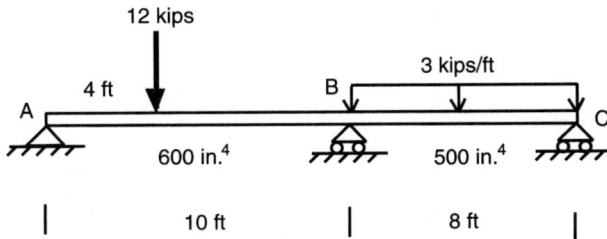

12 kips

3 kips/ft

4 ft

B

A

C

600 in.^{4}

500 in.^{4}

10 ft

8 ft

Problem 4.5

4.6. For the plane frame shown, develop the global load vector. Use the five nodes shown and the four members between them. The distributed loads are normal to the members on which they act. The global x axis passes through nodes 1 and 5; the global y axis passes through nodes 1 and 2.

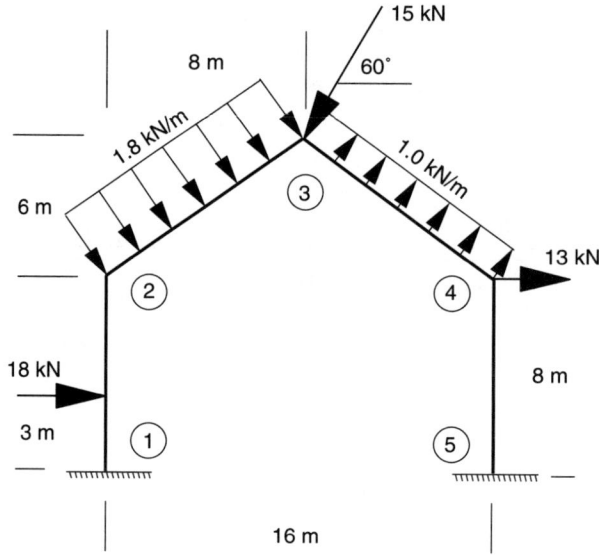

Problem 4.6

4.7. For the plane frame shown, develop the global load vector. Use the six nodes shown and the six members between them. The distributed loads are normal

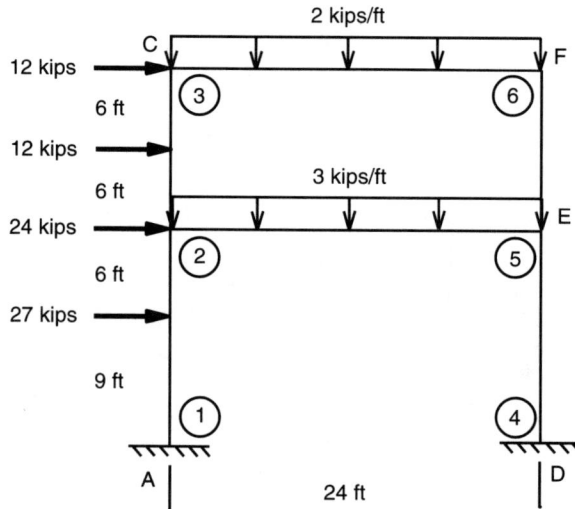

Problem 4.7

to the members on which they act. The global x axis passes through nodes 1 and 4; the global y axis passes through nodes 1, 2, and 3.

4.8. Starting with the global stiffness equations for the truss shown expressed in suitable matrix notation, develop the matrix equations you would use to calculate the reactions at each of the six supports for the given support conditions and restrictions. Show clearly each matrix operation you would use and identify clearly the matrices and submatrices involved and their respective numbers of rows and columns.

The loads (P_i) are applied as shown and the supports provide complete restraint at the points indicated, consistent with what pin and roller supports do. Develop your matrix equations to obtain the reactions so that the number of mathematical operations involved with inversion of any matrices and sub-matrices you use is minimized, consistent with the correct analysis of the truss. You should consider appropriate reduction procedures, but do not attempt to employ banded matrices. Give your estimate of the number of mathematical operations required for matrix inversions in your solution process, first using the Gauss–Jordan algorithm and second using direct matrix inversion.

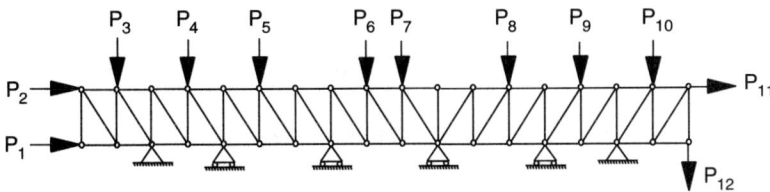

Problem 4.8

4.9. Find the support reactions and member forces for the plane truss shown for $E = 30,000$ ksi.

Use element and global stiffness equations, element transformation matrices, and software which manipulates individual matrices.

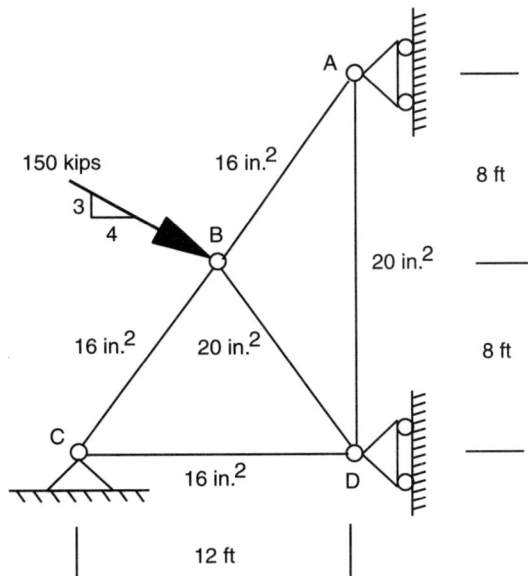

Problem 4.9

4.10. Find the support reactions and member forces for the plate truss shown. The roller support at *C* yields elastically in the vertical direction, 1 in. for each 500 kips of vertical load applied at the support at *C*; $E = 30,000$ ksi.

Use element and global stiffness equations, element transformation matrices, and software which manipulates individual matrices.

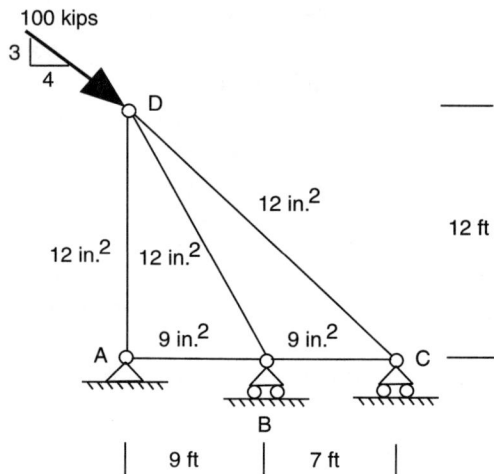

Problem 4.10

4.11. Find the support reactions and member shears and moments for the plane frame shown. For each member,

$$E = 200 \, \text{GPa}, \qquad I = 100 \times 10^{-6} \, \text{m}^4, \qquad A = 20 \times 10^{-3} \, \text{m}^2.$$

Use element and global stiffness equations, element transformation matrices, and software which manipulates individual matrices.

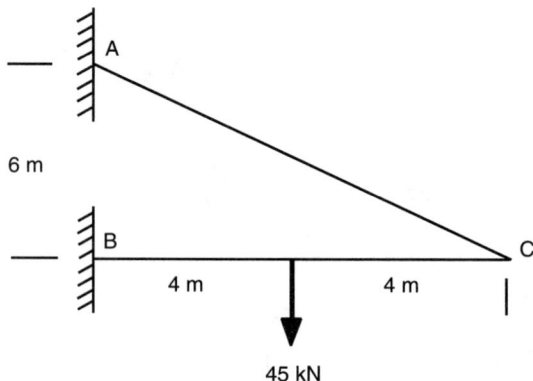

Problem 4.11

4.12. Find the reactions and member shears and moments for the plane frame shown. For each member,

$$E = 30,000 \, \text{ksi,}$$

$$I_{AB} = I_{CD} = 120 \, \text{in.}^4, \qquad A_{AB} = A_{CD} = 12 \, \text{in.}^2,$$

$$I_{BC} = 180 \, \text{in.}^4, \qquad A_{BC} = 20 \, \text{in.}^2$$

Use element and global stiffness equations, element transformation matrices, and software which manipulates individual matrices.

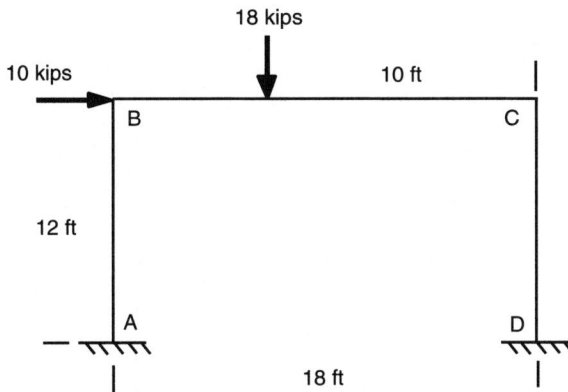

Problem 4.12

4.13. Find the reactions and member shears and moments for the plane frame shown.

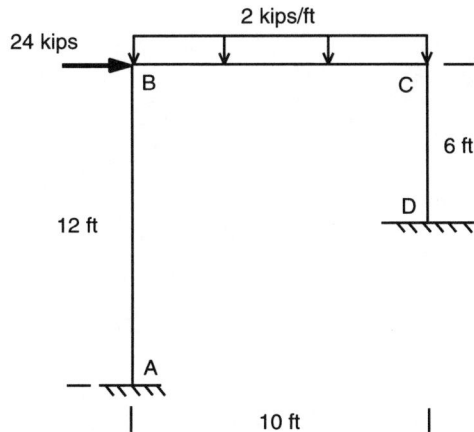

Problem 4.13

For each member

$$E = 30,000 \, \text{ksi},$$

$$I_{AB} = I_{BC} = 800 \, \text{in.}^4, \qquad A_{AB} = A_{BC} = 60 \, \text{in.}^2,$$

$$I_{CD} = 500 \, \text{in.}^4, \qquad A_{CD} = 30 \, \text{in.}^2$$

Use element and global stiffness equations, element transformation matrices, and software which manipulates individual matrices.

CHAPTER 5

ELEMENT STIFFNESS EQUATIONS BY DISPLACED STATE VIRTUAL WORK APPLICATIONS

5.1. DISPLACED STATE OF THE STRUCTURE

In Chapter 2, the displaced state of a structure is defined as the deformed equilibrium position of the structure and its finite elements under the action of its applied loads. In Chapter 3, Figure 3.1 illustrates a truss element in a special displaced state where the nodal displacements, or edof, are specified; Figure 3.2 illustrates a beam element in a similarly specified special displaced state; and Figure 3.3 illustrates a plane frame element in a general displaced state. In each illustration the nodal displacements, or edof, are identified constants. The structural displacements between the nodes vary as a function of local coordinates, conforming to the dof values when the function is at the nodes. This variation of structural displacements may be expressed in terms of the nodal displacements and specially defined displacement variations called global displacement basis functions and element displacement shape functions.

A refinement and extension of the direct determination of the element stiffness equations described in Chapter 3 employs displacement basis and shape functions to obtain element stiffness matrices and load vectors in a general procedure that applies to any structural finite element. The basis and shape functions must be defined in a special way so that they may be used to determine the element stiffness matrices and load vectors.

5.1.1. Global Displacement Basis Functions

A basis function is a special structural variation of displacements across the entire structure. There is one basis function defined for each gdof. Its value at the node of the gdof with which it is identified must equal unity and be zero at the nodes of the other gdof in the structural system. Basis functions have finite value on the elements adjacent to the node of the gdof with which they are identified. They need not be zero on elements more distant, but they usually are.

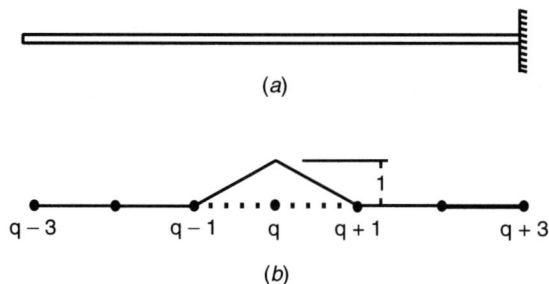

Figure 5.1 Basis function, axial translation

Basis functions must be continuous throughout each finite element of the structure and across each node. The derivatives of basis functions need not be continuous at nodes, but they must be continuous on elements and exist throughout the structure.

Figure 5.1 illustrates the piecewise linear basis function for the gdof at node q for an axially loaded structure. It equals one at the node of its gdof and zero at the other nodes. It is linear across the two finite elements adjacent to the node of its gdof but zero throughout the rest of the structure. It is continuous throughout the structure. Its first derivative exists throughout the structure and is continuous throughout except at node q and the first node to either side of node q.

Figures 5.2 and 5.3 illustrate basis functions for a beam structural system. There are two gdof at each node in a beam, a transverse displacement and a rotation displacement. Thus there are two basis functions at each node, one for each gdof.

The basis function illustrated in Figure 5.2 is for the transverse translation displacement at the node q. The function equals unity at this node and zero at all other nodes. It has finite value on the two elements adjacent to its node and is zero elsewhere in the structure. It is continuous throughout the structure. Since the second gdof at each node is a rotation displacement, the slope of this basis function at each node must be zero. The derivatives of the basis function are continuous throughout the structure.

The basis function illustrated in Figure 5.3 is for the rotation displacement at node q. The second gdof at each node is a rotation, so the slope of the basis function for rotation must be unity at node q and zero at other nodes. Moreover, its transverse translation displacement values must be zero at each node. The basis function

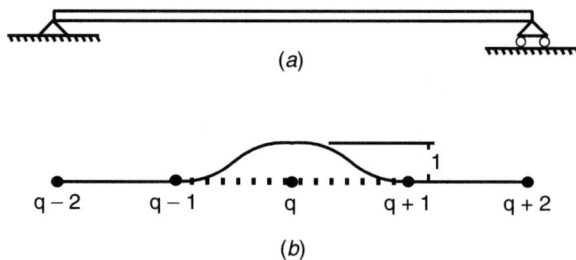

Figure 5.2 Basis function, beam translation

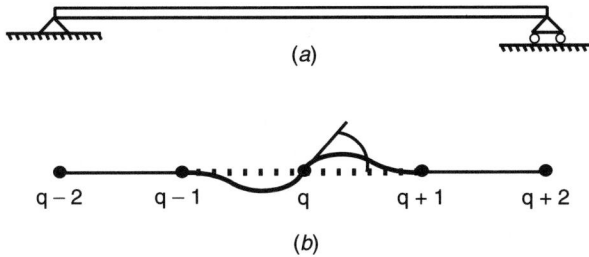

Figure 5.3 Basis function beam rotation

and its derivatives are continuous across the structure, and its transverse translation values have value on the two elements adjacent to its node and zero elsewhere.

Basis functions for beams are transverse translation displacement variations, even when the gdof they are associated with are rotations. The value of the dof identified with the basis function, whether translation or rotation, is unity at its dof nodal location and zero at the other dof locations. As we shall see, this definition of basis functions is necessary to determine the structure's stiffness equations from its displaced state.

The nature of the basis functions is chosen by the analyst. The functions must satisfy the requirements of the definition of basis functions and be differentiable and integrable if they are to be used to develop the structure's stiffness equations. The functions most easily employed are piecewise polynomials of the lowest degree, consistent with the definition requirements placed on basis functions.

Observing Figure 5.1, the simplest polynomial basis function for the axially loaded structure shown is a piecewise linear function. It is zero everywhere except on the two elements to either side of the node and gdof identified with the basis function and is linear on these two elements. On these two elements, there are two conditions which must be satisfied for the function. The function is unity at one node and zero at the other, and two such conditions uniquely determine a straight line.

The simplest polynomial basis functions for the beam structures shown in Figures 5.2 and 5.3 are piecewise cubic polynomials. They are zero everywhere except on the two elements to either side of the node and gdof identified with them and are cubic polynomials on these two elements. On these two elements, there are four conditions that must be satisfied for each function on each element. The function is unity for one dof at one node (either translation or rotation) and zero for the other three gdof, and four such conditions uniquely determine a cubic polynomial.

The basis functions, as defined, permit the representation of the complete displaced state of the structure as the sum of the products of the nodal displacements (or gdof) and their respective basis functions. Such a sum will be a functional variation of the structure's displacements throughout the structure, called its displacement function or displacement field.

The values of the structure's displacement function at the nodes are, of course, the gdof displacement values. The basis functions are unity at their respective gdof and zero elsewhere and are each multiplied by their respective gdof values to construct the structure's displacement function. The variation of displacement values between the nodes will be proportional to and the same nature as the basis functions. The

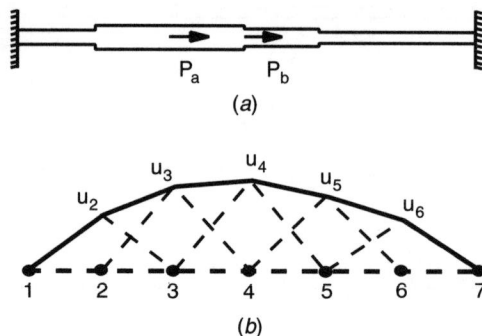

Figure 5.4 Displacement function

accuracy of the displacement function depends on the accuracy with which the nodal displacements are determined and the accuracy with which the functions used for basis functions can represent the actual structural displacement variation.

Figure 5.4 shows piecewise linear basis functions for an axially loaded structure multiplied by their respective gdof displacement values and then summed. This summation is the predicted displacement variation for the structure. It describes the displaced state of the structure.

If the axial loaded structure in Figure 5.4 is composed of linear elastic prismatic elements with concentrated axial loads imposed, then the exact axial displacement variation across these elements will be linear. The internal axial load in each prismatic element will be constant, causing constant axial stress and strain in each element. The axial strain is the first derivative of the axial displacement variation, so that the axial displacement variation must be linear.

The displacement function of a structure is the solution of the structural problem from which internal forces and stresses may be calculated. The analyst chooses the basis functions to be employed, so that only the values of the nodal displacements are needed to complete the displacement function. Recall that these gdof displacement values, or nodal displacements, are what are obtained in the finite element method from the solution of the global stiffness equations.

5.1.2. Element Displacement Shape Functions

The discussion of basis functions above is focused on the structure as a whole. Indeed, basis functions, and their weighted sum, leading to the structure's displacement function, extend throughout the structure. This view of the global structural displaced state provides an understanding of structural displacements, how they are represented in terms of nodal displacements and basis functions, and how they are in essence the solution of the structural problem. It is the basis functions themselves without the nodal displacement values, however, that permit the formulation of stiffness equations.

Though one might work exclusively with global basis functions and obtain global stiffness equations directly, for large structures with many nodes and many gdof it would be a daunting task. It is simpler and more manageable to work with the finite elements of the structure to obtain element stiffness equations. These then are assembled as global stiffness equations by straightforward procedures, such as those discussed in Chapter 4. It is to this end that element displacement shape functions are defined to represent exactly the global displacement basis functions.

Shape functions are continuous displacement variations across a finite element and are subject to the same conditions imposed on global basis functions. There is a shape function for each edof on the finite element. The displacement value, or edof value, of the shape function must be unity at the node of the edof with which it is identified and zero for the other edof at their nodes on the element. The shape function is defined only on the finite element in question and referred to local element coordinate axes.

Figure 5.5. illustrates the two shape functions defined for a 2-node, 2-dof axial finite element, and Figure 5.6 illustrates the four shape functions for a 2-node, 4-dof beam finite element. These functions observe the conditions required for shape functions cited above. Moreover, they and their derivatives are continuous throughout their finite element.

If polynomials are used for the shape functions, as they were for the basis functions, then for the 2-node, 2-dof axial element they are

$$N_1(x) = 1 - x/L, \tag{5.1a}$$

$$N_2(x) = x/L. \tag{5.1b}$$

The length of the element or distance between nodes is L, and x is the local coordinate measured from the element's first node. If the local coordinate is taken as the dimensionless $\zeta = x/L$, the 2-node, 2-dof axial element shape functions are written as

$$N_1(\zeta) = 1 - \zeta, \tag{5.1c}$$

$$N_2(\zeta) = \zeta. \tag{5.1d}$$

The 2-node, 4-dof beam element has cubic polynomial shape functions. They may be determined by writing the general expression for a cubic polynomial as

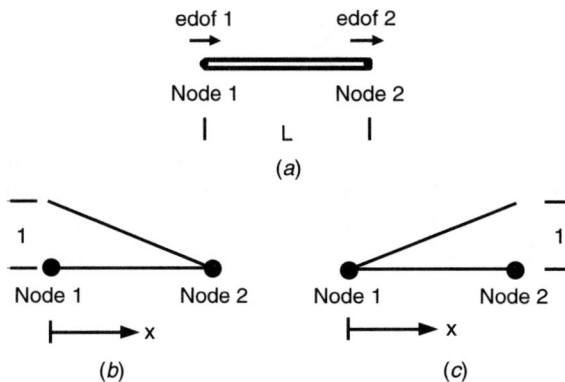

Figure 5.5 Axial shape functions

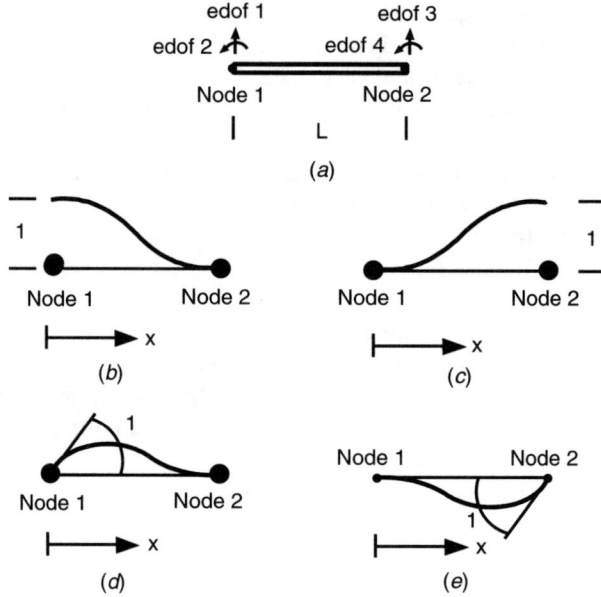

Figure 5.6 Beam shape functions

$$N_i(x) = C_{1i} + C_{2i}x + C_{3i}x^2 + C_{4i}x^3, \tag{5.2a}$$

where $i = 1, 2, 3, 4$, identifying the four shape functions and the four constants of each. The four conditions for a shape function may be imposed on each polynomial to determine its four constants. Using x, L, and ζ as defined above, the four-beam cubic polynomial shape functions may be written as

$$N_1(x) = 1 - 3\zeta^2 + 2\zeta^3, \tag{5.2b}$$

$$N_2(x) = L\zeta(1 - \zeta)^2, \tag{5.2c}$$

$$N_3(x) = \zeta^2(3 - 2\zeta), \tag{5.2d}$$

$$N_4(x) = L\zeta^2(\zeta - 1). \tag{5.2e}$$

Shown in Figure 5.7 are the six linear piecewise basis functions for a 6-node, 5-element axially loaded structure. Observing an individual element in Figure 5.7, say element 2 between nodes 2 and 3, there are two linear variations on it: one from the basis function identified with the element's left node and one from the basis function of its right node. Those two linear variations may also be regarded as the element's shape functions. Clearly the assembly of the structure's basis functions may be constructed by piecing together all of its element shape functions in a node-to-node, dof-to-dof assembly.

A displacement function for an element, which describes its displaced state, may be written so that it is analogous to the structure's displacement function. It is the sum of the products of its edof values and their corresponding element shape func-

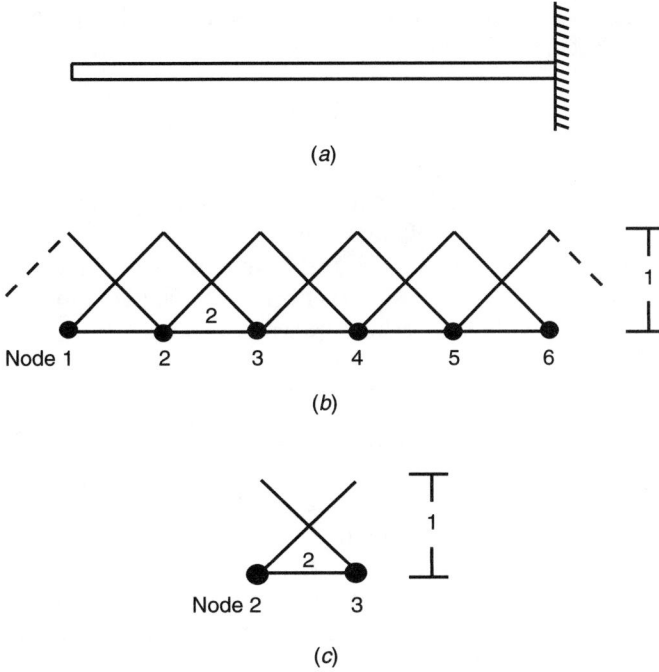

(a)

(b)

(c)

Figure 5.7 Basis functions as shape functions

tions. Because corresponding edof and gdof values (nodal displacements) must be the same and because the assembly of all of the shape functions on elements may be pieced together node to node and dof to dof to construct the entire complex of the structure's basis functions, the structure's displacement function may be obtained by piecing together the element displacement functions node to node and dof to dof throughout the structure.

Equation 5.3a is the expression for the element displacement function for a 2-node, 2-dof axial finite element,

$$u(x) = \lfloor N_1(x) \quad N_2(x) \rfloor \begin{Bmatrix} u_1 \\ u_2 \end{Bmatrix}, \tag{5.3a}$$

where the shape functions $N_i(x)$ are defined in equations 5.1. Equation 5.3b is the expression for the element displacement function for a 2-node, 4-dof beam finite element,

$$v(x) = \lfloor N_1(x) \quad N_2(x) \quad N_3(x) \quad N_4(x) \rfloor \begin{Bmatrix} v_1 \\ \theta_2 \\ v_3 \\ \theta_4 \end{Bmatrix}, \tag{5.3b}$$

where the shape functions $N_i(x)$ are defined in equations 5.2. The general expression for an element displacement function is

$$u(x) = \lfloor \mathbf{N(x)} \rfloor \{\mathbf{q}\}, \qquad (5.3c)$$

where the entries of vector $\mathbf{q}$ are the values of the edof.

The symbol u in equation 5.3c is intended for structural displacement variations of unspecified type. When the type of variation is known, u is translation along an x axis, v is translation along a y axis, w is a translation along a z axis, and θ is rotation. In a 2D setting, θ is a rotation about the axis normal to the plane. In a 3D setting, θ will have subscripts to identify the pertinent axes of rotation.

5.2. VIRTUAL WORK BASIS OF ELEMENT STIFFNESS EQUATIONS

A conservative structural system was discussed in Chapter 3 and is defined as one in which the work done on it by its externally applied forces and moments, as they move through their corresponding displacements, is transformed into elastic potential energy which is completely recovered when the forces and moments are removed. This conservation-of-energy principle is not changed if the forces and moments move through displacements which occur, not because of the forces and moments, but because of some other action, as long as all the work done is accounted for.

One might imagine the structure moving from its displaced state to some other displaced state. The new displacement could be measured with respect to the structure's previous displaced state. If this new displacement is continuous across the structure and is small and consistent with the structure's restraints, it is called an admissible virtual displacement. An admissible structural virtual displacement can be defined as a small, fictitious displacement of a structure which is continuous and consistent with the structure's restraints. The words *virtual* and *fictitious* are included to suggest that the displacement may occur but its reason for occurring need not be known.

Figure 5.8 illustrates a beam element in a displaced state. The element has externally applied loads between its nodes and internal moments and shears at its nodes

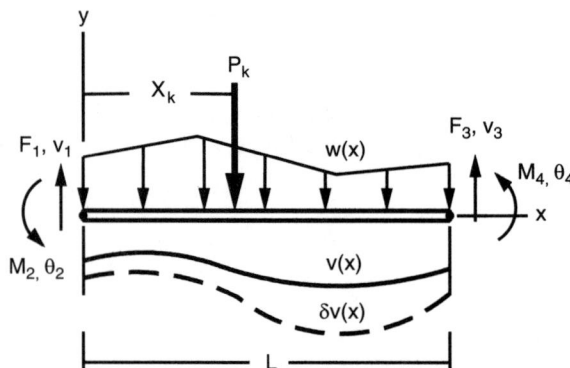

Figure 5.8 Displaced state beam element

and has been given an admissible virtual displacement $\delta v(x)$ from its real displaced state. The latter is expressed by the function $v(x)$. The symbol δ is used to identify quantities whose character is "virtual." The restraints on the element shown occur from the adjacent elements and take the form of internal moments and shears at the nodes and the requirements for continuity of displacements at nodes.

5.2.1. Principle of Virtual Work

The principle of virtual work is a statement of the conditions for equilibrium of a structure and, in a conservative structural system, of the conservation of energy as the structure moves and deforms. It is a fundamental concept in structural analysis which can provide an elegant means for evaluating a structure. In words, for a deformable structure in equilibrium under the action of applied loads and couples, the work done by these applied loads and couples as the structure moves through an admissible virtual displacement is equal to the change in internal strain energy the structure experiences as a consequence of the virtual displacement. In symbols,

$$\delta U = \delta W,$$

where δU is the change in strain energy, or virtual strain energy, the structure experiences as it moves from its displaced state $v(x)$ to its virtual displaced state $\delta v(x)$ and δW is the work done by the applied loads and couples as the structure moves from its displaced state $v(x)$ to its virtual displaced state $\delta v(x)$.

   The principle of virtual work for a rigid body in equilibrium is illustrated in Figure 5.9. The rigid body shown is in equilibrium under the action of the three applied forces F_1, F_2, and F_a and the applied couple C. Equilibrium requires that the sum of the forces be zero and that the sum of the couple and the moments of the forces about a point, say the body's left end, be zero. Thus,

$$F_1 - F_a + F_2 = 0, \tag{5.4a}$$

$$-aF_a + C + LF_2 = 0. \tag{5.4b}$$

If the rigid body is given a virtual displacement, in plane, it may be specified by the small virtual translation δv and the small virtual rotation $\delta\theta$. The work done by the three forces and couple as they move through this virtual displacement is given as

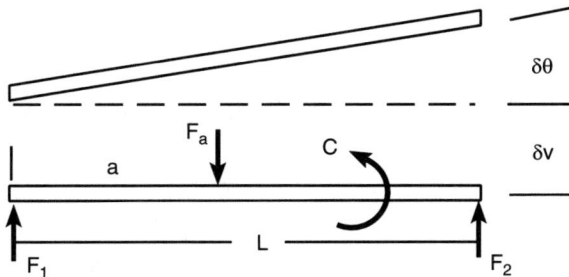

Figure 5.9 Rigid body in equilibrium

$$\delta W = F_1 \, \delta v - F_a \, \delta v - F_a(a \, \delta\theta) + C \, \delta\theta + F_2 \, \delta v + F_2(L \, \delta\theta)$$

$$= (F_1 - F_a + F_2) \, \delta v + (-aF_a + C + LF_2) \, \delta\theta. \tag{5.4c}$$

Substituting equations 5.4a and 5.4b into equation 5.4c, it is clear that

$$\delta W = 0. \tag{5.4d}$$

Equation 5.4d is the statement of the principle of virtual work for this rigid-body structure. Since there can be no strain in a rigid body, its virtual strain energy δU is zero.

The principle of virtual work for a deformable body in equilibrium is illustrated in Figure 5.10. For deformable structures, the virtual work done by the externally applied loads and the internal forces must be clearly distinguished, and the total virtual work done clearly defined. In the figure, P_1 and P_2 are externally applied axial loads at nodes 1 and 2 of the structure. The structure is considered in terms of nodes 1, 2, and 3 and elements a and b between nodes 1 and 2 and nodes 2 and 3, respectively. The internal forces F_{12}, F_{21}, F_{23}, and F_{32} on the elements are as shown. Their equal and opposite reactions F_{12}', F_{21}', and F_{23}' acting on the nodes are also shown. The structure experiences a small virtual axial displacement which is characterized by δu_1 at node 1 and δu_2 at node 2. The restraint at node 3 dictates that there is no virtual displacement there.

The equilibrium of nodes 1 and 2 requires that

$$P_1 - F_{12}' = 0, \tag{5.5a}$$

$$P_2 + F_{21}' - F_{23}' = 0. \tag{5.5b}$$

The total virtual work done by the forces at the nodes is

$$\delta W_{\text{nodes}} = (P_1 - F_{12}') \, \delta u_1 + (P_2 + F_{21}' - F_{23}') \, \delta u_2. \tag{5.5c}$$

Using equations 5.5a and 5.5b in equation 5.5c, clearly,

$$\delta W_{\text{nodes}} = 0. \tag{5.5d}$$

Rearranging equation 5.5c,

$$\delta W_{\text{nodes}} = (P_1 \, \delta u_1 + P_2 \, \delta u_2) + (-F_{12}' \, \delta u_1 + F_{21}' \, \delta u_2 - F_{23}' \, \delta u_2). \tag{5.5e}$$

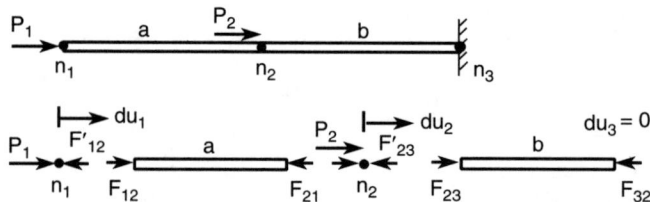

Figure 5.10 Deformable body in equilibrium

The virtual work done by the externally applied forces is

$$\delta W = (P_1 \, \delta u_1 + P_2 \, \delta u_2). \tag{5.5f}$$

The virtual strain energy in the two elements equals the virtual work done by the internal forces acting on the two elements,

$$\delta U = (F_{12} \, \delta u_1 - F_{21} \, \delta u_2 + F_{23} \, \delta u_2)$$
$$= -(-F'_{12} \, \delta u_1 + F'_{21} \, \delta u_2 - F'_{23} \, \delta u_2). \tag{5.5g}$$

Considering equations 5.5d, 5.5e, 5.5f, and 5.5g,

$$0 = \delta W - \delta U,$$

or

$$\delta U = \delta W. \tag{5.5h}$$

Equation 5.5h is the statement of the principle of virtual work for the deformable structure illustrated.

5.2.2. Element Stiffness Equations

Recall that element stiffness equations are systems of linear algebraic equations which relate the unknown element nodal displacements, or edof values, to the element internal nodal forces and moments. The principle of virtual work may be employed to write these equations in the familiar form

$$\{\mathbf{F}\} = [\mathbf{K}]\{\mathbf{q}\}. \tag{5.6}$$

Referring to Figure 5.8, consider first the virtual work δW done by the loads and couples on the element. These include the applied loads between nodes $w(x)$ and P_k and the internal moments and shears at the nodes. The quantity $w(x)$ is a distributed transverse load on the element and P_k is a concentrated transverse load acting at the local coordinate x_k.

Since these loads and couples are acting at their full values at the displaced state of the structure when the virtual displacement begins, the virtual work they do during the virtual displacement is the product of the magnitude of the load or couple and its corresponding virtual displacement. If the load is a force, its virtual displacement is a virtual translation whose line of action is the same as the one for the force. If it is a couple, then the virtual displacement is the virtual rotation whose axis of rotation is the same as the one for the couple.

For the beam element of Figure 5.8, the virtual work of the loads and couples may be expressed as

$$\delta W = \lfloor \delta \mathbf{q} \rfloor \{\mathbf{F}\} + \int_0^L \delta v(x) w(x) \, dx + \delta v(x_k) P_k, \tag{5.7}$$

where

$$\lfloor \delta \mathbf{q} \rfloor = \lfloor \delta v_1 \; \delta \theta_2 \; \delta v_3 \; \delta \theta_4 \rfloor,$$

$$\delta v_1, \; \delta \theta_2, \; \delta v_3, \; \delta \theta_4 = \text{edof virtual values},$$

$$\{\mathbf{F}\} = \begin{Bmatrix} F_1 \\ M_2 \\ F_3 \\ M_4 \end{Bmatrix},$$

$$\delta v(x_k) = \delta v(x) \quad \text{at} \quad x = x_k.$$

The element virtual displacement variation $\delta v(x)$ may be expressed as the product of the edof virtual values and the shape functions in the same manner that its actual displacement variation was expressed, in equation 5.3b. Since $v(x)$ and $\delta v(x)$ are scalar quantities, equation 5.3b may be transposed, so that

$$\delta v(x) = \lfloor \delta v_1 \; \delta \theta_2 \; \delta v_3 \; \delta \theta_4 \rfloor \begin{Bmatrix} N_1(x) \\ N_2(x) \\ N_3(x) \\ N_4(x) \end{Bmatrix}, \tag{5.8a}$$

$$\delta v(x) = \lfloor \delta \mathbf{q} \rfloor \{\mathbf{N}(\mathbf{x})\}. \tag{5.8b}$$

The expression $\delta v(x_k)$ may also be expressed in terms of the edof values and the beam element shape functions. In this circumstance the shape functions assume the values they possess at $x = x_k$. The result is

$$\delta v(x_k) = \lfloor \delta v_1 \; \delta \theta_2 \; \delta v_3 \; \delta \theta_4 \rfloor \begin{Bmatrix} N_1(x_k) \\ N_2(k_x) \\ N_3(x_k) \\ N_4(x_k) \end{Bmatrix}, \tag{5.9a}$$

$$\delta v(x_k) = \lfloor \delta \mathbf{q} \rfloor \{\mathbf{N}(\mathbf{x}_k)\}. \tag{5.9b}$$

Substituting equations 5.8b and 5.9b into equation 5.7 results in the required expression for the virtual work done by the applied loads and couples on the beam finite element. Thus

$$\delta W = \lfloor \delta \mathbf{q} \rfloor \{\mathbf{F}\} + \int_0^L \lfloor \delta \mathbf{q} \rfloor \{\mathbf{N}(\mathbf{x})\} w(x) \; dx + \lfloor \delta \mathbf{q} \rfloor \{\mathbf{N}(\mathbf{x}_k)\} P_k. \tag{5.10a}$$

The expression $\delta\mathbf{q}$ is an array of constant edof virtual values and not a function of x, so that it may be moved out from under the integral as long as the correct order of matrix multiplication is observed. Applying the distributive property of matrix multiplication, the expression for the virtual work done on the beam element (equation 5.10a) may be rewritten as

$$\delta W = \lfloor \delta\mathbf{q} \rfloor \left\{ \{\mathbf{F}\} + \int_0^L \{\mathbf{N}(\mathbf{x})\}w(x)\ dx + \{\mathbf{N}(\mathbf{x}_k)\}P_k \right\}. \tag{5.10b}$$

The column vector braces in equation 5.10b may be reordered as

$$\delta W = \lfloor \delta\mathbf{q} \rfloor \left\{ \{\mathbf{F}\} + \left\{ \int_0^L \mathbf{N}(\mathbf{x})w(x)\ dx \right\} + \{\mathbf{N}(\mathbf{x}_k)P_k\} \right\}. \tag{5.10c}$$

To complete the application of the principle of virtual work, next consider the virtual strain energy δU that the beam element takes on as it moves from its displaced state $v(x)$ to its virtual displaced state $\delta v(x)$.

Figure 5.11 illustrates a differential element from the beam finite element of Figure 5.8. The upper differential element is in its displaced state and the lower one is in its virtual displaced state. It is necessary to consider the action at the differential level because moment and curvature vary across the finite element. When the correct expressions are obtained for the differential element, they may be integrated across the length of the finite element to obtain the required expressions for the finite element.

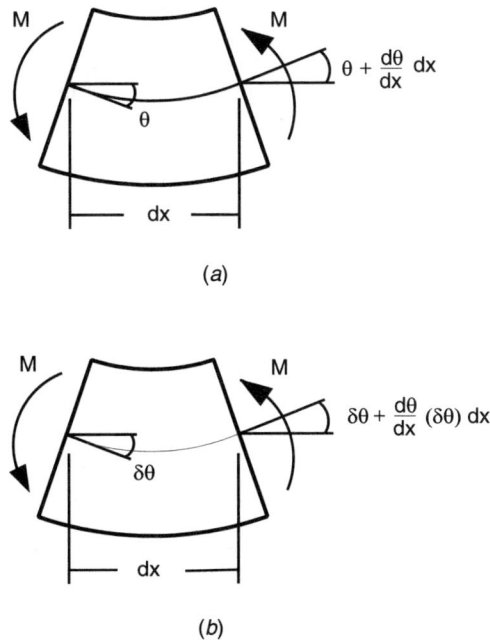

(a)

(b)

Figure 5.11 Differential beam element

The moment shown is the single value of the internal resisting moment at the point on the beam where the differential element is. From basic mechanics, the moment is

$$M = EI \frac{d^2v}{dx^2} = EIv'', \tag{5.11}$$

where, for the beam shown, E is the modulus of elasticity of the beam material, I is the centroidal moment of inertia of the beam cross section, and $v''(x) = d^2v/dx^2$ is the second derivative with respect to x of the transverse displacement variation or the curvature.

The virtual strain energy in the differential element may be obtained by applying the principle of virtual work to it. Only the work done and strain energy change as a consequence of flexure will be considered. Other effects, such as those from shear deformation, are usually negligible and will be ignored. Thus,

$$\delta U_{\text{DE}} = -M \, \delta\theta + M\left(\delta\theta + \frac{d}{dx}(\delta\theta) \, dx\right), \tag{5.12a}$$

$$\delta U_{\text{DE}} = M \frac{d}{dx}(\delta\theta) \, dx. \tag{5.12b}$$

The subscript DE signals that the expression gives the virtual strain energy for the differential element only. Substituting equation 5.11 into equation 5.12b and rearranging,

$$\delta U_{\text{DE}} = \frac{d}{dx}(\delta\theta) EI \frac{d^2v}{dx^2} \, dx. \tag{5.12c}$$

The expression $\delta\theta$ is the change in the virtual slope on the element and clearly is a function of x. It is equal to the first derivative of the virtual transverse displacement,

$$\delta\theta = \frac{d}{dx}(\delta v) = \delta v'. \tag{5.12d}$$

Differentiating equation 5.12d with respect to x,

$$\frac{d}{dx}(\delta\theta) = \frac{d}{dx}(\delta v') = \frac{d^2}{dx^2}(\delta v) = \delta v''. \tag{5.12e}$$

Using equation 5.12e in equation 5.12c,

$$U_{\text{DE}} = \frac{d^2}{dx^2}(\delta v) EI \frac{d^2v}{dx^2} \, dx = \delta v'' \, EIv'' \, dx. \tag{5.12f}$$

To obtain the virtual strain energy for the entire beam element, equation 5.12f must be integrated over the length of the element:

$$\delta U = \int_0^L \delta v'' \, EIv'' \, dx. \tag{5.13}$$

The expressions for $\delta v''(x)$ and $v''(x)$ may be obtained by differentiating equations 5.8 and 5.3b twice with respect to x. Since the edof values in equations 5.8 and 5.3b are constants, one only needs to differentiate the shape functions $N_i(x)$ twice with respect to x. These were chosen as polynomials, so differentiation is easily done:

$$\delta v''(x) = \lfloor \delta \mathbf{q} \rfloor \{\mathbf{N}''(\mathbf{x})\}, \tag{5.14a}$$

$$v''(x) = \lfloor \mathbf{N}''(\mathbf{x}) \rfloor \{\mathbf{q}\}. \tag{5.14b}$$

Substituting equations 5.14 into equation 5.13 results in

$$\delta U = \int_0^L \lfloor \delta \mathbf{q} \rfloor \{\mathbf{N}''(\mathbf{x})\} (EI) \lfloor \mathbf{N}''(\mathbf{x}) \rfloor \{\mathbf{q}\} \ dx. \tag{5.15a}$$

Observing in equation 5.15a that $\lfloor \delta \mathbf{q} \rfloor$ and $\{\mathbf{q}\}$ are arrays of constants which may be moved from under the integral as long as the order of matrix multiplication is not changed and applying the distributive property of matrix multiplication, the required expression for the virtual strain energy of the beam is

$$\delta U = \lfloor \delta \mathbf{q} \rfloor \left[\int_0^L \{\mathbf{N}''(\mathbf{x})\} (EI) \lfloor \mathbf{N}''(\mathbf{x}) \rfloor dx \right] \{\mathbf{q}\}. \tag{5.15b}$$

Equations 5.15b and 5.10c may be equated in conformance with the principle of virtual work, $\delta U = \delta W$, resulting in

$$\lfloor \delta \mathbf{q} \rfloor \left[\int_0^L \{\mathbf{N}''(\mathbf{x})\} (EI) \lfloor \mathbf{N}''(\mathbf{x}) \rfloor \ dx \right] \{\mathbf{q}\}$$
$$= \lfloor \delta \mathbf{q} \rfloor \left\{ \{\mathbf{F}\} + \left\{ \int_0^L \mathbf{N}(\mathbf{x}) w(x) \ dx \right\} + \{\mathbf{N}(\mathbf{x}_k) P_k\} \right\}. \tag{5.16}$$

Since equation 5.16 is a valid matrix equation, so also must be

$$\left[\int_0^L \{\mathbf{N}''(\mathbf{x})\} (EI) \lfloor \mathbf{N}''(\mathbf{x}) \rfloor \ dx \right] \{\mathbf{q}\}$$
$$= \{\mathbf{F}\} + \left\{ \int_0^L \mathbf{N}(\mathbf{x}) w(x) \ dx \right\} + \{\mathbf{N}(\mathbf{x}_k) P_k\}. \tag{5.17a}$$

It will be convenient to represent equation 5.17a in symbols as

$$[\mathbf{K}]\{\mathbf{q}\} = \{\mathbf{F}\} + \{\mathbf{F}_w\} + \{\mathbf{F}_k\}, \tag{5.17b}$$

where

$$[\mathbf{K}] = \left[\int_0^L \{\mathbf{N}''(\mathbf{x})\}(EI)\lfloor \mathbf{N}''(\mathbf{x})\rfloor \, dx \right]$$

$$= \begin{bmatrix} \int_0^L EIN_1''N_1'' \, dx & \int_0^L EIN_1''N_2'' \, dx & \int_0^L EIN_1''N_3'' \, dx & \int_0^L EIN_1''N_4'' \, dx \\ \int_0^L EIN_2''N_1'' \, dx & \int_0^L EIN_2''N_2'' \, dx & \int_0^L EIN_2''N_3'' \, dx & \int_0^L EIN_2''N_4'' \, dx \\ \int_0^L EIN_3''N_1'' \, dx & \int_0^L EIN_3''N_2'' \, dx & \int_0^L EIN_3''N_3'' \, dx & \int_0^L EIN_3''N_4'' \, dx \\ \int_0^L EIN_4''N_1'' \, dx & \int_0^L EIN_4''N_2'' \, dx & \int_0^L EIN_4''N_3'' \, dx & \int_0^L EIN_4''N_4'' \, dx \end{bmatrix}, \quad (5.17\text{c})$$

$$\{\mathbf{q}\} = \begin{Bmatrix} v_1 \\ \theta_2 \\ v_3 \\ \theta_4 \end{Bmatrix}, \qquad (5.17\text{d})$$

$$\{\mathbf{F}\} = \begin{Bmatrix} F_1 \\ M_2 \\ F_3 \\ M_4 \end{Bmatrix}, \qquad (5.17\text{e})$$

$$\{\mathbf{F}_w\} = \left\{ \int_0^L \mathbf{N}(\mathbf{x})w(x) \, dx \right\} = \begin{Bmatrix} \int_0^L N_1(x)w(x) \, dx \\ \int_0^L N_2(x)w(x) \, dx \\ \int_0^L N_3(x)w(x) \, dx \\ \int_0^L N_4(x)w(x) \, dx \end{Bmatrix}, \qquad (5.17\text{f})$$

$$\{\mathbf{F}_k\} = \{\mathbf{N}(\mathbf{x}_k)P_k\} = \begin{Bmatrix} N_1(x_k)P_k \\ N_2(x_k)P_k \\ N_3(x_k)P_k \\ N_4(x_k)P_k \end{Bmatrix}, \qquad (5.17\text{g})$$

The term $[\mathbf{K}]$ is a square matrix whose entries are known constants which may be determined from the definite integrals of the products of the second derivatives of the beam's shape functions (equation 5.17c). The beam's shape functions may be determined from the requirements for shape functions, as was done to obtain equations 5.2. The size of $[\mathbf{K}]$ is $n \times n$, where n is the number of edof. For the beam element studied, the size is 4×4.

The term $\{\mathbf{q}\}$ is the $n \times 1$ vector of edof constant values, or element nodal displacements. They are unknowns in equations 5.17a, 5.17b, and 5.17d. For the beam element, the size is 4×1.

The term $\{\mathbf{F}\}$ is the $n \times 1$ vector of unknown internal moments and shears for the beam element. For the beam element, the size is 4×1.

The terms $\{\mathbf{F}_w\}$ and $\{\mathbf{F}_k\}$ are $n \times 1$ vectors of known equivalent nodal forces and moments which will produce the same nodal displacements in the structure as do the original applied loads between the nodes. They are determined by the expressions in equations 5.17f and 5.17g. The applied loads between the nodes, $w(x)$ and P_k, must be reduced to these equivalent forces and moments at the nodes since the global stiffness equations deal only with quantities at nodes. For the beam element, the size is 4×1.

Comparing equations 5.17b and 5.6, the former clearly are the beam element's stiffness equations. They differ from equation 5.6 in one obvious respect. They include directly the effects of the applied loads between nodes.

An important advantage of equations 5.17 is that they are general. For example, the beam's flexural rigidity EI need not be constant to apply equation 5.17c. If EI is constant, the beam is prismatic, and the term EI may be moved from under the integral. If the integrations indicated are performed using the shape functions cited in equations 5.2, the element stiffness matrix shown in equation 3.5 will be obtained. Readers should perform this exercise to solidify their understanding of the virtual work procedure. Clearly, if EI is a function of x or if shape functions other than those shown in equations 5.2 are used, a different, perhaps better, beam element stiffness matrix will result.

The procedures indicated in equations 5.17f and 5.17g to obtain equivalent nodal forces and moments for applied loads between nodes are general also. It is not necessary to undertake a structural procedure to find fixed end moments and shears. Indeed, except for the simple loads between nodes listed in Figure 3.7, such a procedure may prove cumbersome. The virtual work procedure, however, permits the analyst to determine the shape functions appropriate to the analysis and the loads as suitable functions of x and then to perform the integrations indicated. The equivalent nodal forces and moments for loads between nodes for beam elements tabulated in Figure 3.7 can be duplicated readily using equations 5.17f and 5.17g and the cubic shape functions of equations 5.2. This is left as an exercise for the reader.

The accuracy of equations 5.17 depends on the suitability of the shape functions employed. "Exact" shape functions are seldom known in advance of the analysis, and then only in the most trivial analyses. However, it must be apparent that the variation of structural displacements over a very small finite element will not be great in most instances. If one uses shape functions which can approximate this variation well for small elements, the resulting element stiffness equations, global stiffness equations, and nodal displacements should also be approximated well.

For a beam element, even though cubic polynomials are exact only if the beam is prismatic and not subjected to distributed loads, cubic polynomials will approximate most beam transverse displacements very accurately over reasonably short elements. It is usually simpler, more effective, and more economical to use more and smaller beam finite elements in an analysis than it is to develop and program more refined shape functions and stiffness equations for larger elements. In general, the more finite elements used in the structure, the more accurate should be the analysis, and the more suitable will be the relatively simple shape functions which can be easily manipulated in the finite element method.

What follows in the next two sections are some illustrations of the virtual work procedure in the development of element stiffness matrices for additional 1D structural finite elements. A general approach for multidimensional elements is in Chapter 6.

5.2.3. Beam Finite Element on an Elastic Foundation

The only difference between the beam finite element studied in the last section and one which is supported by an elastic foundation is, of course, the effect which the elastic foundation has. The classical solution to the problem of a beam on an elastic foundation is limited to prismatic beams supporting relatively simple load systems and with either no additional restraints or simple ones at the beam's ends.

The finite element method is not so restricted. Though the beam elements are usually prismatic, the beam structure as a whole may possess several different cross sections. If it is tapered, it may be approximated by numerous short prismatic elements or with fewer larger elements by developing tapered beam element shape functions and element stiffness equations. Additional restraints are easily applied at any gdof in the global stiffness equations. The inclusion of the effects of the elastic foundation is treated easily by the virtual work procedure. The illustration that follows considers the beam element to be prismatic and employs the cubic polynomials of equations 5.2 as element shape functions.

Figure 5.12 is a reproduction of the beam finite element shown in Figure 5.8, except that a subgrade reaction r_s on the beam element from the elastic foundation is shown. This subgrade reaction is taken as a distributed line force which is proportional to the transverse displacement of the beam element. In other words, the subgrade reaction is dependent on and proportional to the element's displaced shape. The inverse is also true, the displaced shape is dependent on the subgrade reaction.

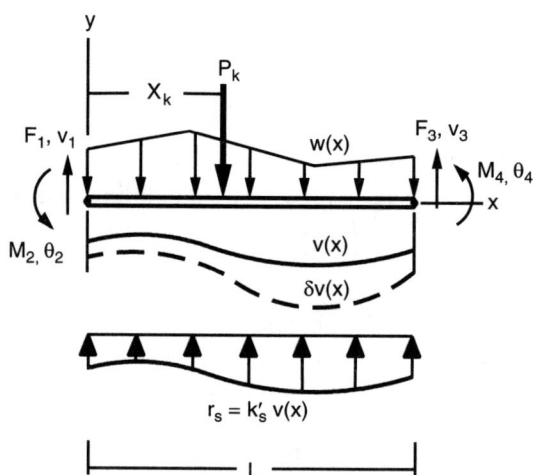

Figure 5.12 Displaced state beam element elastic foundation

The constant of proportionality between the element transverse displacements and the distributed line load subgrade reaction is the beam subgrade modulus k_s'. The beam subgrade modulus is related to the foundation subgrade modulus k_s. This latter modulus is the constant of proportionality between the pressure applied to the foundation (force per unit of area) and the resulting displacement into the foundation. The subgrade modulus k_s is multiplied by the width of the beam in contact with the foundation to give the beam subgrade modulus k_s'.

The virtual work done, δW, by the applied loads on the beam element as it goes through an admissible virtual displacement includes the terms developed for the element in equation 5.10c plus the virtual work δW_{sg} done by the subgrade reaction. The virtual work done by the subgrade reaction will be negative since the reaction must oppose the displacement. It may be expressed as

$$\delta W_{sg} = -\int_0^L \delta v(x) k_s' v(x) \ dx. \tag{5.18a}$$

In equation 5.18a substituting the expressions shown in equations 5.8b and 5.3b for $\delta v(x)$ and $v(x)$ results in

$$\delta W_{sg} = -\int_0^L \lfloor \delta \mathbf{q} \rfloor \{ \mathbf{N(x)} \}(k_s') \lfloor \mathbf{N(x)} \rfloor \{ q \} \ dx. \tag{5.18b}$$

Moving the arrays of constants $\lfloor \delta \mathbf{q} \rfloor$ and $\{ \mathbf{q} \}$ from under the integral without altering the order of matrix multiplication,

$$\delta W_{sg} = -\lfloor \delta \mathbf{q} \rfloor \left[\int_0^L \{ \mathbf{N(x)} \}(k_s') \lfloor \mathbf{N(x)} \rfloor \ dx \right] \{ \mathbf{q} \}. \tag{5.18c}$$

Equation 5.18c is the contribution from the elastic foundation to the virtual work done by the applied loads on the element. In symbols,

$$\delta W_{sg} = -\lfloor \delta \mathbf{q} \rfloor [\mathbf{K}_r] \{ \mathbf{q} \}, \tag{5.18d}$$

where

$$\lfloor \delta \mathbf{q} \rfloor = \lfloor \delta v_1 \ \ \delta \theta_2 \ \ \delta v_3 \ \ \delta \theta_4 \rfloor, \tag{5.18e}$$

$$[\mathbf{K}_r] = \left[\int_0^L \{ \mathbf{N(x)} \}(k_s') \lfloor \mathbf{N(x)} \rfloor \ dx \right],$$

$$= \begin{bmatrix} \int_0^L k_s' N_1 N_1 \ dx & \int_0^L k_s' N_1 N_2 \ dx & \int_0^L k_s' N_1 N_3 \ dx & \int_0^L k_s' N_1 N_4 \ dx \\ \int_0^L k_s' N_2 N_1 \ dx & \int_0^L k_s' N_2 N_2 \ dx & \int_0^L k_s' N_2 N_3 \ dx & \int_0^L k_s' N_2 N_4 \ dx \\ \int_0^L k_s' N_3 N_1 \ dx & \int_0^L k_s' N_3 N_2 \ dx & \int_0^L k_s' N_3 N_3 \ dx & \int_0^L k_s' N_3 N_4 \ dx \\ \int_0^L k_s' N_4 N_1 \ dx & \int_0^L k_s' N_4 N_2 \ dx & \int_0^L k_s' N_4 N_3 \ dx & \int_0^L k_s' N_4 N_4 \ dx \end{bmatrix}, \tag{5.18f}$$

$\{ \mathbf{q} \}$ = see equation 5.17d.

To obtain the total virtual work δW done by the applied loads on the beam element on an elastic foundation, the virtual work done by the subgrade is added algebraically to the virtual work done by the other applied loads. Representing equation 5.10c in symbols and adding equation 5.18d to it result in

$$\delta W = \lfloor \delta \mathbf{q} \rfloor (\{\mathbf{F}\} + \{\mathbf{F}_w\} + \{\mathbf{F}_k\} - [\mathbf{K}_r]\{\mathbf{q}\}). \tag{5.19}$$

The form of the expression for the virtual strain energy taken on by the beam element is not affected by the elastic foundation. It is given by equation 5.15b as

$$\delta U = \lfloor \delta \mathbf{q} \rfloor [\mathbf{K}]\{\mathbf{q}\}. \tag{5.20}$$

Applying the principle of virtual work ($\delta U = \delta W$) by equating equations 5.20 and 5.19 results in

$$\lfloor \delta \mathbf{q} \rfloor [\mathbf{K}]\{\mathbf{q}\} = \lfloor \delta \mathbf{q} \rfloor (\{\mathbf{F}\} + \{\mathbf{F}_w\} + \{\mathbf{F}_k\} - [\mathbf{K}_r]\{\mathbf{q}\}). \tag{5.21}$$

Since equation 5.21 is a valid matrix equation, so also must be

$$[\mathbf{K}]\{\mathbf{q}\} = \{\mathbf{F}\} + \{\mathbf{F}_w\} + \{\mathbf{F}_k\} - [\mathbf{K}_r]\{\mathbf{q}\}. \tag{5.22a}$$

Rearranging yields

$$([\mathbf{K}] + [\mathbf{K}_r])\{\mathbf{q}\} = \{\mathbf{F}\} + \{\mathbf{F}_w\} + \{\mathbf{F}_k\}, \tag{5.22b}$$

or

$$[\mathbf{K}_E]\{\mathbf{q}\} = \{\mathbf{F}\} + \{\mathbf{F}_w\} + \{\mathbf{F}_k\}, \tag{5.22c}$$

where

$$[\mathbf{K}_E] = [\mathbf{K}] + [\mathbf{K}_r]. \tag{5.22d}$$

Thus the effect of the elastic foundation on the beam element's stiffness equations may be accounted for by modifying the element's stiffness matrix using equations 5.17c, 5.18f, and 5.22d.

As an exercise in manipulating these three equations, the reader should develop the 16 stiffness coefficients for the stiffness matrix of a prismatic beam element on an elastic foundation whose beam subgrade modulus k_s' is constant over the length of the beam finite element. Using the cubic polynomial shape functions shown in equation 5.2 and expressing the stiffness coefficients in terms of the beam flexural rigidity EI, its length L, and its beam subgrade modulus k_s' results in

$$k_{E11} = 12EI/L^3 + 0.3714286k_s'L,$$

$$k_{E21} = 6EI/L^2 + 0.0523810k_s'L^2,$$

$$k_{E31} = -12EI/L^3 + 0.1285714k_s'L,$$

$$k_{E41} = 6EI/L^2 - 0.0309524k_s'L^2;$$

$$k_{E12} = k_{E21}$$

$$k_{E22} = 4EI/L + 0.0095238k_s'L^3,$$

$$k_{E32} = -6EI/L^2 + 0.0309524k_s'L^2,$$

$$k_{E42} = 2EI/L - 0.0071429k_s'L^3;$$

$$k_{E13} = k_{E31}$$

$$k_{E23} = k_{E32}$$

$$k_{E33} = 12EI/L^3 + 0.3714286k_s'L,$$

$$k_{E43} = -6EI/L^2 - 0.0523810k_s'L^2;$$

$$k_{E14} = k_{E41}$$

$$k_{E24} = k_{E42}$$

$$k_{E34} = k_{E43}$$

$$k_{E44} = 4EI/L + 0.0095238k_s'L^3.$$

5.2.4. Axial Finite Elements

Axially loaded structures are relatively easy structures to analyze, either with classical procedures or with the finite element method. For this reason they serve to illustrate the finite element method well, especially the virtual work procedure.

Figure 5.13 illustrates a 2-node, 2-edof axial finite element in a displaced and a virtual displaced state. Figure 5.14 similarly illustrates a 3-node, 3-edof axial finite element. Not shown on either is a distributed axial load $p(x)$ which may be present and would be expressed as force per unit of axial length. The virtual work done by the nodal and distributed axial loads as the axial element undergoes an admissible virtual displacement is given as

$$\delta W = \lfloor \delta \mathbf{q} \rfloor \{\mathbf{F}\} + \int_0^L \delta u(x) p(x) \, dx. \tag{5.23}$$

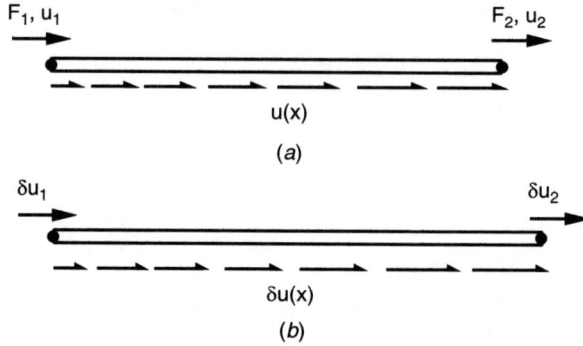

Figure 5.13 Axial element 2 nodes

The vectors $\lfloor \delta \mathbf{q} \rfloor$ and $\{\mathbf{F}\}$ each have two or three entries, depending on whether they are for the 2- or the 3-node element. For the 3-node element they are

$$\lfloor \delta \mathbf{q} \rfloor = \lfloor \delta u_1 \; \delta u_2 \; \delta u_3 \rfloor, \tag{5.24a}$$

$$\{\mathbf{F}\} = \begin{Bmatrix} F_1 \\ F_2 \\ F_3 \end{Bmatrix}. \tag{5.24b}$$

The integral expression in equation 5.23 is treated in the same manner that the beam element was treated earlier. The virtual displacement function $\delta u(x)$ is given as

$$\delta u(x) = \lfloor \delta u_1 \; \delta u_2 \; \delta u_3 \rfloor \begin{Bmatrix} N_1(x) \\ N_2(x) \\ N_3(x) \end{Bmatrix} = \lfloor \delta \mathbf{q} \rfloor \{\mathbf{N(x)}\}. \tag{5.24c}$$

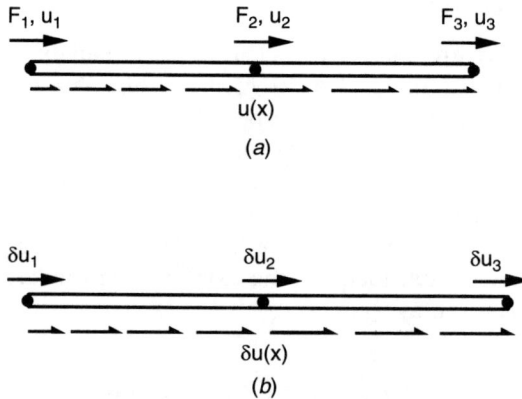

Figure 5.14 Axial element 3 nodes

Using equation 5.24c in equation 5.23, the virtual work done by the nodal and distributed axial loads may be expressed as

$$\delta W = \lfloor \delta \mathbf{q} \rfloor \{\mathbf{F}\} + \lfloor \delta \mathbf{q} \rfloor \left\{ \int_0^L \mathbf{N}(\mathbf{x}) p(x) \; dx \right\}$$

$$= \lfloor \delta \mathbf{q} \rfloor \{\mathbf{F}\} + \lfloor \delta \mathbf{q} \rfloor \{\mathbf{F}_p\}. \tag{5.25}$$

The equivalent nodal axial loads $\{\mathbf{F}_p\}$ of the distributed axial load are computed as indicated in equation 5.25.

Figure 5.15 shows a differential element from an axial finite element in displaced and virtual displaced states. Its virtual strain energy as it passes through an admissible virtual displacement is given as

$$\delta U_{\text{DE}} = -P \, \delta u + P \left(\delta u + \frac{d}{dx} (\delta u) \; dx \right) = P \frac{d}{dx} (\delta u) \; dx. \tag{5.26a}$$

Replacing the axial load P by σA, where σ is the axial stress acting uniformly across the cross-sectional area A of the element,

$$\delta U_{\text{DE}} = \frac{d}{dx} (\delta u)(\sigma A) \; dx. \tag{5.26b}$$

The axial stress $\sigma = E\epsilon$, where E is the modulus of elasticity for the material of the element and ϵ is the axial strain. Axial strain is the first derivative with respect to x of the axial displacement, or $\epsilon = du/dx$. Substituting these quantities into equation 5.26b,

$$\delta U_{\text{DE}} = \frac{d}{dx} (\delta u)(\text{AE}) \frac{du}{dx} dx = \delta u'(\text{AE}) u' \; dx. \tag{5.26c}$$

To obtain the virtual strain energy for the entire axial finite element, equation 5.26c must be integrated over the length of the element:

$$\delta U = \int_0^L \delta u'(\text{AE}) u' \; dx. \tag{5.27}$$

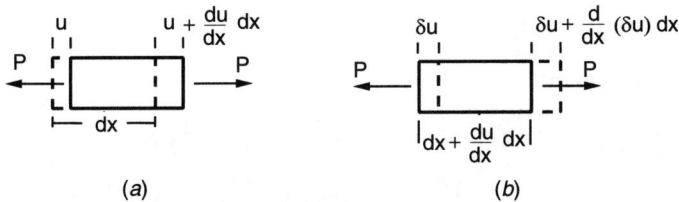

Figure 5.15 Axial differential element

The element displacement function $u(x)$ and virtual displacement function $\delta u(x)$ are the sum of the products of their edof values and their corresponding shape functions:

$$u(x) = \lfloor \mathbf{N(x)} \rfloor \{\mathbf{q}\}, \tag{5.28a}$$

$$\delta u(x) = \lfloor \delta \mathbf{q} \rfloor \{\mathbf{N(x)}\}. \tag{5.28b}$$

Equations 5.28 are differentiated with respect to x and substituted into equation 5.27 to give

$$\delta U = \int_0^L \lfloor \delta \mathbf{q} \rfloor \{\mathbf{N'(x)}\}(AE)\lfloor \mathbf{N'(x)} \rfloor \{\mathbf{q}\} \ dx. \tag{5.29a}$$

Moving the arrays of constants $\lfloor \delta \mathbf{q} \rfloor$ and $\{\mathbf{q}\}$ from under the integral without altering the order of matrix multiplication results in

$$\delta U = \lfloor \delta \mathbf{q} \rfloor \left[\int_0^L \{\mathbf{N'(x)}\}(AE)\lfloor \mathbf{N'(x)} \rfloor \ dx \right] \{\mathbf{q}\}. \tag{5.29b}$$

Applying the principle of virtual work ($\delta U = \delta W$) by equating equations 5.29b and 5.25 results in

$$\lfloor \delta \mathbf{q} \rfloor \left[\int_0^L \{\mathbf{N'(x)}\}(AE)\lfloor \mathbf{N'(x)} \rfloor \ dx \right] \{\mathbf{q}\} = \lfloor \delta \mathbf{q} \rfloor \{\mathbf{F}\} + \lfloor \delta \mathbf{q} \rfloor \{\mathbf{F}_p\}. \tag{5.30a}$$

Equation 5.30a gives rise to

$$\left[\int_0^L \{\mathbf{N'(x)}\}(AE)\lfloor \mathbf{N'(x)} \rfloor \ dx \right] \{\mathbf{q}\} = \{\mathbf{F}\} + \{\mathbf{F}_p\}, \tag{5.30b}$$

or

$$[\mathbf{K}]\{\mathbf{q}\} = \{\mathbf{F}\} + \{\mathbf{F}_p\}. \tag{5.30c}$$

where

$$[\mathbf{K}] = \left[\int_0^L \{\mathbf{N'(x)}\}(AE)\lfloor \mathbf{N'(x)} \rfloor \ dx \right]. \tag{5.30d}$$

For a 2-edof axial element,

$$[\mathbf{K}] = \begin{bmatrix} \int_0^L (AE)N_1'N_1' \ dx & \int_0^L (AE)N_1'N_2' \ dx \\ \int_0^L (AE)N_2'N_1' \ dx & \int_0^L (AE)N_2'N_2' \ dx \end{bmatrix}. \tag{5.30e}$$

For a 3-edof axial element

$$[\mathbf{K}] = \begin{bmatrix} \int_0^L (AE)N_1'N_1' \ dx & \int_0^L (AE)N_1'N_2' \ dx & \int_0^L (AE)N_1'N_3' \ dx \\ \int_0^L (AE)N_2'N_1' \ dx & \int_0^L (AE)N_2'N_2' \ dx & \int_0^L (AE)N_2'N_3' \ dx \\ \int_0^L (AE)N_3'N_1' \ dx & \int_0^L (AE)N_3'N_2' \ dx & \int_0^L (AE)N_3'N_3' \ dx \end{bmatrix}. \tag{5.30f}$$

Here, $\{q\}$ is the array of edof values u_1 and u_2 for a 2-edof element or u_1, u_2, and u_3 for a 3-edof element; $\{F\}$ is the array of internal nodal axial forces F_1 and F_2 for a two 2-edof element or F_1, F_2, and F_3 for a 3-edof element; and

$$\{\mathbf{F}_p\} = \int_0^L \mathbf{N(x)}p(x)\ dx \qquad (5.30g)$$

is the array of equivalent nodal axial forces for the distributed axial force.

For a 2-edof element,

$$\{\mathbf{F}_p\} = \left\{ \begin{array}{c} \int_0^L N_1(x)p(x)\ dx \\ \int_0^L N_2(x)p(x)\ dx \end{array} \right\}, \qquad (5.30h)$$

and for a 3-edof axial element,

$$\{\mathbf{F}_p\} = \left\{ \begin{array}{c} \int_0^L N_1(x)p(x)\ dx \\ \int_0^L N_2(x)p(x)\ dx \\ \int_0^L N_3(x)p(x)\ dx \end{array} \right\}. \qquad (5.30i)$$

Clearly equation 5.30c is the axial element stiffness equation. Equations 5.30d–i, which follow it, indicate procedures to determine the axial element stiffness matrix and equivalent nodal axial loads. Using the linear shape functions shown in equations 5.1 and performing the integrations indicated in equations 5.30e, the stiffness matrix for the 2-edof axial element of equation 3.8 will be obtained.

To obtain the stiffness matrix for the 3-edof axial element, shape functions must be determined first. With three edof on the element there are three conditions that each shape function must satisfy. Thus polynomial shape functions will be quadratic.

Quadratic polynomial shape functions require that the axial strain, or the first derivative of the quadratic axial displacement function, be linear, Thus a 3-edof axial element permits a linear variation of axial stress in it.

The form of the three quadratic shape functions is

$$N_1(x) = c_{11} + c_{21}x + c_{31}x^2, \qquad (5.31a)$$

$$N_2(x) = c_{12} + c_{22}x + x_{32}x^2, \qquad (5.31b)$$

$$N_3(x) = c_{13} + c_{23}x + c_{33}x^2. \qquad (5.31c)$$

Each of equations 5.31 must equal one at the node of its edof and zero at the nodes of the other two edof. Observing Figure 5.14, taking the coordinate of node 1 to be $x = 0$, node 2 to be $x = \frac{1}{2}L$, and node 3 to be $x = L$, and imposing the three conditions on equation 5.31a yields

$$1 = c_{11} + c_{21}(0) + c_{31}(0)^2, \tag{5.32a}$$

$$0 = c_{11} + c_{21}(\tfrac{1}{2}L) + c_{31}(\tfrac{1}{2}L)^2, \tag{5.32b}$$

$$0 = c_{11} + c_{21}(L) + c_{31}(L)^2. \tag{5.32c}$$

In matrix form,

$$\begin{bmatrix} 1 & (0) & (0)^2 \\ 1 & (\tfrac{1}{2}L) & (\tfrac{1}{2}L)^2 \\ 1 & (L) & (L)^2 \end{bmatrix} \begin{Bmatrix} c_{11} \\ c_{21} \\ c_{31} \end{Bmatrix} = \begin{Bmatrix} 1 \\ 0 \\ 0 \end{Bmatrix}. \tag{5.32d}$$

Imposing the three conditions on equation 5.31b yields

$$0 = c_{12} + c_{22}(0) + c_{32}(0)^2, \tag{5.33a}$$

$$1 = c_{12} + c_{22}(\tfrac{1}{2}L) + c_{32}(\tfrac{1}{2}L)^2, \tag{5.33b}$$

$$0 = c_{12} + c_{22}(L) + c_{32}(L)^2. \tag{5.33c}$$

In matrix form

$$\begin{bmatrix} 1 & (0) & (0)^2 \\ 1 & (\tfrac{1}{2}L) & (\tfrac{1}{2}L)^2 \\ 1 & (L) & (L)^2 \end{bmatrix} \begin{Bmatrix} c_{12} \\ c_{22} \\ c_{32} \end{Bmatrix} = \begin{Bmatrix} 0 \\ 1 \\ 0 \end{Bmatrix}. \tag{5.33d}$$

Imposing the three conditions on equation 5.31c yields

$$0 = c_{13} + c_{23}(0) + c_{33}(0)^2, \tag{5.34a}$$

$$0 = c_{13} + c_{23}(\tfrac{1}{2}L) + c_{33}(\tfrac{1}{2}L)^2, \tag{5.34b}$$

$$1 = c_{13} + c_{23}(L) + c_{33}(L)^2. \tag{5.34c}$$

In matrix form

$$\begin{bmatrix} 1 & (0) & (0)^2 \\ 1 & (\tfrac{1}{2}L) & (\tfrac{1}{2}L)^2 \\ 1 & (L) & (L)^2 \end{bmatrix} \begin{Bmatrix} c_{13} \\ c_{23} \\ c_{33} \end{Bmatrix} = \begin{Bmatrix} 0 \\ 0 \\ 1 \end{Bmatrix}. \tag{5.34d}$$

Combining equations 5.32d, 5.33d, and 5.34d as a single matrix equation results in

$$\begin{bmatrix} 1 & (0) & (0)^2 \\ 1 & (\tfrac{1}{2}L) & (\tfrac{1}{2}L)^2 \\ 1 & (L) & (L)^2 \end{bmatrix} \begin{bmatrix} c_{11} & c_{12} & c_{13} \\ c_{21} & c_{22} & c_{23} \\ c_{31} & c_{32} & c_{33} \end{bmatrix} = \begin{bmatrix} 1 & 0 & 0 \\ 0 & 1 & 0 \\ 0 & 0 & 1 \end{bmatrix}. \tag{5.35a}$$

In symbols,

$$[\mathbf{h}][\mathbf{C}] = [\mathbf{I}].\qquad(5.35b)$$

Matrix $\mathbf{C}$ is the array of unknown coefficients of the polynomial shape functions referred to in equations 5.31. By equation 5.35b, it must be equal to the inverse of matrix $\mathbf{h}$, in accordance with the definition of matrix inversion.

Matrix $\mathbf{h}$ is obtained by evaluating the generic functions of the polynomial shape functions, by row, at each corresponding edof. Generic functions of a polynomial are $1, x, x^2, x^3, x^4, \ldots$, which when multiplied by the coefficients c_{ij} and summed result in the polynomial. In the example above, at edof 1, $x = 0$, so that the first row of matrix $\mathbf{h}$ is 1, (0), $(0)^2$. At edof 2, $x = \frac{1}{2}L$, so that the second row of matrix $\mathbf{h}$ is 1, $(\frac{1}{2}L)$, $(\frac{1}{2}L)^2$. Row 3 is formed similarly for edof 3 with $x = L$.

When the edof are simply translations, as in the axial element, they are evaluated directly as indicated above. When some of the edof are rotations, as with a beam element, the first derivatives of the generic functions must be evaluated by row at appropriate positions to form the matrix $\mathbf{h}$.

Thus shape functions may be obtained by forming the matrix $\mathbf{h}$, inverting it, and then identifying the entries from the inverse that apply as coefficients for the shape functions. Ordering the edof numbers as above, column 1 of matrix $\mathbf{h}^{-1}$ gives the coefficients for $N_1(x)$, column 2 the coefficients for $N_2(x)$, and column 3 the coefficients for $N_3(x)$. An efficient way to obtain them is to write equations 5.31 in matrix form and substitute matrix $\mathbf{h}^{-1}$ for the matrix $\mathbf{C}$, as shown below in equations 5.36. Note that if $\mathbf{C} = \mathbf{h}^{-1}$, $\mathbf{C}^T = \mathbf{h}^{-T}$. The symbol $\mathbf{h}^{-T}$ indicates the transpose of matrix $\mathbf{h}^{-1}$:

$$\left\{ \begin{array}{c} N_1(x) \\ N_2(x) \\ N_3(x) \end{array} \right\} = \left[\begin{array}{ccc} C_{11} & C_{21} & C_{31} \\ C_{12} & C_{22} & C_{32} \\ C_{13} & C_{23} & C_{33} \end{array} \right] \left\{ \begin{array}{c} 1 \\ x \\ x^2 \end{array} \right\}, \qquad (5.36a)$$

$$\{\mathbf{N(x)}\} = [\mathbf{C}^T]\{\mathbf{g}\} = [\mathbf{h}^{-T}]\{\mathbf{g}\}. \qquad (5.36b)$$

The shape functions then may be written directly from equations 5.36. The matrix $\mathbf{g}$ is a vector of the generic polynomial functions described above.

If matrix $\mathbf{h}$ from equation 5.35a is inverted using the adjoint method, the result is

$$[\mathbf{h}]^{-1} = \left[\begin{array}{ccc} 1 & 0 & 0 \\ -\frac{3}{L} & \frac{4}{L} & -\frac{1}{L} \\ \frac{2}{L^2} & -\frac{4}{L^2} & \frac{2}{L^2} \end{array} \right].$$

Transposing the matrix $\mathbf{h}^{-1}$ and substituting it into equations 5.36,

$$\left\{ \begin{array}{c} N_1(x) \\ N_2(x) \\ N_3(x) \end{array} \right\} = \left[\begin{array}{ccc} 1 & -\frac{3}{L} & \frac{2}{L^2} \\ 0 & \frac{4}{L} & -\frac{4}{L^2} \\ 0 & -\frac{1}{L} & \frac{2}{L^2} \end{array} \right] \left\{ \begin{array}{c} 1 \\ x \\ x^2 \end{array} \right\}. \qquad (5.37a)$$

The shape functions are written directly from equation 5.37a as

$$N_1(x) = 1 - 3\left(\frac{x}{L}\right) + 2\left(\frac{x}{L}\right)^2, \tag{5.37b}$$

$$N_2(x) = 4\left(\frac{x}{L}\right) - 4\left(\frac{x}{L}\right)^2, \tag{5.37c}$$

$$N_3(x) = -\left(\frac{x}{L}\right) + 2\left(\frac{x}{L}\right)^2. \tag{5.37d}$$

Performing the integrations indicated in equations 5.30f for constant AE results in the stiffness matrix for the 3-node, 3-edof prismatic axial element,

$$[\mathbf{K}] = \left(\frac{AE}{3L}\right)\begin{bmatrix} 7 & -8 & 1 \\ -8 & 16 & -8 \\ 1 & -8 & 7 \end{bmatrix}. \tag{5.38}$$

It is left to the reader to develop the detailed steps which lead to equation 5.38.

5.3. CLOSING REMARKS

The displaced state virtual work procedure for the development of element stiffness equations for the 1D axial and beam finite elements was discussed in detail in an attempt to provide the reader with insights and understandings of the finite element method as it applies to solid structures. One-dimensional elements are easily visualized and illustrated.

Clearly the stiffness equations for the 1D axial and beam finite elements discussed are easily determined by direct procedures, as illustrated in Chapter 3. They are almost as easily determined by the displaced state virtual work procedure, as illustrated in this chapter, where the forms of the stiffness equations developed can be used in more general applications. The fact that each procedure produces the same result should assist the reader in validating each procedure and understanding both.

It remains to generalize the displaced state virtual work procedure for 2D and 3D structural finite elements. As the reader works through the following chapters, where illustrations and visualization of two and three dimensions are more difficult, the insights obtained in this chapter should be recalled.

BIBLIOGRAPHY

Barsoum, F. B., "A Finite Element Approach for a Beam on an Elastic Foundation and Its Application on the Crawlerway Tunnel at the Kennedy Space Center," M.S. Thesis, University of Central Florida, Orlando, 1993.

Barsoum, F. B., and Carroll, W. F., "Evaluating the Serviceability of the Cracked Reinforced Concrete Space Shuttle Crawlerway Tunnel," *Computers and Structures, an International Journal*, Vol. 60, No. 2, 1996, pp. 173–179.

Logan, D. L., *A First Course in the Finite Element Method* (Appendix E), PWS-Kent, Boston, 1992.

McGuire, W., and Gallagher, R. H., *Matrix Structural Analysis* (Chapters 8, 9), Wiley, New York, 1979.

Shames, I. H., and Dym, C. L., *Energy and Finite Elements in Structural Mechanics* (Chapter 3), Hemisphere, (McGraw-Hill), New York, 1985.

Weaver, W. Jr., and Gere, J. M., *Matrix Analysis of Framed Structures* (Chapters 1, 7), 3rd ed., Van Nostrand Reinhold, New York, 1990.

Weaver, W. Jr., and Johnston, P. R., *Finite Elements for Structural Analysis* (Chapter 1), Prentice-Hall, Englewood Cliffs, NJ, 1984.

Yang, T. Y., *Finite Element Structural Analysis* (Chapter 3), Prentice-Hall, Englewood Cliffs, NJ, 1986.

PROBLEMS

5.1. Following the procedure indicated in equations 5.35–5.37 and using software which manipulates individual matrices, develop polynomial shape functions for the 4-node, 4-dof prismatic axial element shown. Sketch each shape function and verify that each satisfies the basic shape function requirements.

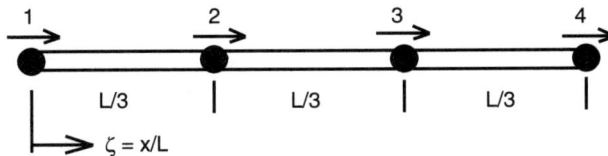

Problem 5.1

5.2. Repeat Problem 5.1 for the 2-node, 4-dof prismatic beam element shown in Figure 5.6. Sketch each shape function and verify that each satisfies the basic shape function requirements.

5.3. Repeat Problem 5.1 for the 3-node, 5-dof prismatic beam element shown. Sketch each shape function and verify that each satisfies the basic shape function requirements.

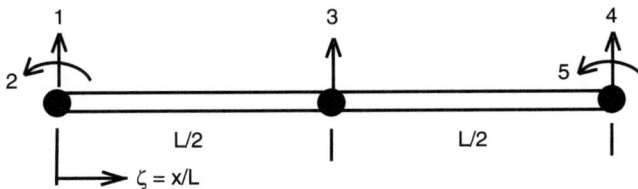

Problem 5.3

5.4. Following the procedure indicated in equations 5.30f, develop the element stiffness matrix displayed in equation 5.38 which is for the 3-node, 3-dof prismatic axial element shown in Figure 5.14.

5.5. Using the results from Problem 5.1 and following the procedure indicated in equations 5.30d, develop the element stiffness matrix for the 4-node, 4-dof prismatic axial element of Problem 5.1.

5.6. Following the procedure indicated in equation 5.17c, develop the element stiffness matrix displayed in equation 3.5 which is for the 2-node, 4-dof prismatic beam element of Problem 5.2.

5.7. Using the results from Problem 5.3 and following the procedure indicated in equation 5.17, develop the element stiffness matrix for the 3-node, 5-dof prismatic beam element of Problem 5.3.

5.8. Following the procedure indicated in equations 5.17c, 5.18f, and 5.22d, develop the element stiffness matrix for the 2-node, 4-dof prismatic beam element on an elastic foundation.

5.9. Using equations 5.17f and 5.17g, obtain the vectors of equivalent nodal forces and moments for the concentrated load, the uniformly distributed load, and the ramp load acting as loads between nodes and displayed in Figure 3.7.

5.10. For a uniformly distributed load w_o acting as a load between nodes and acting across the entire length of the element, determine the vector of equivalent nodal forces and moments for the 3-node, 5-dof prismatic beam element of Problem 5.3.

5.11. Determine the vector of equivalent nodal forces and moments for a 2-node, 4-dof prismatic beam element subjected to the distributed load shown. The exponent n in the function $w(\xi)$ is a nonnegative constant. Verify that when $n = 0$ the results give the correct values for a uniformly distributed load and when $n = 1$ they are correct for a ramp load, as shown in Figure 3.7.

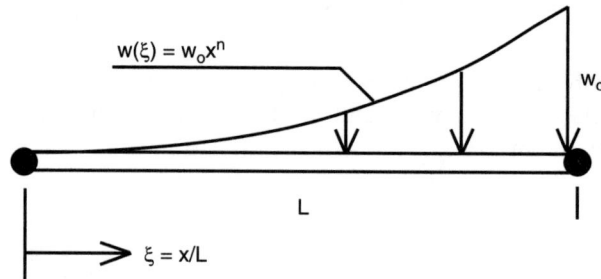

$$w(\xi) = w_o x^n$$

$$w_o$$

$$L$$

$$\xi = x/L$$

Problem 5.11

5.12. Determine the vector of equivalent nodal forces and moments for the 2-node, 6-dof prismatic plane frame element shown.

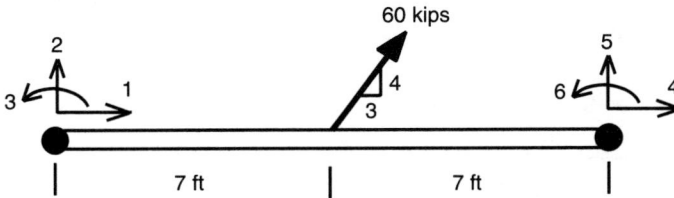

Problem 5.12

5.13. Determine the vector of equivalent nodal forces and moments for the 2-node, 6-dof prismatic plane frame element shown.

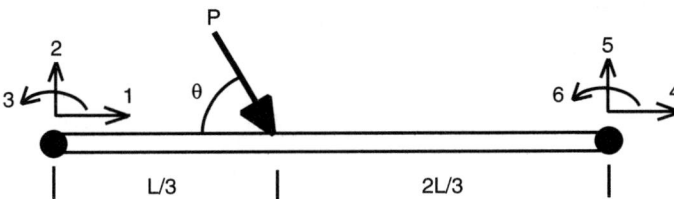

Problem 5.13

5.14. Write the global load vector for the prismatic beam loaded as shown. Determine its support reactions and, at points B and C, its deflections, shears, and moments, Verify that the reactions and applied loads satisfy the conditions for equilibrium of the beam.

Use element and global stiffness equations and software which manipulates individual matrices.

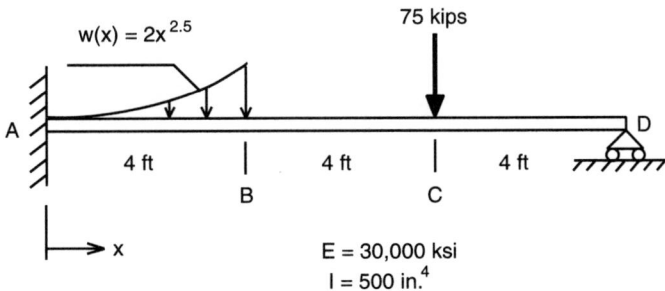

Problem 5.14

5.15. Write the global load vector for the prismatic beam on an elastic foundation loaded as shown and determine its deflections, shears, and moments at points A, B, C, and D.

Use element and global stiffness equations and software which manipulates individual matrices.

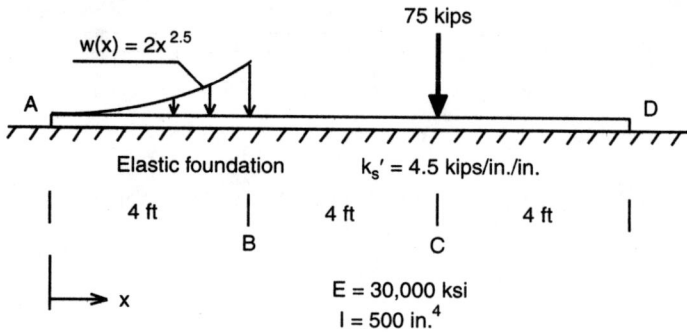

Problem 5.15

CHAPTER 6

GENERAL APPROACH TO ELEMENT STIFFNESS EQUATIONS

6.1. INTRODUCTION

The displaced state virtual work procedure discussed in Chapter 5 is a general approach to the application of the finite element method in structural analysis. The illustrations employed were 1D, 2- and 3-node axial and 2-node beam and plane frame finite elements. There were from one to three edof at each node. The procedure was similar for each of these elements, though the details of its application differed for each.

The procedure may be applied to 2D and 3D finite elements with three, four, or more nodes on each element and with from two to six edof at each node. Moreover, the similarities observed in the procedure for each of the elements may be exploited to develop a single general procedure involving a single set of matrices which is the same for any structural finite element. The differences in application for each element, as we shall see, will be the differences in the size and in the nature of the entries of these matrices.

The general approach using the principle of virtual work and the displaced state of the structure requires the virtual strain energy of the structure to be developed from stress, strain, and displacements in the structure. The relationships among these quantities will be matrix equations which employ nodal displacements, shape functions, and stress–strain relationships.

6.2. STRESS AND STRAIN

Strains in most linear elastic structural systems are regarded as very small quantities. These strains and the stresses corresponding to them are expressed in terms of right-hand orthogonal coordinate systems, such as x–y–z Cartesian coordinates.

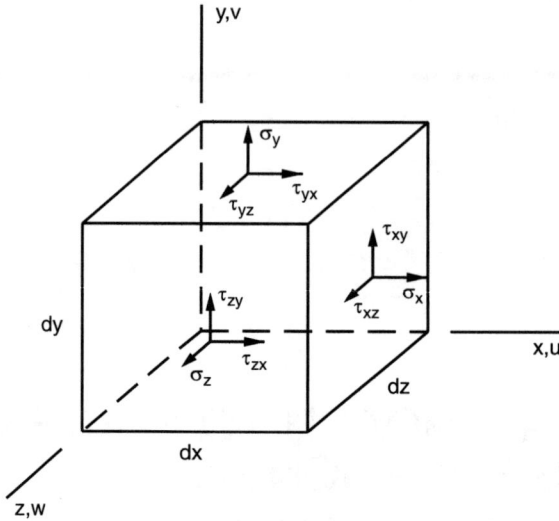

Figure 6.1 Stress displacement

Figure 6.1 shows a differential element from an elastic continuum referenced to Cartesian coordinate axes. Corresponding translation displacements u, v, and w and normal and shear stresses are shown. The symbol σ indicates normal stress. Its subscript indicates its coordinate direction. The symbol τ indicates shear stress. Its first subscript indicates the face of the element on which it acts, and its second subscript indicates its coordinate direction. The face of the differential element is identified by the coordinate axis to which it is perpendicular.

From the moment equilibrium of the differential element,

$$\tau_{xy} = \tau_{yx}, \qquad \tau_{yz} = \tau_{zy}, \qquad \tau_{zx} = \tau_{xz}.$$

Thus only three of the six shear stress components shown on the differential element are independent. There are six independent stress components then that need to be treated, three normal stresses and three shear stresses. These will be represented in vector form as

$$\{\sigma\} = \begin{Bmatrix} \sigma_x \\ \sigma_y \\ \sigma_z \\ \tau_{xy} \\ \tau_{yz} \\ \tau_{zx} \end{Bmatrix}. \tag{6.1}$$

For small strains, normal strains corresponding to the normal stresses in Figure 6.1 and in equation 6.1 are defined in terms of first derivatives of the translation displacements as

$$\epsilon_x = \frac{\partial u}{\partial x}, \qquad \epsilon_y = \frac{\partial v}{\partial y}, \qquad \epsilon_z = \frac{\partial w}{\partial z}. \tag{6.2a}$$

Small shear strains corresponding to the shear stresses in Figure 6.1 and in equation 6.1 are also defined in terms of first derivatives of the translation displacements as

$$\gamma_{xy} = \frac{\partial u}{\partial y} + \frac{\partial v}{\partial x}, \qquad \gamma_{yz} = \frac{\partial v}{\partial z} + \frac{\partial w}{\partial y}, \qquad \gamma_{zx} = \frac{\partial w}{\partial x} + \frac{\partial u}{\partial z}. \tag{6.2b}$$

From a geometric interpretation of shear strain as an angle change, it can be shown that

$$\gamma_{xy} = \gamma_{yx}, \qquad \gamma_{yz} = \gamma_{zy}, \qquad \gamma_{zx} = \gamma_{xz}. \tag{6.2c}$$

Thus only three of the six shear strain components are independent, leaving the three normal strains and three shear strains, which must be treated as independent strain components. These will be represented in vector form as

$$\{\epsilon\} = \begin{Bmatrix} \epsilon_x \\ \epsilon_y \\ \epsilon_z \\ \gamma_{xy} \\ \gamma_{yz} \\ \gamma_{zx} \end{Bmatrix}. \tag{6.2d}$$

In a linear elastic isotropic material, stress is related to strain by a system of linear algebraic equations. The normal strain in the x direction is the algebraic sum of the normal strains in the x direction caused by the three normal stresses σ_x, σ_y, and σ_z. Similarly, the normal strains in the y and z directions are algebraic sums of corresponding normal strains in those directions. The shear strains are independent of one another and of the normal stresses. These relationships may be written as

$$\epsilon_x = \frac{\sigma_x - \nu\sigma_y - \nu\sigma_z}{E}, \qquad \gamma_{xy} = \frac{\tau_{xy}}{G}, \tag{6.3a, b}$$

$$\epsilon_y = \frac{-\nu\sigma_x + \sigma_y - \nu\sigma_z}{E}, \qquad \gamma_{yz} = \frac{\tau_{yz}}{G} \tag{6.3c, d}$$

$$\epsilon_z = \frac{-\nu\sigma_x - \nu\sigma_y + \sigma_z}{E}, \qquad \gamma_{zx} = \frac{\tau_{zx}}{G}. \tag{6.3e, f}$$

The constant E is the modulus of elasticity, the constant G is the shear modulus, and the constant ν is Poisson's ratio. Only two of these constants are independent. It can be shown that

$$G = \frac{E}{2(1 + \nu)}. \tag{6.3g}$$

In matrix form, these equations are

$$
\begin{Bmatrix} \epsilon_x \\ \epsilon_y \\ \epsilon_z \\ \gamma_{xy} \\ \gamma_{yz} \\ \gamma_{zx} \end{Bmatrix} = \frac{1}{E} \begin{bmatrix} 1 & -\nu & -\nu & 0 & 0 & 0 \\ -\nu & 1 & -\nu & 0 & 0 & 0 \\ -\nu & -\nu & 1 & 0 & 0 & 0 \\ 0 & 0 & 0 & 2(1+\nu) & 0 & 0 \\ 0 & 0 & 0 & 0 & 2(1+\nu) & 0 \\ 0 & 0 & 0 & 0 & 0 & 2(1+\nu) \end{bmatrix} \begin{Bmatrix} \sigma_x \\ \sigma_y \\ \sigma_z \\ \tau_{xy} \\ \tau_{yz} \\ \tau_{zx} \end{Bmatrix}, \qquad (6.3h)
$$

or

$$
\{\epsilon\} = [\mathbf{A}]\{\sigma\}. \qquad (6.3i)
$$

The matrix **A** may be inverted so that equation 6.3i is rewritten as

$$
\{\sigma\} = [\mathbf{E}]\{\epsilon\}, \quad \text{where} \quad [\mathbf{E}] = [\mathbf{A}^{-1}]. \qquad (6.3j)
$$

Equation 6.3j is the general stress–strain law used in the finite element method for structures. The matrix **E** is a matrix of elastic constants relating the stress and strain vectors.

The inverse of matrix **A**, or matrix **E**, may be obtained by writing

$$
[\mathbf{A}][\mathbf{E}] = [\mathbf{I}] \qquad (6.4a)
$$

and partitioning each of the 6 × 6 matrices into 3 × 3 submatrices,

$$
\left[\begin{array}{c|c} \mathbf{A}_{11} & \mathbf{A}_{12} \\ \hline \mathbf{A}_{21} & \mathbf{A}_{22} \end{array} \right] \left[\begin{array}{c|c} \mathbf{E}_{11} & \mathbf{E}_{12} \\ \hline \mathbf{E}_{21} & \mathbf{E}_{22} \end{array} \right] = \left[\begin{array}{c|c} \mathbf{I} & \mathbf{0} \\ \hline \mathbf{0} & \mathbf{I} \end{array} \right]. \qquad (6.4b)
$$

Comparing matrix **A** of equations 6.3h and 6.4b,

$$
[\mathbf{A}_{11}] = \frac{1}{E} \begin{bmatrix} 1 & -\nu & -\nu \\ -\nu & 1 & -\nu \\ -\nu & -\nu & 1 \end{bmatrix}, \qquad (6.4c)
$$

$$
[\mathbf{A}_{12}] = [\mathbf{0}], \qquad (6.4d)
$$

$$
[\mathbf{A}_{21}] = [\mathbf{0}], \qquad (6.4e)
$$

$$
[\mathbf{A}_{22}] = \frac{2(1+\nu)}{E}[\mathbf{I}]. \qquad (6.4f)
$$

Four submatrix equations may be written from equation 6.4b:

$$
[\mathbf{A}_{11}][\mathbf{E}_{11}] + [\mathbf{A}_{12}][\mathbf{E}_{21}] = [\mathbf{I}], \qquad (6.4g)
$$

$$
[\mathbf{A}_{11}][\mathbf{E}_{12}] + [\mathbf{A}_{12}][\mathbf{E}_{22}] = [\mathbf{0}]. \qquad (6.4h)
$$

$$[A_{21}][E_{11}] + [A_{22}][E_{21}] = [0]. \tag{6.4i}$$

$$[A_{21}][E_{12}] + [A_{22}][E_{22}] = [I]. \tag{6.4j}$$

Applying equation 6.4d to equation 6.4g,

$$[E_{11}] = [A_{11}^{-1}].$$

Submatrix A_{11} may be inverted using the adjoint method,

$$[E_{11}] = \frac{E}{(1+v)(1-2v)} \begin{bmatrix} 1-v & v & v \\ v & 1-v & v \\ v & v & 1-v \end{bmatrix}. \tag{6.4k}$$

Applying equation 6.4d to equation 6.4h and equation 6.4e to equation 6.4i,

$$[E_{12}] = [E_{21}] = [0]. \tag{6.4l}$$

Applying equation 6.4e to equation 6.4j;

$$[E_{22}] = [A_{22}^{-1}].$$

The submatrix A_{22} is the unit matrix with a scalar multiplier. Its inverse is the unit matrix with the reciprocal of the scalar multiplier,

$$[E_{22}] = \frac{E}{2(1+v)}[I]. \tag{6.4m}$$

Assembling the submatrices E_{11}, E_{12}. E_{21}, and E_{22}, the matrix E is obtained,

$$[E] = \frac{E}{(1+v)(1-2v)} \begin{bmatrix} 1-v & v & v & 0 & 0 & 0 \\ v & 1-v & v & 0 & 0 & 0 \\ v & v & 1-v & 0 & 0 & 0 \\ 0 & 0 & 0 & \frac{1-2v}{2} & 0 & 0 \\ 0 & 0 & 0 & 0 & \frac{1-2v}{2} & 0 \\ 0 & 0 & 0 & 0 & 0 & \frac{1-2v}{2} \end{bmatrix}. \tag{6.4n}$$

The size and details of the matrix E for equation 6.3j will differ from the 3D condition just developed when the conditions in the structure are different. For example, the 2D condition of plane stress illustrated in Figure 6.2 results from a flat plate which is stressed in the plane of the plate so that

$$\sigma_z = \tau_{yz} = \tau_{zx} = 0,$$

$$\gamma_{yz} = \gamma_{zy} = \gamma_{zx} = \gamma_{xz} = 0,$$

$$\tau_{xy} = \tau_{yx},$$

$$\epsilon_z \neq 0.$$

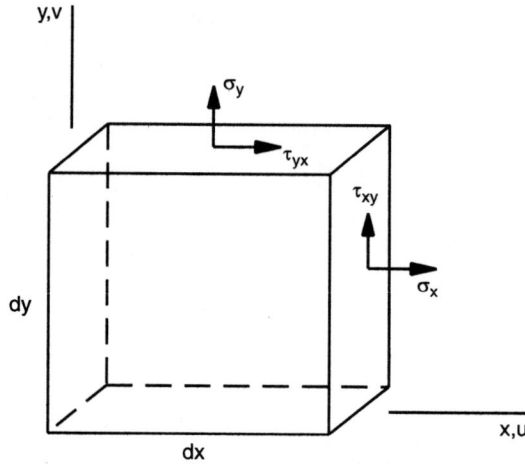

Figure 6.2 Plane stress

For the plane stress condition, then, the stress and strain vectors have three entries each and the matrix **E** is 3×3 in size with different entries and multiplier than for the 3D condition. Modifying equation 6.4n to achieve this condition, equation 6.3j becomes

$$\begin{Bmatrix} \sigma_x \\ \sigma_y \\ \tau_{xy} \end{Bmatrix} = \frac{E}{1 - \nu^2} \begin{bmatrix} 1 & \nu & 0 \\ \nu & 1 & 0 \\ 0 & 0 & \frac{1-\nu}{2} \end{bmatrix} \begin{Bmatrix} \epsilon_x \\ \epsilon_y \\ \gamma_{xy} \end{Bmatrix}. \tag{6.5a}$$

Other stress conditions, such as plane strain, will be different in detail also. The 2D condition of plane strain is similar to but different from plane stress. It results from a long prismatic member with constant stress along its long axis, so that

$$\epsilon_z = \tau_{yz} = \tau_{zx} = 0$$

$$\gamma_{yz} = \gamma_{zy} = \gamma_{zx} = \gamma_{xz} = 0,$$

$$\tau_{xy} = \tau_{yx},$$

$$\sigma_z \neq 0.$$

The stress and strain vectors are the same as for plane stress, but the matrix **E** differs. For plane strain, equation 6.3j becomes

$$\begin{Bmatrix} \sigma_x \\ \sigma_y \\ \tau_{xy} \end{Bmatrix} = \frac{E}{(1 + \nu)(1 - 2\nu)} \begin{bmatrix} 1 - \nu & \nu & 0 \\ \nu & 1 - \nu & 0 \\ 0 & 0 & \frac{1-2\nu}{2} \end{bmatrix} \begin{Bmatrix} \epsilon_x \\ \epsilon_y \\ \gamma_{xy} \end{Bmatrix}. \tag{6.5b}$$

The stress and strain vectors and the elastic matrix **E** for other conditions may be developed in a similar manner as needed.

6.3. STRAIN AND STRESS DISPLACEMENT RELATIONSHIPS

The displacement function for a finite element was shown to be the sum of the products of its edof values, or nodal displacements, and their corresponding shape functions. The translation displacements $u(x, y, z)$, $v(x, y, z)$, and $w(x, y, z)$ in the structural element are functions of the position coordinates and are independent of one another; they will be represented separately in vector form:

$$\{\mathbf{u}\} = \{\mathbf{u}(\mathbf{x}, \ \mathbf{y}, \ \mathbf{z})\} = \left\{ \begin{array}{l} u(x, \ y, \ z) \\ v(x, \ y, \ z) \\ w(x, \ y, \ z) \end{array} \right\}. \tag{6.6}$$

Because of the independence of $u(x, y, z)$, $v(x, y, z)$, and $w(x, y, z)$, a single set of shape functions may be used to represent each. There must be as many shape functions in this set as there are nodes on the element. Only nodal translations in the direction of the x axis and their corresponding shape functions will contribute to $u(x, y, z)$. Only those in the direction of the y axis will contribute to $v(x, y, z)$ and only those in the direction of the z axis to $w(x, y, z)$. The sum of the products of these nodal edof values and their corresponding shape functions form each of the three element displacement functions. For the 8-node, 3D hexahedral element illustrated in Figure 6.3, there are 3 translation edof at each node for a total of 24 edof. Eight edof are in the x direction, 8 are in the y direction, and 8 are in the z direction. One can write a separate equation of the form of equation 5.3c for each displacement function as

$$u(x, \ y, \ z) = \lfloor \mathbf{N}_i(\mathbf{x}, \ \mathbf{y}, \ \mathbf{z}) \rfloor \{\mathbf{q}_u\}, \tag{6.7a}$$

$$v(x, \ y, \ z) = \lfloor \mathbf{N}_i(\mathbf{x}, \ \mathbf{y}, \ \mathbf{z}) \rfloor \{\mathbf{q}_v\}, \tag{6.7b}$$

$$w(x, \ y, \ z) = \lfloor \mathbf{N}_i(\mathbf{x}, \ \mathbf{y}, \ \mathbf{z}) \rfloor \{\mathbf{q}_w\}. \tag{6.7c}$$

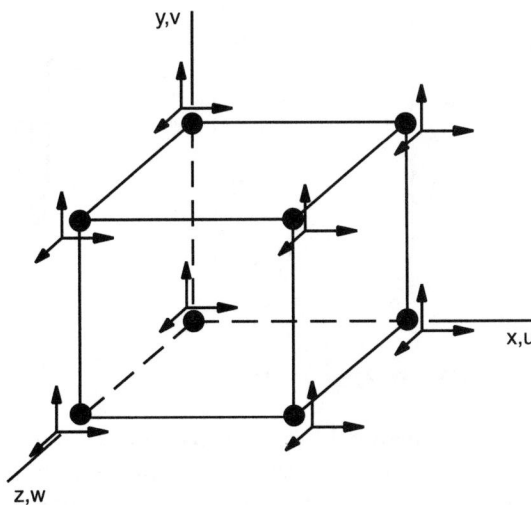

Figure 6.3 Hexahedral element

The row vector $N_i(x, y, z)$ contains eight shape functions, one for each node; it is the same in each of equations 6.7. The column vectors q each contain the eight edof values in their corresponding coordinate directions u, v, or w, which differ in each equation. Equations 6.7 may be combined as a single matrix equation:

$$
\begin{Bmatrix} u(x, y, z) \\ v(x, y, z) \\ w(x, y, z) \end{Bmatrix} = \begin{bmatrix} N_1 & 0 & 0 & | & N_2 & 0 & 0 & | & --- & | & N_8 & 0 & 0 \\ 0 & N_1 & 0 & | & 0 & N_2 & 0 & | & --- & | & 0 & N_8 & 0 \\ 0 & 0 & N_1 & | & 0 & 0 & N_2 & | & --- & | & 0 & 0 & N_8 \end{bmatrix} \begin{Bmatrix} u_1 \\ v_1 \\ w_1 \\ u_2 \\ v_2 \\ w_2 \\ \vdots \\ u_8 \\ v_8 \\ w_8 \end{Bmatrix}, \quad (6.7d)
$$

or

$$\{u(x, \ y, \ z)\} = [N(x, \ y, \ z)]\{q\}. \tag{6.7e}$$

The form of equation 6.7e is general. The matrices u, N, and q are determined by the nature of the finite element being treated.

Strain displacement relationships are derived from the definitions of the strains as derivatives of translation displacements. For the 3D finite element, the strain definition equations 6.2 are applied to the displacement function equation 6.6 to obtain, in matrix form,

$$
\begin{Bmatrix} \epsilon_x(x, y, z) \\ \epsilon_y(x, y, z) \\ \epsilon_z(x, y, z) \\ \gamma_{xy}(x, y, z) \\ \gamma_{yz}(x, y, z) \\ \gamma_{zx}(x, y, z) \end{Bmatrix} = \begin{bmatrix} \frac{\partial}{\partial x} & 0 & 0 \\ 0 & \frac{\partial}{\partial y} & 0 \\ 0 & 0 & \frac{\partial}{\partial z} \\ \frac{\partial}{\partial y} & \frac{\partial}{\partial x} & 0 \\ 0 & \frac{\partial}{\partial z} & \frac{\partial}{\partial y} \\ \frac{\partial}{\partial z} & 0 & \frac{\partial}{\partial x} \end{bmatrix} \begin{Bmatrix} u(x, y, z) \\ v(x, y, z) \\ w(x, y, z) \end{Bmatrix}, \quad (6.8a)
$$

or

$$\{\epsilon(x, \ y, \ z)\} = [d]\{u(x, \ y, \ z)\}. \tag{6.8b}$$

The matrix d is an array of derivative operators. Its entries are determined by the definitions of strain as derivatives of displacements. Its make-up will be different for the 2D plane stress–plane strain condition and different still for other conditions.

Substituting equation 6.7e into equation 6.8b results in

$$\{\epsilon(\mathbf{x}, \ \mathbf{y}, \ \mathbf{z})\} = [\mathbf{d}][\mathbf{N}(\mathbf{x}, \ \mathbf{y}, \ \mathbf{z})]\{\mathbf{q}\}, \tag{6.9a}$$

or

$$\{\epsilon(\mathbf{x}, \ \mathbf{y}, \ \mathbf{z})\} = [\mathbf{B}(\mathbf{x}, \ \mathbf{y}, \ \mathbf{z})]\{\mathbf{q}\}. \tag{6.9b}$$

The matrix **B**, which equals the product of matrices **d** and **N**, is called the strain displacement matrix. It is an array of derivatives of the shape functions. When matrix **B** premultiplies the vector of edof values **q**, the result is the vector of strain functions in the structural element. The form of equations 6.9 is general. The make-up of matrices **d** and **N**, and thus **B**, will depend on the nature of the finite element being treated.

Substituting equation 6.9b into equation 6.3j results in

$$\{\sigma\}(\mathbf{x}, \ \mathbf{y}, \ \mathbf{z})\} = [\mathbf{E}][\mathbf{B}(\mathbf{x}, \ \mathbf{y}, \ \mathbf{z})]\{\mathbf{q}\}. \tag{6.10}$$

Equation 6.10 gives the vector of stress functions in the structural element. They are functions of the position coordinates

6.4. ELEMENT STIFFNESS EQUATIONS

In Section 5.2.1, we saw that the principle of virtual work requires that $\delta U = \delta W$. The term δU is the virtual strain energy in the finite element as the element undergoes an admissable virtual displacement. The term δW is the virtual work of the external actions on the element resulting during the virtual displacement.

6.4.1. Virtual Strain Displacement Relationships

The virtual displacement function of the element is expressed as a vector of functions in the manner that the real displacement functions are expressed in equation 6.6:

$$\{\delta\mathbf{u}\} = \{\delta\mathbf{u}(\mathbf{x}, \ \mathbf{y}, \ \mathbf{z})\} = \left\{ \begin{array}{l} \delta u(x, \ y, \ z) \\ \delta v(x, \ y, \ z) \\ \delta w(x, \ y, \ z) \end{array} \right\}. \tag{6.11}$$

The virtual displacement functions $\delta u(x, y, \ z)$, $\delta v(x, y, \ z)$, and $\delta w(x, \ y, \ z)$ may be expressed in terms of the element shape functions [**N**] and a vector $\delta \mathbf{q}$ of virtual edof values. The matrix representation will be of the form of equations 6.7e, expressed as

$$\{\delta\mathbf{u}(\mathbf{x}, \ \mathbf{y}, \ \mathbf{z})\} = [\mathbf{N}(\mathbf{x}, \ \mathbf{y}, \ \mathbf{z})]\{\delta\mathbf{q}\}. \tag{6.12}$$

Matrix **N** is the array of shape functions defined in equation 6.7d. Vector $\delta \mathbf{q}$ is the same size and form as vector **q**. The former are virtual edof values; the latter are the real edof values.

The virtual strain displacement relationship will be the same form as the real strain displacement relationship (equations 6.9):

$$\{\delta\epsilon(\mathbf{x}, \ \mathbf{y}, \ \mathbf{z})\} = [\mathbf{B}(\mathbf{x}, \ \mathbf{y}, \ \mathbf{z})]\{\delta\mathbf{q}\}. \tag{6.13}$$

Matrix **B** is the array of derivatives of shape functions defined in equation 6.9. Vector $\delta\epsilon$ of virtual strain functions is the same size and form as the vector of strain functions ϵ (equation 6.8).

6.4.2. Virtual Strain Energy

Strain energy for a single stress component is the product of the internal force associated with the stress component and its corresponding displacement. The specific strain energy U'_x (strain energy per unit volume) for the normal stress along the x axis as the stress builds from zero to its current value while passing through a real displacement du is given as

$$U'_x = \frac{1}{2}\frac{F_x\,du}{dx\,dy\,dz}. \tag{6.14a}$$

The prime identifies specific strain energy; the subscript x indicates that the specific strain energy is associated with the stress σ_x. The internal force associated with σ_x is F_x. Since $\sigma_x = F_x/(dy\,dz)$ and $\epsilon_x = du/dx$,

$$u'_x = \tfrac{1}{2}\epsilon_x\sigma_x. \tag{6.14b}$$

Equation 6.14b gives the specific strain energy for σ_x. Similar expressions apply for the other stress components. Since energy is a scalar quantity, the total specific strain energy at a point for all stress components is the algebraic sum of these expressions, or

$$U' = \tfrac{1}{2}(\epsilon_x\sigma_x + \epsilon_y\sigma_y + \epsilon_z\sigma_z + \gamma_{xy}\tau_{xy} + \gamma_{yz}\tau_{yz} + \gamma_{zx}\tau_{zx})$$
$$= \tfrac{1}{2}\lfloor\epsilon\rfloor\{\sigma\}. \tag{6.14c}$$

The virtual strain energy as the differential element undergoes an admissible virtual displacement is expressed similarly. However, since the stress components are at their full values during the virtual displacement, the factor $\tfrac{1}{2}$ does not apply. The virtual strain energy of the entire differential element is its specific virtual strain energy times the volume of the differential element,

$$\delta U_{\mathrm{DE}} = \lfloor\delta\epsilon\rfloor\{\sigma\}\,dV. \tag{6.15a}$$

The subscript DE indicates virtual strain energy of a differential element. The virtual strain energy for the entire finite element is obtained by integrating equation 6.15a over the volume of the finite element:

$$\delta U = \int_v \lfloor\delta\epsilon\rfloor\{\sigma\}\,dV. \tag{6.15b}$$

Equation 6.13 may be rewritten transposed as

$$\lfloor\delta\epsilon\rfloor = \lfloor\delta\mathbf{q}\rfloor[\mathbf{B}^{\mathrm{T}}]. \tag{6.16}$$

Substituting equations 6.16 and 6.10 into equation 6.15b,

$$\delta U = \int_v \lfloor\delta\mathbf{q}\rfloor[\mathbf{B}^{\mathrm{T}}][\mathbf{E}][\mathbf{B}]\{\mathbf{q}\}\,dV. \tag{6.17a}$$

Moving the arrays of constants, $\lfloor \delta \mathbf{q} \rfloor$ and $\{\mathbf{q}\}$, from under the integral without altering the order of matrix multiplication,

$$\delta U = \lfloor \delta \mathbf{q} \rfloor \left[\int_v [\mathbf{B}^T][\mathbf{E}][\mathbf{B}] \, dV \right] \{\mathbf{q}\}. \tag{6.17b}$$

Equation 6.17b is the general expression for the virtual strain energy of a structural finite element as it undergoes an admissible virtual displacement.

6.4.3. Virtual Work of External Actions

The applied actions on a finite element are the unknown nodal internal forces and any externally applied distributed body forces and concentrated loads between nodes.

The unknown nodal internal forces correspond to the edof shown in equations 6.7d. For the 3D condition they are

$$\{\mathbf{F}\} = \begin{Bmatrix} F_{1u} \\ F_{1v} \\ F_{1w} \\ F_{2u} \\ F_{2v} \\ F_{2w} \\ \vdots \\ F_{8u} \\ F_{8v} \\ F_{8w} \end{Bmatrix}. \tag{6.18a}$$

The distributed body forces (force per unit of length, area, or volume) are a vector of body force x, y, and z component functions. For the 3D condition they are

$$\{\mathbf{b}\} = \{\mathbf{b}(\mathbf{x}, \ \mathbf{y}, \ \mathbf{z})\} = \begin{Bmatrix} b_x(x, \ y, \ z) \\ b_y(x, \ y, \ z) \\ b_z(x, \ y, \ z) \end{Bmatrix}. \tag{6.18b}$$

A concentrated load between nodes is a vector of force x, y, and z components located at coordinates $(x_k, \ y_k, \ z_k)$. The vector is

$$\{\mathbf{P}_k\} = \begin{Bmatrix} P_{kx} \\ P_{ky} \\ P_{kz} \end{Bmatrix}. \tag{6.18c}$$

The virtual work of the nodal internal forces, distributed body forces, and concentrated loads between nodes is similar to what was discussed earlier and shown in equation 5.7:

$$\delta W = \lfloor \delta \mathbf{q} \rfloor \{\mathbf{F}\} + \int_{v} \lfloor \delta \mathbf{u}(\mathbf{x}, \ \mathbf{y}, \ \mathbf{z}) \rfloor \{\mathbf{b}(\mathbf{x}, \ \mathbf{y}, \ \mathbf{z})\} \ dV$$
$$+ \lfloor \delta \mathbf{u}(\mathbf{x}_k, \ \mathbf{y}_k, \ \mathbf{z}_k) \rfloor \{\mathbf{P}_k\}. \tag{6.19}$$

Transposing the virtual displacement expression (equation 6.12),

$$\lfloor \delta \mathbf{u}(\mathbf{x}, \ \mathbf{y}, \ \mathbf{z}) \rfloor = \lfloor \delta \mathbf{q} \rfloor [\mathbf{N}^{T}(\mathbf{x}, \ \mathbf{y}, \ \mathbf{z})]. \tag{6.20}$$

Substituting equation 6.20 into equation 6.19,

$$\delta W = \lfloor \delta \mathbf{q} \rfloor \{\mathbf{F}\} + \int_{v} \lfloor \delta \mathbf{q} \rfloor [\mathbf{N}^{T}]\{\mathbf{b}\} \ dV + \lfloor \delta \mathbf{q} \rfloor [\mathbf{N}_{k}^{T}]\{\mathbf{P}_k\}. \tag{6.21a}$$

The symbol N_k indicates the array of shape functions evaluated at the point $(x_k, \ y_k, \ z_k)$ of application of the load P_k. Moving the array of constant edof virtual values $[\delta \mathbf{q}]$ from under the integral without altering the order of matrix multiplication and applying the distributive property of matrix multiplication,

$$\delta W = \lfloor \delta \mathbf{q} \rfloor \left\{ \{\mathbf{F}\} + \int_{v} [\mathbf{N}^{T}]\{\mathbf{b}\} \ dV + [\mathbf{N}_{k}^{T}]\{\mathbf{P}_k\} \right\}. \tag{6.21b}$$

Equation 6.21b is the general expression for the virtual work of the applied actions on a structural finite element as it undergoes an admissible virtual displacement.

6.4.4. General Expressions for Element Stiffness Equations

Equations 6.17b and 6.21b may be equated in accordance with the principle of virtual work:

$$\lfloor \delta \mathbf{q} \rfloor \left[\int_{v} [\mathbf{B}^{T}][\mathbf{E}][\mathbf{B}] \ dV \right] \{\mathbf{q}\}$$
$$= \lfloor \delta \mathbf{q} \rfloor \left\{ \{\mathbf{F}\} + \int_{v} [\mathbf{N}^{T}]\{\mathbf{b}\} \ dV + [\mathbf{N}_{k}^{T}]\{\mathbf{P}_k\} \right\}. \tag{6.22a}$$

Since equation 6.22a is a valid matrix equation, so is

$$\left[\int_{v} [\mathbf{B}^{T}][\mathbf{E}][\mathbf{B}] \ dV \right] \{\mathbf{q}\}$$
$$= \{\mathbf{F}\} + \int_{v} [\mathbf{N}^{T}]\{\mathbf{b}\} \ dV + [\mathbf{N}_{k}^{T}]\{\mathbf{P}_k\}. \tag{6.22b}$$

In symbols,

$$[\mathbf{K}]\{\mathbf{q}\} = \{\mathbf{F}\} + \{\mathbf{F}_b\} + \{\mathbf{F}_k\}. \tag{6.22c}$$

Equation 6.22c is the general form for the element stiffness equations. The matrices $\mathbf{K}$, $\mathbf{F}_b$ and $\mathbf{F}_k$ may be computed from known quantities as

$$[\mathbf{K}] = \left[\int_v [\mathbf{B}^{\mathrm{T}}][\mathbf{E}][\mathbf{B}] \, dV \right], \tag{6.22d}$$

$$\{\mathbf{F}_b\} = \left\{ \int_v [\mathbf{N}^{\mathrm{T}}]\{\mathbf{b}\} \, dV \right\}, \tag{6.22e}$$

$$\{\mathbf{F}_k\} = [\mathbf{N}_k^{\mathrm{T}}]\{\mathbf{P}_k\}. \tag{6.22f}$$

Equations 6.22 provide the means to obtain element stiffness equations for a variety of structural finite elements. A number of the more common elements are discussed in later chapters. The global stiffness equations can be developed from the known element stiffness matrices $[\mathbf{K}]$, the known element load vectors $\{\mathbf{F}_b\}$ and $\{\mathbf{F}_k\}$, and any applied nodal concentrated loads. This assembly of the global stiffness equations from element stiffness equations was discussed in Chapter 4. Solution of the global stiffness equations provides the gdof from which the vectors $\{\mathbf{q}\}$ of edof may be constructed. Calculation of the vector $\{\mathbf{F}\}$ of member internal forces at nodes can then proceed directly from the element stiffness equations.

6.5. GENERALIZED STRESS AND STRAIN

The general form of a structural element's stiffness equations 6.22c–f was developed from conventional expressions for stress and strain (equations 6.3j and 6.9). In some applications, stress may be defined advantageously in a more general sense than force per unit of area on a 3D differential element. In so doing, one must also identify corresponding generalized strains and the relationships among these stresses and strains.

To be useful in the finite element method, generalized stress and strain should reduce the dimensional variations of stress and strain in the structural element, preferably to one dimension. When this is possible, the integrations required in equations 6.22 to develop element stiffness equations are simplified, and the process is seen more clearly.

The generalized stress and strain approach is especially useful in dealing with thin plates and will be so employed in Chapter 10. To illustrate the process here, the familiar stiffness matrix for the linear elastic, straight, thin, prismatic beam element will be developed again.

Figure 6.4a shows the normal stress σ_x acting on the transverse cross section of a beam. As is usually done, the effects of other stresses present are ignored so that a uniaxial state of stress is assumed. The stress σ_x varies with both the x and y position coordinates.

In the conventional approach, the stress–strain relationship is

$$\sigma_x = E\epsilon_x, \tag{6.23a}$$

where $\epsilon_x = \partial u/\partial x$ and u is the translation displacement along the axis of the beam.

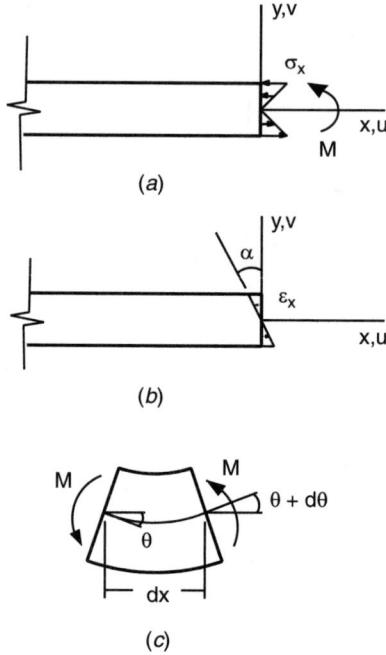

Figure 6.4 Beam element

Figure 6.4*b* shows the normal strain ϵ_x occurring in the beam on the transverse cross section. For a straight beam element and one for which plane transverse sections before bending remain plane and transverse during bending, the distribution of normal strain is linear and in the plane of bending. The rotation θ of the transverse cross section shown in Figure 6.4*c* is the slope $\partial v/\partial x$ of the displaced beam element. The change in slope from one side of the differential element to the other is $d\theta$, which describes a linear variation in the y direction of the change in displacement du on the cross section, or

$$du = \left(\frac{\partial u}{\partial x}\right) dx = (d\theta)y. \tag{6.23b}$$

Thus the variation of the strain ϵ_x in the y direction on the cross section is also linear, or

$$\epsilon_x = \frac{\partial u}{\partial x} = \left(\frac{d\theta}{dx}\right)y. \tag{6.23c}$$

Consequently, the slope of the strain distribution on the transverse cross section is

$$\alpha = \frac{d\theta}{dx} = \frac{\partial}{\partial x}\left(\frac{\partial v}{\partial x}\right) = \frac{\partial^2 v}{\partial x^2}. \tag{6.23d}$$

Clearly this strain distribution slope is the curvature of the displaced beam element. From Figure 6.4*b* it is evident that

$$\epsilon_x = -y\alpha = -y\frac{\partial^2 v}{\partial x^2}. \tag{6.23e}$$

The negative sign is present because ϵ_x is a compressive when y is positive.

Defining moment in the beam as generalized stress and curvature as its corresponding generalized strain provides a simple development and clear view of the process. The generalized stress–strain relationship is

$$M = EI\frac{\partial^2 v}{\partial x^2}, \tag{6.24a}$$

or

$$\bar{\sigma} = \bar{E}\bar{\epsilon}. \tag{6.24b}$$

The overbars signify generalized quantities, where

$$\bar{\sigma} = M, \tag{6.24c}$$

$$\bar{E} = EI, \tag{6.24d}$$

$$\bar{\epsilon} = \frac{\partial^2 v}{\partial x^2}. \tag{6.24e}$$

In the conventional approach,

$$[\mathbf{d}] = -y\left[\frac{\partial^2}{\partial x^2}\right]. \tag{6.25a}$$

Let

$$[\bar{\mathbf{d}}] = \left[\frac{\partial^2}{\partial x^2}\right] \tag{6.25b}$$

be the generalized vector of derivative operators, so that

$$[\mathbf{d}] = -y[\bar{\mathbf{d}}]. \tag{6.25c}$$

The matrix **B** and its generalized counterpart $\bar{\mathbf{B}}$ are related as

$$[\mathbf{B}] = [\mathbf{d}][\mathbf{N}] = -y[\bar{\mathbf{d}}][\mathbf{N}] = -y[\bar{\mathbf{B}}], \tag{6.26a}$$

where

$$[\bar{\mathbf{B}}] = [\bar{\mathbf{d}}][\mathbf{N}]. \tag{6.26b}$$

For the generalized stress–strain approach,

$$[\bar{\mathbf{B}}] = \left[\frac{\partial^2}{\partial x^2}\right] \lfloor N_1 N_2 N_3 N_4 \rfloor = \left\lfloor \frac{\partial^2 N_1}{\partial x^2} \quad \frac{\partial^2 N_2}{\partial x^2} \quad \frac{\partial^2 N_3}{\partial x^2} \quad \frac{\partial^2 N_4}{\partial x^2} \right\rfloor. \tag{6.26c}$$

The element stiffness matrix $[\mathbf{K}]$ is given as

$$
\begin{aligned}
[\mathbf{K}] &= \int_0^L \int_A [\mathbf{B}^{\mathrm{T}}][\mathbf{E}][\mathbf{B}] \ dA \ dx, \\
&= \int_0^L \int_A (-y)[\bar{\mathbf{B}}^{\mathrm{T}}][\mathbf{E}](-y)[\bar{\mathbf{B}}] \ dA \ dx.
\end{aligned}
\tag{6.27a}
$$

Since only y varies over the cross-sectional area A, equation 6.27a may be rewritten as

$$[\mathbf{K}] = \int_0^L [\bar{\mathbf{B}}^{\mathrm{T}}][\mathbf{E}][\bar{\mathbf{B}}] \ dx \left(\int_A y^2 \ dA \right). \tag{6.27b}$$

For the beam element, $[\mathbf{E}]$ is a scalar, the Young's modulus E. The quantity in parentheses in equation 6.27b is a scalar I, the centroidal second moment of the cross-sectional area, which may be combined with E to produce the generalized $[\bar{\mathbf{E}}] = EI$. The element stiffness matrix is thus

$$[\mathbf{K}] = \int_0^L [\bar{\mathbf{B}}^{\mathrm{T}}][\bar{\mathbf{E}}][\bar{\mathbf{B}}] \ dx. \tag{6.27c}$$

Equation 6.27c is evaluated employing the beam element shape functions,

$$[\mathbf{K}] = \int_0^L (EI) \left\{ \begin{array}{c} \frac{\partial^2 N_1}{\partial x^2} \\ \frac{\partial^2 N_2}{\partial x^2} \\ \frac{\partial^2 N_3}{\partial x^2} \\ \frac{\partial^2 N_4}{\partial x^2} \end{array} \right\} \left\lfloor \frac{\partial^2 N_1}{\partial x^2} \quad \frac{\partial^2 N_2}{\partial x^2} \quad \frac{\partial^2 N_3}{\partial x^2} \quad \frac{\partial^2 N_4}{\partial x^2} \right\rfloor dx, \tag{6.27d}$$

which leads directly to equation 5.17c.

6.6. GENERALIZED DOF

When the finite elements of a structure are the same shape, their shape functions are the same. Then the integrals of equations 5.17, 6.22, or 6.27 need to be evaluated only once and the result multiplied by scalars and reused for each element. However, for many finite element analyses, the elements differ in shape so that these integrals must be evaluated individually for each element. This is usually true for multidimensional structures. In these instances, the integrations to develop each element's stiffness equations can be a significant computational effort. The use of generalized degrees of freedom, or generalized displacements, reduce greatly the amount of computation for the integrations.

Earlier the displacement variation for a 1D finite element was represented as

$$u(x) = \lfloor \mathbf{N(x)} \rfloor \{\mathbf{q}\}. \tag{5.3}$$

The edof for the element were the nodal displacements, or the n entries of the vector $\mathbf{q}$. There were as many shape functions in the vector $\mathbf{N(x)}$ as there were edof, and each shape function had as many generic polynomial terms as there were edof.

The determination of each polynomial shape function was illustrated in Chapter 5 (equations 5.31–5.36). The result of this process was the matrix $\mathbf{C}$ of coefficients for all the generic terms of all the shape functions for the element. Matrix $\mathbf{C}$ was arranged so that the entries in a column were the coefficients for the shape function whose number corresponded to that column. Thus,

$$N_1 = c_{11} + c_{21}x + c_{31}x^2 + \cdots,$$
$$N_2 = c_{12} + c_{22}x + c_{32}x^2 + \cdots,$$
$$N_3 = c_{13} + c_{23}x + \cdots,$$

or

$$\{\mathbf{N}\} = [\mathbf{C}^T]\{\mathbf{g}\} = \begin{bmatrix} c_{11} & c_{21} & c_{31} & \cdots & c_{n1} \\ c_{12} & c_{22} & \cdots & \cdots & c_{n2} \\ c_{13} & \ddots & \cdots & \cdots & c_{n3} \\ \vdots & \cdots & \ddots & \cdots & \vdots \\ c_{1n} & \cdots & \cdots & \cdots & c_{nn} \end{bmatrix} \begin{Bmatrix} 1 \\ x \\ x^2 \\ \vdots \\ x^n \end{Bmatrix}. \tag{6.28}$$

The vector $\mathbf{g}$ is a column vector of generic polynomial terms for a 1D finite element.

Recall that (equation 5.35b)

$$[\mathbf{h}][\mathbf{C}] = [\mathbf{I}],$$

so that

$$[\mathbf{C}] = [\mathbf{h}^{-1}]. \tag{6.29}$$

The matrix $\mathbf{h}$ was shown to be an array of constants. It was obtained by row by evaluating the generic functions as a row vector, in succession at each node of the finite element.

Transposing equation 6.28 and substituting equation 6.29 into it,

$$\lfloor \mathbf{N} \rfloor = \lfloor \mathbf{g} \rfloor [\mathbf{C}] = \lfloor \mathbf{g} \rfloor [\mathbf{h}^{-1}]. \tag{6.30}$$

The displacement function $u(x)$ (equation 5.3) is of the same form as its shape functions. If the shape function is a polynomial of n terms, the displacement function is also. The displacement function is the sum of the products of its generic polynomial terms and their coefficients,

$$u(x) = \lfloor \mathbf{g} \rfloor \{\mathbf{D}\}, \tag{6.31}$$

where $\mathbf{g}$ is a row vector of the generic polynomial terms and $\mathbf{D}$ is a column vector of appropriate constant coefficients.

The parallel form of equations 6.31 and 5.3, each representing the displacement function $u(x)$, should be noted. The single-term generic polynomial functions of equation 6.31 play the role of the n-term polynomial shape functions of equations 5.3. The n constant terms D_i in equation 6.31 play the role of the n constant terms q_i in equation 5.3.

We saw earlier that it is the displacement function of the finite element that provides the basis for the determination of element stiffness equations. The development was based on representing $u(x)$ by equation 5.3 and $\mathbf{u(x, y, z)}$ by equation 6.7, along with the application of the principle of virtual work. The displacement function $u(x)$ might also have been represented by equation 6.31. The n entries D_i of vector $\mathbf{D}$ could be the generalized edof or displacements. The generalized edof D_i could then be employed to determine the element stiffness equations, rather than the physical edof q_i. The major advantage of doing so would be that the single-term generic polynomial functions would be manipulated and integrated rather than the n-term polynomial shape functions.

Substituting equation 6.30 into equations 5.3,

$$u(x) = \lfloor \mathbf{g} \rfloor [\mathbf{h}^{-1}] \{\mathbf{q}\}. \tag{6.32}$$

Comparing equation 6.32 to equation 6.31, it is clear that the generalized and physical edof are related by

$$\{\mathbf{D}\} = [\mathbf{h}^{-1}] \{\mathbf{q}\}. \tag{6.33}$$

Equation 6.31 may be extended to three dimensions. What is required is to adapt to the 3D condition where there are three independent translation displacement functions for the finite element and three edof at each node. With eight independent dof in each coordinate direction, the eight generic polynomial terms are $1, x, y, z, xy, yz, zx$, and xyz.

In terms of the generalized edof or displacements, the three independent translation displacement functions would be

$$u(x,\ y, z) = D_{1u} + D_{2u}x + D_{3u}y + D_{4u}z + D_{5u}xy + D_{6u}yz + D_{7u}zx + D_{8u}xyz, \tag{6.34a}$$

$$v(x,\ y,\ z) = D_{1v} + D_{2v}x + D_{3v}y + D_{4v}z + D_{5v}xy + D_{6v}yz + D_{7v}zx + D_{8v}xyz, \tag{6.34b}$$

$$w(x,\ y,\ z) = D_{1w} + D_{2w}x + D_{3w}y + D_{4w}z + D_{5w}xy + D_{6w}yz + D_{7w}zx + D_{8w}xyz. \tag{6.34c}$$

They may be combined as a single matrix equation,

$$\begin{Bmatrix} u(x, \ y, \ z) \\ v(x, \ y, \ z) \\ w(x, \ y, \ z) \end{Bmatrix} = \begin{bmatrix} 1 & 0 & 0 & | & x & 0 & 0 & | & --- & | & xyz & 0 & 0 \\ 0 & 1 & 0 & | & 0 & x & 0 & | & --- & | & 0 & xyz & 0 \\ 0 & 0 & 1 & | & 0 & 0 & x & | & --- & | & 0 & 0 & xyz \end{bmatrix} \begin{Bmatrix} D_{1u} \\ D_{1v} \\ D_{1w} \\ D_{2u} \\ D_{2v} \\ D_{2w} \\ \vdots \\ D_{8u} \\ D_{8v} \\ D_{8w} \end{Bmatrix}, \quad (6.35a)$$

or, more simply,

$$\{\mathbf{u(x, \ y, \ z)}\} = [\mathbf{g}]\{\mathbf{D}\}. \tag{6.35b}$$

The matrix **g** of equation 6.35 is a 3×24 matrix of generic polynomial terms arranged as shown so that when it premultiplies the 24×1 vector of generalized edof, the vector containing the three displacement functions is the result.

The 3D version of equation 6.30 is

$$[\mathbf{N}] = [\mathbf{g}][\mathbf{h}^{-1}] \quad \text{or} \quad [\mathbf{N}^{\mathsf{T}}] = [\mathbf{h}^{-\mathsf{T}}][\mathbf{g}^{\mathsf{T}}]. \tag{6.36}$$

The matrix **N** in equation 6.36 is the 3×24 matrix of shape functions from equation 6.7d. The matrix **g** is the 3×24 matrix of generic polynomial terms from equation 6.35a. The matrix **h** is a 24×24 matrix of constants. It is obtained three rows at a time by evaluating the matrix **g** in succession at each node of the finite element.

The element stiffness matrix was developed earlier (equation 6.22d) as

$$[\mathbf{K}] = \left[\int_v [\mathbf{B}^{\mathsf{T}}][\mathbf{E}][\mathbf{B}] \ dV \right],$$

where

$$[\mathbf{B}] = [\mathbf{d}][\mathbf{N}] \quad \text{and} \quad [\mathbf{B}^{\mathsf{T}}] = [\mathbf{N}^{\mathsf{T}}][\mathbf{d}^{\mathsf{T}}].$$

Thus,

$$[\mathbf{K}] = \left[\int\int_v [\mathbf{N}^T][\mathbf{d}^T][\mathbf{E}][\mathbf{d}][\mathbf{N}] \, dV\right]. \tag{6.37a}$$

Substituting equations 6.36 into equation 6.37a,

$$[\mathbf{K}] = \left[\int\int_v [\mathbf{h}^{-T}][\mathbf{g}^T][\mathbf{d}^T][\mathbf{E}][\mathbf{d}][\mathbf{g}][\mathbf{h}^{-1}] \, dV\right]. \tag{6.37b}$$

The matrix $\mathbf{h}$ is a matrix of constants. Consequently, matrices $\mathbf{h}^{-T}$ and $\mathbf{h}^{-1}$ may be moved from under the integral as long as the order of matrix multiplication is retained. Thus,

$$[\mathbf{K}] = [\mathbf{h}^{-T}]\left[\int\int_v [\mathbf{g}^T][\mathbf{d}^T][\mathbf{E}][\mathbf{d}][\mathbf{g}] \, dV\right][\mathbf{h}^{-1}], \tag{6.37c}$$

$$[\mathbf{K}] = [\mathbf{h}^{-T}]\left[\int\int_v [\mathbf{B}_D^T][\mathbf{E}][\mathbf{B}_D] \, dV\right][\mathbf{h}^{-1}], \tag{6.37d}$$

$$[\mathbf{K}] = [\mathbf{h}^{-T}][\mathbf{K}_D][\mathbf{h}^{-1}]. \tag{6.37e}$$

The matrix $\mathbf{K}_D$ is given as

$$[\mathbf{K}_D] = \left[\int\int_v [\mathbf{B}_D^T][\mathbf{E}][\mathbf{B}_D] \, dV\right]. \tag{6.37f}$$

where

$$[\mathbf{B}_D] = [\mathbf{d}][\mathbf{g}].$$

Substituting equation 6.37e into the element stiffness equation referred to physical displacements (equation 6.22c),

$$[\mathbf{h}^{-T}][\mathbf{K}_D][\mathbf{h}^{-1}]\{\mathbf{q}\} = \{\mathbf{F}\} + \{\mathbf{F}_b\} + \{\mathbf{F}_k\}. \tag{6.38a}$$

Substituting the expression for generalized displacements shown in equation 6.33 into equation 6.38a,

$$[\mathbf{h}^{-T}][\mathbf{K}_D]\{\mathbf{D}\} = \{\mathbf{F}\} + \{\mathbf{F}_b\} + \{\mathbf{F}_k\}. \tag{6.38b}$$

Premutiplying equation 6.38b through by $\mathbf{h}^T$ results in

$$[\mathbf{K}_D]\{\mathbf{D}\} = [\mathbf{h}^T]\{\mathbf{F}\} + [\mathbf{h}^T]\{\mathbf{F}_b\} + [\mathbf{h}^T]\{\mathbf{F}_k\}. \tag{6.38c}$$

Equation 6.38c is more simply shown as

$$[\mathbf{K}_D]\{\mathbf{D}\} = \{\mathbf{F}_D\} + \{\mathbf{F}_{Db}\} + \{\mathbf{F}_{Dk}\}, \tag{6.39a}$$

where

$${\bf \{F_D\} = [h^T]\{F\}} \quad or \quad {\bf \{F\} = [h^{-T}]\{F_D\}}, \tag{6.39b}$$

$${\bf \{F_{Db}\} = [h^T]\{F_b\}} \quad or \quad {\bf \{F_b\} = [h^{-T}]\{F_{Db}\}}, \tag{6.39c}$$

$${\bf \{F_{Dk}\} = [h^T]\{F_k\}} \quad or \quad {\bf \{F_k\} = [h^{-T}]\{F_{Dk}\}}. \tag{6.39d}$$

Equations 6.39 are element stiffness equations, but in terms of generalized displacements. A structural finite element may be treated in terms of generalized displacements, and then when desired, the appropriate matrices may be transformed to reference to physical displacements. The generalized element stiffness matrix is matrix $\mathbf{K}_D$ and may be computed using equation 6.37f. It may then be transformed to the element stiffness matrix referred to physical displacements using equation 6.37e.

As an illustration of the simplifying advantages of using generalized displacements, consider again the linear, elastic, straight, thin prismatic beam element. Generalized stress and strain will be applied also.

The element stiffness matrix is given by

$$[\mathbf{K}_D] = \int_0^L [\bar{\mathbf{B}}_D^T][\bar{\mathbf{E}}][\bar{\mathbf{B}}_D] \, dx, \tag{6.40a}$$

where

$$\lfloor \bar{\mathbf{B}}_D \rfloor = \lfloor \bar{\mathbf{d}} \rfloor \lfloor \bar{\mathbf{g}} \rfloor,$$

$$= \left[\frac{\partial^2}{\partial x^2} \right] \lfloor 1 \quad x \quad x^2 \quad x^3 \rfloor = \lfloor 0 \quad 0 \quad 2 \quad 6x \rfloor$$

and

$$[\bar{\mathbf{E}}] = EI.$$

Thus,

$$[\bar{\mathbf{K}}_D] = \int_0^L \begin{Bmatrix} 0 \\ 0 \\ 2 \\ 6x \end{Bmatrix} (EI) \lfloor 0 \quad 0 \quad 2 \quad 6x \rfloor \, dx,$$

$$\tag{6.40b}$$

$$= EI \int_0^L \begin{bmatrix} 0 & 0 & 0 & 0 \\ 0 & 0 & 0 & 0 \\ 0 & 0 & 4 & 12x \\ 0 & 0 & 12x & 36x^2 \end{bmatrix} dx, = EI \begin{bmatrix} 0 & 0 & 0 & 0 \\ 0 & 0 & 0 & 0 \\ 0 & 0 & 4L & 6L^2 \\ 0 & 0 & 6L^2 & 12L^3 \end{bmatrix}.$$

The manipulations and integrations to develop the element stiffness matrix referred to generalized displacements are much simpler than those for physical displacements (equations 5.17c and 6.27d).

To obtain the element stiffness matrix for physical displacements, equation 6.37e must be applied.

The matrix **h** may be written by evaluating the row vectors

$$\lfloor \mathbf{g} \rfloor = \lfloor 1 \quad x \quad x^2 \quad x^3 \rfloor$$

at $x = 0$ (node 1) for row 1 of matrix **h** and at $x = L$ (node 2) for row 3 of matrix **h** and by evaluating

$$\left\lfloor \frac{\partial \mathbf{g}}{\partial \mathbf{x}} \right\rfloor = \lfloor 0 \quad 1 \quad 2x \quad 3x^2 \rfloor$$

at $x = 0$ (node 1) for row 2 of matrix **h** and at $x = L$ (node 2) for row 4 of matrix **h**.

The row vector **g** was used directly for rows 1 and 3 since edof 1 and 3 are each transverse translations at nodes 1 and 2, respectively. The row vector $\partial \mathbf{g}/\partial \mathbf{x}$ was derived by differentiating the generic polynomial functions of row vector **g** and then used for rows 2 and 4. This was the appropriate operation since edof 2 and 4 are each rotations at nodes 1 and 2, respectively and thus derivatives of the transverse translations.

Thus matrix **h** is given as

$$[\mathbf{h}] = \begin{bmatrix} 1 & 0 & 0 & 0 \\ 0 & 1 & 0 & 0 \\ 1 & L & L^2 & L^3 \\ 0 & 1 & 2L & 3L^2 \end{bmatrix}. \qquad (6.41a)$$

Matrix **h** may be inverted by partitioning it into four 2×2 submatrices and writing the equations

$$[\mathbf{h}][\mathbf{A}] = [\mathbf{I}], \quad \text{where} \quad [\mathbf{A}] = \lfloor \mathbf{h}^{-1} \rfloor,$$

$$\left[\begin{array}{cc|cc} 1 & 0 & 0 & 0 \\ 0 & 1 & 0 & 0 \\ \hline 1 & L & L^2 & L^3 \\ 0 & 1 & 2L & 3L^2 \end{array} \right] \left[\begin{array}{c|c} \mathbf{A}_{11} & \mathbf{A}_{12} \\ \hline \mathbf{A}_{21} & \mathbf{A}_{22} \end{array} \right] = \left[\begin{array}{c|c} \mathbf{I} & \mathbf{0} \\ \hline \mathbf{0} & \mathbf{I} \end{array} \right].$$

Four submatrix equations may be written from this matrix equation and readily solved for the four submatrices $\mathbf{A}_{11}$, $\mathbf{A}_{12}$, $\mathbf{A}_{21}$, and $\mathbf{A}_{22}$:

$$[\mathbf{A}_{11}] = [\mathbf{I}], \qquad [\mathbf{A}_{12}] = [\mathbf{0}],$$

$$[\mathbf{A}_{21}] = \begin{bmatrix} -\frac{3}{L^2} & -\frac{2}{L} \\ \frac{2}{L^3} & \frac{1}{L^2} \end{bmatrix}, \qquad [\mathbf{A}_{22}] = \begin{bmatrix} \frac{3}{L^2} & -\frac{1}{L} \\ -\frac{2}{L^3} & \frac{1}{L^2} \end{bmatrix}.$$

Assembling the four submatrices in the order indicated in the matrix equation above, the inverse of matrix $\mathbf{h}$ may be written as

$$[\mathbf{h}^{-1}] = \begin{bmatrix} 1 & 0 & 0 & 0 \\ 0 & 1 & 0 & 0 \\ -\frac{3}{L^2} & -\frac{2}{L} & \frac{3}{L^2} & -\frac{1}{L} \\ \frac{2}{L^3} & \frac{1}{L^2} & -\frac{2}{L^3} & \frac{1}{L^2} \end{bmatrix}. \tag{6.41b}$$

Matrix $\mathbf{h}^{-1}$ is easily transposed to obtain matrix $\mathbf{h}^{-T}$. Then matrices $\mathbf{h}^{-1}$, $\mathbf{h}^{-T}$, and $\bar{\mathbf{K}}_D$ are substituted into equation 6.37e and the matrix multiplications indicated performed to obtain matrix $\mathbf{K}$. The result will be what is shown in equation 3.5.

The preceding development is the simplest derivation of the beam element stiffness matrix. If done totally numerically, it requires the least computation. For more complicated structural systems comprised of many different 1D, 2D, or 3D finite elements, each element stiffness matrix would be computed numerically. The use of generalized displacements could provide significant savings in computational effort to obtain those element stiffness matrices. One should recognize that computing the matrix $\bar{\mathbf{K}}_D$ for a specific element, writing the numerical matrix $\mathbf{h}$ for the element, inverting and transposing the numerical matrix $\mathbf{h}$, and applying equation 6.37e involve less computation than developing the numerical matrix $\mathbf{K}$ directly from physical displacements.

6.7. CLOSING REMARKS

The general principles for applying the finite element method to linear elastic solid structures are now in place. Their essence is reflected in the discretization of the structure into finite elements, the development of element stiffness equations, the assembly of global stiffness equations from the element stiffness equations, the application of the structural restraints understood in the context of essential, natural, and mixed boundary conditions, and the solution procedures applied to the global stiffness equations identified as the submatrix, solution-without-rearrangement, and penalty methods. The illustrations for the development of these principles have focused on 1D structural elements to facilitate understanding. The chapters to follow will focus on the application of the principles to 2D elements where the finite element method is the most widely used and powerful solution method for such problems.

BIBLIOGRAPHY

Logan, D. L., *A First Course in the Finite Element Method* (Appendix E), PWS-Kent, Boston, 1992.

Shames , I. H., and Dym, C. L., *Energy and Finite Elements in Structural Mechanics* (Chapters 3, 6), Hemisphere (McGraw-Hill), New York, 1985.

Weaver, W. Jr., and Gere, J. M., *Matrix Analysis of Framed Structures* (Chapters 1, 7), 3rd ed., Van Nostrand Reinhold, New York, 1990.

Weaver, W. Jr., and Johnston, P. R., *Finite Elements for Structural Analysis* (Chapter 1), Prentice-Hall, Englewood Cliffs, NJ, 1984.

Zienkiewcz, O. C., and Taylor, R. L., *The Finite Element Method* (Chapter 2), 4th ed., McGraw-Hill, Berkshire, England, 1989.

PROBLEMS

6.1. Starting with equation 6.4n, demonstrate that the constitutive relationships for the plane stress condition are as given by equation 6.5a.

6.2. Starting with equation 6.4n, demonstrate that the constitutive relationships for the plane strain condition are as given by equation 6.5b.

6.3. Write the matrix $\mathbf{B}(\mathbf{x})$ as functions of x for the straight prismatic beam element using the cubic shape functions given as equations 5.2.

6.4. Using the results from Problems 6.3 and equation 6.22d, develop the stiffness matrix for the straight prismatic beam element.

6.5. Using equations 6.40 and 6.41, develop the stiffness matrix for the straight prismatic beam element.

6.6. Starting with equation 6.22e, demonstrate that the element force matrix $\mathbf{F}_{Db}$ in terms of generalized dof for a distributed line load $w(x)$ on a straight prismatic beam element is given as

$$\{\mathbf{F}_{Db}\} = \int_0^L \{\mathbf{g}\} w(x)\, dx,$$

where $\{\mathbf{g}\}$ is the vector of generic functions in x.

6.7. Using the result from Problem 6.6, determine the vector of equivalent forces and moments in terms of physical dof for a straight prismatic beam element for a distributed line load acting across the length of the element for:

a. A uniform load, $w(x) = w_o$.

b. A linearly increasing load, $w(x) = w_o x$.

c. An algebraic power increasing load, $w(x) = w_o x^n$, where n is a nonnegative number.

Why do the results for parts b and c differ from those for Problems 5.9 and 5.11?

6.8. Starting from equation 6.7d, write the displacement function as a matrix equation for the 4-node, 8-dof plane stress flat plate element shown. Express your result in terms of shape functions $N_1(x, y)$, $N_2(x, y)$, $N_3(x, y)$, and $N_4(x, y)$. You need not determine the shape functions themselves.

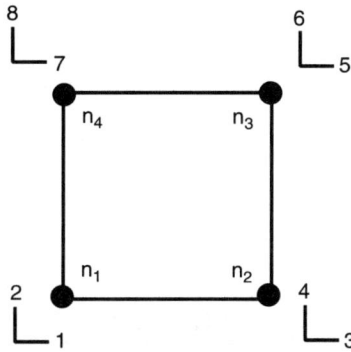

Problem 6.8

6.9. Starting from equations 6.8 and 6.9 and using the result from Problem 6.8, write the matrix $\mathbf{B}(\mathbf{x}, \mathbf{y})$ for the 4-node, 8-dof flat plate plane stress element. Express your result in terms of shape functions $N_1(x, y)$, $N_2(x, y)$, $N_3(x, y)$, and $N_4(x, y)$.

6.10. Using the result from Problem 6.9, show that the matrix $\mathbf{B}(\mathbf{x}, \mathbf{y})$ for the 4-node, 8-dof flat plate plane stress element may be represented as

$$[\mathbf{B}(\mathbf{x}, \mathbf{y})] = [[(\mathbf{B}_1(\mathbf{x}, \mathbf{y})][\mathbf{B}_2(\mathbf{x}, \mathbf{y})][[\mathbf{B}_3(\mathbf{x}, \mathbf{y})][\mathbf{B}_4(\mathbf{x}, \mathbf{y})]],$$

where $\mathbf{B}_i(\mathbf{x}, \mathbf{y})$ for $i = 1, 2, 3, 4$ are 3×2 submatrices.

6.11. Using the result from Problem 6.10, show that the stiffness matrix $\mathbf{K}$ for the flat plate plane stress element may be written as

$$[\mathbf{K}] = t \int_A \begin{bmatrix} \mathbf{B}_1^T\mathbf{EB}_1 & \mathbf{B}_1^T\mathbf{EB}_2 & \mathbf{B}_1^T\mathbf{EB}_3 & \mathbf{B}_1^T\mathbf{EB}_4 \\ \mathbf{B}_2^T\mathbf{EB}_1 & \mathbf{B}_2^T\mathbf{EB}_2 & \mathbf{B}_2^T\mathbf{EB}_3 & \mathbf{B}_2^T\mathbf{EB}_4 \\ \mathbf{B}_3^T\mathbf{EB}_1 & \mathbf{B}_3^T\mathbf{EB}_2 & \mathbf{B}_3^T\mathbf{EB}_3 & \mathbf{B}_3^T\mathbf{EB}_4 \\ \mathbf{B}_4^T\mathbf{EB}_1 & \mathbf{B}_4^T\mathbf{EB}_2 & \mathbf{B}_4^T\mathbf{EB}_3 & \mathbf{B}_4^T\mathbf{EB}_4 \end{bmatrix} dA,$$

where A is the area of the element, t is its thickness, $\mathbf{E}$ is the matrix of elastic constants for the plane stress condition, and $\mathbf{B}_i$ or $\mathbf{B}_j$ for $i, j = 1, 2, 3, 4$ are $\mathbf{B}_i(\mathbf{x}, \mathbf{y})$ for the element.

6.12. Show that the submatrices $\mathbf{B}_i^T\mathbf{EB}_j$ in the integrand of the matrix $\mathbf{K}$ from Problem 6.11 are 2×2 submatrices of functions of x and y which may be represented as

$$\frac{E}{1 - \nu^2} \begin{bmatrix} \dfrac{\partial N_i}{\partial x}\dfrac{\partial N_j}{\partial x} + \dfrac{1-\nu}{2}\dfrac{\partial N_i}{\partial y}\dfrac{\partial N_j}{\partial y} & \nu\dfrac{\partial N_i}{\partial x}\dfrac{\partial N_j}{\partial y} + \dfrac{1-\nu}{2}\dfrac{\partial N_i}{\partial y}\dfrac{\partial N_j}{\partial x} \\ \nu\dfrac{\partial N_i}{\partial y}\dfrac{N_j}{\partial x} + \dfrac{1-\nu}{2}\dfrac{\partial N_i}{\partial x}\dfrac{\partial N_j}{\partial y} & \dfrac{\partial N_i}{\partial y}\dfrac{\partial N_j}{\partial y} + \dfrac{1-\nu}{2}\dfrac{\partial N_i}{\partial x}\dfrac{\partial N_j}{\partial x} \end{bmatrix},$$

CHAPTER 7

PLANE STRESS AND PLANE STRAIN

7.1. INTRODUCTION

Two-dimensional structural finite elements include thin flat plates whose loads are in the plane of the plate and those whose loads are transverse to it. The former are plates in plane stress; the latter are plates in flexure. Thin plates subject to plane stress are discussed in Chapters 8 and 9. Those in flexure are treated in Chapter 10. Axisymmetric solids relate to the plane strain condition and are discussed in Chapter 11. The states of plane stress and its mathematically comparable condition plane strain are discussed below.

7.2. ISOTROPIC PLANE STRESS AND PLANE STRAIN

In Chapter 6, the plane stress condition was described as resulting from a thin flat plate loaded so that its stresses act only on planes normal to the plane of the plate. No stresses at all act on planes parallel to the plate. There is normal strain on planes parallel to the plate, but no shear strain. See Figure 6.2. For an isotropic continuum experiencing small strain, the stress–strain equations for the plane stress condition were shown to be

$$
\left\{ \begin{array}{c} \sigma_x \\ \sigma_y \\ \tau_{xy} \end{array} \right\} = \frac{E}{(1+v)(1-v)} \begin{bmatrix} 1 & v & 0 \\ v & 1 & 0 \\ 0 & 0 & \frac{1-v}{2} \end{bmatrix} \left\{ \begin{array}{c} \epsilon_x \\ \epsilon_y \\ \gamma_{xy} \end{array} \right\}. \tag{7.1a}
$$

The plane strain condition was described in Chapter 6 as resulting from a long prismatic structure loaded so that it can be viewed as a thin transverse slice of the structure. Stresses act on planes normal to the plane of the slice, similar to those for the plane stress condition. On planes parallel to the slice, there is a uniform normal stress, but no shear stress and no normal or shear strain. The stress–strain equations for this condition for an isotropic continuum experiencing small strain were shown to be

$$\left\{ \begin{array}{c} \sigma_x \\ \sigma_y \\ \tau_{xy} \end{array} \right\} = \frac{E}{(1+v)(1-2v)} \left[\begin{array}{ccc} 1-v & v & 0 \\ v & 1-v & 0 \\ 0 & 0 & \frac{1-2v}{2} \end{array} \right] \left\{ \begin{array}{c} \epsilon_x \\ \epsilon_y \\ \gamma_{xy} \end{array} \right\}. \tag{7.1b}$$

Both stress–strain equation systems are represented as

$$\{\sigma\} = [\mathbf{E}]\{\epsilon\}. \tag{7.1c}$$

Though in reality these conditions are 3D, only two coordinate directions in the plane of the plate or slice need to be used to analyze such structures. Stress and strain in the coordinate direction normal to the plate or slice do not affect the analysis. Stress or strain normal to the plane of the plate or slice may be computed after this analysis is carried out.

In the 3D elastic continuum, the strain in the direction of the z axis was written in Chapter 6 as

$$\epsilon_z = \frac{1}{E}(-v\sigma_x - v\sigma_y + \sigma_z). \tag{6.3c}$$

For the plane stress condition, $\sigma_z = 0$, so that equation 6.3c reduces to

$$\epsilon_z = -\frac{v}{E}(\sigma_x + \sigma_y), \tag{7.2a}$$

where the x–y plane is the plane of the plate. Once the stress variation in the x–y plane is computed, the strain in the z direction is easily computed, if it is desired.

For a plane strain condition, $\epsilon_z = 0$, so that σ_z may be computed as

$$\sigma_z = v(\sigma_x + \sigma_y). \tag{7.2b}$$

7.3. BIAXIAL STRESS TRANSFORMATION

Knowing the variation of σ_x, σ_y, and τ_{xy} in the plane of a thin plate subjected to a state of plane stress or plane strain is usually not enough. Stresses also vary with respect to their orientation at a point. The maximum normal and shear stress will rarely be oriented along the x–y coordinate axes chosen for the analysis of a structure.

Stress transformation at a point is dependent on equilibrium alone. It can be dealt with by stress transformation equations or their graphical representation, Mohr's Circle. Figure 7.1 shows an infinitesimal element subjected to a state of plane stress or plane strain. The x–y coordinate directions in the plane of the plate are chosen for the analysis of the structure. They ultimately produce σ_x, σ_y, and τ_{xy}. The x'–y'

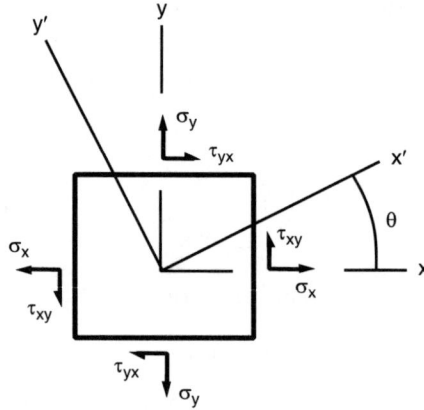

Figure 7.1 Plane stress, or plane strain

coordinate directions are chosen also in the plane of the plate and rotated counter-clockwise through an arbitrary angle θ, measured from the positive x axis.

Figure 7.2 shows a part of the infinitesimal element in the x–y orientation, with a face exposed whose orientation is rotated counterclockwise through the angle θ. The length of the exposed face is taken as unity, as is the dimension of the element perpendicular to the plane of the plate. This piece of the infinitesimal element is shown with the forces acting on each face, rather than the stresses. These forces, of course, must satisfy the conditions for equilibrium.

Setting the sum of the forces in the x' coordinate direction equal to zero results in

$$\sigma_{x'} = \sigma_x \cos^2 \theta + \sigma_y \sin^2 \theta + 2\tau_{xy} \sin \theta \cos \theta. \qquad (7.3a)$$

If θ is replaced in equation 7.3a by $\theta + 90°$, the coordinate direction becomes y' and equation 7.3a becomes

$$\sigma_{y'} = \sigma_x \sin^2 \theta + \sigma_y \cos^2 \theta - 2\tau_{xy} \sin \theta \cos \theta. \qquad (7.3b)$$

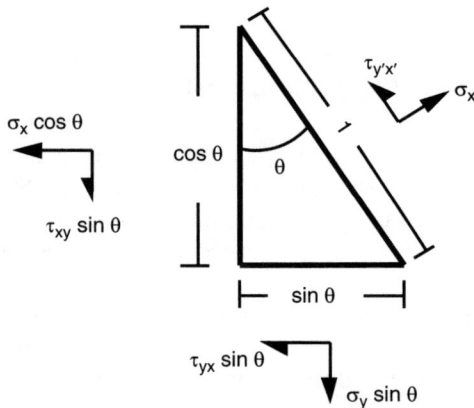

Figure 7.2 Element equilibrium

Setting the sum of the forces in the y' coordinate direction equal to zero results in

$$\tau_{x'y'} = -\sigma_x \sin \theta \cos \theta + \sigma_y \sin \theta \cos \theta + \tau_{xy}(\cos^2 \theta - \sin^2 \theta). \quad (7.3c)$$

Letting $s = \sin \theta$ and $c = \cos \theta$, where θ is understood as the angle measured from the positive x axis to the positive x' axis, equations 7.3 in matrix form are

$$\begin{Bmatrix} \sigma_{x'} \\ \sigma_{y'} \\ \tau_{x'y'} \end{Bmatrix} = \begin{bmatrix} c^2 & s^2 & 2sc \\ s^2 & c^2 & -2sc \\ -sc & sc & c^2 - s^2 \end{bmatrix} \begin{Bmatrix} \sigma_x \\ \sigma_y \\ \tau_{xy} \end{Bmatrix}, \quad (7.3d)$$

or

$$\{\sigma'\} = [T_\theta]\{\sigma\}. \quad (7.3e)$$

Equations 7.3 provide an easy means to compute stresses in the plane of the plate at a point at any orientation θ once the stresses at the point have been determined in the x–y directions. The computation is based on equilibrium alone and is independent of the material properties of the continuum.

If equation 7.3a is differentiated with respect to θ and the result is compared to equation 7.3c,

$$\frac{d\sigma_{x'}}{d\theta} = 2\tau_{x'y'}. \quad (7.4a)$$

Similarly, differentiating equation 7.3b with respect to θ.

$$\frac{d\sigma_{y'}}{d\theta} = -2\tau_{x'y'}. \quad (7.4b)$$

Setting equations 7.3c or 7.4 equal to zero permits the determination of the orientation θ_p for which either σ_x is at an algebraic maximum while σ_y is at an algebraic minimum or vice versa. At this orientation θ_p, obviously the shear stress $\tau_{x'y'} = 0$. The result of these operations determines the orientation θ_p of two principal stresses at the point, both perpendicular to each other and in the plane of the plate. The in-plane principal stresses at a point are the normal stresses, which are algebraically either the largest or the smallest in-plane normal stress at the point and for which the in-plane shear stress is zero. The expression for θ_p is

$$\theta_p = \frac{1}{2}\tan^{-1}\left(\frac{\tau_{xy}}{(\sigma_x - \sigma_y)/2}\right). \quad (7.5a)$$

Differentiating equation 7.3c with respect to θ and setting the result equal to zero permit determining the orientation θ_τ for which $\tau_{x'y'}$ is at a maximum. The simplest expression for θ_τ is

$$\theta_\tau = \theta_p \pm 45°. \quad (7.5b)$$

Equations 7.5 provide an easy determination for the orientations at a point in the continuum for which the in-plane normal or shear stresses are at their maximums. Substituted into equations 7.3d in conjunction with the previously determined values for σ_x, σ_y, and τ_{xy}, these maximum in-plane stress values at the point may be computed.

In the states of plane stress and plane strain, the plane of the plate has zero shear stress acting on it. Thus it is a third principal plane at the point, mutually perpendicular to the other two. One must recognize that although plane stress and plane strain may be analyzed as if they are a 2D condition, in reality there is a third dimension, the one perpendicular to the plane of the plate. There are, at times, important phenomena brought on by that third dimension.

Defining the x and y directions as the in-plane principal directions results in $\theta_p = 0$, and thus, from equations 7.3,

$$\sigma_{x'} = \sigma_x = \sigma_{p1}, \qquad \sigma_{y'} = \sigma_y = \sigma_{p2}, \qquad \tau_{x'y'} = \tau_{xy} = 0,$$

where σ_{p1} and σ_{p2} are principal stresses. The maximum shear stress with respect to σ_{p1} and σ_{p2} occurs at the orientation $\theta_\tau = \pm 45°$ (equation 7.5b). At this orientation, equations 7.3 give

$$\sigma_{x''} = \tfrac{1}{2}(\sigma_{p1} + \sigma_{p2}), \tag{7.6a}$$

$$\sigma_{y''} = \tfrac{1}{2}(\sigma_{p1} + \sigma_{p2}), \tag{7.6b}$$

$$\tau_{x''y''} = \max \ \tau = \tfrac{1}{2}(\sigma_{p1} - \sigma_{p2}). \tag{7.6c}$$

Equation 7.6c is a well-known expression for the maximum in-plane shear stress at a point in terms of the in-plane principal stresses at the point.

Because there are three principal directions at a point, not just two, the maximum shear stress there may not be the in-plane maximum shear stress. Equation 7.6c may be applied using any two of the principal stresses. For plane stress, the third principal stress is $\sigma_z = 0$; for plane strain, it is $\sigma_z = \nu(\sigma_{p1} + \sigma_{p2})$ (equation 7.2b).

In a state of plane stress when the in-plane principal stresses σ_{p1} and σ_{p2} are tension and $\sigma_{p1} > \sigma_{p2}$, the maximum shear stress is determined by σ_{p1} and the out-of-plane principal stress $\sigma_z = 0$. The maximum $\tau = \tfrac{1}{2}\sigma_{p1}$ and is not in-plane. It will occur on a plane at an angle of 45° to the plane of the plate.

In a state of plane strain when the in-plane principal stresses σ_{p1} and σ_{p2} are tension and $\sigma_{p1} = \sigma_{p2}$, the maximum shear stress is determined by an in-plane principal stress and the out-of-plane principal stress σ_z. The maximum $\tau = \tfrac{1}{2}\sigma_{p1}(1 - 2\nu)$. It will occur on a plane at an angle of 45° to the plane of the plate.

There are also other conditions for which the maximum shear stress in a state of plane stress or plane strain is larger than the maximum in-plane shear stress at the point. They may be determined, however, with equation 7.6c when all three principal stresses and directions are examined.

7.4. BIAXIAL STRAIN TRANSFORMATION

Like stresses at a point, strains vary with their orientation at the point. The biaxial strain transformation equations at the point may be obtained through the stress transformation equations, strain energy, and complementary strain energy at the point.

Specific strain energy was discussed in Chapter 6. The specific strain energy with respect to the x coordinate stress and strain was shown to be

$$U_x = \tfrac{1}{2}\sigma_x \epsilon_x. \tag{6.14b}$$

Specific strain energy may be interpreted as the area under the uniaxial stress–strain curve for the stress and strain components under consideration. Figure 7.3 shows the variation in specific strain energy δU_x at a point for $\sigma_x - \epsilon_x$ resulting when a small variation of strain $\delta\epsilon_x$ occurs at the point. Complementary strain energy may be interpreted as the area to the left of the uniaxial stress–strain curve for the stress–strain components under consideration. Figure 7.3 also shows the variation in complementary strain energy δU_{cx} at the point for that same small variation of strain $\delta\epsilon_x$.

For a linear material, if the variation of strain energy δU at one condition is equal to the variation of strain energy at another condition, then the variations of complementary strain energies δU_c must also be equal for the two conditions. Also,

$$\begin{aligned}
\delta U_x &= (\sigma_x)(\delta\epsilon_x) = (E\epsilon_x)(\delta\epsilon_x) = (\epsilon_x)(E\delta\epsilon_x) \\
&= (\epsilon_x)(\delta\sigma_x) \\
&= \delta U_{cx}.
\end{aligned} \tag{7.7}$$

The variation in total strain energy at a point for the element in a state of plane stress or plane strain for a small variation in strain, expressed with respect to the x–y coordinates, is

$$\delta U = \lfloor \sigma_x \quad \sigma_y \quad \tau_{xy} \rfloor \left\{ \begin{array}{c} \delta\epsilon_x \\ \delta\epsilon_y \\ \delta\gamma_{xy} \end{array} \right\} = \lfloor \sigma \rfloor \{\delta\epsilon\}. \tag{7.8a}$$

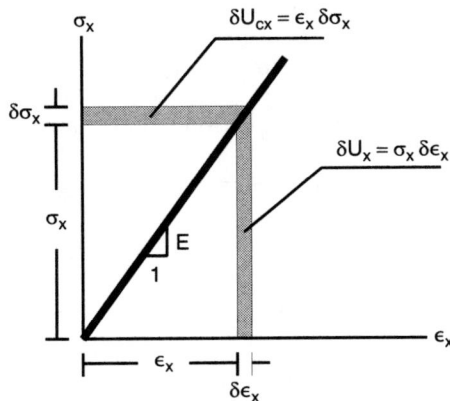

Figure 7.3 Strain energy variation

Referred to the rotated $x'–y'$ axes, it is

$$\delta U' = \lfloor \sigma_{x'} \quad \sigma_{y'} \quad \tau_{x'y'} \rfloor \left\{ \begin{array}{c} \delta\epsilon_{x'} \\ \delta\epsilon_{y'} \\ \delta\gamma_{x'y'} \end{array} \right\} = \lfloor \sigma' \rfloor \{\delta\epsilon'\}. \qquad (7.8b)$$

Strain energy is a scalar quantity, single valued at a point. Thus,

$$\delta U' = \delta U. \qquad (7.8c)$$

The variation in complementary strain energy at the point with respect to the $x–y$ axes is

$$\delta U_c = \lfloor \epsilon_x \quad \epsilon_y \quad \gamma_{xy} \rfloor \left\{ \begin{array}{c} \delta\sigma_x \\ \delta\sigma_y \\ \delta\tau_{xy} \end{array} \right\} = \lfloor \epsilon \rfloor \{\delta\sigma\}. \qquad (7.9a)$$

Referred to the rotated $x'–y'$ axes, it is

$$\delta U_c' = \lfloor \epsilon_{x'} \quad \epsilon_{y'} \quad \gamma_{x'y'} \rfloor \left\{ \begin{array}{c} \delta\sigma_{x'} \\ \delta\sigma_{ypri} \\ \delta\tau_{x'y'} \end{array} \right\} = \lfloor \epsilon' \rfloor \{\delta\sigma'\}. \qquad (7.9b)$$

Considering equations 7.7 and 8.8c,

$$\delta U_c' = \delta U_c, \qquad (7.9c)$$

$$\lfloor \epsilon' \rfloor \{\delta\sigma'\} = \lfloor \epsilon \rfloor \{\delta\sigma\}. \qquad (7.9d)$$

Stresses at a point transform dictated by equation 7.3, so that

$$\{\delta\sigma'\} = [\mathbf{T}_\theta]\{\delta\sigma\}. \qquad (7.10)$$

Substituting equation 7.10 into equation 7.9d,

$$\lfloor \epsilon' \rfloor [\mathbf{T}_\theta]\{\delta\sigma\} = \lfloor \epsilon \rfloor \{\delta\sigma\}. \qquad (7.11a)$$

Since equation 7.11a is a valid matrix equation, so also must be

$$\lfloor \epsilon' \rfloor [\mathbf{T}_\theta] = \lfloor \epsilon \rfloor. \qquad (7.11b)$$

Transposing equation 7.11b,

$$[\mathbf{T}_\theta^T]\{\epsilon'\} = \{\epsilon\}. \qquad (7.11c)$$

Rearranging,

$$\{\epsilon'\} = [\mathbf{T}_\theta^{-T}]\{\epsilon\}. \qquad (7.11d)$$

Equation 7.11d is the biaxial strain transformation equation, the counterpart to equation 7.3e. The strain transformation matrix $\mathbf{T}_\theta^{-T}$ is the transpose of the inverse of the stress transformation matrix $\mathbf{T}_\theta$. It is easily calculated in a numerical manner, but an analytic expression may be obtained for it also.

The inverse of matrix $\mathbf{T}_\theta$ may be obtained by invoking the adjoint method and employing the necessary trigonometric identities. A more elegant approach is to redefine the stress vector as σ_x, σ_y, and $\tau_{xy}\sqrt{2}$, so that equation 7.3d becomes

$$\begin{Bmatrix} \sigma_{x'} \\ \sigma_{y'} \\ \tau_{x'y'}\sqrt{2} \end{Bmatrix} = \begin{bmatrix} c^2 & s^2 & sc\sqrt{2} \\ s^2 & c^2 & -sc\sqrt{2} \\ -sc\sqrt{2} & sc\sqrt{2} & c^2 - s^2 \end{bmatrix} \begin{Bmatrix} \sigma_x \\ \sigma_y \\ \tau_{xy}\sqrt{2} \end{Bmatrix}, \tag{7.12a}$$

$$\{\sigma'_{\text{mod}}\} = [\mathbf{T}_{\theta,\text{mod}}]\{\sigma_{\text{mod}}\}. \tag{7.12b}$$

The matrix $\mathbf{T}_{\theta,\text{mod}}$ is orthogonal, which may be verified by applying equation 1.12 to it. Accordingly, its inverse is its transpose. The inverse of equations 7.12 may be written directly as

$$\{\sigma_{\text{mod}}\} = [\mathbf{T}_{\theta,\text{mod}}^{-1}]\{\sigma'_{\text{mod}}\} = [\mathbf{T}_{\theta,\text{mod}}^{T}]\{\sigma'_{\text{mod}}\}, \tag{7.13a}$$

$$\begin{Bmatrix} \sigma_x \\ \sigma_y \\ \tau_{xy}\sqrt{2} \end{Bmatrix} = \begin{bmatrix} c^2 & s^2 & -sc\sqrt{2} \\ s^2 & c^2 & sc\sqrt{2} \\ sc\sqrt{2} & -sc\sqrt{2} & c^2 - s^2 \end{bmatrix} \begin{Bmatrix} \sigma_{x'} \\ \sigma_{y'} \\ \tau_{x'y'}\sqrt{2} \end{Bmatrix}. \tag{7.13b}$$

Writing equations 7.13b in terms of the original stress vector,

$$\begin{Bmatrix} \sigma_x \\ \sigma_y \\ \tau_{xy} \end{Bmatrix} = \begin{bmatrix} c^2 & s^2 & -2sc \\ s^2 & c^2 & 2sc \\ sc & -sc & c^2 - s^2 \end{bmatrix} \begin{Bmatrix} \sigma_{x'} \\ \sigma_{y'} \\ \tau_{x'y'} \end{Bmatrix}, \tag{7.14a}$$

$$\{\sigma\} = [\mathbf{T}_\theta^{-1}]\{\sigma'\}. \tag{7.14b}$$

Thus the inverse of $\mathbf{T}_\theta$ is obtained.

The strain transformation matrix is the transpose of matrix $\mathbf{T}_\theta^{-1}$, so

$$\begin{Bmatrix} \epsilon_{x'} \\ \epsilon_{y'} \\ \gamma_{x'y'} \end{Bmatrix} = \begin{bmatrix} c^2 & s^2 & sc \\ s^2 & c^2 & -sc \\ -2sc & 2sc & c^2 - s^2 \end{bmatrix} \begin{Bmatrix} \epsilon_x \\ \epsilon_y \\ \gamma_{xy} \end{Bmatrix}. \tag{7.15a}$$

Note that if the strain vector is defined as ϵ_x, ϵ_y, and $\gamma_{xy}/2$, equation 7.15a becomes

$$\begin{Bmatrix} \epsilon_{x'} \\ \epsilon_{y'} \\ \frac{1}{2}\gamma_{x'y'} \end{Bmatrix} = \begin{bmatrix} c^2 & s^2 & 2sc \\ s^2 & c^2 & -2sc \\ -sc & sc & c^2 - s^2 \end{bmatrix} \begin{Bmatrix} \epsilon_x \\ \epsilon_y \\ \frac{1}{2}\gamma_{xy} \end{Bmatrix}, \tag{7.15b}$$

$$\{\epsilon'_{\text{mod}}\} = [\mathbf{T}_\theta]\{\epsilon_{\text{mod}}\}. \tag{7.15c}$$

The modified strain vector transforms as does the stress vector.

Equations 7.3d and 7.15b reveal that the maximum and minimum in-plane normal strains will occur at the orientation θ_p (equation 7.5a), where $\gamma_{x'y'}$ will be zero. Moreover, the maximum in-plane $\gamma_{x'y'}$ will occur at the orientation θ_τ (equation 7.5b).

The strain transformation equations 7.15 are restricted by the material properties of the continuum. The stress–strain components of the continuum material each must possess a linear uniaxial stress–strain relationship so that the equality statement (equation 7.9c) for the complementary strain energy δU_c is true. The stress–strain components need not each have the same modulus, but there must be only one modulus for each.

7.5. ORTHOTROPIC PLANE STRESS

In an isotropic medium, two elastic constants, often Young's modulus E and Poisson's ratio v, govern for all orientations. In an orthotropic medium, however, four elastic constants govern normal stress and strain in two coordinate directions. And to relate shear stress and strain, other constants are needed which involve rotations with respect to the coordinate directions. A mix of all of these values of elastic constants govern in the orientations between the defined mutually perpendicular material orientations. In the discussion that follows, x and y are the orthotropic material orientations which are also in the plane of the plate for the plane stress condition.

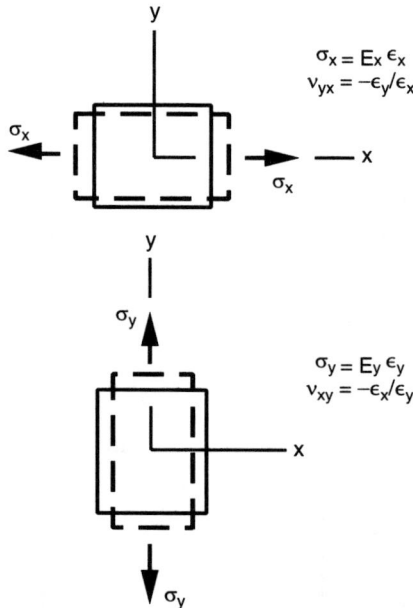

Figure 7.4 Orthotropic uniaxial stress

Figure 7.4 illustrates a linear elastic orthotropic material subjected first to a uniaxial state of stress in the x direction and then to one in the y direction. The four elastic constants relating normal stress and strain are defined as shown. The double subscripted Poisson's ratio ν_{yx} may be interpreted as normal strain in the y direction due to a unit normal strain in the x direction. Similarly, ν_{xy} may be interpreted as normal strain in the x direction due to a unit normal strain in the y direction.

Figure 7.5 illustrates a state of plane stress for a linear elastic orthotropic material. There are normal stresses in both the x and y directions and normal strains in the x, y, and z directions. The dashed rectangle shows the sense of the normal strains ϵ_x and ϵ_y.

Let

$$\epsilon_{x\sigma x} = \text{normal strain in } x \text{ direction caused by } \sigma_x,$$

$$\epsilon_{x\sigma y} = \text{normal strain in } x \text{ direction caused by } \sigma_y,$$

$$\epsilon_{y\sigma x} = \text{normal strain in } y \text{ direction caused by } \sigma_x,$$

$$\epsilon_{y\sigma y} = \text{normal strain in } y \text{ direction caused by } \sigma_y.$$

The normal stress σ_z is zero for plane stress, so there is no contribution from it to strains in the x and y directions. Because the material is linear, strains may be superimposed,

$$\epsilon_x = \epsilon_{x\sigma x} + \epsilon_{x\sigma y} \quad \text{and} \quad \epsilon_y = \epsilon_{y\sigma x} + \epsilon_{y\sigma y}. \tag{7.16}$$

It is apparent from Figure 7.4 that

$$\epsilon_{x\sigma y} = -\nu_{xy}\epsilon_{y\sigma y} \quad \text{and} \quad \sigma_y = E_y\epsilon_{y\sigma y}, \tag{7.17a}$$

$$\epsilon_{y\sigma x} = -\nu_{yx}\epsilon_{x\sigma x} \quad \text{and} \quad \sigma_x = E_x\epsilon_{x\sigma x}. \tag{7.17b}$$

Substituting equations 7.17 into equations 7.16,

$$\epsilon_x = \frac{\sigma_x}{E_x} - \nu_{xy}\frac{\sigma_y}{E_y} \quad \text{and} \quad \epsilon_y = -\nu_{yx}\frac{\sigma_x}{E_x} + \frac{\sigma_y}{E_y}. \tag{7.18a, b}$$

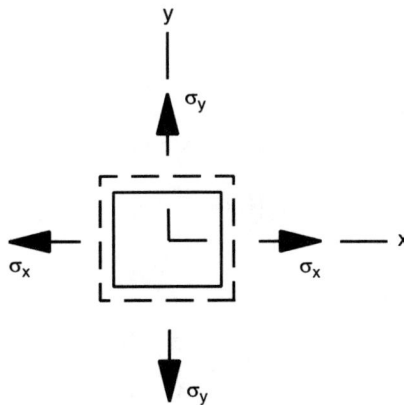

Figure 7.5 Orthotropic plane stress

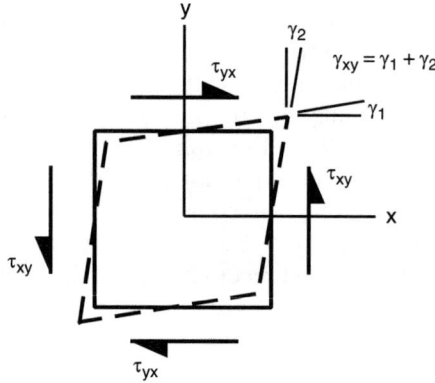

Figure 7.6 Orthotropic shear plane stress

Shear stress and shear strain in a linear elastic orthotropic material in a state of plane stress are shown in Figure 7.6. For any two coordinate directions and for small strains, the relationship is similar to what it is for an isotropic material. That is, it is linear and is uncoupled from the relationships for normal stress and strain. From moment equilibrium of the element shown, $\tau_{xy} = \tau_{yx}$, and by definition, $\gamma_{xy} = \gamma_1 + \gamma_2$. For a linear elastic material, the relationship between the in-plane shear stress and strain is

$$\gamma_{xy} = \frac{\tau_{xy}}{G_{xy}}. \tag{7.18c}$$

Equations 7.18 in matrix form are

$$\begin{Bmatrix} \epsilon_x \\ \epsilon_y \\ \gamma_{xy} \end{Bmatrix} = \begin{bmatrix} \frac{1}{E_x} & -\frac{\nu_{xy}}{E_y} & 0 \\ -\frac{\nu_{yx}}{E_x} & \frac{1}{E_y} & 0 \\ 0 & 0 & \frac{1}{G_{xy}} \end{bmatrix} \begin{Bmatrix} \sigma_x \\ \sigma_y \\ \tau_{xy} \end{Bmatrix}, \tag{7.18d}$$

$$\{\epsilon\} = [C]\{\sigma\}. \tag{7.18e}$$

The matrix **C** is a matrix of elastic constants relating the stress vector to the strain vector. Its inverse is the matrix **E** of elastic constants, which is employed in the expressions for the element stiffness matrix, such as equation 6.22d. Matrix **C** may be inverted by partitioning it and writing submatrix equations:

$$[C][E] = [I], \tag{7.19a}$$

$$\begin{bmatrix} \frac{1}{E_x} & -\frac{\nu_{xy}}{E_y} & 0 \\ -\frac{\nu_{yx}}{E_x} & \frac{1}{E_y} & 0 \\ 0 & 0 & \frac{1}{G_{xy}} \end{bmatrix} \begin{bmatrix} A & B \\ C & D \end{bmatrix} = \begin{bmatrix} I & 0 \\ 0 & I \end{bmatrix}. \tag{7.19b}$$

Four submatrix equations may be written from equation 7.19b and the four sub-matrices **A**, **B**, **C**, and **D** obtained:

$$[\mathbf{A}] = \frac{1}{1 - \nu_{xy}\nu_{yx}} \begin{bmatrix} E_x & \nu_{xy}E_x \\ \nu_{yx}E_y & E_y \end{bmatrix}, \tag{7.19c}$$

$$[\mathbf{B}] = \begin{bmatrix} 0 \\ 0 \end{bmatrix}, \tag{7.19d}$$

$$[\mathbf{C}] = [0 \quad 0], \tag{7.19e}$$

$$[\mathbf{D}] = [G_{xy}]. \tag{7.19f}$$

Assembling these four submatrices in the order indicated in equation 7.19b, the matrix **E** is obtained, and the plane stress orthotropic stress–strain equation is written as

$$\begin{Bmatrix} \sigma_x \\ \sigma_y \\ \tau_{xy} \end{Bmatrix} = \frac{1}{1 - \nu_{xy}\nu_{yx}} \begin{bmatrix} E_x & \nu_{xy}E_x & 0 \\ \nu_{yx}E_y & E_y & 0 \\ 0 & 0 & (1 - \nu_{xy}\nu_{yx})G_{xy} \end{bmatrix} \begin{Bmatrix} \epsilon_x \\ \epsilon_y \\ \gamma_{xy} \end{Bmatrix}, \tag{7.20a}$$

$$\{\boldsymbol{\sigma}\} = [\mathbf{E}]\{\boldsymbol{\epsilon}\}. \tag{7.20b}$$

Note that there are five elastic constants in the matrix **E** (equation 7.20a), compared to only two for the isotropic plane stress condition (equation 7.1a). Recognize also that the stresses and strains in equations 7.20a are along the orthotropic material orientations only, whereas those in equations 7.1a may be along any orientation.

Only four of the constant entries in the matrix **E** of equations 7.20a are independent, as may be shown by considering the strain energy of the structural element. The strain energy at the point of the infinitesimal element shown in Figure 7.5 may be obtained by first applying the normal stress σ_x to the element, resulting in the following component of strain energy:

$$U_{1x} = \tfrac{1}{2}\epsilon_{x\sigma x}\sigma_x. \tag{7.21a}$$

Then the other normal stress σ_y is applied to the element, causing the following additional strain energy in the element:

$$U_{1y} = \tfrac{1}{2}\epsilon_{y\sigma y}\sigma_y + \epsilon_{x\sigma y}\sigma_x. \tag{7.21b}$$

The total strain energy in the element developed during this stressing is

$$U_1 = U_{1x} + U_{1y} = \tfrac{1}{2}\epsilon_{x\sigma x}\sigma_x + \tfrac{1}{2}\epsilon_{y\sigma y}\sigma_y + \epsilon_{x\sigma y}\sigma_x. \tag{7.21c}$$

Alternatively, the strain energy at the point may be obtained by first applying the normal stress σ_y to the element, resulting in the following component of strain energy:

$$U_{2y} = \tfrac{1}{2}\epsilon_{y\sigma y}\sigma_y. \tag{7.22a}$$

Then the normal stress σ_x is applied to the element, causing the following additional strain energy:

$$U_{2x} = \tfrac{1}{2}\epsilon_{x\sigma x}\sigma_x + \epsilon_{y\sigma x}\sigma_y. \tag{7.22b}$$

The strain energy during the stressing of the element this time is

$$U_2 = U_{2x} + U_2 y = \tfrac{1}{2}\epsilon_{x\sigma x}\sigma_x + \tfrac{1}{2}\epsilon_{y\sigma y}\sigma_y + \epsilon_{y\sigma x}\sigma_y. \tag{7.22c}$$

Since strain energy is a scalar quantity with a single value at a point, $U_1 = U_2$. Comparing equations 7.21c and 7.22c,

$$\epsilon_{x\sigma y}\sigma_x = \epsilon_{y\sigma x}\sigma_y. \tag{7.23}$$

Substituting equations 7.17 into equation 7.23,

$$(-\nu_{xy}\epsilon_{y\sigma y})(\sigma_x) = (-\nu_{yx}\epsilon_{x\sigma x})(\sigma_y), \tag{7.24a}$$

$$\nu_{xy}\left(\frac{\sigma_x}{\epsilon_{x\sigma x}}\right) = \nu_{yx}\left(\frac{\sigma_y}{\epsilon_{y\sigma y}}\right), \tag{7.24b}$$

$$\nu_{xy}E_x = \nu_{yx}E_y. \tag{7.24c}$$

Equation 7.24c is an expression of Maxwell's reciprocal relations. It reduces the number of independent elastic constants in equations 7.20a from five to four.

There is no unique expression for G_{xy} in terms of the other elastic constants, such as exists for an isotropic material (equation 6.3g). However, a reasonably good approximate relationship is

$$\frac{1}{G_{xy}} \approx \frac{1 + \nu_{yx}}{E_x} + \frac{1 + \nu_{xy}}{E_y}. \tag{7.25}$$

Equation 7.25 in conjunction with equation 7.24c allows the treatment of linear elastic orthotropic materials in the plane stress condition with just three independent elastic constants.

When the orthotropic material axes are aligned with the axes used for the analysis of the structure, the analysis may proceed directly using the matrix **E** of equations 7.20a. On the other hand, when the orthotropic material axes are rotated with respect to the structural axes, the elastic constants needed for the matrix **E**, such as in equation 6.22d, must be referred to the structural axes. They may be computed for the structural axes from the elastic constants defined for the orthotropic axes.

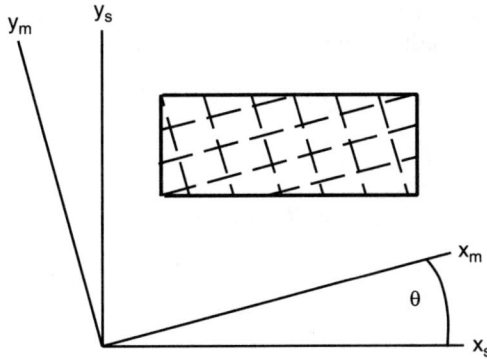

Figure 7.7 Orthyotropic structural element

Figure 7.7 illustrates an orthotropic structural element in plane stress for which the orthotropic material axes are rotated from the structural axes. The subscript s is for the axes of the structure; the subscript m designates the orthotropic axes.

Stresses transform independent of material properties as

$$\{\sigma_m\} = [\mathbf{T}_\theta]\{\sigma_s\}. \tag{7.26a}$$

Because the material is linear, strains transform in accordance with

$$\{\epsilon_m\} = [\mathbf{T}_\theta^{-T}]\{\epsilon_s\}. \tag{7.26b}$$

The orthotropic stress–strain relationship is

$$\{\sigma_m\} = [\mathbf{E}_m]\{\epsilon_m\}, \tag{7.26c}$$

where matrix $\mathbf{E}_m$ is the orthotropic matrix of elastic constants (equations 7.20a, 7.24c, and 7.25).

Substituting equations 7.26a and 7.26b into equation 7.26c,

$$[\mathbf{T}_\theta]\{\sigma_s\} = [\mathbf{E}_m][\mathbf{T}_\theta^{-T}]\{\epsilon_s\}, \tag{7.27a}$$

$$\{\sigma_s\} = [\mathbf{T}_\theta^{-1}][\mathbf{E}_m][\mathbf{T}_\theta^{-T}]\{\epsilon_s\}, \tag{7.27b}$$

$$\{\sigma_s\} = [\mathbf{E}_s]\{\epsilon_s\}, \tag{7.27c}$$

where

$$[\mathbf{E}_s] = [\mathbf{T}_\theta^{-1}][\mathbf{E}_m][\mathbf{T}_\theta^{-T}]. \tag{7.27d}$$

The matrix $\mathbf{E}_s$ is the matrix of elastic constants for the axes of the structure. It is used to compute the element stiffness matrix, such as in equations 6.22d. For a linear orthotropic material in the plane stress condition, computation of the element stiffness matrix requires four (or three) elastic constants for the matrix $\mathbf{E}_m$ and the orientation θ of the orthotropic axes with respect to the structural axes for the matrix $\mathbf{T}_\theta$ for each structural element.

Note that if the strain vector of equation 7.15b is employed, for which shear strains are treated as one-half of their values, equation 7.26b becomes

$$\{\epsilon_{m(\text{mod})}\} = [\mathbf{T}_\theta]\{\epsilon_{s(\text{mod})}\}. \tag{7.28a}$$

As a consequence, equation 7.27d becomes

$$[\mathbf{E}_s] = [\mathbf{T}_\theta^{-1}][\mathbf{E}_m][\mathbf{T}_\theta]. \tag{7.28b}$$

However, if the shear strain is so modified, the matrix **B** of equations 6.22 must also be modified to reflect this changed definition of shear strain.

7.6. ORTHOTROPIC PLANE STRAIN

The plane strain condition in an orthotropic material has many similarities to the plane stress condition, just as it did in an isotropic material. In plane strain, however, the normal stress in the transverse direction σ_z is not zero. Consequently, there are three normal stresses that contribute to each normal strain. Equations 7.16, 7.18a, and 7.18b must include the effects of σ_z if they are to apply for plane strain. From Figure 7.4 visualize a third imposition of uniaxial stress on the element in the z direction, in addition to those shown for the x and y directions. Consider also the interactions of normal stress and strain between the x–z pair of coordinate directions and those between the y–z pair of coordinate directions. Extending the rationale that led to equations 7.16,

$$\epsilon_x = \epsilon_{x\sigma x} + \epsilon_{x\sigma y} + \epsilon_{x\sigma z}, \tag{7.29a}$$

$$\epsilon_y = \epsilon_{y\sigma x} + \epsilon_{y\sigma y} + \epsilon_{y\sigma z}, \tag{7.29b}$$

$$\epsilon_z = \epsilon_{z\sigma x} + \epsilon_{z\sigma y} + \epsilon_{z\sigma z}. \tag{7.29c}$$

Equations 7.17, 7.18a, and 7.18b may then be extended, so that

$$\epsilon_x = \frac{\sigma_x}{E_x} - \nu_{xy}\frac{\sigma_y}{E_y} - \nu_{xz}\frac{\sigma_z}{E_z}E_z, \tag{7.30a}$$

$$\epsilon_y = -\nu_{yx}\frac{\sigma_x}{E_x} + \frac{\sigma_y}{E_y} - \nu_{yz}\frac{\sigma_z}{E_z}, \tag{7.30b}$$

$$\epsilon_z = -\nu_{zx}\frac{\sigma_x}{E_x} - \nu_{zy}\frac{\sigma_y}{E_y} + \frac{\sigma_z}{E_z} = 0. \tag{7.30c}$$

Note that five additional elastic constants have been introduced: an elastic modulus E_z for the transverse direction, Poisson's ratios ν_{xz} and ν_{zx} for the x–z coordinate directions, and Poisson's ratios ν_{yz} and ν_{zy} for the y–z coordinate directions. For plane strain, ϵ_z is zero, so that equation 7.30c may be rewritten as

$$\frac{\sigma_z}{E_z} = \nu_{zx}\frac{\sigma_x}{E_x} + \nu_{zy}\frac{\sigma_y}{E_y}. \tag{7.30d}$$

When this result is substituted into equations 7.30a and 7.30b,

$$\epsilon_x = \left(\frac{1 - v_{xz}v_{zx}}{E_x}\right)\sigma_x - \left(\frac{v_{xy} + v_{xz}v_{zy}}{E_y}\right)\sigma_y, \tag{7.31a}$$

$$\epsilon_y = -\left(\frac{v_{yx} + v_{yz}v_{zx}}{E_x}\right)\sigma_x + \left(\frac{1 - v_{yz}v_{zy}}{E_y}\right)\sigma_y. \tag{7.31b}$$

Equations 7.31 are the counterparts to the plane stress equations 7.18a and 7.18b.

The relationship for in-plane shear stress and shear strain is the same as it is for plane stress (equation 7.18c).

In matrix form, equations 7.31 and 7.18c may be expressed as

$$\begin{Bmatrix} \epsilon_x \\ \epsilon_y \\ \gamma_{xy} \end{Bmatrix} = \begin{bmatrix} \frac{1-v_{xz}v_{zx}}{E_x} & -\frac{v_{xy}+v_{xz}v_{zy}}{E_y} & 0 \\ -\frac{v_{yx}+v_{yz}v_{zx}}{E_x} & \frac{1-v_{yz}v_{zy}}{E_y} & 0 \\ 0 & 0 & \frac{1}{G_{xy}} \end{bmatrix} \begin{Bmatrix} \sigma_x \\ \sigma_y \\ \tau_{xy} \end{Bmatrix}, \tag{7.31c}$$

$$\{\epsilon\} = [\mathbf{C}]\{\boldsymbol{\sigma}\}. \tag{7.31d}$$

Clearly a plane strain analysis in an orthotropic material may be done for the in-plane stresses and strains independently of the transverse stress. However, it is also clear that the elastic constants contained in equations 7.31c are influenced by the material properties in the transverse coordinate direction.

The matrix **C** for plane strain (equations 7.31c) may be inverted using the same method employed for the inversion of the matrix **C** for plane stress (equations 7.19). The result is the matrix **E** of elastic constants for plane strain. The plane strain orthotropic stress–strain equations may then be written as

$$\begin{Bmatrix} \sigma_x \\ \sigma_y \\ \tau_{xy} \end{Bmatrix} = \frac{1}{D} \begin{bmatrix} (1 - v_{yz}v_{zy})E_x & (v_{xy} + v_{xz}v_{zy})E_x & 0 \\ (v_{yx} + v_{yz}v_{zx})E_y & (1 - v_{xz}v_{zx})E_y & 0 \\ 0 & 0 & DG_{xy} \end{bmatrix} \begin{Bmatrix} \epsilon_x \\ \epsilon_y \\ \gamma_{xy} \end{Bmatrix}, \tag{7.32a}$$

where

$$D = (1 - v_{xz}v_{zx})(1 - v_{yz}v_{zy}) - (v_{xy} + v_{xz}v_{zy})(v_{yx} + v_{yz}v_{zx})$$

and

$$\{\boldsymbol{\sigma}\} = [\mathbf{E}]\{\boldsymbol{\epsilon}\}. \tag{7.32b}$$

Maxwell's reciprocal relations may be developed for use with equations 7.32 in a manner similar to the method employed for plane stress (equations 7.21–7.24). The development leads to

$$v_{xy}E_x = v_{yx}E_y, \qquad v_{xz}E_x = v_{zx}E_z, \qquad v_{yz}E_y = v_{zy}E_z. \tag{7.33a}$$

Equations 7.33a may be used to show that

$$(\nu_{xy} + \nu_{xz}\nu_{zy})E_x = (\nu_{yx} + \nu_{yz}\nu_{zx})E_y. \tag{7.33b}$$

As with the plane stress condition, the stresses and strains in equations 7.32a are along the orthotropic material orientations only. When these axes are aligned with the axes used for the analysis of the structure, the analysis and computation of element stiffness matrices may proceed directly using the matrix **E** of equation 7.32a. If the orthotropic axes are rotated with respect to the structural axes, then the appropriate matrix $\mathbf{E}_s$ must be computed from the matrix $\mathbf{E}_m$ of equation 7.32a using equation 7.27d.

7.7. CLOSING REMARKS

Plane stress and plane strain are characterized by their stress and strain vectors

$$\{\boldsymbol{\sigma}\} = \begin{Bmatrix} \sigma_x \\ \sigma_y \\ \tau_{xy} \end{Bmatrix}, \qquad \{\boldsymbol{\epsilon}\} = \begin{Bmatrix} \epsilon_x \\ \epsilon_y \\ \gamma_{xy} \end{Bmatrix} \tag{7.34a}$$

and their elastic matrix

$$[\mathbf{E}] = \begin{bmatrix} E_{11} & E_{12} & 0 \\ E_{21} & E_{22} & 0 \\ 0 & 0 & E_{33} \end{bmatrix}. \tag{7.34b}$$

Only three in-plane stresses and strains need to be dealt with in plane stress or plane strain for either isotropic or orthotropic media. Because of Maxwell's reciprocal relations, which require that $E_{12} = E_{21}$, there are at most four independent elastic constants which must be determined for the matrix **E**: E_{11}, E_{12}, E_{22}, and E_{33}.

In an isotropic medium, only two constants such as Young's modulus E and Poisson's ratio ν are needed, which then may be used to calculate the four E_{ij} using equations 7.1.

In an orthotropic medium subjected to plane stress, either the four constants E_{ij} might be determined directly by some means or the moduli E_x and E_y and one Poisson's ratio ν_{xy} might be obtained. These then could be used to calculate the four E_{ij} using equations 7.20a, 7.24c, and 7.25.

In an orthotropic medium subjected to plane strain, the four constants E_{ij} should be determined directly. The alternative is to determine E_x, E_y, E_z, ν_{xy}, ν_{xz}, ν_{yz}, and G_{xy} and then calculate the four E_{ij} using equations 7.32a and 7.33.

In an orthotropic medium, when the orthotropic axes are rotated from the structural axes, the elastic matrix used for the calculation of the element stiffness matrix must be referred to the structural axes. This may be accomplished from the orthotropic elastic matrix and its four constants E_{ij} (equation 7.34b) and transforming the matrix using equation 7.27d.

Finally for either plane stress or plane strain, once the stresses are determined at a point in the structure, it is often necessary to determine the principal stresses and maximum shear stress at that point. The principal stresses may be calculated using equations 7.3 and 7.5a. The maximum shear stresses, in-plane and out of plane, may be determined using equation 7.6c.

BIBLIOGRAPHY

Beer, F. P., and Johnson, E. R. Jr., *Mechanics of Materials* (Chapters 1, 2, 6), 2nd ed., McGraw-Hill, New York, 1992.

Boresi, A. P., and Lynn, P. O., *Elasticity in Engineering Mechanics* (Chapters 2, 3), Prentice-Hall, Englewood Cliffs, NJ, 1974.

Boresi, A. P., Sidebottom, O. M., Seely, F. B., and Smith, J. O., *Advanced Mechanics of Materials* (Chapter 1), Wiley, New York, 1978.

Logan, D. L., *A First Course in the Finite Element Method* (Chapter 7), PWS-Kent, Boston, 1992.

McGuire, W., and Gallagher, R. H., *Matrix Structural Analysis* (Chapter 12), Wiley, New York, 1979.

Prathap, G., *The Finite Element Method in Structural Mechanics* (Chapter 7), Kluwer Academic, Dordrecht, The Netherlands, 1993.

Weaver, W. Jr., and Johnston, P. R., *Finite Elements for Structural Analysis* (Chapter 2), Prentice-Hall, Englewood Cliffs, NJ, 1984.

Yang, T. Y., *Finite Element Structural Analysis* (Chapter 9), Prentice-Hall, Englewood Cliffs, NJ, 1986.

PROBLEMS

The use of computer software such as MATLAB or a spreadsheet which manipulates individual matrices and evaluates functions at a point in their space will facilitate solution of many of these problems.

7.1. Show that equation 7.3a can be written as

$$\sigma_{x'} = \tfrac{1}{2}(\sigma_x + \sigma_y) + \tfrac{1}{2}(\sigma_x - \sigma_y) \cos 2\theta + \tau_{xy} \sin 2\theta.$$

7.2. Show that equation 7.3c can be written as

$$\tau_{x'y'} = -\tfrac{1}{2}(\sigma_x - \sigma_y) \sin 2\theta + \tau_{xy} \cos 2\theta.$$

7.3. Derive the expression for the angle θ_p (equation 7.5a) which gives the orientation of the in-plane principal stresses of a structure subjected to plane stress or plane strain.

7.4. Derive the expression for the angle θ_τ (equation 7.5b) which gives the orientation of the in-plane maximum shear stress of a structure subjected to plane stress or plane strain.

7.5. Determine the maximum shear stress in a structure subjected to plane stress when

   a. $\sigma_{p1} = 10 \, \text{ksi} \, (T)$, $\sigma_{p2} = 8 \, \text{ksi} \, (C)$;
   b. $\sigma_{p1} = 16 \, \text{ksi} \, (T)$, $\sigma_{p2} = 6 \, \text{ksi} \, (T)$;
   c. $\sigma_{p1} = 2 \, \text{ksi} \, (C)$, $\sigma_{p2} = 8 \, \text{ksi} \, (C)$; and
   d. $\sigma_{p1} = 8 \, \text{ksi} \, (T)$, $\sigma_{p2} = 8 \, \text{ksi} \, (T)$.

7.6. Repeat Problem 7.5 but for a structure subjected to plane strain whose Poisson's ratio $\nu = 0.25$.

7.7. A thin plate is subjected to plane stress. Its elastic constants $E = 30,000 \, \text{ksi}$ and $\nu = 0.25$. Its stresses are $\sigma_x = 20 \, \text{ksi} \, (T)$, $\sigma_y = 4 \, \text{ksi} \, (T)$, and $\tau_{xy} = 6 \, \text{ksi}$. Use the matrix equations of this chapter to determine:

   a. The strains corresponding to the stresses given.
   b. The in-plane principal stresses and strains.
   c. The in-plane maximum shear stress and strains.
   d. The maximum shear stress in the structure.
   e. The stress and strain normal to the plane of the plate.

7.8. Repeat Problem 7.7 but for a structure subjected to plane strain.

7.9. Repeat Problem 7.7 if $\sigma_x = 10 \, \text{ksi} \, (T)$, $\sigma_y = 2 \, \text{ksi} \, (C)$, and $\tau_{xy} = 8 \, \text{ksi}$.

7.10. Repeat Problem 7.9 but for a structure subjected to plane strain.

7.11. The structure shown is subjected to plane stress and is fabricated from an orthotropic material. Its elastic properties with respect to the x_m–y_m material axes and stresses with respect to the x–y structure axes are

$$
\begin{aligned}
E_x &= 30,000 \, \text{ksi}, & \sigma_x &= 80 \, \text{ksi}, \\
E_y &= 20,000 \, \text{ksi}, & \sigma_y &= 50 \, \text{ksi}, \\
\nu_{xy} &= 0.22, & \tau_{xy} &= 27 \, \text{ksi}.
\end{aligned}
$$

Determine its strains with respect to the x–y structure axes.

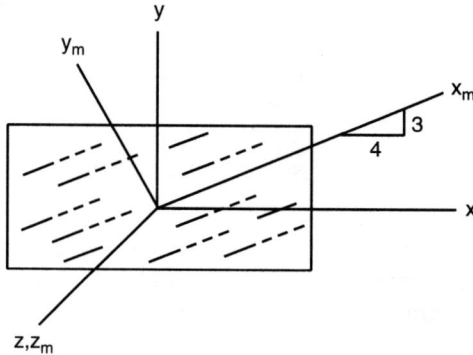

Problem 7.11

7.12. Repeat Problem 7.11 for a structure subjected to plane strain whose elastic properties are

$$
\begin{aligned}
E_x &= 30{,}000 \text{ ksi}, & \nu_{xy} &= 0.20, \\
E_y &= 20{,}000 \text{ ksi}, & \nu_{xz} &= 0.24, \\
E_z &= 25{,}000 \text{ ksi}, & \nu_{yz} &= 0.22. \\
G_{xy} &= 10{,}000 \text{ ksi},
\end{aligned}
$$

CHAPTER 8

PLANE STRESS STRUCTURAL TRIANGULAR FINITE ELEMENTS

8.1. INTRODUCTION

Structural elements subjected to plane stress are treated by the finite element method similarly to the structural elements discussed in earlier chapters. The important difference is the incorporation of two spatial dimensions instead of just one. Recall that in plane stress the thickness of the structure must be constant.

The process begins by discretizing the structure. The finite elements employed are plane shapes characterized by nodes located at key points on the shape. For plane stress, two independent physical in-plane displacements can occur at any point. Thus each node will possess two dof. Shape functions must be determined for the element which make possible the computation of the matrices $\mathbf{B}$, $\mathbf{F}_b$, and $\mathbf{F}_k$ of the element stiffness equations 6.22. Once the element stiffness equations are determined for the plane stress finite elements, the finite element method proceeds identically, as described earlier. That is, the global stiffness equations are assembled from the element stiffness equations, the boundary conditions are imposed, the global stiffness equations are solved for the nodal displacements, the edof–gdof correspondence is invoked, and internal forces and stresses are calculated.

8.2. CONSTANT-STRAIN TRIANGULAR ELEMENT (3-NODE, 6-DOF)

The simplest plane shape one might elect for a 2D finite element is a triangle. An obvious location of nodes on the element is one at each vertex. Almost any plane structural shape may be discretized with such triangular finite elements, though individual triangular finite elements may be different in size and shape. It is this ability to discretize a structural shape so readily that led to the early development of the 3-node triangular finite element. With only three nodes, the pertinent shape functions will be linear, as we shall see, leading to simplified mathematics for this element, an additional advantage. The linear shape functions, on the other hand, dictate a constant value of strain, and therefore stress, throughout the element. This latter circumstance is a disadvantage when stress and strain vary sharply throughout the structure. To achieve a reasonable analysis of such structures, a large number of 3-node triangular finite elements must be used to obtain an accurate representation of the stresses in the structure. Large numbers of elements mean large numbers of stiffness equations to develop and solve with the consequent large computational effort.

8.2.1. Shape and Displacement Functions

Figure 8.1 illustrates a 3-node, 6-dof triangular finite element referred to x–y reference axes, the global coordinate axes for the structure. For this finite element, local reference axes are not needed. The element stiffness equations may be developed and referred directly to the global axes, obviating the need for transformations among axes.

Since the in-plane displacements in the element are independent of one another, the displacement functions may be written as

$$u(x, \ y) = N_1(x, \ y)q_1 + N_2(x, \ y)q_3 + N_3(x, \ y)q_5, \tag{8.1a}$$

$$v(x, \ y) = N_1(x, \ y)q_2 + N_2(x, \ y)q_4 + N_3(x, \ y)q_6. \tag{8.1b}$$

Only three shape functions are needed. The two displacement functions differ only in the nodal displacements: those in the x direction (q_1, q_3, q_5) for the displacement function $u(x, \ y)$, which is also in the x direction, and those in the y direction (q_2, q_4, q_6) for the displacement function $v(x, \ y)$, which is in the y direction.

The shape functions are polynomials in x and y. Each relates to three dof, one dof at each node, so that each will be a three-term polynomial. They may be written as

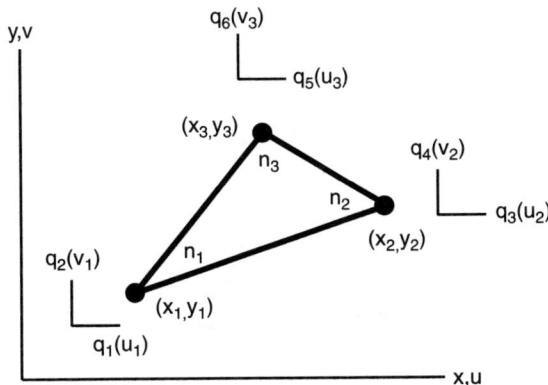

Figure 8.1 Constant-strain triangle

$$N_1(x, \ y) = C_{11} + C_{21}x + C_{31}y, \tag{8.2a}$$

$$N_2(x, \ y) = C_{12} + C_{22}x + C_{32}y, \tag{8.2b}$$

$$N_3(x, \ y) = C_{13} + C_{23}x + C_{33}y. \tag{8.2c}$$

In matrix form, they are

$$\begin{Bmatrix} N_1 \\ N_2 \\ N_3 \end{Bmatrix} = \begin{bmatrix} C_{11} & C_{21} & C_{31} \\ C_{12} & C_{22} & C_{32} \\ C_{13} & C_{23} & C_{33} \end{bmatrix} \begin{Bmatrix} 1 \\ x \\ y \end{Bmatrix}, \tag{8.2d}$$

$$\{\mathbf{N}\} = [\mathbf{C}^{\mathrm{T}}]\{\mathbf{g}\}. \tag{8.2e}$$

The matrix $\mathbf{C}$ is a matrix of shape function constants and the vector $\mathbf{g}$ is a column vector of generic polynomial terms similar to what was shown in equations 6.28. In this instance, however, they are for the 3-node 2D triangular plane stress finite element rather than for a general 1D finite element.

Though these shape functions describe in-plane displacements, they provide insight on plotting them as surfaces over the finite element. The ordinates from the element to the surface represent the values of the in-plane displacements given by the shape functions. Since these shape functions are linear, each may be plotted over the finite element as a plane whose ordinate equals one at its node and zero at the other two nodes, as shown in Figure 8.2.

The displacement functions are linear combinations of the shape functions; thus they also are linear functions of x and y. They may each be plotted as planes over the triangular finite element. Ordinates to the planes at each node will equal the corresponding nodal displacement q_i or q_{Gi}.

Figure 8.3 illustrates an element displacement function $u(x, \ y)$ or $v(x, \ y)$ plotted for three adjacent triangular finite elements (e_1, e_2, and e_3) of a plane stress structure. The figure shows that displacements must be continuous, not only at nodes, but also along the edges between nodes. It also shows that the slopes of the planes are constant for each element in either the global x or y direction. The slopes of these

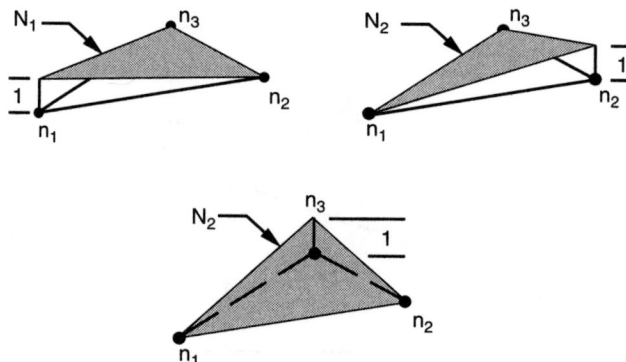

Figure 8.2 Shape functions

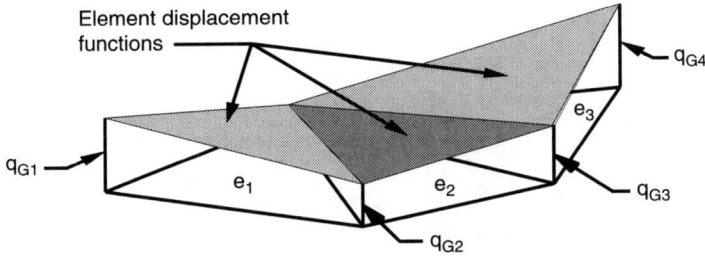

Figure 8.3 Displacement functions

planes in the global x and y directions relate directly to the strains in the elements. Hence each finite element possesses constant strain throughout the element and therefore constant stress. The strain and stress values will be different on different finite elements and of course discontinuous at the element edges.

The shape function constants (the entries in matrix **C**) may be determined from equations developed earlier:

$$[\mathbf{h}][\mathbf{C}] = [\mathbf{I}] \quad \text{and} \quad [\mathbf{C}] = [\mathbf{h}^{-1}]. \tag{6.29}$$

The matrix **h** was shown earlier to be an array of constants obtained by evaluating the generic polynomial terms of each shape function, by row, at each corresponding node of the finite element. Thus,

$$\begin{bmatrix} c_{11} & c_{12} & c_{13} \\ c_{21} & c_{22} & c_{23} \\ c_{31} & c_{32} & c_{33} \end{bmatrix} = \begin{bmatrix} 1 & x_1 & y_1 \\ 1 & x_2 & y_2 \\ 1 & x_3 & y_3 \end{bmatrix}^{-1}. \tag{8.3a}$$

The matrix **h** is inverted using the adjoint method, so that

$$[\mathbf{C}] = \frac{\begin{bmatrix} x_2 y_3 - x_3 y_2 & -(x_1 y_3 - x_3 y_1) & x_1 y_2 - x_2 y_1 \\ -(y_3 - y_2) & y_3 - y_1 & -(y_2 - y_1) \\ x_3 - x_2 & -(x_3 - x_1) & x_2 - x_1 \end{bmatrix}}{\begin{vmatrix} 1 & x_1 & y_1 \\ 1 & x_2 & y_2 \\ 1 & x_3 & y_3 \end{vmatrix}}. \tag{8.3b}$$

Equation 8.3b shows that the shape function constants are determined completely by the triangular finite element's global x–y coordinates of nodes: $(x_1,\ y_1)$, $(x_2,\ y_2)$, and $(x_3,\ y_3)$. The denominator of equation 8.3b, det **h**, may be interpreted as twice the area of the triangular finite element, or $2A_e$, as shown below. The node-numbering sequence on the element must be in a counterclockwise direction. Otherwise det $\mathbf{h} = -2A_e$.

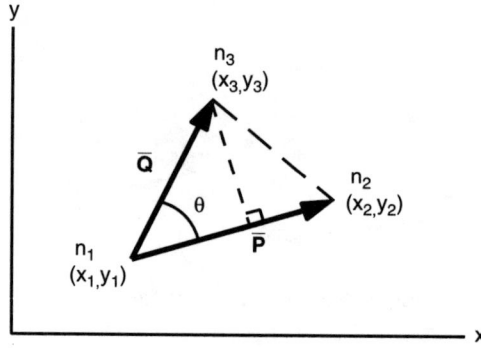

Figure 8.4 Area of triangle

To obtain the expression for twice the area of the triangle, let two vectors $\bar{\mathbf{P}}$ and $\bar{\mathbf{Q}}$ represent two sides of a triangle in x–y space, as shown in Figure 8.4. The magnitude of the cross product of the two vectors is defined as the product of the lengths of the two vectors and the sine of the angle between them:

$$|\bar{\mathbf{P}} \times \bar{\mathbf{Q}}| = PQ \ \sin \ \theta.$$

If P is regarded as the base of the triangle, $Q \sin \theta$ is its altitude. Clearly the area of the triangle is

$$A_e = \tfrac{1}{2}P(Q \ \sin \ \theta) = \tfrac{1}{2}|\bar{\mathbf{P}} \times \bar{\mathbf{Q}}|.$$

The cross product may also be developed in terms of the rectangular components of the vectors,

$$\bar{\mathbf{P}} \times \bar{\mathbf{Q}} = (P_x\mathbf{i} + P_y\mathbf{j}) \times (Q_x\mathbf{i} + Q_y\mathbf{j}),$$

where $\mathbf{i}$, $\mathbf{j}$, and $\mathbf{k}$ are unit vectors in the x, y, and z directions, respectively, and P_x, P_y, Q_x, and Q_y are the x and y components of the vectors $\bar{\mathbf{P}}$ and $\bar{\mathbf{Q}}$ respectively. Expanding this expression,

$$\bar{\mathbf{P}} \times \bar{\mathbf{Q}} = P_xQ_x\mathbf{i} \times \mathbf{i} + P_xQ_y\mathbf{i} \times \mathbf{j} + P_yQ_x\mathbf{j} \times \mathbf{i} + P_yQ_y\mathbf{i} \times \mathbf{j}.$$

Applying the cross-product definition to the unit vectors,

$$\mathbf{i} \times \mathbf{i} = \mathbf{j} \times \mathbf{j} = 0 \quad \text{and} \quad \mathbf{i} \times \mathbf{j} = -\mathbf{j} \times \mathbf{i} = \mathbf{k}.$$

Thus

$$\bar{\mathbf{P}} \times \bar{\mathbf{Q}} = (P_XQ_y - Q_xP_y)\mathbf{k}.$$

The magnitude of the cross product is

$$2A_e = P_xQ_y - Q_xP_y.$$

Observing Figure 8.4, the x and y components of the two vectors may be expressed in terms of the coordinates of the vertices of the triangle:

$$2A_e = (x_2 - x_1)(y_3 - y_1) - (x_3 - x_1)(y_2 - y_1)$$
$$= x_2y_3 - x_3y_2 - x_1y_3 + x_3y_1 + x_1y_2 - x_2y_1.$$

The right-hand side of this equation may be conveniently expressed as a 3×3 determinant, so that

$$2A_e = \begin{vmatrix} 1 & x_1 & y_1 \\ 1 & x_2 & y_2 \\ 1 & x_3 & y_3 \end{vmatrix} = \det \mathbf{h}. \tag{8.4}$$

Applying equations 8.3b and 8.4 to equation 8.2d, the three shape functions may be written in terms of the triangular element's global nodal coordinates and its area:

$$N_1(x, y) = \frac{1}{2A_e}[(x_2y_3 - x_3y_2) - (y_3 - y_2)x + (x_3 - x_2)y], \tag{8.5a}$$

$$N_2(x, y) = \frac{1}{2A_e}[(x_3y_1 - x_1y_3) - (y_1 - y_3)x + (x_1 - x_3)y], \tag{8.5b}$$

$$N_3(x, y) = \frac{1}{2A_e}[(x_1y_2 - x_2y_1) - (y_2 - y_1)x + (x_2 - x_1)y]. \tag{8.5c}$$

The properties of shape functions (illustrated in Figure 8.2) may be verified from equations 8.5 by substituting for x and y the coordinates of the nodes (x_1, y_1), (x_2, y_2), or (x_3, y_3).

8.2.2. Element Stiffness Matrix

To obtain the stiffness matrix $\mathbf{K}$ of the triangular finite element, equation 6.22d is used:

$$[\mathbf{K}] = \left[\int_v [\mathbf{B}^T][\mathbf{E}][\mathbf{B}] \, dv\right]. \tag{6.22d}$$

The matrix $\mathbf{B}$ is an array of derivatives of the shape functions, differentiated in accordance with the strain displacement relationships. For plane stress,

$$\epsilon_x = \frac{\partial u}{\partial x}, \qquad \epsilon_y = \frac{\partial v}{\partial y}, \qquad \gamma_{xy} = \frac{\partial u}{\partial y} + \frac{\partial v}{\partial x}; \tag{8.6a}$$

$$\left\{\begin{array}{c} \epsilon_x \\ \epsilon_y \\ \gamma_{xy} \end{array}\right\} = \begin{bmatrix} \frac{\partial}{\partial x} & 0 \\ 0 & \frac{\partial}{\partial y} \\ \frac{\partial}{\partial y} & \frac{\partial}{\partial x} \end{bmatrix} \left\{\begin{array}{c} u \\ v \end{array}\right\}; \tag{8.6b}$$

$$\{\epsilon\} = [\mathbf{d}]\{\mathbf{u}\}. \tag{8.6c}$$

Equations 8.1 in matrix form are

$$
\left\{ \begin{array}{c} u \\ v \end{array} \right\} = \left[\begin{array}{ccc|ccc} N_1 & 0 & N_2 & 0 & N_3 & 0 \\ 0 & N_1 & 0 & N_2 & 0 & N_3 \end{array} \right] \left\{ \begin{array}{c} q_1 \\ q_2 \\ q_3 \\ q_4 \\ q_5 \\ q_6 \end{array} \right\},
\tag{8.7a}
$$

$$
\{\mathbf{u}\} = [\mathbf{N}]\{\mathbf{q}\}.
\tag{8.7b}
$$

Substituting equation 8.7b into equation 8.6c,

$$
\{\epsilon\} = [\mathbf{d}][\mathbf{N}]\{\mathbf{q}\},
\tag{8.8a}
$$

$$
\{\epsilon\} = [\mathbf{B}]\{\mathbf{q}\}.
\tag{8.8b}
$$

The matrix **B** for plane stress using the 3-node triangular finite element therefore is

$$
[\mathbf{B}] = [\mathbf{d}][\mathbf{N}]
\tag{8.9a}
$$

$$
= \left[\begin{array}{cc} \frac{\partial}{\partial x} & 0 \\ 0 & \frac{\partial}{\partial y} \\ \frac{\partial}{\partial y} & \frac{\partial}{\partial x} \end{array} \right] \left[\begin{array}{ccc|ccc} N_1 & 0 & N_2 & 0 & N_3 & 0 \\ 0 & N_1 & 0 & N_2 & 0 & N_3 \end{array} \right]
\tag{8.9b}
$$

$$
= \left[\begin{array}{ccc|ccc|ccc} \frac{\partial N_1}{\partial x} & 0 & \frac{\partial N_2}{\partial x} & 0 & \frac{\partial N_3}{\partial x} & 0 \\ 0 & \frac{\partial N_1}{\partial y} & 0 & \frac{\partial N_2}{\partial y} & 0 & \frac{\partial N_3}{\partial y} \\ \frac{\partial N_1}{\partial y} & \frac{\partial N_1}{\partial x} & \frac{\partial N_2}{\partial y} & \frac{\partial N_2}{\partial x} & \frac{\partial N_3}{\partial y} & \frac{\partial N_3}{\partial x} \end{array} \right].
\tag{8.9c}
$$

Performing the differentiations indicated by equation 8.9c on equations 8.5 and substituting the result into equation 8.9c,

$$
[\mathbf{B}] = \frac{1}{2A_e} \left[\begin{array}{ccc|ccc|ccc} -(y_3 - y_2) & 0 & -(y_1 - y_3) & 0 & -(y_2 - y_1) & 0 \\ 0 & x_3 - x_2 & 0 & x_1 - x_3 & 0 & x_2 - x_1 \\ x_3 - x_2 & -(y_3 - y_2) & x_1 - x_3 & -(y_1 - y_3) & x_2 - x_1 & -(y_2 - y_1) \end{array} \right].
\tag{8.9d}
$$

For the plane stress 3-node triangular finite element, the matrix **B** is a matrix of constants, determined entirely by the global x–y coordinates of the nodes on the finite element. Since the matrix **E** is also a matrix of constants, the integrand of equation 6.22d is constant and may be moved outside of the integral. The remaining integral is the volume of the finite element or its area times its thickness, $A_e t$. The stiffness matrix for this finite element may be computed directly from matrices whose entries are expressions of the element's nodal global coordinates, its elastic constants, and its area and thickness.

The stiffness matrix is given as

$$[\mathbf{K}] = [\mathbf{B}^T][\mathbf{E}][\mathbf{B}]A_e t. \tag{8.10a}$$

For plane stress, equation 8.10a becomes

$$[\mathbf{K}] = \frac{Et}{4A_e(1-v^2)}
\begin{bmatrix}
-(y_3 - y_2) & 0 & x_3 - x_2 \\
0 & x_3 - x_2 & -(y_3 - y_2) \\
\hline
-(y_1 - y_3) & 0 & x_1 - x_3 \\
0 & x_1 - x_3 & -(y_1 - y_3) \\
\hline
-(y_2 - y_1) & 0 & x_2 - x_1 \\
0 & x_2 - x_1 & -(y_2 - y_1)
\end{bmatrix}
\begin{bmatrix}
1 & v & 0 \\
v & 1 & 0 \\
0 & 0 & \frac{1-v}{2}
\end{bmatrix}$$

$$\times
\begin{bmatrix}
-(y_3 - y_2) & 0 & \mid & -(y_1 - y_3) & 0 & \mid & -(y_2 - y_1) & 0 \\
0 & x_3 - x_2 & \mid & 0 & x_1 - x_3 & \mid & 0 & x_2 - x_1 \\
x_3 - x_2 & -(y_3 - y_2) & \mid & x_1 - x_3 & -(y_1 - y_3) & \mid & x_2 - x_1 & -(y_2 - y_1)
\end{bmatrix}. \tag{8.10b}$$

As an illustration of the use of equation 8.10b, the stiffness matrix for the plane stress finite element shown in Figure 8.5 will be computed. The global coordinates shown are in inches, the material is isotropic steel for which Young's modulus is 30,000 ksi and Poisson's ratio is 0.25. The element is $\frac{9}{16}$ in. thick.

The area of the element is

$$A_e = \frac{1}{2} \det[\mathbf{h}] = \frac{1}{2}
\begin{vmatrix}
1 & 3 & 8 \\
1 & 9 & 5 \\
1 & 7 & 11
\end{vmatrix} = 15.000 \text{ in}^2.$$

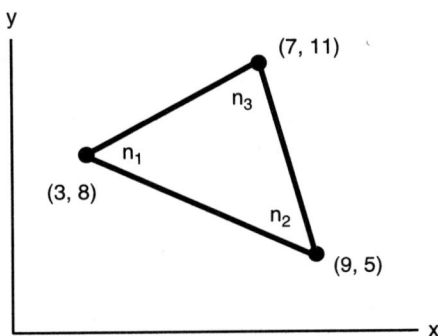

Figure 8.5 Example element

The scalar multiplier of equation 8.10b is

$$S = \frac{Et}{4A_e(1 - v^2)} = \frac{(30,000)(9/16)}{4(15)(1 - 0.25^2)} = 300.00\,\text{kips/in}^3.$$

The matrices **B** and **E** without their scalar multipliers are

$$[\mathbf{B}] = \begin{bmatrix} -6 & 0 & 3 & 0 & 3 & 0 \\ 0 & -2 & 0 & -4 & 0 & 6 \\ -2 & -6 & -4 & 3 & 6 & 3 \end{bmatrix} \text{in.,}$$

$$[\mathbf{E}] = \begin{bmatrix} 1 & 0.25 & 0 \\ 0.25 & 1 & 0 \\ 0 & 0 & 0.375 \end{bmatrix}.$$

Thus

$$[\mathbf{K}] = 300.00 \begin{bmatrix} -6 & 0 & -2 \\ 0 & -2 & -6 \\ 3 & 0 & -4 \\ 0 & -4 & 3 \\ 3 & 0 & 6 \\ 0 & 6 & 3 \end{bmatrix} \begin{bmatrix} 1 & 0.25 & 0 \\ 0.25 & 1 & 0 \\ 0 & 0 & 0.375 \end{bmatrix} \begin{bmatrix} -6 & 0 & 3 & 0 & 3 & 0 \\ 0 & -2 & 0 & -4 & 0 & 6 \\ -2 & -6 & -4 & 3 & 6 & 3 \end{bmatrix}$$

$$= \begin{bmatrix} 11250.0 & 2250.0 & -4500.0 & 1125.0 & -6750.0 & -3375.0 \\ 2250.0 & 5250.0 & 2250.0 & 375.0 & -4500.0 & -5625.0 \\ -4500.0 & 2250.0 & 4500.0 & -2250.0 & 0 & 0 \\ 1125.0 & 375.0 & -2250.0 & 5812.5 & 1125.0 & -6187.5 \\ -6750.0 & -4500.0 & 0 & 1125.0 & 6750.0 & 3375.0 \\ -3375.0 & -5625.0 & 0 & -6187.5 & 3375.0 & 11812.5 \end{bmatrix} \text{kips/in.}$$

Note that it is usually simpler to write the **B** and **E** matrices numerically with their scalar multipliers and then perform the numerical manipulations indicated by equation 8.10b than it would be to expand equation 8.10b in terms of the global coordinates x_i and y_i and enter the values for the coordinates.

8.2.3. Loads between Nodes

In a 2D plane stress structure, externally applied loads are in-plane and may be concentrated at a point, distributed throughout portions of the plane structure, or distributed along a line on the structure.

An in-plane concentrated load that occurs at a node is treated as described earlier by inserting the values of its global x and y components directly into the global load vector positions corresponding to the gdof of the node.

A concentrated load applied within the element is treated in accordance with equations 6.22f, developed earlier:

$$\{\mathbf{F}_k\} = [\mathbf{N}_k^T]\{\mathbf{P}_k\}. \tag{6.22f}$$

The load P_k is at global coordinates $(x_k,\ y_k)$ and the vector $\mathbf{P}_k$ comprises the x and y components of load P_k. Matrix $\mathbf{N}_k$ contains the values of the shape functions at coordinates $(x_k,\ y_k)$, arrayed as shown in equation 8.7a but transposed. The vector $\mathbf{F}_k$ is the vector of loads at each edof which produce the same nodal displacements that the load P_k produces, as computed by equations 6.22f. Equations 6.22f applied for plane stress are

$$\{\mathbf{F}_k\} = \begin{Bmatrix} F_{1k} \\ F_{2k} \\ F_{3k} \\ F_{4k} \\ F_{5k} \\ F_{6k} \end{Bmatrix} = \begin{bmatrix} N_{1k} & 0 \\ 0 & N_{1k} \\ N_{2k} & 0 \\ 0 & N_{2k} \\ N_{3k} & 0 \\ 0 & N_{3k} \end{bmatrix} \begin{Bmatrix} P_{kx} \\ P_{ky} \end{Bmatrix} = \begin{Bmatrix} N_{1k}P_{kx} \\ N_{1k}P_{ky} \\ N_{2k}P_{kx} \\ N_{2k}P_{ky} \\ N_{3k}P_{kx} \\ N_{3k}P_{ky} \end{Bmatrix}. \tag{8.11}$$

The use of equation 8.11 will be illustrated by computing the vector $\mathbf{F}_k$ for a concentrated in-plane load applied to the plane stress finite element illustrated in Figures 8.5 and 8.6. The load is applied at global coordinates (5, 9). Its magnitude is 50 kips and its line of action slopes up and to the right at 3 vertical on 4 horizontal. The area of the finite element was previously computed as $15\,\text{in}^2$.

The element's shape functions are computed from equations 8.5:

$$N_1(x,\ y) = \tfrac{1}{30}[(9 \times 11 - 7 \times 5) - (11 - 5)x + (7 - 9)y] = \tfrac{1}{15}(32 - 3x - y),$$

$$N_2(x,\ y) = \tfrac{1}{30}[(7 \times 8 - 3 \times 11) - (8 - 11)x + (3 - 7)y] = \tfrac{1}{30}(23 + 3x - 4y),$$

$$N_3(x,\ y) = \tfrac{1}{30}[(3 \times 5 - 9 \times 8) - (5 - 8)x + (9 - 3)y] = \tfrac{1}{10}(-19 + x + 2y).$$

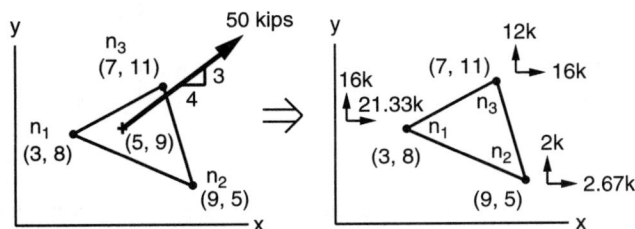

Figure 8.6 Example equivalent nodal loads

The shape functions evaluated at the coordinates (5, 9) are

$$N_1(5,\ 9) = \tfrac{8}{15}, \qquad N_2(5,\ 9) = \tfrac{1}{15}, \qquad N_3(5,\ 9) = \tfrac{6}{15}.$$

The x and y components of the 50-kip load are

$$P_{kx} = +40\ \text{kips} \quad \text{and} \quad P_{ky} = +30\ \text{kips}.$$

Thus the equivalent nodal loads are

$$\{\mathbf{F}_k\} = \begin{Bmatrix} F_{1k} \\ F_{2k} \\ F_{3k} \\ F_{4k} \\ F_{5k} \\ F_{6k} \end{Bmatrix} = \frac{1}{15} \begin{Bmatrix} 8 \times 40 \\ 8 \times 30 \\ 1 \times 40 \\ 1 \times 30 \\ 6 \times 40 \\ 6 \times 30 \end{Bmatrix} = \begin{Bmatrix} 21.333 \\ 16.000 \\ 2.667 \\ 2.000 \\ 16.000 \\ 12.000 \end{Bmatrix} \text{kips}.$$

Figure 8.6 shows the 50-kip in-plane concentrated load and its in-plane equivalent nodal loads.

An in-plane load distributed throughout the triangular finite element is treated with equations 6.22e, developed earlier:

$$\{\mathbf{F}_b\} = \int_v [\mathbf{N}^T(\mathbf{x},\ \mathbf{y})]\{\mathbf{b}(\mathbf{x},\ \mathbf{y})\}\ dV. \tag{6.22e}$$

The distributed body force vector $\mathbf{b}(\mathbf{x},\ \mathbf{y})$ is force per unit of volume and contains two entries, the in-plane x and y components $b_x(x,\ y)$ and $b_y(x,\ y)$. The matrix $\mathbf{N}(\mathbf{x},\ \mathbf{y})$ is a matrix of shape functions arrayed as shown in equation 8.7a and transposed. Vector $\mathbf{F}_b$ contains the loads at each edof which produce the same nodal displacements that the body force $\mathbf{b}(\mathbf{x},\ \mathbf{y})$ produces. Equations 6.22e applied for plane stress triangular finite elements of thickness t are

$$\{\mathbf{F}_b\} = \begin{Bmatrix} F_{1b} \\ F_{2b} \\ F_{3b} \\ F_{4b} \\ F_{5b} \\ F_{6b} \end{Bmatrix} = \int_v \begin{bmatrix} N_1(x,\ y) & 0 \\ 0 & N_1(x,\ y) \\ N_2(x,\ y) & 0 \\ 0 & N_2(x,\ y) \\ N_3(x,\ y) & 0 \\ 0 & N_3(x,\ y) \end{bmatrix} \begin{Bmatrix} b_x(x,\ y) \\ b_y(x,\ y) \end{Bmatrix} dV = \begin{Bmatrix} \int_A N_1(x,\ y)b_x(x,\ y)t\ dA \\ \int_A N_1(x,\ y)b_y(x,\ y)t\ dA \\ \int_A N_2(x,\ y)b_x(x,\ y)t\ dA \\ \int_A N_2(x,\ y)b_y(x,\ y)t\ dA \\ \int_A N_3(x,\ y)b_x(x,\ y)t\ dA \\ \int_A N_3(x,\ y)b_y(x,\ y)t\ dA \end{Bmatrix}.$$
$$\tag{8.12a}$$

In a plane stress condition, the thickness t of the element is constant and may be placed outside of each integral. If the body force is also constant throughout the element, the b_x or b_y of each integrand in equations 8.12a may be placed outside its integral. The remaining integral for each entry then is the integral of the shape function over the triangular area of the finite element.

Referring to Figure 8.2, the integral of a shape function over the element area is the volume of the pyramid formed by the plane which represents the shape function and the element's area. The altitude of the pyramid is one and its base is A_e, so the volume of each pyramid is $\frac{1}{3}(1)(A_e)$. Thus the equivalent nodal loads for a constant body force on a plane stress triangular finite element are

$$
\{\mathbf{F}_b\} = \begin{Bmatrix} F_{1b} \\ F_{2b} \\ F_{3b} \\ F_{4b} \\ F_{5b} \\ F_{6b} \end{Bmatrix} = \frac{A_e t}{3} \begin{Bmatrix} b_x \\ b_y \\ b_x \\ b_y \\ b_x \\ b_y \end{Bmatrix}. \tag{8.12b}
$$

When the body force is variable, each entry of $\mathbf{F}_b$ is the integral over the element's triangular area A_e of the product of the element thickness t, an x or y component of the body force $b_x(x, y)$ or $b_y(x, y)$, and a shape function $N_i(x, y)$.

Figure 8.7 shows a distributed in-plane load applied along the edge of a triangular finite element of a plane stress structure. This load is part of a distributed line load on the structure which the analyst ensures will occur on the edge of one or more of the finite elements. Its magnitude may vary along the length of the edge, but it must be constant across the thickness t of the element. The load is expressed as a function of a line variable L along the edge where it is applied. If this load $b(L)$ is force per unit of area, the applied line load is $b(L)t$ and is force per unit of length.

Equations 6.22e apply; however, the integrals must be along the edge of the element where the load is actually applied. This is because the line load does the virtual work there, which is the basis for equations 6.22e. For the same reason, the shape functions $N_i(x, y)$, which express the displacements participating in the virtual work, apply only along the edge of the element where the load is applied.

Taking the variable of integration as the line variable L and expressing the shape functions as functions of L, equations 6.22e may be written for the distributed edge load shown in Figure 8.7 as

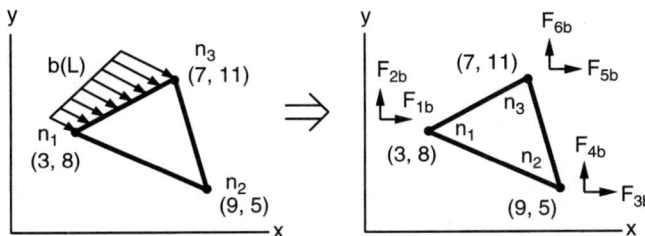

Figure 8.7 Equivalent nodal loads

$$\{\mathbf{F}_b\} = \begin{Bmatrix} F_{1b} \\ F_{2b} \\ F_{3b} \\ F_{4b} \\ F_{5b} \\ F_{6b} \end{Bmatrix} = \int_0^{L_{13}} \begin{bmatrix} N_1(L) & 0 \\ 0 & N_1(L) \\ 0 & 0 \\ 0 & 0 \\ N_3(L) & 0 \\ 0 & N_3(L) \end{bmatrix} \begin{Bmatrix} b_x(L)t \\ b_y(L)t \end{Bmatrix} dL = \begin{Bmatrix} \int_0^{L_{13}} N_1(L)b_x(L)t\, dL \\ \int_0^{L_{13}} N_1(L)b_y(L)t\, dL \\ 0 \\ 0 \\ \int_0^{L_{13}} N_3(L)b_x(L)t\, dL \\ \int_0^{L_{13}} N_3(L)b_y(L)t\, dL \end{Bmatrix}.$$

$$(8.12c)$$

Equations 8.12c are for a line load applied along edge 1–3 of the element where shape function N_2 equals zero. For a load along edge 1–2, shape function N_3 would be zero along edge 1–2 and N_1 and N_2 would be nonzero. The corresponding entries for N_3 in equation 8.12c would be zero. Similarly for a load along edge 2–3, shape function N_1 would be zero, N_2 and N_3 would be nonzero, and the corresponding entries for N_1 in equation 8.12c would be zero.

Again the thickness of the element is constant and may be placed outside of each integral. If the line load $b(L)$ is also constant along the edge of the element, the b_x or b_y portion of each integrand in equations 8.12c may be placed outside of its integral. The remaining integral for each entry then is simply the integral of the shape function as it varies along the edge of the finite element where the line load is applied.

Referring to Figure 8.2, it is apparent that each shape function varies linearly along the edges of the triangular finite element, from zero at one end of the edge to one at the other end. Its integral therefore is the area of a triangle whose base is the length of the edge and whose altitude is one. Thus the equivalent nodal loads for a constant line load on the edge between nodes 1 and 3 of a plane stress triangular finite element are

$$\{\mathbf{F}_b\} = \begin{Bmatrix} F_{1b} \\ F_{2b} \\ F_{3b} \\ F_{4b} \\ F_{5b} \\ F_{6b} \end{Bmatrix} = \frac{L_{13}t}{2} \begin{Bmatrix} b_x \\ b_y \\ 0 \\ 0 \\ b_x \\ b_y \end{Bmatrix}.$$

$$(8.12d)$$

The term L_{13} is the length of edge 1–3. If the line load were along edge 1–2, the length of edge 1–2 would be used and the zeros in the vector $\mathbf{F}_b$ would appear at the edof positions of node 3 (edof 5 and 6), rather than at node 2. Similarly, if edge 2–3 were loaded, its length would be used and the zeros would appear at the edof positions of node 1.

Clearly if the line load is not constant along the edge of the triangular finite element, equations 8.12d do not apply. Instead, the integrals of equations 8.12c must be evaluated, with integrands which are the product of the element's thickness t, an x or y component of the variable line load $b_x(L)$ or $b_y(L)$, and a shape function $N_i(L)$.

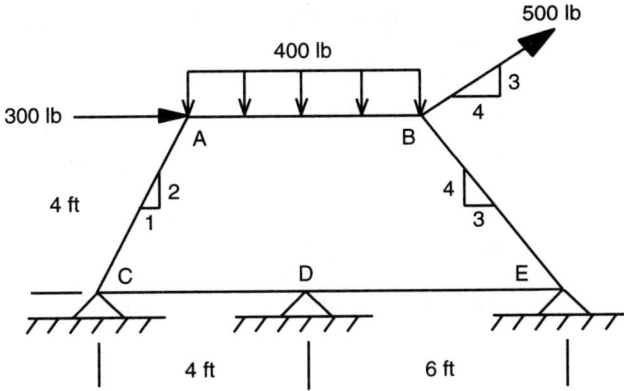

Figure 8.8 Plane stress plate

8.2.4. Plane Stress Plate Example Analysis

To illustrate the use of the 3-node, 6-dof plane stress triangular finite element, the support reactions will be determined for the statically indeterminate thin plate structure shown in Figure 8.8. The structure has three pin supports: two in-plane concentrated loads, a distributed line load, and an in-plane body force (its weight due to gravity). The modulus of elasticity $E = 30,000$ ksi, Poisson's ratio $v = \frac{1}{4}$, the plate thickness $t = 1.00$ in, and the unit weight for the plate material $\gamma = 500$ lb/ft^3.

The discretized structure is shown in Figure 8.9 with 3 elements (a, b, and c), 5 nodes, and 10 gdof. Element nodes are numbered inside each element from 1 to 3. Global x–y axes and the global coordinates of nodes are shown also.

The area of element a is obtained from equation 8.4:

$$A_a = \frac{1}{2} \begin{vmatrix} 1 & 0 & 0 \\ 1 & 4 & 0 \\ 1 & 2 & 4 \end{vmatrix} = 8.00 \text{ ft}^2.$$

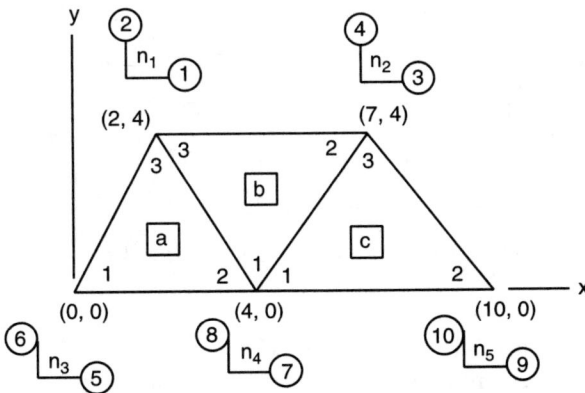

Figure 8.9 Discretized plate

The strain–displacement matrix $\mathbf{B}_a$ for element a is obtained from equation 8.9d:

$$[\mathbf{B}_a] = \frac{1}{16} \begin{bmatrix} -4 & 0 & 4 & 0 & 0 & 0 \\ 0 & -2 & 0 & -2 & 0 & 4 \\ -2 & -4 & -2 & 4 & 4 & 0 \end{bmatrix} \quad \text{rad/ft.}$$

The scalar multiplier of equation 8.10b is evaluated for element a as

$$S = \frac{30 \times 10^6 \times 144 \times (1/12)}{4 \times 8(1 - 0.25^2)} = 12 \times 10^6 \, \text{lb/ft}^3.$$

The stiffness matrix $\mathbf{K}_a$ for the element may be written from equation 8.10b as

$$[\mathbf{K}_a] = 12 \times 10^6 \begin{bmatrix} -4 & 0 & -2 \\ 0 & -2 & -4 \\ 4 & 0 & -2 \\ 0 & -2 & 4 \\ 0 & 0 & 4 \\ 0 & 4 & 0 \end{bmatrix} \begin{bmatrix} 1 & 0.25 & 0 \\ 0.25 & 1 & 0 \\ 0 & 0 & 0.375 \end{bmatrix}$$

$$\times \begin{bmatrix} -4 & 0 & 4 & 0 & 0 & 0 \\ 0 & -2 & 0 & -2 & 0 & 4 \\ -2 & -4 & -2 & 4 & 4 & 0 \end{bmatrix}.$$

$$= \begin{array}{c} \quad \\ ⑤ \\ ⑥ \\ ⑦ \\ ⑧ \\ ① \\ ② \end{array} \begin{array}{cccccc} ⑤ & ⑥ & ⑦ & ⑧ & ① & ② \end{array} \\ \begin{bmatrix} 210 & 60 & -174 & -12 & -36 & -48 \\ 60 & 120 & 12 & -24 & -72 & -96 \\ -174 & 12 & 210 & -60 & -36 & 48 \\ -12 & -24 & -60 & 120 & 72 & -96 \\ -36 & -72 & -36 & 72 & 72 & 0 \\ -48 & -96 & 48 & -96 & 0 & 192 \end{bmatrix} \times 10^6 \, \text{lb/ft.}$$

The circled numbers on matrix $\mathbf{K}_a$ refer to corresponding gdof.

The weight of the element must be discretized as equivalent nodal loads along the lines of action of the pertinent dof. Applying equation 8.12b,

$$\{\mathbf{F}_{ba}\} = \frac{8 \times (1/12)}{3} \begin{Bmatrix} 0 \\ -500 \\ 0 \\ -500 \\ 0 \\ -500 \end{Bmatrix} = \begin{matrix} ⑤ \\ ⑥ \\ ⑦ \\ ⑧ \\ ① \\ ② \end{matrix} \begin{Bmatrix} 0 \\ -111.11 \\ 0 \\ -111.11 \\ 0 \\ -111.11 \end{Bmatrix} \text{ lb.}$$

Load and stiffness matrices for element b are developed similarly. However, element b has the added contribution of the distributed line load across its top from gdof 1–4. Its equivalent nodal forces are added to those due to the element's weight.

Thus the stiffness matrix and equivalent load vector for element b are written as

$$[\mathbf{K}_b] = \begin{matrix} ⑦ \\ ⑧ \\ ③ \\ ④ \\ ① \\ ② \end{matrix} \begin{bmatrix} \overset{⑦}{90} & \overset{⑧}{0} & \overset{③}{-36} & \overset{④}{-72} & \overset{①}{-54} & \overset{②}{72} \\ 0 & 240 & -48 & -96 & 48 & -144 \\ -36 & -48 & 168 & 48 & -132 & 0 \\ -72 & -96 & 48 & 96 & 24 & 0 \\ -54 & 48 & -132 & 24 & 186 & -72 \\ 72 & -144 & 0 & 0 & -72 & 144 \end{bmatrix} \times 10^6 \text{ lb/ft.}$$

$$\{\mathbf{F}_{bb}\} = \begin{Bmatrix} 0 \\ -138.89 \\ 0 \\ -138.89 \\ 0 \\ -138.89 \end{Bmatrix} + \begin{Bmatrix} 0 \\ 0 \\ 0 \\ -1000 \\ 0 \\ -1000 \end{Bmatrix} = \begin{matrix} ⑦ \\ ⑧ \\ ③ \\ ④ \\ ① \\ ② \end{matrix} \begin{Bmatrix} 0 \\ -138.89 \\ 0 \\ -1138.89 \\ 0 \\ -1138.89 \end{Bmatrix} \text{ lb}$$

The stiffness matrix and load vector for element c are developed in the same manner and are

$$[\mathbf{K}_c] = \begin{matrix} ⑦ \\ ⑧ \\ ⑨ \\ ⑩ \\ ③ \\ ④ \end{matrix} \begin{bmatrix} \overset{⑦}{155} & \overset{⑧}{60} & \overset{⑨}{-101} & \overset{⑩}{-12} & \overset{③}{-54} & \overset{④}{-48} \\ 60 & 120 & 12 & 24 & -72 & -144 \\ -101 & 12 & 155 & -60 & -54 & 48 \\ -12 & 24 & -60 & 120 & 72 & -144 \\ -54 & -72 & -54 & 72 & 108 & 0 \\ -48 & -144 & 48 & -144 & 0 & 288 \end{bmatrix} \times 10^6 \text{ lb/ft}, \quad \{\mathbf{F}_{cc}\} = \begin{matrix} ⑦ \\ ⑧ \\ ⑨ \\ ⑩ \\ ③ \\ ④ \end{matrix} \begin{Bmatrix} 0 \\ -166.67 \\ 0 \\ -166.67 \\ 0 \\ -166.67 \end{Bmatrix} \text{ lb.}$$

The global stiffness equations are assembled following the procedure described in Chapter 4, and the restraints at the supports (gdof 5–10) are imposed:

$$[\mathbf{K}_G]\{\mathbf{q}_G\} = \{\mathbf{F}_G\},$$

$$\left[\begin{array}{c|c} \mathbf{K}_{ff} & \mathbf{K}_{fs} \\ \hline \mathbf{K}_{sf} & \mathbf{K}_{ss} \end{array}\right] \left\{\begin{array}{c} \mathbf{q}_f \\ \mathbf{q}_s \end{array}\right\} = \left\{\begin{array}{c} \mathbf{F}_f \\ \mathbf{R}_s \end{array}\right\},$$

	①	②	③	④	⑤	⑥	⑦	⑧	⑨	⑩				
①	258	−72	−132	24	−36	−72	−90	120	0	0		q_1		300.00
②	−72	336	0	0	−48	−96	120	−240	0	0		q_2		−1250.00
③	−132	0	276	48	0	0	−90	−120	−54	72		q_3		400.00
④	24	0	48	384	0	0	−120	−240	48	−144		q_4		−1005.56
⑤	−36	−48	0	0	210	60	−174	−12	0	0	$\times 10^6$	0	=	R_5
⑥	−72	−96	0	0	60	120	12	−24	0	0		0		$R_6 - 111.11$
⑦	−90	120	−90	−120	−174	12	455	0	−101	−12		0		R_7
⑧	120	−240	−120	−240	−12	−24	0	480	12	24		0		$R_8 - 416.67$
⑨	0	0	−54	48	0	0	−101	12	155	−60		0		R_9
⑩	0	0	72	−144	0	0	−12	24	−60	120		0		$R_{10} - 166.67$

Note that the concentrated loads for gdof 1, 3, and 4 are added directly into the global load vector at positions 1, 3, and 4.

The global stiffness equations are readily solved by any of the three methods discussed in Chapter 1. Using the submatrix method,

$$\{\mathbf{q}_f\} = [\mathbf{K}_{ff}^{-1}]\{\mathbf{F}_f\} = \begin{array}{c} ① \\ ② \\ ③ \\ ④ \end{array} \left\{\begin{array}{c} 2.0616 \\ -3.2785 \\ 2.9778 \\ 3.1197 \end{array}\right\} \times 10^{-6}\,\text{ft.}$$

The initial calculation for the support reactions is

$$\{\mathbf{R}_s\} = [\mathbf{K}_{sf}]\{\mathbf{q}_f\} = \begin{array}{c} ⑤ \\ ⑥ \\ ⑦ \\ ⑧ \\ ⑨ \\ ⑩ \end{array} \left\{\begin{array}{c} 83.15 \\ 166.30 \\ -472.60 \\ 1425.62 \\ -310.55 \\ 663.64 \end{array}\right\}\,\text{lb.}$$

The vector $\mathbf{R}_s$ must be corrected for equivalent nodal loads by subtracting from it the equivalent loads acting at the supports:

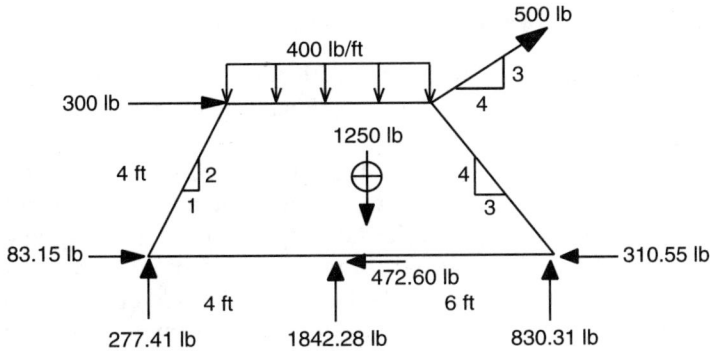

Figure 8.10 Plate's reactions

$$\{\mathbf{R}_{s,\text{corr}}\} = \begin{Bmatrix} 83.15 \\ 166.30 \\ -472.60 \\ 1425.62 \\ -310.55 \\ 633.64 \end{Bmatrix} - \begin{Bmatrix} 0.00 \\ -111.11 \\ 0.00 \\ -416.67 \\ 0.00 \\ -166.67 \end{Bmatrix} = \begin{matrix} ⑤ \\ ⑥ \\ ⑦ \\ ⑧ \\ ⑨ \\ ⑩ \end{matrix} \begin{Bmatrix} 83.15 \\ 277.41 \\ -472.60 \\ 1842.28 \\ -310.55 \\ 830.31 \end{Bmatrix} \text{ lb.}$$

Figure 8.10 shows the structure with its applied loads, weight, and support reactions. The weight of the structure is calculated as 1250 lb, the sum of the weights of the three elements. The center of gravity is calculated as 4.7778 ft to the right of the left pin support. It is easily shown that the loads, weight, and reactions satisfy the equations of equilibrium for the structure.

It is left as an exercise for the reader to complete the calculations for the matrices of elements b and c, the global matrices, the solution of the global stiffness equations, and the weight and center of gravity of the structure and to verify that the structure's loads, weight, and reactions satisfy the conditions for equilibrium.

8.3. NATURAL COORDINATES AND QUADRATURE FOR TRIANGULAR ELEMENTS

Three-node, 6-dof triangular finite elements for plane stress structures may be dealt with explicitly using equations 8.9d, 8.10, 8.11, 8.12b, and 8.12d provided that any body force $b(x, y)$ and any applied line load $b(L)$ are constant. If these applied loads are variable, then the integrals of equations 8.12a and 8.12c must be evaluated. Moreover, if higher order triangular finite elements are employed, such as a 6-node, 12-dof element, its stiffness matrix $\mathbf{K}$ and internal force vector $\mathbf{F}_b$ must be evaluated over the triangular area of the element employing integrals of functions of higher degree than those for the 3-node, 6-dof element.

Equations 8.9–8.12 would not apply, but comparable equations can be developed from equations 6.22.

8.3.1. Natural Coordinates for Triangular Elements

Natural or local coordinates locate points within the element by referring to key geometric attributes of the element. These coordinates lend themselves to convenient integration over the area of the element.

Figure 8.11 illustrates a point within a triangle located by measuring its perpendicular distance from each side of the triangle. The distance measured from the side opposite node 1 is designated s_1, from the side opposite node 2 it is s_2, and from the side opposite node 3 it is s_3. If the altitude of the triangle from the side opposite node 1 is h_1, from the side opposite node 2 is h_2, and from the side opposite node 3 is h_3, the following dimensionless local coordinates of the point may be defined:

$$\zeta_1 = \frac{s_1}{h_1}, \qquad \zeta_2 = \frac{s_2}{h_2}, \qquad \zeta_3 = \frac{s_3}{h_3}. \tag{8.13}$$

The value of each local coordinate will be from zero to one. Each point on the triangle may be specified as $(\zeta_1, \zeta_2, \zeta_3)$. The coordinates of node 1 are $(1, 0, 0)$, of node 2 are $(0, 1, 0)$, and of node 3 are $(0, 0, 1)$.

Another interpretation of these coordinates is that they are ratios of areas. The coordinate s_1 may be considered as the altitude of a triangle whose base is the length of side 2–3, the side opposite node 1. The ratio of the area of this triangle to the area of the triangular element is

$$\frac{A_1}{A_e} = \frac{(1/2)s_1 L_{23}}{(1/2)h_1 L_{23}} = \frac{s_1}{h_1} = \zeta_1. \tag{8.14a}$$

Similarly a second triangle within the triangular finite element may be identified whose altitude is s_2 and base is the length of side 1–3. The ratio of its area to the area of the triangular element will be ζ_2. The area ratio of the third triangle with altitude s_3 and base the length of side 1–2 gives rise to the coordinate ζ_3. These local or natural coordinates $(\zeta_1, \zeta_2, \zeta_3)$ are often called area coordinates and are illustrated in Figure 8.12.

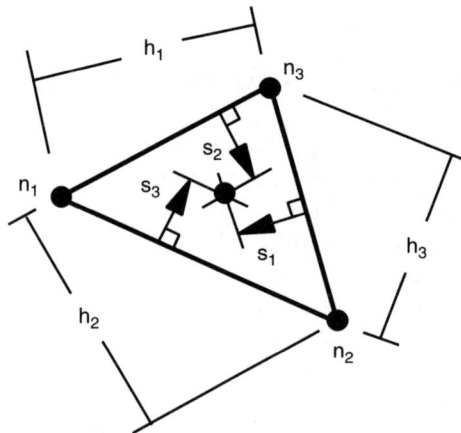

Figure 8.11 Local triangular coordinates

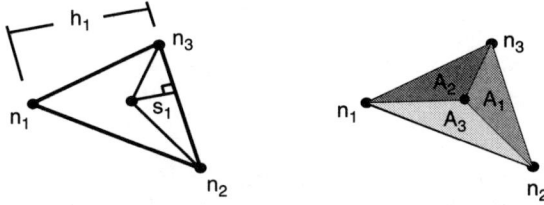

Figure 8.12 Area triangular coordinates

Letting A_1, A_2, and A_3 be the areas of the triangles with altitudes s_1, s_2, and s_3, respectively,

$$\zeta_1 = \frac{A_1}{A_e}, \qquad \zeta_2 = \frac{A_2}{A_e}, \qquad \zeta_3 = \frac{A_3}{A_e}. \tag{8.14b}$$

Though three area coordinates are specified, only two are independent:

$$A_1 + A_2 + A_3 = A_e \tag{8.15a}$$

or

$$\frac{A_1}{A_e} + \frac{A_2}{A_e} + \frac{A_3}{A_e} = 1, \tag{8.15b}$$

and thus

$$\zeta_1 + \zeta_2 + \zeta_3 = 1. \tag{8.15c}$$

Area coordinates so defined are dependent on the attributes of the triangle alone and as we shall see give rise to a convenient mechanism for integration in terms of the area coordinates over the triangular area of the finite element. Since the integrands of equations 6.22 are usually developed in terms of Cartesian x and y coordinates, transformations must be found to convert integrands developed as $I(x, y)$ to integrands as $I(\zeta_1, \zeta_2, \zeta_3)$.

Equations 8.14b may be rewritten as

$$2A_1 = 2A_e\zeta_1 = \begin{vmatrix} 1 & x & y \\ 1 & x_2 & y_2 \\ 1 & x_3 & y_3 \end{vmatrix}, \tag{8.16a}$$

$$2A_2 = 2A_e\zeta_2 = \begin{vmatrix} 1 & x_1 & y_1 \\ 1 & x & y \\ 1 & x_3 & y_3 \end{vmatrix}, \tag{8.16b}$$

$$2A_3 = 2A_e\zeta_3 = \begin{vmatrix} 1 & x_1 & y_1 \\ 1 & x_2 & y_2 \\ 1 & x & y \end{vmatrix}. \tag{8.16c}$$

If one multiplies equation 8.16a by x_1, equation 8.16b by x_2, and equation 8.16c by x_3, adds the three resulting equations, expands the three determinants contained in this sum, and simplifies, the result is

$$x = \zeta_1 x_1 + \zeta_2 x_2 + \zeta_3 x_3. \tag{8.17a}$$

The process can be repeated using y_1, y_2, and y_3 as multipliers of equations 8.16 instead of x_1, x_2, and x_3, to give

$$y = \zeta_1 y_1 + \zeta_2 y_2 + \zeta_3 y_3. \tag{8.17b}$$

Equations 8.17 are the transformation from triangular area coordinates (ζ_1, ζ_2, ζ_3) to Cartesian coordinates (x, y). Substituting them into a function $I(x, y)$ will transform it to $I(\zeta_1, \zeta_2, \zeta_3)$.

Writing equations 8.15c and 8.17 in matrix form,

$$\begin{bmatrix} 1 & 1 & 1 \\ x_1 & x_2 & x_3 \\ y_1 & y_2 & y_3 \end{bmatrix} \begin{Bmatrix} \zeta_1 \\ \zeta_2 \\ \zeta_3 \end{Bmatrix} = \begin{Bmatrix} 1 \\ x \\ y \end{Bmatrix}. \tag{8.18a}$$

or

$$[\mathbf{h}^T]\{\zeta\} = \{\mathbf{g}\}. \tag{8.18b}$$

Transposing and rearranging,

$$\lfloor \zeta \rfloor [\mathbf{h}] = \lfloor \mathbf{g} \rfloor, \tag{8.18c}$$

$$\lfloor \zeta \rfloor = \lfloor \mathbf{g} \rfloor [\mathbf{h}^{-1}]. \tag{8.18d}$$

Substituting the matrix $\mathbf{C}$ for matrix $\mathbf{h}^{-1}$ (equation 6.29) in equation 8.18d,

$$\lfloor \zeta \rfloor = \lfloor \mathbf{g} \rfloor [\mathbf{C}], \tag{8.18e}$$

and transposing,

$$\{\zeta\} = [\mathbf{C}^T]\{\mathbf{g}\}, \tag{8.18f}$$

$$\begin{Bmatrix} \zeta_1 \\ \zeta_2 \\ \zeta_3 \end{Bmatrix} = \begin{bmatrix} C_{11} & C_{21} & C_{31} \\ C_{12} & C_{22} & C_{32} \\ C_{13} & C_{23} & C_{33} \end{bmatrix} \begin{Bmatrix} 1 \\ x \\ y \end{Bmatrix}. \tag{8.18g}$$

Comparing equation 8.18g to equation 8.2d,

$$\begin{Bmatrix} \zeta_1 \\ \zeta_2 \\ \zeta_3 \end{Bmatrix} = \begin{Bmatrix} N_1(x, \ y) \\ N_2(x, \ y) \\ N_3(x, \ y) \end{Bmatrix}. \tag{8.19}$$

Equation 8.19 is the inverse of the transformation portrayed by equations 8.17.

8.3.2. Quadrature for Triangular Elements

Functions which are polynomials of the variables ζ_1, ζ_2, and ζ_3 may be expressed as

$$f(\zeta_1, \ \zeta_2, \ \zeta_3) = \sum_{i=1}^{n} B_i \zeta_1^{a_i} \zeta_2^{b_i} \zeta_3^{c_i}, \tag{8.20}$$

where i identifies the term of the polynomial function; n is the number of terms in the polynomial function; B_i is a constant multiplier of term i of the polynomial function; and a_i, b_i, c_i, are the power exponents of ζ_1, ζ_2, and ζ_3, respectively, in term i of the polynomial function. They must be nonnegative integer values $(0, \ 1, 2, \ 3, \ldots)$.

The exact integral over a triangular area of a single term of a polynomial function of the area coordinates of the triangle was developed by Eisenberg and Malvern (see bibliography at chapter's end) as

$$\int_{A_e} B\zeta_1^a \zeta_2^b \zeta_3^c \ dA_e = 2A_e \frac{Ba!b!c!}{(a+b+c+2)!}. \tag{8.21a}$$

The area of the triangle is A_e. The constants B, a, b, and c are as defined for equation 8.20. The exclamation marks in equation 8.21a indicate factorials.

Combining equations 8.20 and 8.21a, the exact integral of a polynomial function of area coordinates of a triangle over the triangular area is

$$\int_{A_e} (\zeta_1, \ \zeta_2, \ \zeta_3) \ dA_e = \int_{A_e} \left(\sum_{i=1}^{n} B_i \zeta_1^{a_i} \zeta_2^{b_i} \zeta_3^{c_i} \right) dA_e$$

$$= 2A_e \sum_{i=1}^{n} \left(\frac{B_i a_i! b_i! c_i!}{(a_i + b_i + c_i + 2)!} \right). \tag{8.21b}$$

8.3.3. Example Body Force

As an illustration of the use of equations 8.17, 8.19 and 8.21b, an in-plane body force is imposed on the plane stress triangular finite element illustrated in Figure 8.5 whose area was determined earlier as $15.0 \, \text{in}^2$. The magnitude of the body force is proportional to its x and y global Cartesian coordinates, or

$$b_x(x, \ y) = k_1 x \quad \text{and} \quad b_y(x, \ y) = k_2 y,$$

where k_1 and k_2 are constants of proportionality.

The vector of in-plane equivalent nodal forces is

$$\{F_b\} = \begin{Bmatrix} k_1 \int_{A_e} x N_1(x,\ y)\ dA_e \\ k_2 \int_{A_e} y N_1(x,\ y)\ dA_e \\ k_1 \int_{A_e} x N_2(x,\ y)\ dA_e \\ k_2 \int_{A_e} y N_2(x,\ y)\ dA_e \\ k_1 \int_{A_e} x N_3(x,\ y)\ dA_e \\ k_2 \int_{A_e} y N_3(x,\ y)\ dA_e \end{Bmatrix}.$$

Equations 8.17 give

$$x = 3\zeta_1 + 9\zeta_2 + 7\zeta_3 \quad \text{and} \quad y = 8\zeta_1 + 5\zeta_2 + 11\zeta_3.$$

Equation 8.19 gives

$$N_1(x,\ y) = \zeta_1, \quad N_2(x,\ y) = \zeta_2, \quad N_3(x,\ y) = \zeta_3.$$

Thus the vector $\mathbf{F}_b$ is given as

$$\{F_b\} = \begin{Bmatrix} k_1 \int_{A_e} (3\zeta_1^2 + 9\zeta_1\zeta_2 + 7\zeta_1\zeta_3)\ dA_e \\ k_2 \int_{A_e} (8\zeta_1^2 + 5\zeta_1\zeta_2 + 11\zeta_1\zeta_3)\ dA_e \\ k_1 \int_{A_e} (3\zeta_1\zeta_2 + 9\zeta_2^2 + 7\zeta_2\zeta_3)\ dA_e \\ k_2 \int_{A_e} (8\zeta_1\zeta_2 + 5\zeta_2^2 + 11\zeta_2\zeta_3)\ dA_e \\ k_1 \int_{A_e} (3\zeta_1\zeta_3 + 9\zeta_2\zeta_3 + 7\zeta_3^2)\ dA_e \\ k_2 \int_{A_e} (8\zeta_1\zeta_3 + 5\zeta_2\zeta_3 + 11\zeta_3^2)\ dA_e \end{Bmatrix}.$$

The first integral of vector $\mathbf{F}_b$ evaluated using equation 8.21b is

$$F_{1b} = (2)(15)(k_1)\left(\frac{3 \cdot 2! \cdot 0! \cdot 0!}{(2+0+0+2)!} + \frac{9 \cdot 1! \cdot 1! \cdot 0!}{(1+1+0+2)!} + \frac{7 \cdot 1! \cdot 0! \cdot 1!}{(1+0+1+2)!} \right)$$

$$= 30 k_1 \left(\frac{3 \cdot 2 \cdot 1 \cdot 1}{24} + \frac{9 \cdot 1 \cdot 1 \cdot 1}{24} + \frac{7 \cdot 1 \cdot 1 \cdot 1}{24} \right) = 27.5 k_1.$$

The other five integrals are evaluated similarly, so that

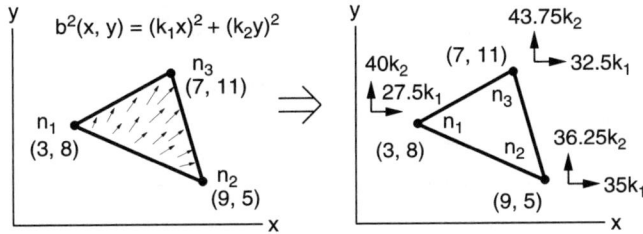

Figure 8.13 Example equivalent nodal loads

$$\{\mathbf{F}_b\} = \begin{Bmatrix} 27.50k_1 \\ 40.00k_2 \\ 35.00k_1 \\ 36.25k_2 \\ 32.50k_1 \\ 43.75k_2 \end{Bmatrix}.$$

Figure 8.13 shows the in-plane body force on the triangular finite element and its equivalent in-plane nodal loads.

8.4. NATURAL COORDINATES AND QUADRATURE FOR LINE SEGMENTS

Triangular finite elements often require that a product of a variable line load and a shape function be integrated along the edge of the element to determine equivalent nodal loads. Natural coordinates for a straight line segment may be developed in a manner similar to what was used for the natural coordinates for triangular areas. These natural line coordinates lead to a convenient mechanism for integration along the line, similar to the mechanism for integration over a triangular area.

8.4.1. Natural Coordinates for Lines

Figure 8.14 illustrates a point p within a line segment located by measuring its distance from each end. The distance measured from the end opposite node 1 is s_1 and from the end opposite node 2 is s_2. If the length of the line segment is L, the following dimensionless natural or local coordinates of the point may be defined:

$$\zeta_1 = \frac{s_1}{L} \quad \text{and} \quad \zeta_2 = \frac{s_2}{L}. \tag{8.22}$$

The value of each local coordinate will be from zero to one. The location of each point on the line may be specified as (ζ_1, ζ_2). The coordinates of node 1 are $(1, 0)$ and of node 2 are $(0, 1)$. Though two local coordinates are specified, only one is independent

$$s_1 + s_2 = L \quad \text{or} \quad \frac{s_1}{L} + \frac{s_2}{L} = 1$$

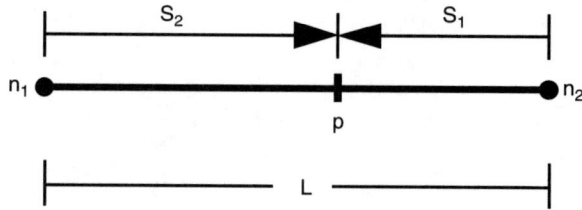

Figure 8.14 Local line coordinates

and thus

$$\zeta_1 + \zeta_2 = 1. \tag{8.23}$$

Figure 8.15 illustrates a line segment referred to global Cartesian coordinates. It is convenient to transform the global x–y axes to rotated x'–y' axes, where x' is parallel to the line segment. Equation 3.15b may be applied for this transformation, so that

$$\begin{Bmatrix} x_1' \\ x_2' \end{Bmatrix} = \begin{bmatrix} c & s & 0 & 0 \\ 0 & 0 & c & s \end{bmatrix} \begin{Bmatrix} x_1 \\ y_1 \\ x_2 \\ y_2 \end{Bmatrix}. \tag{8.24a}$$

The length of the line segment L is $x_2' - x_1'$, which may be computed from equation 8.24a:

$$\begin{aligned} L = x_2' - x_1' &= x_2 \cos \theta + y_2 \sin \theta - x_1 \cos \theta - y_1 \sin \theta \\ &= (x_2 - x_1) \cos \theta + (y_2 - y_1) \sin \theta. \end{aligned} \tag{8.24b}$$

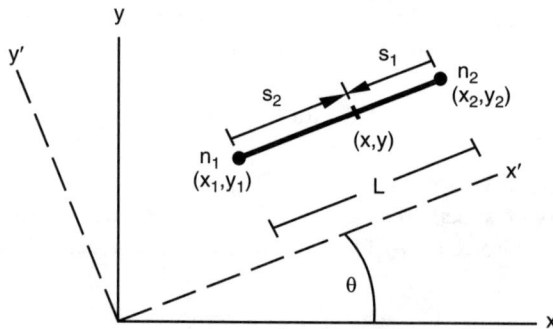

Figure 8.15 Line coordinate transformation

Equation 8.24b reduces to the Pythagorean expression for the length of the segment by squaring both sides of the equation and substituting in it trigonometric identities. The Pythagorean expression is

$$L = \sqrt{(x_2 - x_1)^2 + (y_2 - y_1)^2}. \tag{8.24c}$$

Equations 8.22 may be rewritten as

$$s_1 = \zeta_1 L = x_2' - x', \tag{8.25a}$$

$$s_2 = \zeta_2 L = x' - x_1'. \tag{8.25b}$$

Multiplying equation 8.25a by x_1' and equation 8.25b by x_2', adding the resulting equations, and simplifying,

$$x' = \zeta_1 x_1' + \zeta_2 x_2'. \tag{8.26}$$

Equation 8.26 is the transformation from local line coordinates (ζ_1, ζ_2) to the rotated global coordinate (x'). Writing equations 8.23 and 8.26 in matrix form,

$$\begin{bmatrix} 1 & 1 \\ x_1' & x_2' \end{bmatrix} \begin{Bmatrix} \zeta_1 \\ \zeta_2 \end{Bmatrix} = \begin{Bmatrix} 1 \\ x' \end{Bmatrix}, \tag{8.27a}$$

or

$$[\mathbf{h}^T]\{\zeta\} = \{\mathbf{g}\}. \tag{8.27b}$$

The development followed by equations 8.18 applies, so that

$$\begin{Bmatrix} \zeta_1 \\ \zeta_2 \end{Bmatrix} = \begin{Bmatrix} N_1(x') \\ N_2(x') \end{Bmatrix}. \tag{8.27c}$$

Equations 8.27c are the inverse of the transformation portrayed by equations 8.25.

The shape functions for a 2-node line segment referred to rotated global Cartesian coordinates may be written from equations 8.2 and 6.30 as

$$\lfloor \mathbf{N}_i(\mathbf{x}') \rfloor = \lfloor \mathbf{g} \rfloor [\mathbf{h}^{-1}], \tag{8.28a}$$

or

$$\lfloor N_1(x') \quad N_2(x') \rfloor = \lfloor 1 \quad x' \rfloor \begin{bmatrix} 1 & x_1' \\ 1 & x_2' \end{bmatrix}^{-1} \tag{8.28b}$$

$$= \left\lfloor \frac{x_2' - x'}{L} \quad \frac{-x_1' + x'}{L} \right\rfloor. \tag{8.28c}$$

A special case is when $x_1' = 0$, so that $x_2' = L$:

$$\lfloor N_1(x') \quad N_2(x') \rfloor = \left\lfloor 1 - \frac{x'}{L} \quad \frac{x'}{L} \right\rfloor. \tag{8.28d}$$

Equation 8.28d gives the shape functions developed earlier for the 2-node axial finite element (equations 5.1).

8.4.2. Quadrature for a Line Segment

The quadrature formula for the exact integral along a line of a single-term polynomial function of local line coordinates ζ_1 and ζ_2 is

$$\int_L B\zeta_1^a \zeta_2^b \, dL = L \frac{Ba!b!}{(a+b+1)!}. \tag{8.29a}$$

Equation 8.29a may be developed by employing substitution techniques and integration by parts.

The exact integral of a polynomial function of the local line coordinates along the line is

$$\begin{aligned}
\int_L f(\zeta_1, \ \zeta_2) \, dL &= \int_L \left(\sum_{i=1}^n B_i \zeta_1^{a_i} \zeta_2^{b_i} \right) dL \\
&= L \sum_{i=1}^n \left(\frac{B_i a_i! b_i!}{(a_i + b_i + 1)!} \right).
\end{aligned} \tag{8.29b}$$

A polynomial function of the variables ζ_1 and ζ_2 is similar to what is expressed in equation 8.20 except that the third variable ζ_3 is omitted.

8.4.3. Example Line Load

As an illustration of the use of equations 8.24, 8.26, 8.27c, and 8.29b, the triangular element shown in Figure 8.5 is impressed with an in-plane line load, varying in a linear manner from 10 lb/in. at node 1 to 40 lb/in. at node 3. The load is directed to the right and down at an angle of 30° clockwise from the positive x axis and is shown in Figure 8.16.

The vector $\mathbf{F}_b$ of equivalent nodal loads is given by equation 8.12c. Because a definite integral is a function of its limits only, the coordinate axes employed to develop the integrand may be chosen to facilitate that development. To this end, a convenient set of axes are Cartesian coordinate axes, rotated so the x' axis is parallel to side 1–3 and the origin is at node 1. For these axes, the magnitude of $b(L)$ may be written as

$$|b(L)| = |b(x')| = \frac{40 - 10}{L_{13}} x' + 10.$$

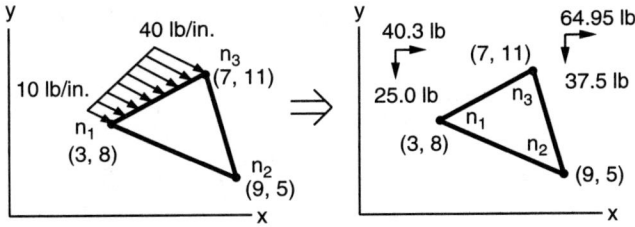

Figure 8.16 Example equivalent nodal loads

The length of side 1–3 is given by equation 8.24c,

$$L_{13} = \sqrt{(7-3)^2 + (11-8)^2} = 5.00 \, \text{in.}$$

Thus

$$|b(L)| = 6x' + 10,$$

and

$$b_x(L) = (6x' + 10) \, \cos \, 30° = 3x'\sqrt{3} + 5\sqrt{3},$$
$$b_y(L) = -(6x' + 10) \, \sin \, 30° = -3x' - 5.$$

with $x_1' = 0$ and $x_3' = 5.00$, equation 8.26 gives

$$x' = \zeta_1 \cdot 0 + \zeta_3 \cdot 5 = 5\zeta_3.$$

Substituting with this result,

$$b_x(L) = 5\sqrt{3}(3\zeta_3 + 1), \qquad b_y(L) = -5(3\zeta_3 + 1).$$

Equation 8.27c indicates

$$N_1(L) = \zeta_1, \qquad N_3(L) = \zeta_3.$$

Thus the integral expression of vector $\mathbf{F_b}$ may be written as

$$\{\mathbf{F_b}\} = \begin{Bmatrix} 5\sqrt{3} \int_L (3\zeta_1\zeta_3 + \zeta_1) \, dL \\ -5 \int_L (3\zeta_1\zeta_3 + \zeta_1) \, dL \\ 0 \\ 0 \\ 5\sqrt{3} \int_L (3\zeta_3^2 + \zeta_3) \, dL \\ -5 \int_L (3\zeta_3^2 + \zeta_3) \, dL \end{Bmatrix}.$$

The first integral of the vector $\mathbf{F}_b$ evaluated using equation 8.29b is

$$F_{1b} = (5\sqrt{3})(5)\left(\frac{3 \cdot 1! \cdot 1!}{(1+1+1)!} + \frac{1 \cdot 1! \cdot 0!}{(1+0+1)!}\right)$$

$$= 25\sqrt{3}\left(\frac{3 \cdot 1 \cdot 1}{6} + \frac{1 \cdot 1 \cdot 1}{2}\right) = 43.30 \text{ lb.}$$

The other three integrals are evaluated similarly, so that

$$\{\mathbf{F}_b\} = \begin{Bmatrix} 43.30 \\ -25.00 \\ 0 \\ 0 \\ 64.95 \\ -37.50 \end{Bmatrix} \text{ lb.}$$

Figure 8.16 shows the in-plane line load on the triangular finite element and its equivalent in-plane nodal loads.

8.5. LINEAR STRAIN TRIANGULAR ELEMENT (6-NODE, 12-DOF)

The accuracy of the 3-node, 6-dof constant strain triangular finite element for stress calculation is restricted because the strain and stress cannot vary over the element. To achieve acceptable results with this element when stresses vary sharply in a plane stress structure, a large number of elements must be used. An alternative is to use an element which permits a variation of strain and stress within it. Such an element must possess displacement and shape functions which are second-degree polynomials or higher, so that the strains within the element, computed as first partial derivatives of these polynomials, will be variable.

For a plane stress triangular finite element to possess second-degree or higher polynomial shape functions, it must have more than three nodes. A 6-node, 12-dof triangular finite element is shown in Figure 8.17. There is a node at each vertex of the triangle and one at the center of each side. It needs six shape functions, each relating to one of the dof at each node.

8.5.1. Shape and Displacement Functions

The displacement functions for the element of Figure 8.17 are

$$u(x, y) = N_1(x, y)q_1 + N_2(x, y)q_3 + N_3(x, y)q_5 + N_4(x, y)q_7$$
$$+ N_5(x, y)q_9 + N_6(x, y)q_{11}, \tag{8.30a}$$

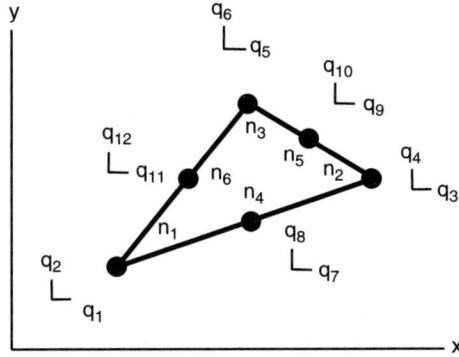

Figure 8.17 Linear strain trinagle

$$v(x,\ y) = N_1(x,\ y)q_2 + N_2(x,\ y)q_4 + N_3(x,\ y)q_6 + N_4(x,\ y)q_8$$
$$+ N_5(x,\ y)q_{10} + N_6(x,\ y)q_{12}. \tag{8.30b}$$

The shape functions are polynomials in x and y. To meet the requirements for shape functions ($N_i = 1$ at node i, $N_i = 0$ at the other five nodes), there must be six polynomial terms in each shape function, each with a constant multiplier. The lowest degree six-term polynomial in x and y is a complete quadratic polynomial with terms 1, x, y, x^2, xy, and y^2. Thus the six shape functions are

$$\begin{Bmatrix} N_1 \\ N_2 \\ N_3 \\ N_4 \\ N_5 \\ N_6 \end{Bmatrix} = \begin{bmatrix} C_{11} & C_{21} & C_{31} & C_{41} & C_{51} & C_{61} \\ C_{12} & C_{22} & C_{32} & C_{42} & C_{52} & C_{62} \\ C_{13} & C_{23} & C_{33} & C_{43} & C_{53} & C_{63} \\ C_{14} & C_{24} & C_{34} & C_{44} & C_{54} & C_{64} \\ C_{15} & C_{25} & C_{35} & C_{45} & C_{55} & C_{65} \\ C_{16} & C_{26} & C_{36} & C_{46} & C_{56} & C_{66} \end{bmatrix} \begin{Bmatrix} 1 \\ x \\ y \\ x^2 \\ xy \\ y^2 \end{Bmatrix}, \tag{8.31a}$$

$$\{\mathbf{N}\} = [\mathbf{C}^T]\{\mathbf{g}\}. \tag{8.31b}$$

The first partial derivatives of these shape functions with respect to x and y will be linear polynomials in x and y. Thus the strains and stresses within this finite element vary linearly.

Applying equation 6.29 to equation 8.31b,

$$\{\mathbf{N}\} = [\mathbf{h}^{-T}]\{\mathbf{g}\}. \tag{8.31c}$$

The matrix $\mathbf{h}$, as before, is an array of constants obtained by evaluating the generic polynomial terms of each shape function by row at each corresponding node of the finite element. It may be written directly from the global coordinates of the six nodes:

$$[\mathbf{h}] = \begin{bmatrix} 1 & x_1 & y_1 & x_1^2 & x_1y_1 & y_1^2 \\ 1 & x_2 & y_2 & x_2^2 & x_2y_2 & y_2^2 \\ 1 & x_3 & y_3 & x_3^2 & x_3y_3 & y_3^2 \\ 1 & x_4 & y_4 & x_4^2 & x_4y_4 & y_4^2 \\ 1 & x_5 & y_5 & x_5^2 & x_5y_5 & y_5^2 \\ 1 & x_6 & y_6 & x_6^2 & x_6y_6 & y_6^2 \end{bmatrix}. \tag{8.32}$$

The transpose and inverse of matrix **h** are easily determined numerically, assisted by a computer, with known values for the nodal coordinates on the element. The six row entries of matrix $\mathbf{h}^{-T}$ provide, respectively, the constant multipliers for each term of the six shape functions for the element (equations 8.31a).

8.5.2. Element Stiffness Matrix

The element stiffness matrix for the triangular finite element is obtained, as before, by starting with equation 6.22d:

$$[\mathbf{K}] = \left[\int_v [\mathbf{B}^{\mathrm{T}}][\mathbf{E}][\mathbf{B}] \, dV \right]. \tag{6.22d}$$

The strain displacement matrix **B** is the array of derivatives of shape functions, differentiated in accordance with the strain displacement relationships, equations 8.6b, 8.6c, 8.8, and 8.9a:

$$[\mathbf{B}] = [\mathbf{d}][\mathbf{N}]. \tag{8.9a}$$

The correct form for matrix **N** is obtained by writing equations 8.30 in matrix form as

$$\begin{Bmatrix} u \\ v \end{Bmatrix} = \begin{bmatrix} N_1 & 0 & | & N_2 & 0 & | & \cdots & \cdots & | & N_6 & 0 \\ 0 & N_1 & | & 0 & N_2 & | & \cdots & \cdots & | & 0 & N_6 \end{bmatrix} \begin{Bmatrix} q_1 \\ q_2 \\ q_3 \\ \vdots \\ q_{11} \\ q_{12} \end{Bmatrix}, \tag{8.33a}$$

or

$$\{\mathbf{u}\} = [\mathbf{N}]\{\mathbf{q}\}. \tag{8.33b}$$

Applying equation 8.9a,

$$[\mathbf{B}] = \begin{bmatrix} \frac{\partial}{\partial x} & 0 \\ 0 & \frac{\partial}{\partial y} \\ \frac{\partial}{\partial y} & \frac{\partial}{\partial x} \end{bmatrix} \begin{bmatrix} N_1 & 0 & | & N_2 & 0 & | & \cdots & \cdots & | & N_6 & 0 \\ 0 & N_1 & | & 0 & N_2 & | & \cdots & \cdots & | & 0 & N_6 \end{bmatrix}. \tag{8.34a}$$

It is convenient to represent the matrix **B** as a row of submatrices $\mathbf{B}_i$, where the index i corresponds to the node numbering sequence on the triangular finite element. Thus,

$$[\mathbf{B}] = [\,\mathbf{B}_1 \quad \mathbf{B}_2 \quad \mathbf{B}_3 \quad \mathbf{B}_4 \quad \mathbf{B}_5 \quad \mathbf{B}_6\,] \tag{8.34b}$$

and

$$[\mathbf{B}_i] = \begin{bmatrix} \frac{\partial}{\partial x} & 0 \\ 0 & \frac{\partial}{\partial y} \\ \frac{\partial}{\partial y} & \frac{\partial}{\partial x} \end{bmatrix} \begin{bmatrix} N_i & 0 \\ 0 & N_i \end{bmatrix} = \begin{bmatrix} \frac{\partial N_i}{\partial x} & 0 \\ 0 & \frac{\partial N_i}{\partial y} \\ \frac{\partial N_i}{\partial y} & \frac{\partial N_i}{\partial x} \end{bmatrix}. \tag{8.34c}$$

To implement equations 6.22d, the expressions for the x and y partial derivatives of the six shape functions must be written as

$$\frac{\partial N_i}{\partial x} = \frac{\partial}{\partial x}(C_{1i} + C_{2i}x + C_{3i}y + C_{4i}x^2 + C_{5i}xy + C_{6i}y^2)$$

$$= C_{2i} + 2C_{4i}x + C_{5i}y, \tag{8.34d}$$

$$\frac{\partial N_i}{\partial y} = \frac{\partial}{\partial y}(C_{1i} + C_{2i}x + C_{3i}y + C_{4i}x^2 + C_{5i}xy + C_{6i}y^2)$$

$$= C_{3i} + C_{5i}x + 2C_{6i}y. \tag{8.34e}$$

The submatrix $\mathbf{B}_i$ is given as

$$[\mathbf{B}_i] = \begin{bmatrix} C_{2i} + 2C_{4i}x + C_{5i}y & 0 \\ 0 & C_{3i} + C_{5i}x + 2C_{6i}y \\ C_{3i} + C_{5i}x + 2C_{6i}y & C_{2i} + 2C_{4i}x + C_{5i}y \end{bmatrix}, \tag{8.34f}$$

where the constants C_{ji} are the entries of the transposed inverse of matrix **h** (equation 8.32).

The matrix **K** is given as

$$[\mathbf{K}] = \int_v [\mathbf{B}^{\mathrm{T}}][\mathbf{E}][\mathbf{B}]\,dV = \int_V \begin{bmatrix} \mathbf{B}_1^{\mathrm{T}} \\ \mathbf{B}_2^{\mathrm{T}} \\ \mathbf{B}_3^{\mathrm{T}} \\ \mathbf{B}_4^{\mathrm{T}} \\ \mathbf{B}_5^{\mathrm{T}} \\ \mathbf{B}_6^{\mathrm{T}} \end{bmatrix} [\mathbf{E}][\,\mathbf{B}_1 \quad \mathbf{B}_2 \quad \mathbf{B}_3 \quad \mathbf{B}_4 \quad \mathbf{B}_5 \quad \mathbf{B}_6\,]\,dV. \tag{8.35a}$$

Alternatively, by expanding the integrand of equation 8.35a,

$$[K] = \int_v \begin{bmatrix} \mathbf{B}_1^T \mathbf{E} \mathbf{B}_1 & \mathbf{B}_1^T \mathbf{E} \mathbf{B}_2 & \mathbf{B}_1^T \mathbf{E} \mathbf{B}_3 & \mathbf{B}_1^T \mathbf{E} \mathbf{B}_4 & \mathbf{B}_1^T \mathbf{E} \mathbf{B}_5 & \mathbf{B}_1^T \mathbf{E} \mathbf{B}_6 \\ \mathbf{B}_2^T \mathbf{E} \mathbf{B}_1 & \mathbf{B}_2^T \mathbf{E} \mathbf{B}_2 & \mathbf{B}_2^T \mathbf{E} \mathbf{B}_3 & \mathbf{B}_2^T \mathbf{E} \mathbf{B}_4 & \mathbf{B}_2^T \mathbf{E} \mathbf{B}_5 & \mathbf{B}_2^T \mathbf{E} \mathbf{B}_6 \\ \mathbf{B}_3^T \mathbf{E} \mathbf{B}_1 & \mathbf{B}_3^T \mathbf{E} \mathbf{B}_2 & \mathbf{B}_3^T \mathbf{E} \mathbf{B}_3 & \mathbf{B}_3^T \mathbf{E} \mathbf{B}_4 & \mathbf{B}_3^T \mathbf{E} \mathbf{B}_5 & \mathbf{B}_3^T \mathbf{E} \mathbf{B}_6 \\ \mathbf{B}_4^T \mathbf{E} \mathbf{B}_1 & \mathbf{B}_4^T \mathbf{E} \mathbf{B}_2 & \mathbf{B}_4^T \mathbf{E} \mathbf{B}_3 & \mathbf{B}_4^T \mathbf{E} \mathbf{B}_4 & \mathbf{B}_4^T \mathbf{E} \mathbf{B}_5 & \mathbf{B}_4^T \mathbf{E} \mathbf{B}_6 \\ \mathbf{B}_5^T \mathbf{E} \mathbf{B}_1 & \mathbf{B}_5^T \mathbf{E} \mathbf{B}_2 & \mathbf{B}_5^T \mathbf{E} \mathbf{B}_3 & \mathbf{B}_5^T \mathbf{E} \mathbf{B}_4 & \mathbf{B}_5^T \mathbf{E} \mathbf{B}_5 & \mathbf{B}_5^T \mathbf{E} \mathbf{B}_6 \\ \mathbf{B}_6^T \mathbf{E} \mathbf{B}_1 & \mathbf{B}_6^T \mathbf{E} \mathbf{B}_2 & \mathbf{B}_6^T \mathbf{E} \mathbf{B}_3 & \mathbf{B}_6^T \mathbf{E} \mathbf{B}_4 & \mathbf{B}_6^T \mathbf{E} \mathbf{B}_5 & \mathbf{B}_6^T \mathbf{E} \mathbf{B}_6 \end{bmatrix} dV. \qquad (8.35b)$$

The indices i and j in equations 8.35b are in the nodal numbering sequence. Each integrand $\mathbf{B}_i^T \mathbf{E} \mathbf{B}_j$ is a 2×2 submatrix whose entries are the products of the linear polynomial functions shown in equations 8.34f combined with the entries of the matrix $\mathbf{E}$ for the element. There are 36 of these submatrices, resulting in a 12×12 element stiffness matrix with 144 integrands, each a six-term second-degree polynomial in x and y.

8.5.3. Numerical Quadrature over a Triangular Area

The process of developing each polynomial integrand and performing the integrations over the triangular area of the element is tedious at best, even employing the methods presented in Section 8.3. These integrations are more efficiently done without transformation to area coordinates, in a purely numerical manner with an appropriate quadrature (or numerical integration) formula, and with respect to the global Cartesian coordinates of the structure.

The form of the quadrature formula is

$$\int_{A_e} f(x, y) \, dA_e = \sum_{k=1}^{N_k} f(x_k, y_k) H_k, \qquad (8.36)$$

where $f(x, y)$ is the integrand; A_e is the area of the triangle over which the integration is performed; (x_k, y_k) are the coordinates of points on the triangular area, called integration points; k is an index identifying the integration points; N_k is the number of integration points employed; and H_k is the weight factor for integration point k.

Equation 8.36 is sometimes referred to as Gaussian quadrature for triangles, because the integration points are located to optimize the accuracy of the evaluation, as do Gauss derivations for lines, rectangles, and 3D parallelopipeds. Gaussian quadrature is discussed in detail in Chapter 9.

Equation 8.36 will integrate exactly a quadratic polynomial in x and y over a triangular area using three integration points. The integration points should be located at the center of each side of the triangle at area coordinates $(\frac{1}{2}, \frac{1}{2}, 0)$, $(0, \frac{1}{2}, \frac{1}{2})$, and $(\frac{1}{2}, 0, \frac{1}{2})$. Note that on the 6-node, 12-dof linear strain triangle, these three integration points are located at nodes 4, 5, and 6. The weight factors for each of these three integration points are the same and equal $\frac{1}{3} A_e$.

As an illustration of the use of equation 8.36, the integral

$$\int_{A_e} (x^2 + xy + y^2) \, dA_e$$

will be evaluated over the area of the triangle in Figure 8.5. For this example, the methods of Section 8.3 may be employed without difficulty to obtain an exact value for the integral by transforming the integrand to a function of the triangle's area coordinates (equations 8.17) and then applying equation 8.21b. It is left as an exercise for the reader to do this. An explicit exact integral in terms of the Cartesian coordinates is possible but tedious. The numerical evaluation by equation 8.36 is exact with three integration points and is the simplest to carry out.

The coordinates of the three integration points are determined with equations 8.17:

$$x_{k1} = \tfrac{1}{2} \cdot 3 + \tfrac{1}{2} \cdot 9 + 0 \cdot 7 = 6.0 \text{ in.}, \quad y_{k1} = \tfrac{1}{2} \cdot 8 + \tfrac{1}{2} \cdot 5 + 0 \cdot 11 = 6.5 \text{ in.},$$
$$x_{k2} = 0 \cdot 3 + \tfrac{1}{2} \cdot 9 + \tfrac{1}{2} \cdot 7 = 8.0 \text{ in.}, \quad y_{k2} = 0 \cdot 8 + \tfrac{1}{2} \cdot 5 + \tfrac{1}{2} \cdot 11 = 8.0 \text{ in.},$$
$$x_{k3} = \tfrac{1}{2} \cdot 3 + 0 \cdot 9 + \tfrac{1}{2} \cdot 7 = 5.0 \text{ in.}, \quad y_{k3} = \tfrac{1}{2} \cdot 8 + 0 \cdot 5 + \tfrac{1}{2} \cdot 11 = 9.5 \text{ in.}$$

The area of the triangle was determined earlier as 15.0 in^2.

Thus

$$\int_{A_e} (x^2 + xy + y^2) \, dA_e$$
$$= \tfrac{1}{3} \cdot 15[(6^2 + 6 \cdot 6.5 + 6.5^2) + (8^2 + 8 \cdot 8 + 8^2) + (5^2 + 5 \cdot 9.5 + 9.5^2)] = 2360.0 \text{ in}^4.$$

The application of equation 8.36 to perform the integrations required for matrix **K** in equations 8.35 is straightforward. Three integration points will give exact results since each integrand is a polynomial in x and y, not higher than second degree. For plane stress, where $dV = t \, dA_e$, equations 6.22d and 8.35 become

$$[\mathbf{K}] = \frac{A_e t}{3} \sum_{k=1}^{3} ([\mathbf{B}_k^\mathrm{T}(x_k, \ y_k)][\mathbf{E}][\mathbf{B}_k(x_k, \ y_k)]). \qquad (8.37a)$$

The coordinates $(x_k, \ y_k)$ of each of the three integration points are determined for the element using equations 8.17 or are known already as the coordinates of nodes 4, 5, and 6. These coordinate values and the appropriate elastic constants are substituted into equation 8.37a to form the numerical matrices $\mathbf{B}_k(x_{k1}, \ y_{k1})$, $\mathbf{B}_k(x_{k2}, \ y_{k2})$, $\mathbf{B}_k(x_{k3}, \ y_{k3})$, and **E**, and the matrix multiplications indicated are performed. The result is the 12×12 numerical matrix **K** for the element. These steps are illustrated as

$$[\mathbf{K}] = [\mathbf{K}_{k=1}] + [\mathbf{K}_{k=2}] + [\mathbf{K}_{k=3}]$$
$$= \tfrac{1}{3} A_e t [\mathbf{B}_k^\mathrm{T}(x_{k1}, \ y_{k1})][\mathbf{E}][\mathbf{B}_k(x_{k1}, \ y_{k1})]$$
$$+ \tfrac{1}{3} A_e t [\mathbf{B}_k^\mathrm{T}(x_{k2}, \ y_{k2})][\mathbf{E}][\mathbf{B}_k(x_{k2}, \ y_{k2})] \qquad (8.37b)$$
$$+ \tfrac{1}{3} A_e t [\mathbf{B}_k^\mathrm{T}(x_{k3}, \ y_{k3})][\mathbf{E}][\mathbf{B}_k(x_{k3}, \ y_{k3})].$$

8.5.4. Loads between Nodes

The discussion of loads between nodes for the 3-node, 6-dof constant strain triangular finite element, discussed in Section 8.2.3, applies directly for the 6-node, 12-dof linear strain triangular finite element.

The equations developed for body forces for the constant strain element are the same form as for the linear strain element. For the linear strain element, however, the vector of equivalent nodal loads has 12 entries instead of 6, and there are 6 shape functions instead of 3 included in the formulas. The integrations required in the formulas may be performed by the methods of either Section 8.3 or Section 8.5.3.

The equations developed for line loads on the edge of a constant strain element are also of the same form as for the linear strain element. Again there are 12 entries instead of 6 in the vector of equivalent nodal loads for the linear strain element. There will be 3 shape functions which have value on the edge of the linear strain element instead of 2, so that 6 entries in the 12×1 vector of equivalent nodal loads will be nonzero instead of 4 in the 6×1 vector for the constant strain element.

The shape functions on the edge of the linear strain element will be second-degree functions of the line variable L, rather than linear functions, since there are three nodes on the edge of the linear strain element. If rotated Cartesian axes are used with an end node as the origin, the edge shape functions will be the same as shown for equations 5.37 for the 3-node 1D axial finite element. In equation 5.37, node 1 identifies the origin of the rotated Cartesian axes, node 2 the center of the edge of the element, and node 3 the end of the edge opposite the origin.

8.5.5. Linear Strain Element Example Analysis

To illustrate the use of the 6-node, 12-dof linear strain triangular finite element, the reactions of the plane stress constant strain element plate example of Section 8.2.4 will be recalculated using the linear strain element. The loads, supports, and geometry of the structure are shown in Figure 8.8. The plate discretized for three 6-node, 12-dof linear strain elements is in Figure 8.18.

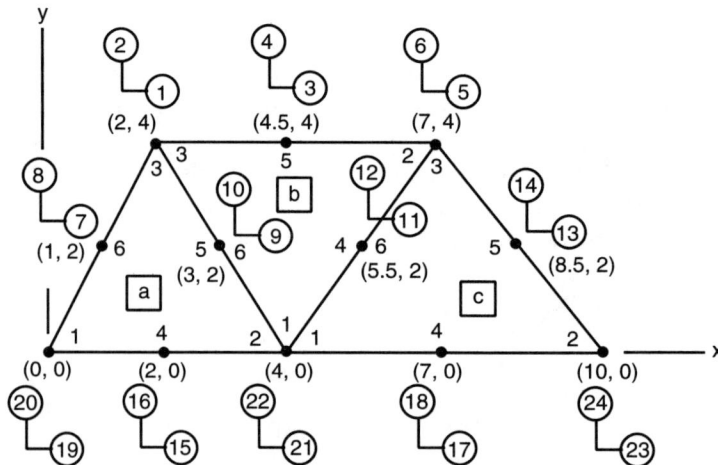

Figure 8.18 Discretized plate

Equations 6.22d, 8.31, 8.32, 8.34f, 8.35, and 8.37 indicate the procedure to obtain the element's matrix $\mathbf{K}_a$.

The element's matrix $\mathbf{h}_a$ is written from equation 8.32 using the global coordinates nodes and then inverted:

$$[\mathbf{h}_a] = \begin{bmatrix} 1 & 0 & 0 & 0 & 0 & 0 \\ 1 & 4 & 0 & 16 & 0 & 0 \\ 1 & 2 & 4 & 4 & 8 & 16 \\ 1 & 2 & 0 & 4 & 0 & 0 \\ 1 & 3 & 2 & 9 & 6 & 4 \\ 1 & 1 & 2 & 1 & 2 & 4 \end{bmatrix},$$

$$[\mathbf{h}_a^{-1}] = \begin{bmatrix} 1 & 0 & 0 & 0 & 0 & 0 \\ -0.75 & -0.25 & 0 & 1 & 0 & 0 \\ -0.375 & 0.125 & -0.25 & -0.5 & 0 & 1 \\ 0.125 & 0.125 & 0 & -0.25 & 0 & 0 \\ 0.125 & -0.125 & 0 & 0 & 0.25 & -0.25 \\ 0.03125 & 0.03125 & 0.125 & 0.0625 & -0.125 & -0.125 \end{bmatrix}.$$

The columns of matrix $\mathbf{h}_a^{-1}$ are the constants for the shape functions of element a (equation 8.31).

The product matrix $[\mathbf{B}_a^{\mathrm{T}}(x, y)][\mathbf{E}][\mathbf{B}_a(x, y)]$ of equation 8.35a, a 12×12 array of functions of x and y, must be integrated over the triangular area of the element. The simplest approach is to implement the quadrature rule expressed in equation 8.37. The integration points are at nodes 4, 5, and 6 of element a.

The matrix $\mathbf{B}_a(x, y)$, a 3×12 array of functions of x and y, is written out in accordance with equation 8.34 using the shape function constants obtained above in matrix $\mathbf{h}_a^{-1}$. It is then evaluated at the first integration point, or node 4 at $(2, 0)$, as

$$[\mathbf{B}_a(2, 0)] = \begin{bmatrix} -0.25 & 0 & 0.25 & 0 & 0 & 0 & 0 & 0 & 0 & 0 & 0 & 0 \\ 0 & -0.125 & 0 & -0.125 & 0 & -0.25 & 0 & -0.5 & 0 & 0.5 & 0 & 0.5 \\ -0.125 & -0.25 & -0.125 & 0.25 & -0.25 & 0 & -0.5 & 0 & 0.5 & 0 & 0.5 & 0 \end{bmatrix}.$$

Next the product matrix $[\mathbf{B}_a^{\mathrm{T}}(2, 0)][\mathbf{E}][\mathbf{B}_a(2, 0)]$ is formed. The elastic matrix $\mathbf{E}$ was written out in Section 8.2.4. The scalar associated with this product matrix is the product of the scalar of the elastic matrix, $E/(1 - v^2)$, and the weight factor, $\frac{1}{3}A_e t$. The area of element a is $8\,\mathrm{ft}^2$ (Section 8.2.4). Thus the scalar is

$$S = \left(\frac{30,000,000 \times 144}{1 - 0.25^2} \right) \left(\frac{8 \times (1/12)}{3} \right) = 1024 \times 10^6\,\mathrm{lb/ft}.$$

The product matrix evaluated at (2, 0) thus is

$$
[\mathbf{B}_a^\mathrm{T}(2,\ 0)\mathbf{E}\mathbf{B}_a(2,\ 0)] =
\begin{bmatrix}
70 & 20 & -58 & -4 & 12 & 16 & 24 & 32 & -24 & -32 & -24 & -32 \\
20 & 40 & 4 & -8 & 24 & 32 & 48 & 64 & -48 & -64 & -48 & -64 \\
-58 & 4 & 70 & -20 & 12 & -16 & 24 & -32 & -24 & 32 & -24 & 32 \\
-4 & -8 & -20 & 40 & -24 & 32 & -48 & 64 & 48 & -64 & 48 & -64 \\
12 & 24 & 12 & -24 & 24 & 0 & 48 & 0 & -48 & 0 & -48 & 0 \\
16 & 32 & -16 & 32 & 0 & 64 & 0 & 128 & 0 & -128 & 0 & -128 \\
24 & 48 & 24 & -48 & 48 & 0 & 96 & 0 & -96 & 0 & -96 & 0 \\
32 & 64 & -32 & 64 & 0 & 128 & 0 & 256 & 0 & -256 & 0 & -256 \\
-24 & -48 & -24 & 48 & -48 & 0 & -96 & 0 & 96 & 0 & 96 & 0 \\
-32 & -64 & 32 & -64 & 0 & -128 & 0 & -256 & 0 & 256 & 0 & 256 \\
-24 & -48 & -24 & 48 & -48 & 0 & -96 & 0 & 96 & 0 & 96 & 0 \\
-32 & -64 & 32 & -64 & 0 & -128 & 0 & -256 & 0 & 256 & 0 & 256
\end{bmatrix}
\ \text{lb/ft.}
$$

The process is repeated for the second integration point, or node 5 at (3, 2), to give $[\mathbf{B}_a^\mathrm{T}(3,\ 2)][\mathbf{E}][\mathbf{B}_a(3,\ 2)]$ and then for the third integration point, or node 6 at (1, 2), to give $[\mathbf{B}_a^\mathrm{T}(1,\ 2)][\mathbf{E}][\mathbf{B}_a(1,\ 2)]$. The three product matrices are added and the result is the stiffness matrix $\mathbf{K}_a$ for element a,

	⑲	⑳	㉑	㉒	①	②	⑮	⑯	⑨	⑩	⑦	⑧
⑲	210	60	58	4	12	16	-232	-16	0	0	-48	-64
⑳	60	120	-4	8	24	32	16	-32	0	0	-96	-128
㉑	58	-4	210	-60	12	-16	-232	16	-48	64	0	0
㉒	4	8	-60	120	-24	32	-16	-32	96	-128	0	0
①	12	24	12	-24	72	0	0	0	-48	96	-48	-96
②	16	32	-16	32	0	192	0	0	64	-128	-64	-128
⑮	-232	16	-232	-16	0	0	656	0	-96	-160	-96	160
⑯	-16	-32	16	-32	0	0	0	576	-160	-256	160	-256
⑨	0	0	-48	96	-48	64	-96	-160	656	0	-464	0
⑩	0	0	64	-128	96	-128	-160	-256	0	576	0	-64
⑦	-48	-96	0	0	-48	-64	-96	160	-464	0	656	0
⑧	-64	-128	0	0	-96	-128	160	-256	0	-64	0	576

$[\mathbf{K}_a] = $ (the matrix above), lb/ft.

Numbers in circles are the corresponding gdof.

The stiffness matrices for elements b and c are obtained similarly. The plate's unit weight, $\gamma = 500\ \text{lb/ft}^3$, is the body force for element a. Its equivalent nodal loads are obtained by adapting equation 8.12a:

$$
\{\mathbf{F}_{ba}\} = tb_y
\begin{Bmatrix}
0 \\
\int_{A_a} N_{a1}\, dA \\
0 \\
\int_{A_a} N_{a2}\, dA \\
0 \\
\int_{A_a} N_{a3}\, dA \\
0 \\
\int_{A_a} N_{a4}\, dA \\
0 \\
\int_{A_a} N_{a5}\, dA \\
0 \\
\int_{A_a} N_{a6}\, dA
\end{Bmatrix}.
$$

The shape functions for element a are written from the matrix $\mathbf{h}_a^{-1}$ above and then integrated over the triangular area of the element. In the application of the quadrature rule, the integration points are also nodes 4, 5, and 6. Thus the values of the shape functions are either unity or zero:

$$
\{\mathbf{F}_{ba}\} = tb_y
\begin{Bmatrix}
0 \\
0 \\
0 \\
0 \\
0 \\
0 \\
0 \\
\frac{A_e \times 1}{3} \\
0 \\
\frac{A_e \times 1}{3} \\
0 \\
\frac{A_e \times 1}{3}
\end{Bmatrix}
= \frac{1(-500)}{12}
\begin{Bmatrix}
0 \\
0 \\
0 \\
0 \\
0 \\
0 \\
0 \\
\frac{8 \times 1}{3} \\
0 \\
\frac{8 \times 1}{3} \\
0 \\
\frac{8 \times 1}{3}
\end{Bmatrix}
=
\begin{matrix}
⑲ \\ ⑳ \\ ㉑ \\ ㉒ \\ ① \\ ② \\ ⑮ \\ ⑯ \\ ⑨ \\ ⑩ \\ ⑦ \\ ⑧
\end{matrix}
\begin{Bmatrix}
0 \\
0 \\
0 \\
0 \\
0 \\
0 \\
0 \\
-111.11 \\
0 \\
-111.11 \\
0 \\
-111.11
\end{Bmatrix}
\quad \text{lb.}
$$

The vectors of equivalent nodal forces for the gravitational body force for elements b and c are developed in like manner.

The line load on element b is reduced to equivalent nodal loads by adapting equation 8.12c. Note that only nodes 3, 5, and 2 of element b, and thus gdof 2, 4, and 6, are involved. Nodes 1, 4, and 6 of element b are not where the load acts, and since the load is vertical, gdof 1, 3, and 5 have no load component. Thus,

$$
\{\mathbf{F}_{bb11}\} = tb_y
\begin{Bmatrix}
0 \\
0 \\
0 \\
\int_L N_{a2} \, dL \\
0 \\
\int_L N_{a3} \, dL \\
0 \\
0 \\
0 \\
\int_L N_{a5} \, dL \\
0 \\
0
\end{Bmatrix}.
$$

The shape functions are written from matrix $\mathbf{h}_b^{-1}$ evaluated along the line $y = 4$ where the load acts and then integrated from $x = 2$ where the load starts to $x = 7$ where the load ends:

$$[\mathbf{h}_b] = \begin{bmatrix} 1 & 4 & 0 & 16 & 0 & 0 \\ 1 & 7 & 4 & 49 & 28 & 16 \\ 1 & 2 & 4 & 4 & 8 & 16 \\ 1 & 5.5 & 2 & 30.25 & 11 & 4 \\ 1 & 4.5 & 4 & 20.25 & 18 & 16 \\ 1 & 3 & 2 & 9 & 6 & 4 \end{bmatrix},$$

$$[\mathbf{h}_b^{-1}] = \begin{bmatrix} 1 & 2.08 & 0.48 & -3.2 & -2.56 & 3.2 \\ 0 & -0.84 & -0.44 & 0.8 & 1.28 & -0.8 \\ -0.75 & -0.42 & 0.33 & 1.2 & -0.16 & -0.02 \\ 0 & 0.08 & 0.08 & 0 & -0.16 & 0 \\ 0 & 0.08 & -0.12 & -0.2 & 0.04 & 0.2 \\ 0.125 & 0.02 & 0.045 & -0.1 & 0.06 & -0.15 \end{bmatrix}.$$

Thus

$$N_{b2}(x, \ 4) = 2.08 - 0.84x - 0.42(4) + 0.08x^2 + 0.08x(4) + 0.02(4^2)$$
$$= 0.72 - 0.52x + 0.08x^2,$$

$$\int_L N_{b2}(L)b_y t \ dL = -400 \int_2^7 (0.72 - 0.52x + 0.08x^2) \ dx = -333.33 \, \text{lb}.$$

Similarly

$$\int_L N_{b3}(L)b_y t \ dL = -333.33 \, \text{lb}, \qquad \int_L L_{b5}(L)b_y t \ dL = -1333.33 \, \text{lb}.$$

These load values are inserted in the vector of equivalent nodal forces for the line load on element b as

$$\{\mathbf{F}_{bbLL}\} = \begin{array}{c} ㉑ \\ ㉒ \\ ⑤ \\ ⑥ \\ ① \\ ② \\ ⑪ \\ ⑫ \\ ③ \\ ④ \\ ⑨ \\ ⑩ \end{array} \left\{ \begin{array}{c} 0 \\ 0 \\ 0 \\ -333.33 \\ 0 \\ -333.33 \\ 0 \\ 0 \\ 0 \\ -1333.33 \\ 0 \\ 0 \end{array} \right\} \quad \text{lb.}$$

Finally, the three element stiffness matrices are assembled as the global stiffness matrix following the procedures discussed in Chapter 4. The three element equivalent nodal force vectors for the gravitational load, the element b equivalent nodal force vector for the line load, the applied concentrated loads, and the unknown support reactions are assembled as the global load vector using the procedures of Chapter 4. The support restraints are imposed by specifying zero values for the displacements at positions 19–24 in the global displacement vector. Thus the 24 global stiffness equations are ready for solution:

$$\left[\begin{array}{c|c} \mathbf{K}_{ff} & \mathbf{K}_{fs} \\ \hline \mathbf{K}_{sf} & \mathbf{K}_{ss} \end{array} \right] \left\{ \begin{array}{c} \mathbf{q}_f \\ \mathbf{0} \end{array} \right\} = \left\{ \begin{array}{c} \mathbf{F}_f \\ \mathbf{R}_s \end{array} \right\}.$$

Using the submatrix solution procedure,

$$[\mathbf{K}_{ff}]\{\mathbf{q}_f\} = \{\mathbf{F}_f\},$$

235.2	−69.1	−121.6	74.2	25.6	−30.7	−48.0	−96.0	−124.8	98.6	19.2	48.6	0.0	0.0	0.0	0.0	0.0	0.0	q_1		300
−69.1	332.8	−41.0	−51.2	20.5	25.6	−64.0	−128.0	181.7	−230.4	−41.0	−51.2	0.0	0.0	0.0	0.0	0.0	0.0	q_2		−333
−121.6	−41.0	499.2	5.1	−153.6	10.2	0.0	0.0	−102.4	−158.7	−89.6	158.7	0.0	0.0	0.0	0.0	0.0	0.0	q_3		0
74.2	−51.2	5.1	614.4	−15.4	−51.2	0.0	0.0	−158.7	−307.2	158.7	−204.8	0.0	0.0	0.0	0.0	0.0	0.0	q_4		−1472
25.6	20.58	−153.6	−15.4	212.8	46.1	0.0	0.0	12.8	15.4	−99.2	−130.6	−48.0	64.0	0.0	0.0	0.0	0.0	q_5		400
−30.7	25.6	10.2	−51.2	46.1	294.4	0.0	0.0	15.4	51.2	−134.8	−281.6	42.7	−128.0	0.0	0.0	0.0	0.0	q_6		−33
−48.0	−64.0	0.0	0.0	0.0	0.0	656.0	0.0	−464.0	0.0	0.0	0.0	0.0	0.0	−96.0	160.0	0.0	0.0	q_7		0
−96.0	−128.0	0.0	0.0	0.0	0.0	0.0	576.0	0.0	−64.0	0.0	0.0	0.0	0.0	160.0	−256.0	0.0	0.0	q_8		−111
−124.8	181.8	−102.4	−158.7	12.8	15.4	−464.0	0.0	1155.2	5.1	−307.2	−5.1	0.0	0.0	−96.0	−160.0	0.0	0.0	q_9	=	0
98.6	−230.4	−158.7	−307.2	15.4	51.2	0.0	−64.0	5.1	1190.4	−5.1	−102.4	0.0	0.0	−160.0	−256.0	0.0	0.0	q_{10}		−250
19.2	−41.0	−89.6	158.7	−99.2	−134.8	0.0	0.0	−307.2	−5.1	870.7	5.1	−179.6	−0.0	0.0	0.0	−96.0	106.7	q_{11}		0
48.6	−51.2	158.7	−204.8	−130.6	−281.6	0.0	0.0	−5.1	−102.4	5.1	1083.7	0.0	42.7	0.0	0.0	106.7	−256.0	q_{12}		−306
0.0	0.0	0.0	0.0	−48.0	42.7	0.0	0.0	0.0	0.0	−179.6	0.0	371.6	0.0	0.0	0.0	−96.0	−106.7	q_{13}		0
0.0	0.0	0.0	0.0	64.0	−128.0	0.0	0.0	0.0	0.0	0.0	42.7	0.0	469.3	0.0	0.0	−106.7	−256.0	q_{14}		−167
0.0	0.0	0.0	0.0	0.0	0.0	−96.0	160.0	−96.0	−160.0	0.0	0.0	0.0	0.0	656.0	0.0	0.0	0.0	q_{15}		0
0.0	0.0	0.0	0.0	0.0	0.0	160.0	−256.0	−160.0	−256.0	0.0	0.0	0.0	0.0	0.0	576.0	0.0	0.0	q_{16}		−111
0.0	0.0	0.0	0.0	0.0	0.0	0.0	0.0	0.0	0.0	−96.0	106.7	−96.0	−106.7	0.0	0.0	371.6	0.0	q_{17}		0
0.0	0.0	0.0	0.0	0.0	0.0	0.0	0.0	0.0	0.0	106.7	−256.0	−106.7	−256.0	0.0	0.0	0.0	469.3	q_{18}		−167

$$[\mathbf{K}_{sf}]\{\mathbf{q}_f\} = \{\mathbf{R}_s\},$$

$$
\begin{bmatrix}
12.0 & 16.0 & 0.0 & 0.0 & 0.0 & 0.0 & -48.0 & -64.0 & 0.0 & 0.0 & 0.0 & 0.0 & 0.0 & 0.0 & -232.0 & -16.0 & 0.0 & 0.0 \\
24.0 & 32.0 & 0.0 & 0.0 & 0.0 & 0.0 & -96.0 & -128.0 & 0.0 & 0.0 & 0.0 & 0.0 & 0.0 & 0.0 & 16.0 & -32.0 & 0.0 & 0.0 \\
2.4 & -3.2 & -32.0 & -64.0 & 37.6 & 61.9 & 0.0 & 0.0 & -73.6 & 204.8 & -118.4 & -183.5 & 0.0 & 0.0 & -232.0 & 16.0 & -89.8 & -10.7 \\
-49.6 & 70.4 & 25.6 & 0.0 & 16.0 & 57.6 & 0.0 & 0.0 & 121.6 & -281.6 & -89.6 & -230.4 & 0.0 & 0.0 & -16.0 & -32.0 & 10.7 & 21.3 \\
0.0 & 0.0 & 0.0 & 0.0 & 12.0 & -1.67 & 0.0 & 0.0 & 0.0 & 0.0 & 0.0 & 0.0 & -48.0 & 42.7 & 0.0 & 0.0 & -89.8 & 10.7 \\
0.0 & 0.0 & 0.0 & 0.0 & -16.0 & 32.0 & 0.0 & 0.0 & 0.0 & 0.0 & 0.0 & 0.0 & 64.0 & -128.0 & 0.0 & 0.0 & -10.7 & 21.3
\end{bmatrix}
\begin{Bmatrix} q_1 \\ q_2 \\ q_3 \\ q_4 \\ q_5 \\ q_6 \\ q_7 \\ q_8 \\ q_9 \\ q_{10} \\ q_{11} \\ q_{12} \\ q_{13} \\ q_{14} \\ q_{15} \\ q_{16} \\ q_{17} \\ q_{18} \end{Bmatrix}
= \begin{Bmatrix} R_{19} \\ R_{20} \\ R_{21} \\ R_{22} \\ R_{23} \\ R_{24} \end{Bmatrix},
$$

where the unit for $\mathbf{K}_{ff}$ and $\mathbf{K}_{sf}$ is kips per foot $\times 10^6$, for vector $\mathbf{F}_f$ is pounds, for vector $\mathbf{q}_f$ is feet $\times 10^{-6}$, and for vector $\mathbf{R}_s$ is lb:

$$
\{\mathbf{q}_f\} = [\mathbf{K}_{ff}]^{-1}\{\mathbf{F}_f\} = \begin{Bmatrix} 10.857 \\ -5.101 \\ 8.954 \\ -11.358 \\ 9.809 \\ -14.519 \\ 3.613 \\ -1.472 \\ 3.597 \\ -4.915 \\ 4.809 \\ -10.244 \\ 1.951 \\ -11.380 \\ 0.216 \\ -3.036 \\ 1.421 \\ -12.800 \end{Bmatrix}, \quad
\{\mathbf{R}_s\} = [\mathbf{K}_{sf}]\{\mathbf{q}_f\} = \begin{Bmatrix} -31.96 \\ 39.52 \\ -97.38 \\ 2238.76 \\ -570.66 \\ 671.72 \end{Bmatrix}.
$$

Note that the calculated support reactions need not be corrected for equivalent nodal forces, since no equivalent forces act along the lines of action of these reactions.

Figure 8.19 shows the structure with its applied loads, weight, and reactions. The center of gravity of the plate is 4.7778 ft to the right of the left pin support, as in the earlier example.

It is left as an exercise for the reader to complete the calculations for this example, to include showing that the loads, weight, and reactions satisfy the conditions for equilibrium.

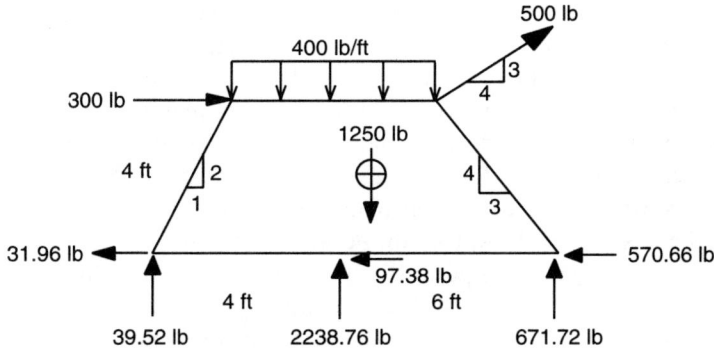

Figure 8.19 Plate's reactions

On comparing Figure 8.10 and 8.19, it is clear that the two triangular strain finite elements produce different results, even though both sets of results satisfy the equilibrium and support restraint conditions. The two different elements possess different stiffness equations. With only three elements in each analysis, the global stiffness equations for the structure reflect strongly the character of the element stiffness equations. With more elements in the analysis, the global stiffness equations will exhibit in a truer manner the behavior of the continuum of the plate and be less affected by the character of the finite element used. A large number of either type of element will produce support reactions which closely agree with the behavior of the plate as a continuum.

8.6. CLOSING REMARKS

The straight-sided 3- and 6-node triangular finite elements discussed in this chapter provide a means for analysis of plane stress structures. The inherent geometric advantage of the triangular shape for modeling 2D structures provides an ability to treat plane structures with a great variety of shapes. The 3-node constant strain element is the simplest to use and program but requires many elements for analysis of stress when stresses in the structure vary sharply. The 6-node linear strain element is more complicated to use. Its matrices are twice as large as those for the constant strain element, and the entries for its matrices must be obtained by a quadrature procedure. However, comparably sized 6-node linear strain elements are more accurate than constant strain elements.

If a plane stress structure were analyzed twice, once using the constant strain element and once using the linear strain element, and the same pattern of nodes were used in each analysis, the same number of global stiffness equations would result from each. In a space occupied by the linear strain element, the stresses would vary linearly. That same space would be occupied by four constant strain elements and thus would exhibit four values of each stress, one set of values for each constant strain element. The result would likely be less accurate for the constant strain element analysis, in stress prediction, than for the linear strain element analysis.

The computations for the constant strain element would be simpler and carried out faster. There would be four times as many element stiffness matrices to determine and account for but the same number of element stiffness coefficients since the linear strain element stiffness matrix has four times as many entries as does a constant strain element stiffness matrix. Element stiffness matrices computed for the linear strain element would be done numerically as suggested in Section 8.5.3, so that to obtain each stiffness coefficient, approximately three times as many computations would be needed as compared to the computations required to obtain each constant strain stiffness coefficient. Computations for equivalent nodal load vectors likely would be somewhat greater also for the linear strain element. Whether the greater computational effort for a linear strain analysis is justified must be resolved by the analyst in terms of greater accuracy achieved in the results.

BIBLIOGRAPHY

Cook, R. D., Malkus, D. S., and Plesha, M. E., *Concepts and Applications of Finite Element Analysis* (Chapter 5), Wiley, New York, 1989.

Eisenberg, M. A., and Malvern, L. E., "On Finite Element Integration in Natural Coordinates," *International Journal of Numerical Methods in Engineering*, Vol. 7, No. 4, 1973, pp. 574–575.

Logan, D. L., *A First Course in the Finite Element Method* (Chapters 7–9), PWS-Kent, Boston, 1992.

McGuire, W., and Gallagher, R. H., *Matrix Structural Analysis* (Chapter 12), Wiley, New York, 1979.

Prathap, G., *The Finite Element Method in Structural Mechanics* (Chapter 7), Kluwer Academic, Dordrecht, The Netherlands, 1993.

Shames, I. H., and Dym, C. L., *Energy and Finite Elements in Structural Mechanics* (Chapter 12), Hemisphere, (McGraw-Hill), New York, 1985.

Weaver, W. Jr., and Johnston, P. R., *Finite Elements for Structural Analysis* (Chapters 2, 3, Appendix B), Prentice-Hall, Englewood Cliffs, NJ, 1984.

Yang, T. Y., *Finite Element Structural Analysis* (Chapters 9, 11), Prentice-Hall, Englewood Cliffs, NJ, 1986.

Zienkiewcz, O. C., and Taylor, R. L., *The Finite Element Method* (Chapter 3), 4th ed., McGraw-Hill, Berkshire, England, 1989.

PROBLEMS

The use of computer software such as MATLAB or a spreadsheet which manipulates individual matrices and evaluates functions at a point in their space will facilitate solution for many of these problems.

8.1. The nodal coordinates for the 3-node, 6-dof plane stress thin plate triangular finite element shown are inches. Its properties are $t = 0.25$ in., $E = 30,000$ ksi and $v = 0.25$.

a. Write the element's shape functions.
b. Write its stiffness matrix.

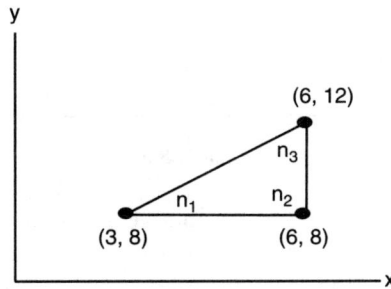

Problem 8.1

8.2. The 3-node, 6-dof plane stress thin plate triangular finite element shown has its nodal coordinates expressed in feet. Write the vector of equivalent nodal loads for:

a. A concentrated 500-kip load acting from a point at (6, 4) and sloping up and to the right at 3 vertical on 4 horizontal.
b. A distributed line load acting along edge 1–3 with an intensity of 25 lb/ft of edge length and sloping down and to the right at 4 vertical and 3 horizontal.
c. A distributed line load acting horizontally to the right along edge 1–3 with a linearly increasing intensity from 0 lb/ft of edge length at node 1 to 20 lb/ft of edge length at node 3.

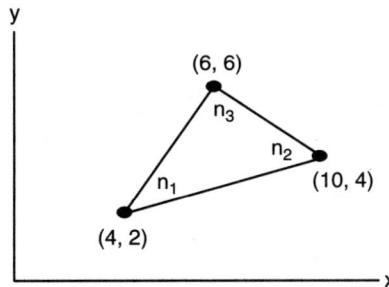

Problem 8.2

8.3. Write the stiffness matrix for the plane stress thin plate triangular finite element shown. The global coordinates of nodes are indicated in feet. The plate is isotropic steel. Its modulus of elasticity $E = 30,000\,\text{ksi}$, Poisson's ratio $\nu = 0.3$, and thickness $t = 1\frac{1}{4}\,\text{in}$.

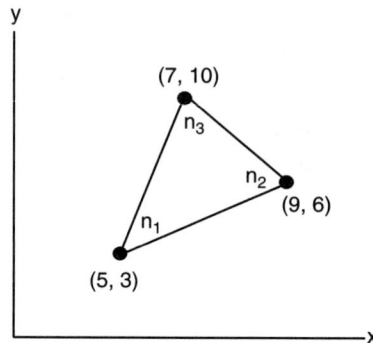

Problem 8.3

8.4. The element of Problem 8.3 has a unit weight of $485\,\text{lb/ft}^3$ and is subjected to an in-plane concentrated load of $1000\,\text{lb}$ acting at coordinates $(7.5, 6.5)$ and sloping down and to the right at an angle of $25°$ from the horizontal. Write the vector of equivalent nodal loads for the element.

8.5. The element loaded as described in Problem 8.4 is also subjected to an in-plane line load of $200\,\text{lb/ft}$ of edge length which acts normal to side 1–2 and out from the element. Write the vector of equivalent nodal loads for the element.

8.6. Determine the support reactions of the plane stress plate in the example of Section 8.2.4. The plate remains loaded and configured as shown in Figure 8.8, except that the pin support at C is changed to a roller support which rolls horizontally. Verify that these reactions, the loads, and the plate weight satisfy the equations of equilibrium.

8.7. Repeat Problem 8.6, but with pin supports at C and E and a roller support at D which rolls horizontally.

8.8. The structure shown is a thin plate with two in-plane concentrated loads and pin supports at C and D. Using the two 3-node, 6-dof triangular plane stress finite elements indicated, determine the displacement at point B and the support reactions. Verify that the reactions satisfy the equations of equilibrium. Neglect the plate's weight. Its elastic modulus $E = 30,000$ ksi, Poisson's ratio $v = \frac{1}{4}$, and thickness $t = \frac{7}{8}$ in.

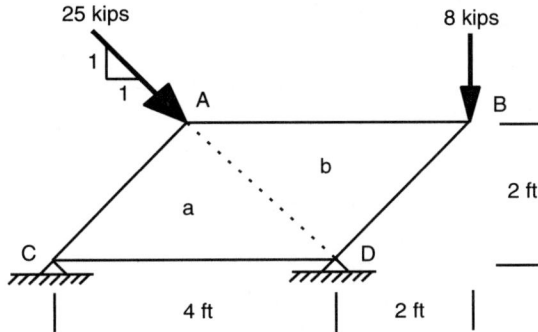

Problem 8.8

8.9. Repeat Problem 8.8, but use four 3-node, 6-dof triangular plane stress finite elements. The four elements should be obtained by dividing element a of Problem 8.8 into two elements with a vertical line from point A to side CD and element b into two elements with a vertical line from point D to side AB.

8.10. The structure shown is a thin plate with a single in-plane concentrated load at point D. It is supported with a pin support at A, a roller at C, and lateral support along its entire length. Determine the support reactions at points A and C and the displacement of point B using the eight 3-node, 6-dof plane stress triangular finite elements indicated. Neglect the weight of the plate. Its elastic modulus $E = 30,000$ ksi, Poisson's ratio $v = \frac{1}{4}$, and thickness $t = \frac{3}{4}$ in.

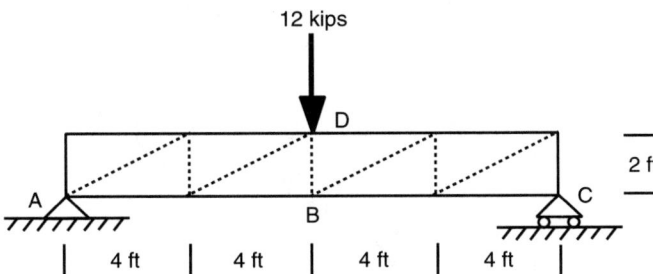

Problem 8.10

8.11. Repeat Problem 8.10 but change the roller at C to a pin support.

8.12. Repeat Problem 8.10 using the eight 3-node, 6-dof plane stress triangular finite elements oriented as shown. Why do these results differ from those of Problem 8.10?

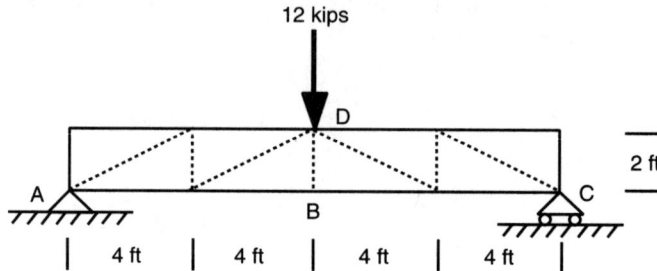

Problem 8.12

8.13. Repeat Problem 8.12 changing the roller support at C to a pin support. Why do these results differ from those of Problem 8.11?

8.14. The 3-node, 6-dof plane stress thin plate triangular finite element shown has a thickness $t = 1.20$ in., and its nodal coordinates are expressed in feet.

a. Write the shape functions $N_i(x, y)$ for the element by writing the area coordinates of a point on the element in terms of x, y, and the nodal coordinates.

b. Integrate each shape function over the area of the element.

c. Write the vector of equivalent nodal loads for the weight of the element. Its unit weight $\gamma = 485$ lb/ft^3 and acts in-plane.

d. An in-plane body force on the element is $b_x(x, y) = 0$, $b_y(x, y) = y$, where the unit is pounds per square feet. Write the element's vector of equivalent nodal loads for this body force.

e. An in-plane distributed line load acts horizontally right along edge 1–3 with a linearly increasing intensity from 0 lb/ft of edge length at node 1 to 20 lb/ft of edge length at node 3. Using the methods described in Section 8.4, write the vector of equivalent nodal loads for this line load.

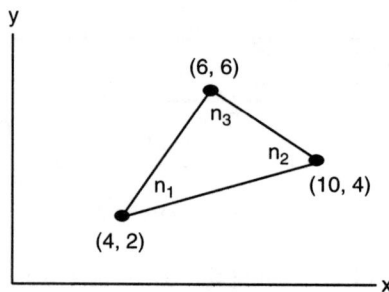

Problem 8.14

8.15. A steel plate structure rests on two pin supports and is subjected to in-plane loads, as shown. The line load from D to A acts horizontally; its intensity varies linearly from 0 to 500 lb/ft of edge length. The line load from A to C acts vertically; its intensity varies parabolically from 0 to 300 to 0 lb/ft of edge length. Using the four 3-node, 6-dof plane stress thin plate triangular finite elements indicated, determine the reactions at the two pin supports. Use software which manipulates individual matrices, and verify that loads, plate weight, and reactions satisfy the equations of equilibrium. The elastic modulus $E = 30,000$ ksi, Poisson's ratio $v = \frac{1}{4}$, thickness $t = 0.800$ in., and unit weight $\gamma = 500$ lb/ft^3.

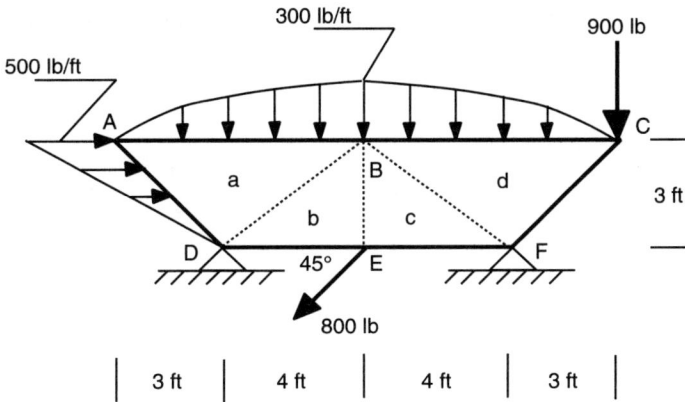

Problem 8.15

8.16. The 6-node, 12-dof plane stress thin plate triangular finite element shown has a thickness $t = 1.20$ in., and its nodal coordinates are expressed in feet.

a. Write the shape function $N_i(x, y)$ for the element.
b. Integrate each shape function over the area of the element.
c. Write the vector of equivalent nodal loads for the weight of the element. Its unit weight $\gamma = 485$ lb/ft^3 and acts in-plane.

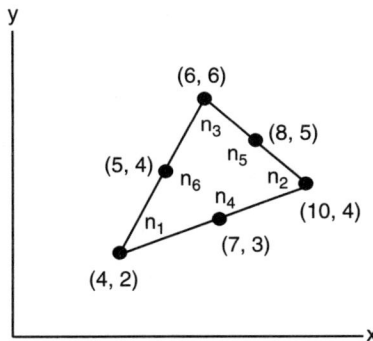

Problem 8.16

d. A body force acting in-plane and throughout the element is $b_x(x, y) = 0$ and $b_y(x, y) = y$, where the unit is pounds per square feet. Write the element's vector of equivalent nodal loads for this body force.

e. A distributed line load acts horizontally to the right along edge 1–3 with a linearly increasing intensity from 0 lb/ft of edge length at node 1 to 20 lb/ft of edge length at node 3. Write the vector of equivalent nodal loads for this line load,

    (1) Use the method of the example in Section 8.5.5 to write appropriate integrands.

    (2) Use equations 5.37 and the line load as a function of the length variable $(0 - L_{13})$ to write appropriate integrands.

8.17. Repeat Problem 8.8 using two 6-node, 12-dof plane stress thin plate triangular finite elements.

CHAPTER 9

ISOPARAMETRIC PLANE STRESS STRUCTURAL QUADRILATERAL FINITE ELEMENTS

9.1. INTRODUCTION

Quadrilateral plane stress finite elements offer an approach slightly different from triangular elements in the modeling of structures subjected to the conditions of plane stress. They possess more nodes and dof than comparable triangular elements and therefore require shape functions of higher polynomial degree. This allows more variation of strain and stress within the element with the consequent opportunities for better accuracy. The quadrilateral is as adept at modeling complex plane shapes as is the triangle.

The mathematics of developing element stiffness equations is more complicated for quadrilateral finite elements. Explicit expressions for stiffness coefficients will be possible only for the simplest quadrilateral element, the rectangle. The more useful quadrilateral finite elements must have their stiffness coefficients developed as integral expressions, and these integrals must be evaluated by numerical means. This approach was employed for the 6-node, 12-dof linear strain triangular finite element in the last chapter, achieving simplifications and efficiencies over what would have occurred if explicit expressions were attempted.

Three plane stress quadrilateral finite elements are common. They are the 4-node, 8-dof bilinear element, the 8-node, 16-dof serendipity element, and the 9-node, 18-dof Lagrange (tensor) element. Each is illustrated in Figure 9.1.

The generic polynomial functions used in the displacement and shape functions for each are

$$\lfloor g \rfloor = \lfloor 1 \quad x \quad y \quad xy \rfloor \quad \text{(bilinear)}, \tag{9.1a}$$

$$\lfloor g \rfloor = \lfloor 1 \quad x \quad y \quad x^2 \quad xy \quad y^2 \quad x^2y \quad xy^2 \rfloor \quad \text{(serendipity)}, \tag{9.1b}$$

$$\lfloor g \rfloor = \lfloor 1 \quad x \quad y \quad x^2 \quad xy \quad y^2 \quad x^2y \quad xy^2 \quad x^2y^2 \rfloor \quad \text{(Lagrange)}. \tag{9.1c}$$

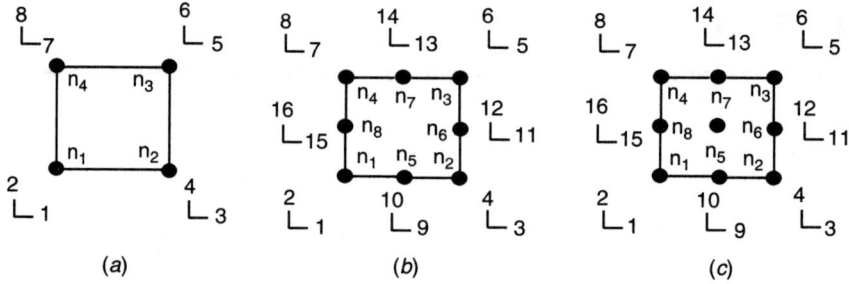

Figure 9.1. Quadrilateral finite elements: (a) bilinear; (b) serendipity; (c) Lagrange (center node is n_9 with edof 17, 18).

The simplest is the 4-node bilinear element, but it is the least accurate. It affords less variation of strain and stress within the element and, in conjunction with isoparametric transformation, allows geometric modeling only with straight-sided quadrilaterals. Isoparametric transformation will be described subsequently. The 8-node serendipity and 9-node Lagrange elements, on the other hand, allow greater variation of strain and stress within their elements and allow geometric modeling with quadratic curve sided quadrilaterals if isoparametric transformation is employed.

9.2. BILINEAR RECTANGULAR ELEMENT (4-NODE, 8-DOF)

Though rectangular elements offer no advantage over 3-node, 6-dof triangular elements, they do provide a convenient opening for discussion of more general quadrilateral finite elements. A 4-node, 8-dof rectangular finite element is shown in Figure 9.2.

It is convenient to define local x–y reference axes with their origin at the center of the rectangle. Local natural dimensionless coordinates are taken as $\zeta = x/a$ and $\eta = y/b$. The length of the rectangle in the x direction is $2a$, and in the y direction it is $2b$.

9.2.1. Shape and Displacement Functions

With four nodes and the two dof for each node chosen as translation displacements in the x and y directions, the element displacement functions u and v will each contain four generic polynomial terms comparable to what is shown in equation

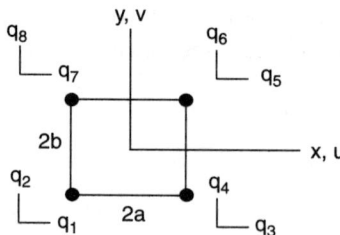

Figure 9.2. Rectangular finite element.

9.1a. They may be expressed as functions of ζ and η. The shape functions $N_i(\zeta,\ \eta)$ therefore will contain the same four generic polynomial terms. The displacement functions are written as the sum of the products of the shape functions and their corresponding nodal displacements:

$$u(\zeta,\ \eta) = N_1(\zeta,\ \eta)q_1 + N_2(\zeta,\ \eta)q_3 + N_3(\zeta,\ \eta)q_5 + N_4(\zeta,\ \eta)q_7, \qquad (9.2\text{a})$$

$$v(\zeta,\ \eta) = N_1(\zeta,\ \eta)q_2 + N_2(\zeta,\ \eta)q_4 + N_3(\zeta,\ \eta)q_6 + N_4(\zeta,\ \eta)q_8. \qquad (9.2\text{b})$$

In matrix form they are

$$
\left\{ \begin{array}{c} u(\zeta,\ \eta) \\ v(\zeta,\ \eta \end{array} \right\} =
\begin{bmatrix}
N_1(\zeta,\ \eta) & 0 & N_2(\zeta,\ \eta) & 0 & N_3(\zeta,\ \eta) & 0 & N_4(\zeta,\ \eta) & 0 \\
0 & N_1(\zeta,\ \eta) & 0 & N_2(\zeta,\ \eta) & 0 & N_3(\zeta,\ \eta) & 0 & N_4(\zeta,\ \eta)
\end{bmatrix}
\left\{ \begin{array}{c} q_1 \\ q_2 \\ q_3 \\ q_4 \\ q_5 \\ q_6 \\ q_7 \\ q_8 \end{array} \right\},
$$

$$(9.2\text{c})$$

$$\{\mathbf{u}(\zeta,\ \eta)\} = [\mathbf{N}(\zeta,\ \eta)]\{\mathbf{q}\}. \qquad (9.2\text{d})$$

The shape functions $N_i(\zeta,\ \eta)$ are

$$N_1(\zeta,\ \eta) = C_{11} + C_{21}\zeta + C_{31}\eta + C_{41}\zeta\eta, \qquad (9.3\text{a})$$

$$N_2(\zeta,\ \eta) = C_{12} + C_{22}\zeta + C_{32}\eta + C_{42}\zeta\eta, \qquad (9.3\text{b})$$

$$N_3(\zeta,\ \eta) = C_{13} + C_{23}\zeta + C_{33}\eta + C_{43}\zeta\eta, \qquad (9.3\text{c})$$

$$N_4(\zeta,\ \eta) = C_{14} + C_{24}\zeta + C_{34}\eta + C_{44}\zeta\eta. \qquad (9.3\text{d})$$

In matrix form,

$$
\left\{ \begin{array}{c} N_1(\zeta,\ \eta) \\ N_2(\zeta,\ \eta) \\ N_3(\zeta,\ \eta) \\ N_4(\zeta,\ \eta) \end{array} \right\} =
\begin{bmatrix}
C_{11} & C_{21} & C_{31} & C_{41} \\
C_{12} & C_{22} & C_{32} & C_{42} \\
C_{13} & C_{23} & C_{33} & C_{43} \\
C_{14} & C_{24} & C_{34} & C_{41}
\end{bmatrix}
\left\{ \begin{array}{c} 1 \\ \zeta \\ \eta \\ \zeta\eta \end{array} \right\},
\qquad (9.3\text{e})
$$

$$\{\mathbf{N}(\zeta,\ \eta)\} = \lfloor \mathbf{C}^{\mathrm{T}} \rfloor \{\mathbf{g}\}, \qquad (9.3\text{f})$$

or transposing,

$$\lfloor \mathbf{N}(\zeta,\ \eta) \rfloor = \lfloor \mathbf{g} \rfloor [\mathbf{C}]. \qquad (9.3\text{g})$$

Evaluating equation 9.3g successively at each node allows the formation by row of the matrix equation:

$$
\begin{bmatrix} 1 & 0 & 0 & 0 \\ 0 & 1 & 0 & 0 \\ 0 & 0 & 1 & 0 \\ 0 & 0 & 0 & 1 \end{bmatrix} = \begin{bmatrix} 1 & -1 & -1 & 1 \\ 1 & 1 & -1 & -1 \\ 1 & 1 & 1 & 1 \\ 1 & -1 & 1 & -1 \end{bmatrix} \begin{bmatrix} C_{11} & C_{12} & C_{13} & C_{14} \\ C_{21} & C_{22} & C_{23} & C_{24} \\ C_{31} & C_{32} & C_{33} & C_{34} \\ C_{41} & C_{42} & C_{43} & C_{44} \end{bmatrix},
\tag{9.4a}
$$

$$
[\mathbf{I}] = [\mathbf{h}][\mathbf{C}].
\tag{9.4b}
$$

Thus

$$
[\mathbf{C}] = [\mathbf{h}^{-1}].
\tag{9.4c}
$$

To obtain the constant multipliers for the shape function polynomial terms, that is, the entries of matrix $\mathbf{C}$, the inverse of matrix $\mathbf{h}$ must be determined. Recalling equations 1.12, one notes that the matrix $\mathbf{h}$ multiplied by the scalar $\frac{1}{2}$ is an orthogonal matrix:

$$
\frac{1}{2}[\mathbf{h}] = \begin{bmatrix} \frac{1}{2} & -\frac{1}{2} & -\frac{1}{2} & \frac{1}{2} \\ \frac{1}{2} & \frac{1}{2} & -\frac{1}{2} & -\frac{1}{2} \\ \frac{1}{2} & \frac{1}{2} & \frac{1}{2} & \frac{1}{2} \\ \frac{1}{2} & -\frac{1}{2} & \frac{1}{2} & -\frac{1}{2} \end{bmatrix}.
\tag{9.5a}
$$

Therefore

$$
[\tfrac{1}{2}\mathbf{h}]^{-1} = [\tfrac{1}{2}\mathbf{h}]^{\mathrm{T}}.
\tag{9.5b}
$$

Since

$$
[\mathbf{h}] = 2[\tfrac{1}{2}\mathbf{h}],
$$

$$
[\mathbf{h}^{-1}] = \tfrac{1}{2}[\tfrac{1}{2}\mathbf{h}]^{-1} = \tfrac{1}{2}[\tfrac{1}{2}\mathbf{h}]^{\mathrm{T}},
$$

and

$$
[\mathbf{h}^{-1}] = \tfrac{1}{4}[\mathbf{h}^{\mathrm{T}}].
\tag{9.5c}
$$

Substituting equation 9.5c into equation 9.4c,

$$
[\mathbf{C}] = \tfrac{1}{4}[\mathbf{h}^{\mathrm{T}}] \quad \text{or} \quad [\mathbf{C}^{\mathrm{T}}] = \tfrac{1}{4}[\mathbf{h}].
\tag{9.5d}
$$

Substituting equation 9.5d into equation 9.3f,

$$
\{\mathbf{N}(\zeta,\ \eta)\} = \tfrac{1}{4}[\mathbf{h}]\{\mathbf{g}\},
\tag{9.6a}
$$

$$
\begin{Bmatrix} N_1(\zeta,\ \eta) \\ N_2(\zeta,\ \eta) \\ N_3(\zeta,\ \eta) \\ N_4(\zeta,\ \eta) \end{Bmatrix} = \frac{1}{4} \begin{bmatrix} 1 & -1 & -1 & 1 \\ 1 & 1 & -1 & -1 \\ 1 & 1 & 1 & 1 \\ 1 & -1 & 1 & -1 \end{bmatrix} \begin{Bmatrix} 1 \\ \zeta \\ \eta \\ \zeta\eta \end{Bmatrix},
\tag{9.6b}
$$

and

$$
N_1(\zeta,\ \eta) = \tfrac{1}{4}(1 - \zeta - \eta + \zeta\eta) = \tfrac{1}{4}(1 - \zeta)(1 - \eta),
\tag{9.6c}
$$

$$
N_2(\zeta,\ \eta) = \tfrac{1}{4}(1 + \zeta - \eta - \zeta\eta) = \tfrac{1}{4}(1 + \zeta)(1 - \eta),
\tag{9.6d}
$$

$$
N_3(\zeta,\ \eta) = \tfrac{1}{4}(1 + \zeta + \eta + \zeta\eta) = \tfrac{1}{4}(1 + \zeta)(1 + \eta),
\tag{9.6e}
$$

$$
N_4(\zeta,\ \eta) = \tfrac{1}{4}(1 - \zeta + \eta - \zeta\eta) = \tfrac{1}{4}(1 - \zeta)(1 + \eta).
\tag{9.6f}
$$

If the local natural nodal coordinates are represented by $(\zeta_i,\ \eta_i)$, where i is the node number, $(\zeta_i,\ \eta_i)$ is $(-1,\ -1)$ at node 1, $(-1,\ 1)$ at node 2, $(1,\ 1)$ at node 3, and $(-1,\ 1)$ at node 4, the shape functions may then be represented by the expression

$$
N_i(\zeta,\ \eta\} = \tfrac{1}{4}(1 + \zeta_i\zeta)(1 + \eta_i\eta).
\tag{9.6g}
$$

A sketch of each shape function plotted as a curved 3D surface over the element in ζ–η space is shown in Figure 9.3. The surfaces are bilinear. As can be seen from the figure, they are linear along the edges of the element. Examination of equations 9.6 indicates that along any line parallel to the ζ axis (constant η) the shape functions and surfaces are linear with respect to ζ. Similarly, along lines parallel to the η axis (constant ζ) the shape functions and surfaces are linear with respect to η. In other directions, however, they are second degree. For example, along the diagonal line from node 1 to node 3 ($\zeta = \eta$),

$$
N_i(\zeta,\ \eta) = N_i(\zeta) = \tfrac{1}{4}(1 + \zeta_i\zeta)(1 + \eta_i\zeta)
$$

$$
= \tfrac{1}{4}(1 + (\zeta_i + \eta_i)\zeta + \zeta_i\eta_i\zeta^2).
$$

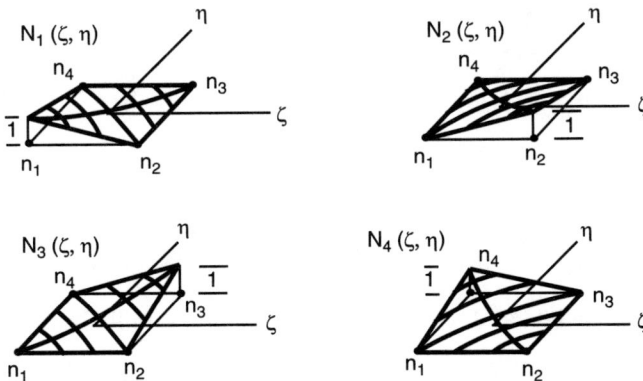

Figure 9.3. Shape functions.

9.2.2. Element Stiffness Matrix

Equation 6.22d provides the means to compute the element stiffness matrix

$$[\mathbf{K}] = \left[\int_{\text{vol}} [\mathbf{B}^{\mathrm{T}}][\mathbf{E}][\mathbf{B}] \, dV \right],$$

(6.22d)

where the matrix **B** is an array of derivatives of shape functions differentiated in accordance with the strain displacement relationships for plane stress (equations 8.6–8.9a and 9.2):

$$[\mathbf{B}] = [\mathbf{d}][\mathbf{N}],$$

(9.7a)

$$[\mathbf{B}] = \begin{bmatrix} \frac{\partial}{\partial x} & 0 \\ 0 & \frac{\partial}{\partial y} \\ \frac{\partial}{\partial y} & \frac{\partial}{\partial x} \end{bmatrix} \begin{bmatrix} N_1 & 0 & | & N_2 & 0 & | & N_3 & 0 & | & N_4 & 0 \\ 0 & N_1 & | & 0 & N_2 & | & 0 & N_3 & | & 0 & N_4 \end{bmatrix},$$

(9.7b)

$$[\mathbf{B}] = \begin{bmatrix} \mathbf{B}_1 & \mathbf{B}_2 & \mathbf{B}_3 & \mathbf{B}_4 \end{bmatrix},$$

(9.7c)

where

$$[\mathbf{B}_i] = \begin{bmatrix} \frac{\partial N_i}{\partial x} & 0 \\ 0 & \frac{\partial N_i}{\partial y} \\ \frac{\partial N_i}{\partial y} & \frac{\partial N_i}{\partial x} \end{bmatrix}.$$

(9.7d)

The shape functions must be differentiated with respect to x and y, not ζ and η, since the strains are defined as derivatives with respect to x and y. The shape functions were developed as functions of ζ and η, so the chain rule for differentiation must be employed:

$$\frac{\partial N_i(\zeta, \eta)}{\partial x} = \frac{\partial N_i(\zeta, \eta)}{\partial \zeta} \cdot \frac{\partial \zeta}{\partial x} + \frac{\partial N_i(\zeta, \eta)}{\partial \eta} \cdot \frac{\partial \eta}{\partial x},$$

(9.8a)

$$\frac{\partial N_i(\zeta, \eta)}{\partial y} = \frac{\partial N_i(\zeta, \eta)}{\partial \zeta} \cdot \frac{\partial \zeta}{\partial y} + \frac{\partial N_i(\zeta, \eta)}{\partial \eta} \cdot \frac{\partial \eta}{\partial y}.$$

(9.8b)

The natural coordinates ζ and η may be differentiated with respect to x and y to give

$$\zeta = \frac{x}{a}, \qquad \frac{\partial \zeta}{\partial x} = \frac{1}{a}, \qquad \frac{\partial \zeta}{\partial y} = 0;$$

(9.8c)

$$\eta = \frac{y}{b}, \qquad \frac{\partial \eta}{\partial x} = 0, \qquad \frac{\partial \eta}{\partial y} = \frac{1}{b}.$$

(9.8d)

Substituting equations 9.8c and 9.8d into equations 9.8a and 9.8b,

$$\frac{\partial N_i(\zeta, \eta)}{\partial x} = \frac{1}{a} \frac{\partial N_i(\zeta, \eta)}{\partial \zeta},$$

(9.8e)

$$\frac{\partial N_i(\zeta,\ \eta)}{\partial y} = \frac{1}{b}\frac{\partial N_i(\zeta,\ \eta)}{\partial \eta}. \tag{9.8f}$$

Substituting equations 9.8e and 9.8f into equation 9.7d,

$$[\mathbf{B}_i] = \begin{bmatrix} \frac{1}{a}\frac{\partial N_i}{\partial \zeta} & 0 \\ 0 & \frac{1}{b}\frac{\partial N_i}{\partial \eta} \\ \frac{1}{b}\frac{\partial N_i}{\partial \eta} & \frac{1}{a}\frac{\partial N_i}{\partial \zeta} \end{bmatrix}. \tag{9.9a}$$

differentiating equations 9.6g and recalling that the area of the rectangular element $A_e = 4ab$,

$$\frac{1}{a}\frac{\partial N_i}{\partial \zeta} = \frac{1}{4a}(\zeta_i)(1 + \eta_i\eta) = \frac{1}{A_e}(b\zeta_i)(1 + \eta_i\eta), \tag{9.9b}$$

$$\frac{1}{b}\frac{\partial N_i}{\partial \eta} = \frac{1}{4b}(\eta_i)(1 + \zeta_i\zeta) = \frac{1}{A_e}(a\eta_i)(1 + \zeta_i\zeta). \tag{9.9c}$$

Substituting equations 9.9b and 9.9c into equation 9.9a,

$$[\mathbf{B}_i] = \frac{1}{A_e} \begin{bmatrix} b\zeta_i(1 + \eta_i\eta) & 0 \\ 0 & a\eta_i(1 + \zeta_i\zeta) \\ a\eta_i(1 + \zeta_i\zeta) & b\zeta_i(1 + \eta_i\eta) \end{bmatrix}. \tag{9.9d}$$

The terms ζ_i and η_i are the coordinates of node i, $(-1,\ -1)$, $(1,\ -1)$, $(1,\ 1)$, and $(-1,\ 1)$ as $i = 1,\ 2,\ 3,\ 4$.

The complete matrix $\mathbf{B}$ for the 4-node rectangular plane stress finite element (Figure 9.2) is given as

$$[\mathbf{B}] = \frac{1}{A_3} \begin{bmatrix} -b(1-\eta) & 0 & | & b(1-\eta) & 0 & | & b(1+\eta) & 0 & | & -b(1+\eta) & 0 \\ 0 & -a(1-\zeta) & | & 0 & -a(1+\zeta) & | & 0 & a(1+\zeta) & | & 0 & a(1-\zeta) \\ -a(1-\zeta) & -b(1-\eta) & | & -a(1+\zeta) & b(1-\eta) & | & a(1+\zeta) & b(1+\eta) & | & a(1-\zeta) & -b(1+\eta) \end{bmatrix}. \tag{9.9e}$$

It contains linear functions of the local natural coordinates ζ and η, the dimensions of the rectangle a and b, and the area A_e of the rectangle which equals $4ab$.

The integral expression for the element's stiffness matrix, cited above as equation 6.22d, may be adapted for the plane stress condition by changing the variable of integration from V (volume) to x and y and then to ζ and η. The thickness t of a plane stress plate is constant, so that

$$dV = t\ dA = t\ dx\ dy. \tag{9.10a}$$

From the natural and rectilinear coordinate relationships,

$$x = a\zeta, \qquad dx = a\ d\zeta, \tag{9.10b}$$

$$y = b\eta, \qquad dy = b\ d\eta. \tag{9.10c}$$

Substituting equations 9.10b and 9.10c into equation 9.10a,

$$dV = (abt)\ d\zeta\ d\eta = (\tfrac{1}{4} A_e t)\ d\zeta\ d\eta. \tag{9.10d}$$

The matrix **E** was written earlier for an isotropic material in plane stress. As contained in equation 6.5a,

$$[\mathbf{E}] = \frac{E}{1 - \nu^2} \begin{bmatrix} 1 & \nu & 0 \\ \nu & 1 & 0 \\ 0 & 0 & \frac{1-\nu}{2} \end{bmatrix}. \tag{9.10e}$$

The othotropic expression for matrix **E** for plane stress is contained in equation 7.20.

Letting the symbols **B**$'$ and **E**$'$ refer to the matrices **B** and **E** without the scalar multipliers, the expression for matrix **K** may be written as

$$[\mathbf{K}] = \frac{Et}{4A_e(1 - \nu^2)} \int_{-1}^{1} \int_{-1}^{1} \left[\mathbf{B}'^{\mathrm{T}}(\zeta,\ \eta) \right][\mathbf{E}'][\mathbf{B}'(\zeta,\ \eta)]\ d\zeta\ d\eta,$$

$$= \frac{Et}{4A_e(1 - \nu^2)} \int_{-1}^{1} \int_{-1}^{1} [\mathbf{I}(\zeta,\ \eta)]\ d\zeta\ d\eta. \tag{9.10f}$$

The integrand $\mathbf{I}(\zeta,\ \eta)$ is an 8×8 matrix whose entries are quadratic polynomials in ζ and η which incorporate the elastic constant ν. Each entry must be integrated over the area of the rectangular element in ζ–η space.

Further insights into the matrix **K** and the integrand $\mathbf{I}(\zeta,\ \eta)$ may be gained by treating them in terms of submatrices, as was done for the linear strain triangular finite element (equations 8.35):

$$[\mathbf{I}(\zeta, \eta)] = [\mathbf{B}'^{\mathrm{T}}(\zeta,\ \eta)][\mathbf{E}'][\mathbf{B}'(\zeta,\ \eta)] \tag{9.11a}$$

$$= \begin{bmatrix} \mathbf{B}_1'^{\mathrm{T}} \\ \mathbf{B}_2'^{\mathrm{T}} \\ \mathbf{B}_3'^{\mathrm{T}} \\ \mathbf{B}_4'^{\mathrm{T}} \end{bmatrix} [\mathbf{E}'][\mathbf{B}_1'\ \ \mathbf{B}_2'\ \ \mathbf{B}_3'\ \ \mathbf{B}_4']. \tag{9.11b}$$

Alternatively, by expanding equation 9.11b,

$$[\mathbf{I}(\zeta,\ \eta)] = \begin{bmatrix} \mathbf{B}_1'^{\mathrm{T}}\mathbf{E}'\mathbf{B}_1' & \mathbf{B}_1'^{\mathrm{T}}\mathbf{E}'\mathbf{B}_2' & \mathbf{B}_1'^{\mathrm{T}}\mathbf{E}'\mathbf{B}_3' & \mathbf{B}_1'^{\mathrm{T}}\mathbf{E}'\mathbf{B}_4' \\ \mathbf{B}_2'^{\mathrm{T}}\mathbf{E}'\mathbf{B}_1' & \mathbf{B}_2'^{\mathrm{T}}\mathbf{E}'\mathbf{B}_2' & \mathbf{B}_2'^{\mathrm{T}}\mathbf{E}'\mathbf{B}_3' & \mathbf{B}_2'^{\mathrm{T}}\mathbf{E}'\mathbf{B}_4' \\ \mathbf{B}_3'^{\mathrm{T}}\mathbf{E}'\mathbf{B}_1' & \mathbf{B}_3'^{\mathrm{T}}\mathbf{E}'\mathbf{B}_2' & \mathbf{B}_3'^{\mathrm{T}}\mathbf{E}'\mathbf{B}_3' & \mathbf{B}_3'^{\mathrm{T}}\mathbf{E}'\mathbf{B}_4' \\ \mathbf{B}_4'^{\mathrm{T}}\mathbf{E}'\mathbf{B}_1' & \mathbf{B}_4'^{\mathrm{T}}\mathbf{E}'\mathbf{B}_2' & \mathbf{B}_4'^{\mathrm{T}}\mathbf{E}'\mathbf{B}_3' & \mathbf{B}_4'^{\mathrm{T}}\mathbf{E}'\mathbf{B}_4' \end{bmatrix}. \tag{9.11c}$$

The entries $\mathbf{B}_i'^{\mathrm{T}}\mathbf{E}'\mathbf{B}_j'$ are each 2×2 submatrices, consistent with the fact that the integrand $\mathbf{I}(\zeta, \eta)$ must be an 8×8 array of functions of ζ and η. As an example, examine the submatrix $\mathbf{B}_2'^T\mathbf{E}'\mathbf{B}_1'$, located in matrix $\mathbf{I}(\zeta, \eta)$ in the shaded space shown below in equation 9.11d. Note it comprises the integrands for the 4-element stiffness coefficients K_{31}, k_{32}, k_{41}, and k_{42}. The circled numbers refer to the node numbers which identify the submatrices; the other numbers are dof numbers which identify the integrands for individual stiffness coefficients.

$$(9.11d)$$

The submatrix $\mathbf{B}_2'^T\mathbf{E}'\mathbf{B}_1'$ may be evaluated as

$$
[\mathbf{B}_2'^T\mathbf{E}'\mathbf{B}_1'] =
\begin{bmatrix} b(1-\eta) & 0 & -a(1+\zeta) \\ 0 & -a(1+\zeta) & b(1-\eta) \end{bmatrix}
\begin{bmatrix} 1 & \nu & 0 \\ \nu & 1 & 0 \\ 0 & 0 & \frac{1-\nu}{2} \end{bmatrix}
\begin{bmatrix} -b(1-\eta) & 0 \\ 0 & -a(1-\zeta) \\ -a(1-\zeta) & -b(1-\eta) \end{bmatrix}
$$

$$
= \begin{bmatrix} -b^2(1-\eta)^2 + \frac{(1-\nu)a^2}{2}(1-\zeta^2) & \frac{4_c}{4}\left(-\nu(1-\zeta)(1-\eta) + \frac{1-\nu}{2}(1+\zeta)(1-\eta)\right) \\ \frac{4_c}{4}\left(\nu(1+\zeta)(1-\eta) - \frac{1-\nu}{2}(1-\zeta)(1-\eta)\right) & a^2(1-\zeta^2) - \frac{(1-\nu)b^2}{2}(1-\eta)^2 \end{bmatrix}.
$$

$$(9.11e)$$

To evaluate a specific stiffness coefficient from among the four whose integrands are in submatrix $\mathbf{B}_2'^T\mathbf{E}'\mathbf{B}_1'$, the integrand with its scalar multiplier is assembled as an integral over the area of the rectangular element in ζ–η space. For example.

$$
k_{31} = \frac{Et}{4A_e(1-\nu^2)} \int_{-1}^{1}\int_{-1}^{1} \left(-b^2(1-\eta)^2 + \frac{(1-\nu)a^2}{2}(1-\zeta^2)\right) d\zeta\, d\eta.
$$

$$(9.11f)$$

The integrals for the stiffness coefficients k_{32}, k_{41}, and k_{42} may be written in like manner using the appropriate integrands from submatrix $\mathbf{B}_2'^T\mathbf{E}'\mathbf{B}_1'$. The other integrals required for the remaining stiffness coefficients of the element stiffness matrix may also be evaluated in this manner. The other submatrices would be formed, the integrands of the stiffness coefficients identified, and the integral expressions written.

Explicit evaluations of these integrals are certainly possible but not necessary. An efficient way to evaluate the integrals for the rectangular element is to evaluate the integrals of equation 9.10f by numerical means for a specific element whose quantities a, b, t, E, and ν have known numerical values. The process of numerical integration over quadrilateral areas will be discussed in detail later in this chapter.

9.3. QUADRILATERAL ELEMENT (4-NODE, 8-DOF)

Figure 9.4 illustrates a 4-node quadrilateral plane stress finite element. Global recti-linear x–y coordinate axes and local natural ζ–η axes are shown. The natural coor-dinate axes are "skewed" coordinate lines (curvilinear coordinate axes). These axes each bisect the angle between their corresponding sides of the quadrilateral. Lines of constant ζ or η proportionally bisect the angles between their corresponding sides of the quadrilateral and their ζ or η axis.

The ranges of values for the ζ–η natural coordinates are defined as

$$-1 \leq \zeta \leq 1, \qquad -1 \leq \eta \leq 1.$$

The natural coordinate values for the element's nodes are shown in Figure 9.4. The origin of the natural coordinate axes is defined as

$$x_g = \frac{(x_1 + x_2)/2 + (x_3 + x_4)/2}{2} = \frac{1}{4}(x_1 + x_2 + x_3 + x_4), \qquad (9.12a)$$

$$y_g = \frac{(y_1 + y_2)/2 + (y_3 + y_4)/2}{2} = \frac{1}{4}(y_1 + y_2 + y_3 + y_4). \qquad (9.12b)$$

The quadrilateral element is shown in ζ–η space in Figure 9.5. In ζ–η space, the quadrilateral is a 2×2 square. This ζ–η square also represents the rectangular finite element discussed earlier. Thus the shape functions expressed in terms of natural coordinates ζ–η for the rectangular element are the same as for the quadrilateral element as long as they remain expressed in terms of the natural coordinates ζ–η defined for it. These shape functions $N_i(\zeta, \eta)$ are given by equations 9.6 and are illustrated in Figure 9.3.

The integral expressions for the stiffness coefficients of the stiffness matrix for the quadrilateral element, however, will be different from those for the rectangular element. The transformations from ζ–η to x–y space are different, causing both

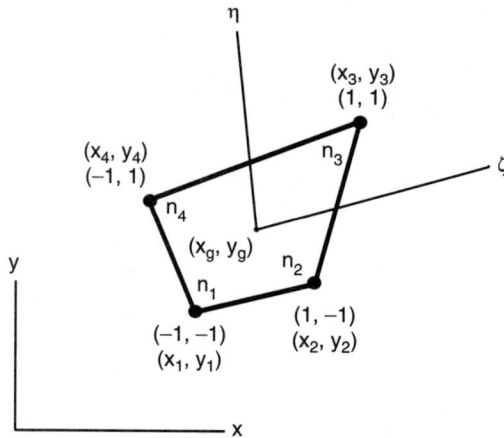

Figure 9.4. Quadrilateral finite element x–y space.

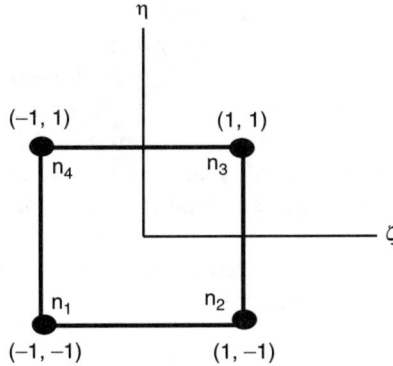

Figure 9.5. Quadrilateral finite element ζ–η space.

the derivatives of the matrix $\mathbf{B}(\zeta, \eta)$ with respect to x and y and the integration over the volume to differ for the two elements.

9.3.1. Isoparametric Transformation

A fundamental requirement for shape functions is that they equal one at their respective dof and zero at the other dof. It is this property of shape functions that permits the displacement functions for the element to be represented by the sum of the products of the shape functions and their corresponding nodal displacements. That same property permits the coordinate variation to be represented in a manner entirely analogous to the displacement variation, but using the corresponding values of nodal coordinates instead of nodal displacements. Note the x–y coordinate variations are independent of one another, similar to the u–v displacement variation. Thus,

$$x = x_1 N_1(\zeta, \eta) + x_2 N_2(\zeta, \eta) + x_3 N_3(\zeta, \eta) + x_4 N_4(\zeta, \eta), \tag{9.13a}$$

$$y = y_1 N_1(\zeta, \eta) + y_2 N_2(\zeta, \eta) + y_3 N_3(\zeta, \eta) + y_4 N_4(\zeta, \eta). \tag{9.13b}$$

In matrix form,

$$x(\zeta, \eta) = \lfloor \mathbf{N}(\zeta, \eta) \rfloor \{\mathbf{x}\}, \tag{9.13c}$$

$$y(\zeta, \eta) = \lfloor \mathbf{N}(\zeta, \eta) \rfloor \{\mathbf{y}\}. \tag{9.13d}$$

Equations 9.13 are the isoparametric transformation from ζ–η space to x–y space for the quadrilateral finite element, so named because this coordinate transformation is of the same mathematical form as the displacement interpolation. The transformation may be viewed as mapping the square element in ζ–η space (Figure 9.5) to the quadrilateral element in x–y space (Figure 9.4). Clearly the square element of Figure 9.5 with shape functions $N_i(\zeta, \eta)$ (equations 9.6) may be used to map any quadrilateral element by employing these shape functions with the corresponding global rectilinear coordinates (x_i, y_i) of the four nodes of the quadrilateral element. The element of Figure 9.5 is thus called a master element.

The transformation defined by equations 9.13 does not have an explicit inverse. That is, $\zeta(x, y)$ and $\eta(x, y)$ cannot be found directly. However, what is needed is not an explicit inverse transformation, but an ability to differentiate the shape functions with respect to x and y in order to formulate the entries in the matrix **B** in the expression for the element stiffness matrix equation 6.22d. In addition, the integration scheme to calculate stiffness coefficients must be done in ζ–η space, so that the variables of integration must be transformed from the volume $V(x, y, z)$, to x and y, to ζ to η

The matrix $\mathbf{B}_i$ for plane stress is given by equations 9.7d. Applying the differentiation chain rule to the shape functions,

$$\frac{\partial N_i(\zeta, \eta)}{\partial x} = \frac{\partial N_i(\zeta, \eta)}{\partial \zeta} \cdot \frac{\partial \zeta}{\partial x} + \frac{\partial N_i(\zeta, \eta)}{\partial \eta} \cdot \frac{\partial \eta}{\partial x}, \tag{9.14a}$$

$$\frac{\partial N_i(\zeta, \eta)}{\partial y} = \frac{\partial N_i(\zeta, \eta)}{\partial \zeta} \cdot \frac{\partial \zeta}{\partial y} + \frac{\partial N_i(\zeta, \eta)}{\partial \eta} \cdot \frac{\partial \eta}{\partial y}. \tag{9.14b}$$

In matrix form

$$\left\{ \begin{array}{c} \frac{\partial N_i(\zeta, \eta)}{\partial x} \\ \frac{\partial N_i(\zeta, \eta)}{\partial y} \end{array} \right\} = \begin{bmatrix} \frac{\partial \zeta}{\partial x} & \frac{\partial \eta}{\partial x} \\ \frac{\partial \zeta}{\partial y} & \frac{\partial \eta}{\partial y} \end{bmatrix} \left\{ \begin{array}{c} \frac{\partial N_i(\zeta, \eta)}{\partial \zeta} \\ \frac{\partial N_i(\zeta, \eta)}{\partial \eta} \end{array} \right\}. \tag{9.14c}$$

There is no way to obtain directly the derivatives $\partial \zeta/\partial x$, $\partial \eta/\partial x$, $\partial \zeta/\partial y$, and $\partial \eta/\partial y$ since the inverse functions $\zeta(x, y)$ and $\eta(x, y)$ cannot be found. On the other hand, the differentiation chain rule can be invoked in an inverse process as

$$\frac{\partial N_i(\zeta, \eta)}{\partial \zeta} = \frac{\partial N_i(\zeta, \eta)}{\partial x} \cdot \frac{\partial x}{\partial \zeta} + \frac{\partial N_i(\zeta, \eta)}{\partial y} \cdot \frac{\partial y}{\partial \zeta}, \tag{9.14d}$$

$$\frac{\partial N_i(\zeta, \eta)}{\partial \eta} = \frac{\partial N_i(\zeta, \eta)}{\partial x} \cdot \frac{\partial x}{\partial \eta} + \frac{\partial N_i(\zeta, \eta)}{\partial y} \cdot \frac{\partial y}{\partial \eta}. \tag{9.14e}$$

In matrix form,

$$\left\{ \begin{array}{c} \frac{\partial N_i(\zeta, \eta)}{\partial \zeta} \\ \frac{\partial N_i(\zeta, \eta)}{\partial \eta} \end{array} \right\} = \begin{bmatrix} \frac{\partial x}{\partial \zeta} & \frac{\partial y}{\partial \zeta} \\ \frac{\partial x}{\partial \eta} & \frac{\partial y}{\partial \eta} \end{bmatrix} \left\{ \begin{array}{c} \frac{\partial N_i(\zeta, \eta)}{\partial x} \\ \frac{\partial N_i(\zeta, \eta)}{\partial y} \end{array} \right\}. \tag{9.14f}$$

The derivatives $\partial x/\partial \zeta$, $\partial y/\partial \zeta$, $\partial x/\partial \eta$, and $\partial y/\partial \eta$ in matrix array in equation 9.14f may be obtained directly from the transformation from ζ–η space to x–y space (equation 9.13). This matrix of derivatives, called the Jacobian matrix $\mathbf{J}(\zeta, \eta)$ of the transformation, is then inverted, so that

$$\left\{ \begin{array}{c} \frac{\partial N_i(\zeta, \eta)}{\partial x} \\ \frac{\partial N_i(\zeta, \eta)}{\partial y} \end{array} \right\} = \begin{bmatrix} \frac{\partial x}{\partial \zeta} & \frac{\partial y}{\partial \zeta} \\ \frac{\partial x}{\partial \eta} & \frac{\partial y}{\partial \eta} \end{bmatrix}^{-1} \left\{ \begin{array}{c} \frac{\partial N_i(\zeta, \eta)}{\partial \zeta} \\ \frac{\partial N_i(\zeta, \eta)}{\partial \eta} \end{array} \right\}. \tag{9.14g}$$

Equation 9.14g provides the derivatives of shape functions with respect to x and y needed for matrix **B** in the stiffness matrix equation 6.22d. The derivatives of shape functions with respect to ζ and η are written directly from equations 9.6:

$$\frac{\partial N_i(\zeta,\ \eta)}{\partial \zeta} = \frac{\zeta_i}{4}(1 + \eta_i \eta), \qquad \frac{\partial N_i(\zeta,\ \eta)}{\partial \eta} = \frac{\eta_i}{4}(1 + \zeta_i \zeta), \tag{9.15a}$$

or

$$\frac{\partial N_1(\zeta,\ \eta)}{\partial \zeta} = -\frac{1}{4}(1 - \eta), \qquad \frac{\partial N_1(\zeta,\ \eta)}{\partial \eta} = -\frac{1}{4}(1 - \zeta); \tag{9.15b}$$

$$\frac{\partial N_2(\zeta,\ \eta)}{\partial \zeta} = \frac{1}{4}(1 - \eta), \qquad \frac{\partial N_2(\zeta,\ \eta)}{\partial \eta} = -\frac{1}{4}(1 + \zeta); \tag{9.15c}$$

$$\frac{\partial N_3(\zeta,\ \eta)}{\partial \zeta} = \frac{1}{4}(1 + \eta), \qquad \frac{\partial N_3(\zeta,\ \eta)}{\partial \eta} = \frac{1}{4}(1 + \zeta); \tag{9.15d}$$

$$\frac{\partial N_4(\zeta,\ \eta)}{\partial \zeta} = -\frac{1}{4}(1 + \eta), \qquad \frac{\partial N_4(\zeta,\ \eta)}{\partial \eta} = \frac{1}{4}(1 - \zeta). \tag{9.15e}$$

The derivative expressions for the Jacobian matrix $\mathbf{J}(\zeta,\ \eta)$, obtained by differentiating the transformation equations 9.13, are

$$\frac{\partial x}{\partial \zeta} = x_1 \frac{\partial N_1}{\partial \zeta} + x_2 \frac{\partial N_2}{\partial \zeta} + x_3 \frac{\partial N_3}{\partial \zeta} + x_4 \frac{\partial N_4}{\partial \zeta}, \tag{9.16a}$$

$$\frac{\partial y}{\partial \zeta} = y_1 \frac{\partial N_1}{\partial \zeta} + y_2 \frac{\partial N_2}{\partial \zeta} + y_3 \frac{\partial N_3}{\partial \zeta} + y_4 \frac{\partial N_4}{\partial \zeta}, \tag{9.16b}$$

$$\frac{\partial x}{\partial \eta} = x_1 \frac{\partial N_1}{\partial \eta} + x_2 \frac{\partial N_2}{\partial \eta} + x_3 \frac{\partial N_3}{\partial \eta} + x_4 \frac{\partial N_4}{\partial \eta}, \tag{9.16c}$$

$$\frac{\partial y}{\partial \eta} = y_1 \frac{\partial N_1}{\partial \eta} + y_2 \frac{\partial N_2}{\partial \eta} + y_3 \frac{\partial N_3}{\partial \eta} + y_4 \frac{\partial N_4}{\partial \eta}. \tag{9.16d}$$

Each of the derivatives, given by the expressions in equations 9.16, are linear functions of ζ and η, since the derivatives of the shape functions with respect to ζ and η are the linear functions of ζ and η shown in equations 9.15. The inverse of the Jacobian matrix $\mathbf{J}(\zeta,\ \eta)$ may be obtained by application of the adjoint method of matrix inversion:

$$[\mathbf{J}(\zeta,\ \eta)]^{-1} = \begin{bmatrix} \frac{\partial x}{\partial \zeta} & \frac{\partial y}{\partial \zeta} \\ \frac{\partial x}{\partial \eta} & \frac{\partial y}{\partial \eta} \end{bmatrix}^{-1} \tag{9.17a}$$

$$= \frac{1}{|J(\zeta, \eta)|} \begin{bmatrix} \frac{\partial y}{\partial \eta} & -\frac{\partial y}{\partial \zeta} \\ -\frac{\partial x}{\partial \eta} & \frac{\partial x}{\partial \zeta} \end{bmatrix}, \tag{9.17b}$$

where the determinant of the Jacobian matrix, called simply the Jacobian of the transformation, is given by

$$|J(\zeta, \ \eta)| = \frac{\partial x}{\partial \zeta} \cdot \frac{\partial y}{\partial \eta} - \frac{\partial x}{\partial \eta} \cdot \frac{\partial y}{\partial \zeta}. \tag{9.17c}$$

The Jacobian is a linear polynomial in ζ and η. Though it is formed as products of the derivatives given in equations 9.16, which would seem to make it bilinear, it can be shown that the way it is formed causes the possible $\zeta\eta$ term to vanish.

The expressions for the derivatives of the shape functions with respect to x and y may now be written by substituting equations 9.16 into 9.17, and then 9.15 and 9.17 into equations 9.14g, as illustrated in the equations

$$\left\{ \begin{array}{c} \frac{\partial N_i(\zeta, \ \eta)}{\partial x} \\ \frac{\partial N_i(\zeta, \ \eta)}{\partial y} \end{array} \right\} = \frac{1}{|J(\zeta, \ \eta)|} \left[\begin{array}{cc} \frac{\partial y}{\partial \eta} & -\frac{\partial y}{\partial \zeta} \\ -\frac{\partial x}{\partial \eta} & \frac{\partial x}{\partial \zeta} \end{array} \right] \left\{ \begin{array}{c} \frac{\partial N_i(\zeta, \ \eta)}{\partial \zeta} \\ \frac{\partial N_i(\zeta, \ \eta)}{\partial \eta} \end{array} \right\}. \tag{9.18a}$$

where

$$\frac{\partial N_i(\zeta, \ \eta)}{\partial x} = \frac{1}{|J(\zeta, \ \eta)|} \left(\frac{\partial y}{\partial \eta} \cdot \frac{\partial N_i(\zeta, \ \eta)}{\partial \zeta} - \frac{\partial y}{\partial \zeta} \cdot \frac{\partial N_i(\zeta, \ \eta)}{\partial \eta} \right), \tag{9.18b}$$

$$\frac{\partial N_i(\zeta, \ \eta)}{\partial y} = \frac{1}{|J(\zeta, \ \eta)|} \left(-\frac{\partial x}{\partial \eta} \cdot \frac{\partial N_i(\zeta, \ \eta)}{\partial \zeta} + \frac{\partial x}{\partial \zeta} \cdot \frac{\partial N_i(\zeta, \ \eta)}{\partial \eta} \right). \tag{9.18c}$$

To transform the variable of integration in the stiffness matrix equation 6.22d from $V(x, \ y, \ z)$ to x and y,

$$dV = t \ dx \ dy. \tag{9.10a}$$

The term t is the constant thickness of the plane stress plate.

The transformation of the variables of integration x and y to ζ and η may be achieved with an application of Green's theorem in 2D space. Green's theorem relates a line integral of two functions around the perimeter of an area A to the area integral of derivatives of the two functions over the area A. In symbols,

$$\oint [P(x, \ y) \ dx + Q(x, \ y) \ dy] = \int_A \left(\frac{\partial Q(x, \ y)}{\partial x} - \frac{\partial P(x, \ y)}{\partial y} \right) dx \ dy. \tag{9.19a}$$

Let $P(x, \ y) = 0$ and $Q(x, \ y) = x$ so that $\partial P(x, \ y)/\partial y = 0$ and $\partial Q(x, \ y)/\partial x = 1$. Substituting into equation 9.19a,

$$\oint x \ dy = \int_A dx \ dy. \tag{9.19b}$$

Let x and y define a transformation $x = x(\zeta, \eta)$ and $y = (\zeta, \ \eta)$. Therefore,

$$dy = \frac{\partial y(\zeta, \ \eta)}{\partial \zeta} d\zeta + \frac{\partial y(\zeta, \ \eta)}{\partial \eta} d\eta.$$

Substituting into equation 9.19b,

$$\int_A dx\,dy = \oint\left(x(\zeta,\ \eta)\frac{\partial y(\zeta,\ \eta)}{\partial \zeta}d\zeta + x(\zeta,\ \eta)\frac{\partial y(\zeta,\ \eta)}{\partial \eta}d\eta\right)$$

$$= \oint[p(\zeta,\ \eta)\,d\zeta + q(\zeta,\ \eta)d\eta],$$

(9.19c)

where

$$p(\zeta,\ \eta) = x(\zeta,\ \eta)\frac{\partial y(\zeta,\ \eta)}{\partial \zeta},$$

(9.19d)

$$q(\zeta,\ \eta) = x(\zeta,\ \eta)\frac{\partial y(\zeta,\ \eta)}{\partial \eta}.$$

(9.19e)

Applying Green's theorem to the right-hand side of equation 9.19c,

$$\int_A dx\,dy = \int_A \left(\frac{\partial q(\zeta,\ \eta)}{\partial \zeta} - \frac{\partial p(\zeta,\ \eta)}{\partial \eta}\right) d\zeta\,d\eta.$$

(9.19f)

The integrand of equations 9.19f may be evaluated by differentiating equations 9.19d and 9.19e:

$$\frac{\partial q}{\partial \zeta} = \frac{\partial x}{\partial \zeta}\cdot\frac{\partial y}{\partial \eta} + x\frac{\partial^2 y}{\partial \zeta\,\partial \eta},$$

(9.19g)

$$\frac{\partial p}{\partial \eta} = \frac{\partial x}{\partial \eta}\cdot\frac{\partial y}{\partial \zeta} + x\frac{\partial^2 y}{\partial \eta\,\partial \zeta}.$$

(9.19h)

Subtracting equation 9.19h from equation 9.19g and comparing the result to equation 9.17c, equation 9.19f may be written as

$$\int_A dx\,dy = \int_A |J(\zeta,\ \eta)|\,d\zeta\,d\eta$$

(9.19i)

or

$$dx\,dy = |J(\zeta,\ \eta)|\,d\zeta\,d\eta.$$

(9.19j)

Equation 9.19j expresses the transformation from ζ–η space to x–y space. It maps the differential area in ζ–η space to a corresponding differential area in x–y space, as shown in Figure 9.6. The Jacobian $|J(\zeta,\ \eta)|$ of the transformation must not change sign in the region over which integration will occur. Otherwise there will not be a one-to-one correspondence between points of the region in ζ–η space and its mapped region in x–y space. If the Jacobian is positive throughout the region, the direction of traverse on the perimeter of the region will be the same in both spaces. If it is negative, those directions will be opposite.

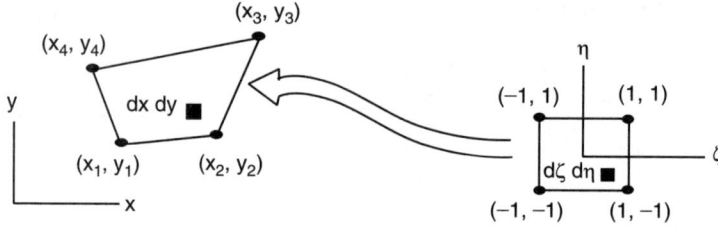

Figure 9.6. Element mapping.

9.3.2. Element Stiffness Matrix

Equation 9.19j in conjunction with equations 9.18 permits the calculation in ζ–η space of the stiffness coefficients of a finite element which is a quadrilateral in x–y space and which has been transformed isoparametrically to the square master element in ζ–η space. Equation 6.22d, modified for the plane stress condition to achieve this, is

$$[\mathbf{K}] = t \int_A [\mathbf{B}^\mathrm{T}(\zeta,\ \eta)][\mathbf{E}][\mathbf{B}(\zeta,\ \eta)]\ dA \tag{9.20a}$$

$$= t \int_{-1}^{1} \int_{-1}^{1} [\mathbf{B}^\mathrm{T}(\zeta,\ \eta)][\mathbf{E}][\mathbf{B}(\zeta,\ \eta)]|J(\zeta,\ \eta)|\ d\zeta\ d\eta \tag{9.20b}$$

$$= t \int_{-1}^{1} \int_{-1}^{1} [\mathbf{I}(\zeta,\ \eta)]\ d\zeta\ d\eta. \tag{9.20c}$$

The integrand $\mathbf{I}(\zeta,\ \eta)$ is a matrix of functions of ζ and η, 8×8 in size, reflecting the eight dof on the 4-node quadrilateral plane stress finite element. Each of its 64 entries must be determined, integrated over the area of the master element in ζ–η space, and multiplied by the constant thickness t of the element. The determination of matrix $\mathbf{I}(\zeta,\ \eta)$ may proceed as follows:

$$[\mathbf{I}(\zeta,\ \eta)] = [\mathbf{B}^\mathrm{T}(\zeta,\ \eta)][\mathbf{E}][\mathbf{B}(\zeta,\ \eta)]|J(\zeta,\ \eta)|. \tag{9.21a}$$

The matrix $\mathbf{B}(\zeta,\ \eta)$ is given by equations 9.7 for the plane stress condition. The derivatives of the shape functions $N_i(\zeta,\ \eta)$ with respect to x and y, called for in equations 9.7 are expressed in equations 9.18. Each of the derivatives in equations 9.18 is divided by the scalar Jacobian $|J(\zeta,\ \eta)|$. Let the symbol $\mathbf{B}'(\zeta,\ \eta)$ represent matrix $\mathbf{B}(\zeta,\ \eta)$ without division by the Jacobian; then

$$[\mathbf{I}(\zeta,\ \eta)] = \frac{[\mathbf{B}'^\mathrm{T}(\zeta,\ \eta)][\mathbf{E}][\mathbf{B}'(\zeta,\ \eta)]}{|J(\zeta,\ \eta)|} \tag{9.21b}$$

The derivatives of $N_i(\zeta,\ \eta)$, x, and y, each with respect to ζ and η in equations 9.17 and 9.18 for the matrix $\mathbf{B}(\zeta,\ \eta)$ and the Jacobian $|J(\zeta,\ \eta)|$, are written out in equations 9.15 and 9.16.

It is clear that each entry of the matrix $\mathbf{I}(\zeta,\ \eta)$ is a polynomial fraction. The numerator of each fraction consists of polynomial terms resulting from the products of entries of the matrices $\mathbf{B}'^T(\zeta, \eta)$, $\mathbf{E}$, and $\mathbf{B}'(\zeta, \eta)$. They will be products of bilinear polynomials, resulting in quadratic polynomials in ζ and η and incorporating the elastic constants of matrix $\mathbf{E}$ and the element's nodal coordinates (x_i, y_i). The denominator of each fraction is the Jacobian, a liner polynomial in ζ and η, as stated earlier.

Though one might persist in determining each entry of $\mathbf{I}(\zeta, \eta)$ as a polynomial fraction in ζ and η and the analytical integration of each over the area of the master element, the process would be extremely tedious and would result in the stiffness matrix for only one finite element. A numerical integration scheme, such as described for the linear strain triangular finite element in Section 8.5.3, requires only that the integrand function be numerically evaluated using the coordinates of selected points in the area over which the integration is to occur. A weighted sum of these evaluations then equals the value of the integral over that area.

The integrand $\mathbf{I}(\zeta, \eta)$ of equation 9.21b is easily evaluated for a known set of coordinates (ζ_k, η_k) of a point, the known values of the element's nodal rectilinear coordinates (x_i, y_i), and the known elastic constants. The simplest procedure is to evaluate numerically each derivative of the shape functions $N_i(\zeta, \eta)$ with respect to ζ and η at (ζ_k, η_k), as indicated in equations 9.15. Then the derivatives of x and y with respect to ζ and η are evaluated at (ζ_k, η_k) using the values of shape function derivatives with respect to ζ and η just computed, as indicated in equations 9.16. Next the values of the Jacobian at (ζ_k, η_k) are computed using the values of the derivatives of x and y just computed, as indicated in equations 9.17c. The derivatives of the shape functions $N_i(\zeta, \eta)$ with respect to x and y at (ζ_k, η_k) are then computed using derivative values just obtained and equations 9.18, but without division by the Jacobian. Finally the matrix $\mathbf{B}'(\zeta_k, \eta_k)$ is assembled using equation 9.7 as a guide, and the numerical matrix $\mathbf{I}(\zeta_k, \eta_k)$ is obtained by performing the matrix operations indicated in equation 9.21b.

9.4. QUADRATURE RULES

The need for a numerical integration scheme to evaluate the stiffness coefficients for the quadrilateral finite element was just discussed. The scheme suggested, the one employed earlier for the linear strain triangular finite element, is known as a quadrature rule. The mathematical form for the process, applied to a function of one variable integrated along a line segment, is expressed as

$$\int_{\zeta_L}^{\zeta_U} f(\zeta)\, d\zeta = \sum_{k=1}^{n} f(\zeta_k) H_k. \tag{9.22a}$$

Applied to a function of two variables integrated over a rectangular area, the form is

$$\int_{\eta_L}^{\eta_U} \int_{\zeta_L}^{\zeta_U} f(\zeta, \eta)\, d\zeta\, d\eta = \sum_{k=1}^{n} f(\zeta_k, \eta_k) H_k. \tag{9.22b}$$

The terms ζ_U, ζ_L, η_U, and η_L are simply the upper and lower limits of the integrals. The terms ζ_k and η_k are coordinates of points within the region over which integration will occur. These points are called integration points. The term H_k is a weight factor. There is one weight factor associated with each integration point.

Thus to develop and use a quadrature rule, the number and locations of the integration points where the integrand must be evaluated and the weight factors associated with each integration point must be determined.

9.4.1. Nodal Point Quadrature

In its simplest form, the quadrature rule for the integration of a function of one variable along a line segment of length h specifies the locations of the integration points ζ_k at the two ends of the segment, leaving the weight factors H_k to be determined. Since a definite integral is a function of its limits only, it may be evaluated with respect to any convenient reference axes as long as the integrand function, the variable of integration, and the limits of integration are properly transformed. Figure 9.7 illustrates the integrand plotted over its segment and transformed so that it is a function of the variable ζ whose origin is at the center of the segment and whose ends are at $\zeta_1 = -\frac{1}{2}h$ and $\zeta_2 = \frac{1}{2}h$. The length of the segment is h. The quadrature rule may be written from equation 9.22a as

$$\int_{-h/2}^{h/2} f(\zeta)\, d\zeta = \sum_{k=1}^{2} f(\zeta_k) H_k$$

$$= f\left(\frac{-h}{2}\right) H_1 + f\left(\frac{h}{2}\right) H_2. \tag{9.23a}$$

The weight factors may be determined by assuming that the $f(\zeta)$ is approximated along the length of the segment by a polynomial $g(\zeta)$:

$$f(\zeta) \approx g(\zeta) = C_1 + C_2 \zeta. \tag{9.23b}$$

For its complete determination, it will be a two-term polynomial, or a linear function, since only two conditions are available to impose on it for the determination of the constants C_1 and C_2. The conditions are that, at ζ_1, the function $f(\zeta_1)$ must equal $g(\zeta_1)$ and, at ζ_2, $f(\zeta_2)$ must equal $g(\zeta_2)$.

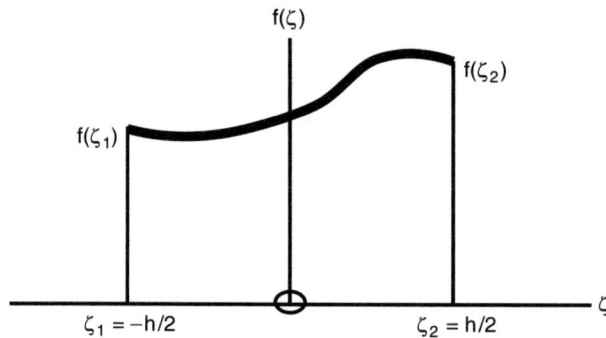

Figure 9.7. Integrand.

Substituting equation 9.23b into equation 9.23a,

$$\int_{-h/2}^{h/2} (C_1 + C_2\zeta)\, d\zeta = \left(C_1 - C_2\frac{h}{2}\right)H_1 + \left(C_1 + C_2\frac{h}{2}\right)H_2$$

$$= (H_1 + H_2)C_1 + \frac{h}{2}(-H_1 + H_2)C_2.$$

(9.23c)

Integrating the polynomial of equation 9.23b term by term,

$$\int_{-h/2}^{h/2} (C_1 + C_2\zeta)\, d\zeta = C_1\left(\frac{h}{2} + \frac{h}{2}\right) + C_2\left(\frac{h^2}{8} - \frac{h^2}{8}\right)$$

$$= h \cdot C_1 + 0 \cdot C_2.$$

(9.23d)

Comparing the multipliers of the polynomial constants C_1 and C_2 in equations 9.23c and 9.23d,

$$H_1 + H_2 = h \quad \text{and} \quad -H_1 + H_2 = 0.$$

(9.23e)

Equations 9.23e may be solved for the weight factors:

$$H_1 = H_2 = \tfrac{1}{2}h.$$

(9.23f)

The nodal two-point quadrature rule may be written as

$$\int_{-h/2}^{h/2} f(\zeta)\, d\zeta = \sum_{k=1}^{2} f(\zeta_k)H_k$$

$$= f\left(\frac{-h}{2}\right) \cdot \frac{h}{2} + f\left(\frac{h}{2}\right) \cdot \frac{h}{2}$$

(9.24)

The accuracy of the rule depends entirely on how well the linear polynomial $g(\zeta)$ approximates the integrand $f(\zeta)$ over the length of the segment h. Equation 9.24 is a formalized statement of the trapezoid method of numerical integration.

Simpson's rule is another nodal point quadrature rule for the integration of a function of one variable along a line segment of length h. In it, integration points are specified at each end and at the center of the segment. It approximates the integrand with a three-term polynomial, or a quadratic function, since three known conditions are available to be imposed on it. The three weight factors for the three integration points may be determined by following the procedure just used:

$$H_1 = H_3 = \tfrac{1}{6}h \quad \text{and} \quad H_2 = \tfrac{2}{3}h.$$

The nodal three-point quadrature rule (Simpson's rule) is

$$\int_{-h/2}^{h/2} f(\zeta)\, d\zeta = \sum_{k=1}^{3} f(\zeta_k) H_k$$

$$= f\left(\frac{-h}{2}\right) \cdot \frac{h}{6} + f(0) \cdot \frac{2h}{3} + f\left(\frac{h}{2}\right) \cdot \frac{h}{6}. \tag{9.25}$$

The three-point nodal rule (equation 9.25) is more accurate than the two-point rule (equation 9.24) since quadratic polynomials approximate most integrands better than linear polynomials. The three-point rule, however, requires more effort to implement.

9.4.2. Gaussian Quadrature: Functions of One Variable

By specifying the location of the integration points, nodal point quadrature rules approximate the integrand function with a polynomial possessing the same number of terms as there are integration points. If n is the number of integration points, the degree of the polynomial used for the approximation will be $n - 1$.

Gaussian quadrature does not specify the location of the integration points. Instead it leaves their location along with their weight factors to be determined. A two-point Gauss quadrature rule will require a four-term polynomial with four constants to deal with the four unknowns: the coordinates of the two integration points and the two weight factors.

Following the procedure employed above,

$$\int_{-h/2}^{h/2} f(\zeta)\, d\zeta = \sum_{k=1}^{2} f(\zeta_k) H_k$$

$$= f(\zeta_1) H_1 + f(\zeta_2) H_2. \tag{9.26a}$$

Let

$$f(\zeta) \approx g(\zeta) = C_1 + C_2\zeta + C_3\zeta^2 + C_4\zeta^3. \tag{9.26b}$$

Substituting equation 9.26b into equation 9.26a,

$$\int_{-h/2}^{h/2} (C_1 + C_2\zeta + C_3\zeta^2 + C_4^3)\, d\zeta$$

$$= (C_1 + C_2\zeta_1 + C_3\zeta_1^2 + C_4\zeta_1^3)H_1 + (C_1 + C_2\zeta_2 + C_3\zeta_2^2 + C_4\zeta_2^3)H_2$$

$$= (H_1 + H_2)C_1 + (\zeta_1 H_1 + \zeta_2 H_2)C_2 + (\zeta_1^2 H_1 + \zeta_2^2 H_2)C_3 + (\zeta_1^3 H_1 + \zeta_2^3 H_2)C_4. \tag{9.26c}$$

Integrating the polynomial of equation 9.26b term by term,

$$\int_{-h/2}^{h/2} (C_1 + C_2\zeta + C_3\zeta^2 + C_4^3)\, d\zeta = h \cdot C_1 + 0 \cdot C_2 + \frac{h^3}{12} \cdot C_3 + 0 \cdot C_4. \tag{9.26d}$$

Comparing the multipliers of the polynomial constants C_1, C_2, C_3, and C_4 in equations 9.26c and 9.26d,

$$H_1 + H_2 = h, \tag{9.26e}$$

$$\zeta_1 H_1 + \zeta_2 H_2 = 0, \tag{9.26f}$$

$$\zeta_1^2 H_1 + \zeta_2^2 H_2 = h^3/12, \tag{9.26g}$$

$$\zeta_1^3 H_1 + \zeta_2^3 H_2 = 0. \tag{9.26h}$$

Solving equations 9.26e–h for the coordinates of the integration points ζ_k and the weight factors H_k,

$$H_1 = H_2 = \frac{h}{2}, \qquad \zeta_1 = -\frac{h}{2\sqrt{3}}, \qquad \zeta_2 = \frac{h}{2\sqrt{3}}. \tag{9.26i}$$

The Gaussian quadrature rule for the integration of a function of one variable along a line segment of length h, using two integration points, may be written as

$$\int_{-h/2}^{h/2} f(\zeta)\, d\zeta = \sum_{k=1}^{2} f(\zeta_k) H_k$$

$$= f\left(\frac{-h}{2\sqrt{3}}\right) \cdot \frac{h}{2} + f\left(\frac{h}{2\sqrt{3}}\right) \cdot \frac{h}{2}. \tag{9.27}$$

The accuracy of the two-point Gauss rule is limited by the ability of a cubic polynomial $g(\zeta)$ to approximate the integrand $f(\zeta)$ over the length h of the line segment. It will do better than either of the nodal point quadrature rules discussed, since a cubic polynomial will usually approximate an integrand better than either a linear or a quadratic polynomial. Moreover, it will take less or comparable effort to implement.

A Gaussian quadrature rule using three integration points would require a six-term, or fifth-degree, polynomial to deal with the six unknown quantities: the coordinates of the three integration points and their three corresponding weight factors. The procedure illustrated above can be used to determine these as

$$\zeta_1 = -\frac{h\sqrt{0.6}}{2}, \qquad \zeta_2 = 0, \qquad \zeta_3 = \frac{h\sqrt{0.6}}{2}, \tag{9.28a}$$

$$H_1 = H_3 = \frac{5h}{18}, \qquad H_2 = \frac{4h}{9}. \tag{9.28b}$$

Integration point 1 is left of the origin, which is at the center of the segment, integration point 2 is at the origin, and integration point 3 is right of the origin. The three-point Gauss quadrature rule may be written as

$$\int_{-h/2}^{h/2} f(\zeta) \, d\zeta = \sum_{k=1}^{3} f(\zeta_k) H_k,$$

$$= f\left(\frac{-h\sqrt{0.6}}{2}\right) \cdot \frac{5h}{18} + f(0) \cdot \frac{4h}{9} + f\left(\frac{h\sqrt{0.6}}{2}\right) \cdot \frac{5h}{18}.$$

(9.29)

The accuracy of the three-point Gauss rule is restricted by the ability of a fifth-degree polynomial to approximate the integrand over the length h of the line segment. It should provide sufficient accuracy for our purposes.

In not specifying locations for integration points, Gaussian quadrature approximates the integrand with a polynomial having twice as many terms as integration points. If the number of integration points is n, the degree of that polynomial is $2n - 1$. The coordinates of integration points and their corresponding weight factors for Gaussian quadrature are available for in excess of 10 integration points.

As an illustration of an application of the two-point Gauss rule, the following integral will be evaluated:

$$\int_0^4 (x + 1)(x - 4)(x - 6) \, dx.$$

The integrand is a cubic polynomial and can be integrated explicitly for its exact value without difficulty. The two-point Gauss quadrature rule (equation 9.27) will evaluate it exactly, since it uses a cubic polynomial to approximate the integrand.

The integrand shown is referenced to a variable x and must be evaluated over the range of x from 0 to 4. The length of the line segment then is 4.

The quadrature rule was established for coordinates referenced to an origin located at the center of the line segment. Consequently, the coordinates of the integration points given for equation 9.27 must be transformed to the x coordinates used in this example integral and the integrand evaluated with those coordinates. Alternatively, the integrand, its variable of integration x, and its limits may be transformed to the variable ζ with its origin at the center of the segment. The integrand, so transformed, would then be evaluated at the coordinates ζ_1 and ζ_2 given for equation 9.27. In the finite element method, this latter approach is done almost exclusively. Either approach will yield the same results, though in this example the latter approach is illustrated.

The center of the segment is at $x = 2$. If the length of the segment remains 4 with respect to the ζ coordinate for equation 9.27,

$$x = \zeta + 2.$$

Transforming the integral to ζ coordinates,

$$dx = d\zeta,$$

$$\zeta_L = 0 - 2 = -2, \quad \zeta_U = 4 - 2 = 2,$$

$$g(\zeta) = f(\zeta + 2) = (\zeta + 2 + 1)(\zeta + 2 - 4)(\zeta + 2 - 6)$$

$$= (\zeta + 3)(\zeta - 2)(\zeta - 4).$$

The transformed integral is

$$\int_0^4 (x+1)(x-4)(x-6)\,dx = \int_{-2}^2 (\zeta+3)(\zeta-2)(\zeta-4)\,d\zeta$$

$$= \sum_{k=1}^2 g(\zeta_k)H_k.$$

The weight factors are

$$H_1 = H_2 = \frac{4}{2} = 2.$$

The integration points are at

$$\zeta_1 = -\frac{4}{2\sqrt{3}}, \quad \text{and} \quad \zeta_2 = \frac{4}{2\sqrt{3}}.$$

The integrand evaluated at ζ_1 and ζ_2 is

$$g(\zeta_1) = \left(-\frac{2}{\sqrt{3}}+3\right)\left(-\frac{2}{\sqrt{3}}-2\right)(-\frac{2}{\sqrt{3}}-4)$$

$$= 30.007404666,$$

$$g(\zeta_2) = \left(\frac{2}{\sqrt{3}}+3\right)\left(\frac{2}{\sqrt{3}}-2\right)\left(\frac{2}{\sqrt{3}}-4\right)$$

$$= 9.992595344.$$

Applying equation 9.27, the value of the integral is

$$(30.007404666)\,(2) + (9.992595344)\,(2) = 80.000000000.$$

It is left as an exercise for the reader to evaluate the integral explicitly and to apply the alternative approach by transforming the integration points from ζ to x and evaluating the integral as a function of x using the quadrature rule.

9.4.3. Gaussian Quadrature: Functions of Two Variables

The quadrature rules for functions of two variables acting over a rectangular area whose dimensions are $2a$ in width and $2b$ in height are easily developed from those for one variable acting over a line segment whose length is h. For convenience, the coordinate axes will be ζ and η, with their origin at the center of the rectangular area over which the integration must occur. The coordinate ζ will be parallel to the side whose length is $2a$; the coordinate η will be parallel to the side whose length is $2b$. Figure 9.8 illustrates the rectangular area and shows the presence of integration points.

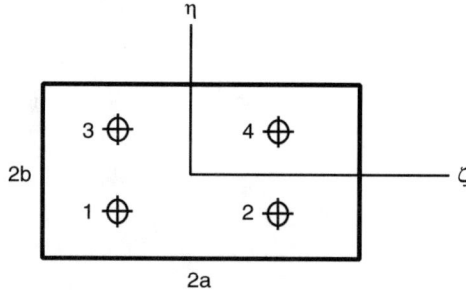

Figure 9.8. Rectangular area.

The integral of a function of two variables over a rectangular area is expressed as a double integral:

$$\int_A f(\zeta, \eta)\, dA = \int_{-b}^{b} \int_{-a}^{a} f(\zeta, \eta)\, d\zeta\, d\eta. \tag{9.30a}$$

Integration begins with the variable ζ, the inner integral of equation 9.30a, which by the quadrature rule (equation 9.22a) may be written as

$$\int_{-a}^{a} f(\zeta, \eta)\, d\zeta = \sum_{i=1}^{L} f(\zeta_i, \eta) H_i = g(\eta). \tag{9.30b}$$

Rewriting equation 9.30a, using equation 9.30b, and again applying the quadrature rule,

$$\int_A f(\zeta, \eta)\, dA = \int_{-b}^{b} g(\eta)\, d\eta = \sum_{j=1}^{M} g(\eta_j) H_j. \tag{9.30c}$$

Replacing $g(\eta_j)$ in equation 9.30c with its summation form shown in equation 9.30b, equation 9.30c may be written as

$$\int_A f(\zeta, \eta)\, dA = \sum_{j=1}^{M} \left(\sum_{i=1}^{L} f(\zeta_i, \eta_j) H_i \right) H_j$$

$$= \sum_{j=1}^{M} \sum_{i=1}^{L} f(\zeta_i, \eta_j) H_i\, H_j. \tag{9.30d}$$

Equation 9.30d is a double summation, with L terms in the ζ direction and M terms in the η direction for a total of $L \times M$ terms. It may be conveniently represented by a single summation:

$$\int_A f(\zeta, \eta)\, dA = \sum_{k=1}^{n} f(\zeta_k, \eta_k) H, \tag{9.30e}$$

where k is the single-sum index ($k = 1, 2, 3, \ldots$); $n = L \times M$; (ζ_k, η_k) are the coordinates of the integration points, sequencing from left to right in rows which run from the bottom to the top; and $H_k = H_i \times H_j$, where H_i and H_j are the weight factors for the quadrature in the ζ and η directions, respectively, corresponding to (ζ_k, η_k).

If two integration points are used in each direction over a rectangular area, the two-point Gauss quadrature rule for functions of one variable may be applied in each direction to develop the 2×2 (four-point) Gauss quadrature rule for a function of two variables over the rectangular area. The table below and Figure 9.9 illustrate the rectangular area, its integration points, and its weight factors:

Integration Point	ζ_k	η_k	H_k
1	$-a/\sqrt{3}$	$-b/\sqrt{3}$	ab
2	$a/\sqrt{3}$	$-b/\sqrt{3}$	ab
3	$-a/\sqrt{3}$	$b/\sqrt{3}$	ab
4	$a/\sqrt{3}$	$b/\sqrt{3}$	ab

This 2×2 Gauss quadrature rule for integration of a function of two variables over a rectangular area may be written as

$$\int_A f(\zeta,\ \eta)\ dA = f\left(-\frac{a}{\sqrt{3}}, -\frac{b}{\sqrt{3}}\right)ab + f\left(\frac{a}{\sqrt{3}}, -\frac{b}{\sqrt{3}}\right)ab$$
$$+ f\left(-\frac{a}{\sqrt{3}}, \frac{b}{\sqrt{3}}\right)ab + f\left(\frac{a}{\sqrt{3}}, \frac{b}{\sqrt{3}}\right)ab. \tag{9.31a}$$

The 2×2 Gauss rule is more often stated for a square area whose sides are each two units long. It may be written from equation 9.31a by setting $a = 1$ and $b = 1$:

$$\int_A f(\zeta,\ \eta)\ dA = f\left(-\frac{1}{\sqrt{3}}, -\frac{1}{\sqrt{3}}\right) + f\left(\frac{1}{\sqrt{3}}, -\frac{1}{\sqrt{3}}\right) + f\left(-\frac{1}{\sqrt{3}}, \frac{1}{\sqrt{3}}\right) + f\left(\frac{1}{\sqrt{3}}, \frac{1}{\sqrt{3}}\right).$$
$$\tag{9.31b}$$

The 2×2 square area is identical to what is portrayed as the quadrilateral finite element in ζ–η space (Figures 9.5 and 9.6).

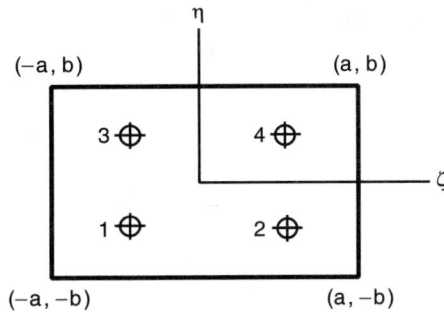

Figure 9.9. Gauss 2×2 rule.

If the 2×2 Gauss rule is applied, as expressed in equation 9.31a, to a function over a rectangular area where the origin of the variables of integration is not at the center of the rectangle, the coordinates of the integration points must be translated from the (ζ_k, η_k) values used in equation 9.31a to values in the variables of integration. If the rule is applied, as stated in equation 9.31b, where the origin of the variables of integration is not at the center of the rectangular area and the rectangular area is not a 2×2 square, the coordinates of the integration points (ζ_k, η_k) must be scaled by the factors a and b and then translated. The weight factors must be scaled by ab.

The alternative approach, the one used exclusively in the finite element method, is to transform the $f(x, y)$, the differential area dA, and the rectangular area to the 2×2 square and its corresponding quadrature rule. Transformations $x = x(\zeta, \eta)$, $y = y(\zeta, \eta)$, and $dA = |J(\zeta, \eta)| \, d\zeta \, d\eta$ must be written so that

$$
\int_A f(x, y) \, dA = \int_{-1}^{1} \int_{-1}^{1} f(x(\zeta, \eta), y(\zeta, \eta))|J(\zeta, \eta)| \, d\zeta \, d\eta
$$

$$
= \int_{-1}^{1} \int_{-1}^{1} I(\zeta, \eta) \, d\zeta \, d\eta.
$$

(9.31c)

The rule of equation 9.31b may then be used directly as

$$
\int_A f(\zeta, \eta) \, dA = \sum_{k=1}^{4} I(\zeta_k, \eta_k)
$$

$$
= I\left(-\frac{1}{\sqrt{3}}, -\frac{1}{\sqrt{3}}\right) + I\left(\frac{1}{\sqrt{3}}, -\frac{1}{\sqrt{3}}\right) + I\left(-\frac{1}{\sqrt{3}}, \frac{1}{\sqrt{3}}\right) + I\left(\frac{1}{\sqrt{3}}, \frac{1}{\sqrt{3}}\right).
$$

(9.31d)

The 2×2 Gauss rule uses cubic polynomials in ζ and η to approximate the integrand. It will evaluate exactly polynomial integrals with terms of $\zeta^3 \eta^3$ or lower degree in ζ, η, or both. If the integrand contains higher degree terms or is not a simple polynomial, the evaluation will not be exact, though it may be sufficiently accurate for the process used.

The three-point Gauss rule for quadrature of a function of a single variable along a line segment (equation 9.29) may be applied in the manner described above to develop a 3×3 (nine-point) Gauss quadrature rule for integration of a function of two variables over a rectangular area whose dimensions are $2a$ in width and $2b$ in height. The table below and Figure 9.10 illustrate the rectangular area, its integration points, and its weight factors:

Integration Points	ζ_k	η_k	H_k
1, 3, 7, 9	$\pm a\sqrt{0.6}$	$\pm b\sqrt{0.6}$	$25ab/81$
2, 8	0	$\pm b\sqrt{0.6}$	$40ab/81$
4, 6	$\pm a\sqrt{0.6}$	0	$40ab/81$
5	0	0	$64ab/81$

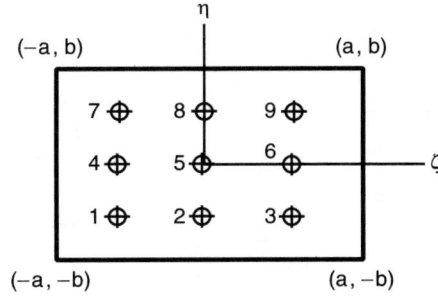

Figure 9.10. Gauss 3×3 rule.

The 3×3 Gauss quadrature rule is given as

$$\int_A f(\zeta, \eta) \, dA = f(-a\sqrt{0.6}, -b\sqrt{0.6})\frac{25ab}{81} + f(0, -b\sqrt{0.6})\frac{40ab}{81}$$

$$+ f(a\sqrt{0.6}, -b\sqrt{0.6})\frac{25ab}{81} + f(-a\sqrt{0.6}, 0)\frac{40ab}{81}$$

$$+ f(0, 0)\frac{64ab}{81} + f(a\sqrt{0.6}, 0)\frac{40ab}{81} + f(-a\sqrt{0.6}, b\sqrt{0.6})\frac{25ab}{81}$$

$$+ f(0, b\sqrt{0.6})\frac{40ab}{81} + f(a\sqrt{0.6}, b\sqrt{0.6})\frac{25ab}{81}. \tag{9.32a}$$

The 3×3 Gauss rule for a square whose sides are each two units long is written, from equation 9.32a, by setting $a = 1$ and $b = 1$:

$$\int_A f(\zeta, \eta) \, dA = f(-\sqrt{0.6}, -\sqrt{0.6})\frac{25}{81} + f(0, -\sqrt{0.6})\frac{40}{81} + f(\sqrt{0.6}, -\sqrt{0.6})\frac{25}{81}$$

$$+ f(-\sqrt{0.6}, 0)\frac{40}{81} + f(0, 0)\frac{64}{81} + f(\sqrt{0.6}, 0)\frac{40}{81}$$

$$+ f(-\sqrt{0.6}, \sqrt{0.6})\frac{25}{81} + f(0, \sqrt{0.6})\frac{40}{81} + f(\sqrt{0.6}, \sqrt{0.6})\frac{25}{81}. \tag{9.32b}$$

The 3×3 Gauss rule uses fifth-degree polynomials in ζ and η to approximate the integrand. It will evaluate exactly polynomial integrals with terms $\zeta^5\eta^5$ or lower degree in ζ, η, or both.

Clearly Gauss quadrature rules for rectangular areas can be devised as 2×3, 3×2, 3×4, 4×3, 4×4, ..., as desired. The integration points in the rectangular area must be located in accordance with the appropriate Gauss rules for single-variable functions. Then the corresponding weight factors for the rectangular area rule must be calculated from the weight factors of the single-variable rules.

As an example of the application of the 2×2 Gauss rule, the integral $\int_A [x^2 y/(x + 1)] \, dA$ will be evaluated over a rectangular area whose width is eight units and height is four units. The x axis parallels the long side and the y axis parallels the short side. The origin of the x–y axes is at the lower left corner of

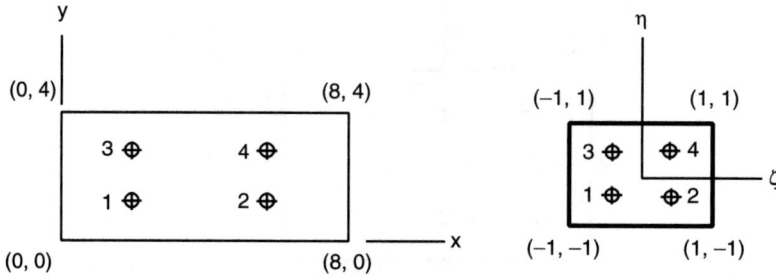

Figure 9.11. Example area quadrature.

the area. Figure 9.11 shows the rectangle, its integration points, and transformed square area whose sides are each two units long.

The transformation of the integral and its rectangle to the 2×2 square requires a transformation which translates the origin of the rectangle to its center and scales its dimensions to those of the square. This transformation may be written using the methods for isoparametric transformation (equations 9.13) and the shape functions for a rectangle in terms of its local natural coordinates (equations 9.6):

$$x = 0 \cdot N_i(\zeta,\ \eta) + 8 \cdot N_2(\zeta,\ \eta) + 8 \cdot N_3(\zeta,\ \eta) + 0 \cdot N_4(\zeta,\ \eta)x_4$$

$$= 8\left(\frac{1}{4}(1+\zeta)(1-\eta)\right) + 8\left(\frac{1}{4}(1+\zeta)(1+\eta)\right)$$

$$= 4(1+\zeta),$$

$$y = 0 \cdot N_i(\zeta,\ \eta) + 0 \cdot N_2(\zeta,\ \eta) + 4 \cdot N_3(\zeta,\ \eta) + 4 \cdot N_4(\zeta,\ \eta)x_4$$

$$= 4\left(\frac{1}{4}(1+\zeta)(1+\eta)\right) + 4\left(\frac{1}{4}(1-\zeta)(1+\eta)\right)$$

$$= 2(1+\eta).$$

From these expressions for $x(\zeta,\ \eta)$ and $y(\zeta,\ \eta)$,

$$|J(\zeta,\ \eta)| = \frac{\partial x}{\partial \zeta} \cdot \frac{\partial y}{\partial \eta} - \frac{\partial x}{\partial \eta} \cdot \frac{\partial y}{\partial \zeta} = 4 \cdot 2 - 0 \cdot 0 = 8,$$

$$I(\zeta,\ \eta) = f(4(1+\zeta),\ 2(1+\eta))|J(\zeta,\ \eta)|$$

$$= \frac{(4(1+\zeta))^2(2(1+\eta))}{(4(1+\zeta))+1}8 = \frac{256(1+\zeta)^2(1+\eta)}{5+4\zeta}.$$

The transformed integral (equation 9.31c) is

$$\int_A \frac{x^2 y}{x+1}\, dA = \int_{-1}^{1}\int_{-1}^{1} \frac{256(1+\zeta)^2(1+\eta)}{5+4\zeta}\, d\zeta\, d\eta = \sum_{k=1}^{4} \frac{256(1+\zeta_k)^2(1+\eta_k)}{5+4\zeta_k} H_k.$$

The coordinates of the four integration points for the 2×2 square area are obtained from equation 9.31d. The integrand evaluated at each integration point is

$$I\left(-\frac{1}{\sqrt{3}}, -\frac{1}{\sqrt{3}}\right) = 7.1834452, \qquad I\left(\frac{1}{\sqrt{3}}, -\frac{1}{\sqrt{3}}\right) = 36.8294351,$$

$$I\left(-\frac{1}{\sqrt{3}}, \frac{1}{\sqrt{3}}\right) = 26.8089830, \qquad I\left(\frac{1}{\sqrt{3}}, \frac{1}{\sqrt{3}}\right) = 137.4493230.$$

Using the weight factors for the 2×2 square ($H_1 = H_2 = H_3 = H_4 = 1$) and applying the quadrature rule (equation 9.31d),

$$\int_A \frac{x^2 y}{x+1} \, dA = (7.1834452) \cdot 1 + (36.8294351) \cdot 1$$

$$+ (26.8089830) \cdot 1 + (137.4493230) \cdot 1$$

$$= 208.271186$$

It is left as an exercise for the reader to evaluate the integral first explicitly and second to transform the integration points from (ζ_k, η_k) to (x_k, y_k) to evaluate the integral as a function of x and y first using the 2×2 and then the 3×3 Gauss quadrature rules.

9.5. ELEMENT MAPPING AND MASTER ELEMENT CALCULATIONS

The strength and utility of the finite element method is a result of its ability to analyze structures with irregular geometric shapes. We saw earlier that the straight-sided triangular elements possessed this ability because of the inherent versatility of the triangle to model irregular shapes as assemblages of triangles. A disadvantage is that the triangles are straight sided while the irregular structure may not be, thus introducing error. The 3-node constant strain straight-sided triangular finite element (Section 8.2) employs linear shape functions to approximate displacement variation within the element. This results in a single set of values for the stress and strain vectors for each element, further compounding its inaccuracies as a structural model. The straight-sided 6-node linear strain triangular finite element (Section 8.5) and the straight-sided 4-node quadrilateral finite element (Section 9.3) are improvements. Both use higher degree polynomial shape functions for displacement variations with improved accuracy. However, both must still model irregular structures with straight-sided finite elements, even though the boundaries of the structure may be curved.

The development of the isoparametric transformation scheme coupled with Gaussian quadrature (Sections 9.3 and 9.4) makes it possible to calculate the element stiffness matrix for the straight-sided 4-node quadrilateral finite element. Unlike the developments for the triangular finite elements, the isoparametric transformation scheme is not limited to either the number of nodes on the finite element or its shape.

If an 8-node quadrilateral finite element is employed in the isoparametric transformation scheme, it must use shape functions with eight polynomial terms, or quadratic polynomials. Clearly these will approximate better the structural displacement variation within the finite element than will the bilinear shape functions of the 4-node quadrilateral element. Equally important, however, the transformation of the 2×2 square master element in ζ–η space using quadratic shape functions results in curved-sided quadrilaterals in x–y space. Their sides will be quadratic polynomials which can approximate most irregularly shaped structures better than the straight-sided 4-node quadrilateral can.

The trade-off, of course, is that 8-node quadrilateral elements possess 16 edof and 16×16 element stiffness matrices, whereas 4-node quadrilateral elements possess 8 edof and 8×8 element stiffness matrices. More computation must be done per element to achieve the improved accuracy, though the fewer elements are usually needed.

A 9-node quadrilateral element may be a further improvement in accuracy over the 8-node quadrilateral element. However, in many instances, the element stiffness coefficients for the 8-node element can be calculated with sufficient accuracy using the 2×2 Gauss quadrature rule. On the other hand, the 3×3 Gauss quadrature rule is almost always needed for the 9-node element. The 3×3 Gauss rule, as compared to the 2×2 Gauss rule, imposes over a twofold increase in the computation effort for the integration to obtain the element stiffness matrix.

The 6-node straight-sided triangular finite element, discussed earlier in Section 8.5, may be formulated in the isoparametric transformation scheme in the same way that is done for the quadrilateral finite element. In so doing, the triangular finite element will be mapped from a straight-sided master triangular element in ζ–η space to a curved-sided triangular finite element in x–y space. The curved sides will be quadratic polynomials. A 7-node triangular finite element, with the seventh node at the center of the triangle, could be developed similarly.

9.5.1. Element Mapping

To better understand the isoparametric algorithm, it is helpful to view the assemblage of elements in x–y space as a series of transformations T_e, where $e = 1$, $2, \ldots, N$ (N being the number of finite elements). The finite elements in x–y space are images of the master element in ζ–η space, projected by their isoparametric transformations T_e. The transformations are the sum of the products of the nodal coordinates of each x–y element and the corresponding shape functions of the ζ–η master element. Figure 9.12 illustrates the concept.

For the transformation process to be possible and practical, several conditions must obtain. The element transformation T_e must be easy to construct from the geometry of the structure in x–y space. They are. All that is needed are the coordinates of the nodes of the elements in x–y space and the shape functions of the master element in ζ–η space.

In addition, the calculation of element stiffness coefficients must be relatively easy to do. It can be. The description of the stiffness matrix calculation of the 4-node quadrilateral element using equations 9.21 applies directly. The shape functions for the master element are written and differentiated with respect to ζ and η. These derivatives are evaluated at each Gaussian integration point and, with the x–y coordinates of each node of the element, substituted into equations 9.17 and 9.18 to develop the entries for matrices in equation 9.21b. Then the Gaussian quadrature rule is completed to produce the complete element stiffness matrix.

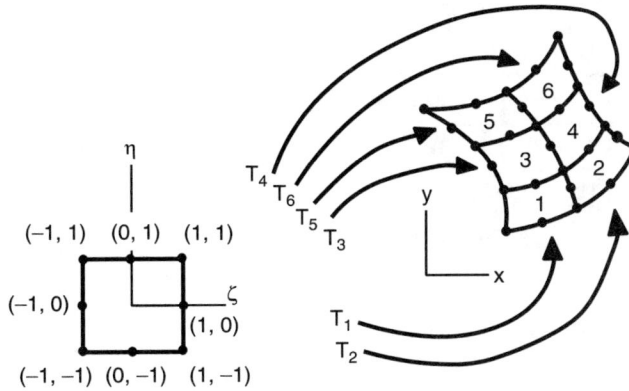

Figure 9.12. Element mapping.

The Jacobian of the transformation $|J(\zeta, \eta)|$ (equation 9.17c) must be positive throughout the element for the transformation to yield a one-to-one correspondence of all points between the x–y element and its ζ–η master element. If it were positive in one region of the element and negative in another, differential areas of the master element would map differently in the two regions, as may be surmised from equation 9.19j. Moreover, the Jacobian would be zero as its value changes from positive to negative so that the integrand matrix $\mathbf{I}(\zeta, \eta)$ of equation 9.21b would be undefined there.

The value and sign of the Jacobian at any point in the element can be verified using equation 9.17c. In general, however, transformations such as the isoparametric transformation will result in positive values for their Jacobians throughout each element if reasonable care is exercised in locating the nodes on the x–y elements. They should be at each corner of the x–y quadrilateral or triangular elements. If additional nodes are used, there should be one near the center of each side. If an added node is used, it should be near the center of the x–y element. The interior angles of the x–y element should be greater than zero and less than 180°, preferably much greater than zero and much less than 180°. In addition, the lengths of the sides of the x–y element should not be greatly different. As a guide, the ratio of the longest to the shortest side should not exceed 5.

A final condition is that the assemblage of x–y elements must map into x–y space without gaps or overlap between them. Observing the sides between adjacent elements in x–y space shown in Figure 9.12, adjacent elements each share three nodes. Since the curved sides of each are quadratic polynomials in an isoparametric transformation, there can be no gap or overlap between them. This is because the coordinates of the three shared nodes determine uniquely the one quadratic polynomial curve that is the boundary for each.

If the elements had nodes only at corners, adjacent elements would share two nodes. Their sides would be straight, and the coordinates of the two shared nodes would determine uniquely the one straight line that is the boundary for each.

An example of the transformation for a well-behaved 4-node element is shown in Figure 9.13. Shown are the element in x–y space and its ζ–η master element.

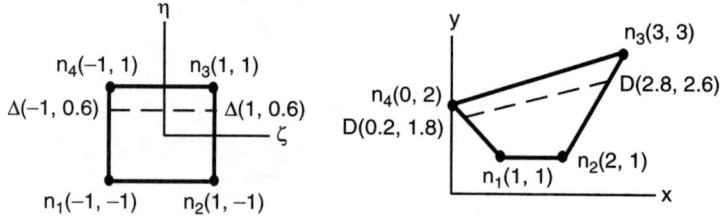

Figure 9.13. Example transformation.

The transformation is given as

$$x = 1 \cdot N_1(\zeta, \eta) + 2 \cdot N_2(\zeta, \eta) + 3 \cdot N_3(\zeta, \eta) + 0 \cdot N_4(\zeta, \eta)$$
$$= \tfrac{1}{4}(1 - \zeta)(1 - \eta) + \tfrac{2}{4}(1 + \zeta)(1 - \eta) + \tfrac{3}{4}(1 + \zeta)(1 + \eta)$$
$$= \tfrac{1}{4}(6 + 4\zeta + 2\zeta\eta),$$
$$y = 1 \cdot N_1(\zeta, \eta) + 1 \cdot N_2(\zeta, \eta) + 3 \cdot N_3(\zeta, \eta) + 2 \cdot N_4(\zeta, \eta)$$
$$= \tfrac{1}{4}(1 - \zeta)(1 - \eta) + \tfrac{1}{4}(1 + \zeta)(1 - \eta) + \tfrac{3}{4}(1 + \zeta)(1 + \eta) + \tfrac{2}{4}(1 - \zeta)(1 + \eta)$$
$$= \tfrac{1}{4}(7 + \zeta + 3\eta + \zeta\eta).$$

The Jacobian of the transformation is

$$|J(\zeta, \eta)| = \begin{vmatrix} \frac{\partial x}{\partial \zeta} & \frac{\partial y}{\partial \zeta} \\ \frac{\partial x}{\partial \eta} & \frac{\partial y}{\partial \eta} \end{vmatrix} = \begin{vmatrix} \frac{4+2\eta}{4} & \frac{1+\eta}{4} \\ \frac{2\zeta}{4} & \frac{3+\zeta}{4} \end{vmatrix} = \frac{6 + \zeta + 3\eta}{8}.$$

The Jacobian is a linear polynomial in ζ and η and is positive throughout the range of ζ and η. It plots as a plane over the ζ–η master element (Figure 9.14).

The line Δ–Δ on the master element maps to the line D–D on the x–y element, as shown in Figure 9.13. On the master element its equation is $\eta = 0.6$. Its parametric equations in x–y space may be written from the transformation:

$$x(\zeta, \eta) = \tfrac{1}{4}[6 + 4\zeta + 2\zeta(0.6)] = 1.5 + 1.3\zeta,$$
$$y(\zeta, \eta) = \tfrac{1}{4}[7 + \zeta + 3(0.6) + \zeta(0.6)] = 2.2 + 0.4\zeta.$$

Since x and y are linear with respect to ζ, line D–D plots as a straight line on the x–y

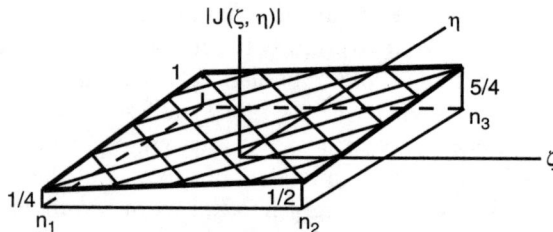

Figure 9.14. Example Jacobian.

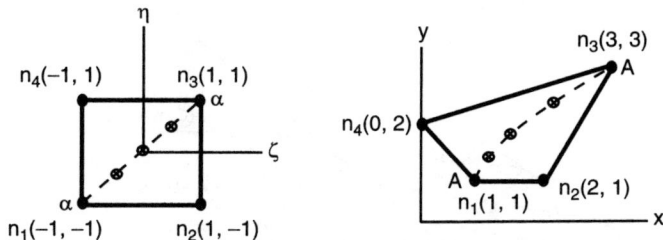

Figure 9.15. Example transformation.

element. Its equation in x–y space may be obtained by eliminating ζ from the parametric equations:

$$y = \tfrac{1}{13}(4x + 22.6).$$

The line α–α on the master element is a diagonal from node 1 to node 3, as shown in Figure 9.15. On the master element its equation is $\zeta = \eta$. Its parametric equations in x–y space may be written from the transformation:

$$x = \tfrac{1}{4}(6 + 4\zeta + 2\zeta \cdot \zeta) = \tfrac{1}{4}(6 + 4\zeta + 2\zeta^2),$$

$$y = \tfrac{1}{4}(7 + \zeta + 3\zeta + \zeta \cdot \zeta) = \tfrac{1}{4}(7 + 4\zeta + \zeta^2).$$

Since x and y are not linear with respect to ζ, the straight diagonal line α–α on the ζ–η master element plots as a curved line A–A on the x–y element. Some coordinates for line A–A are easily computed from its parametric equations and are tabulated below. Lines α–α and A–A are shown in Figure 9.15:

ζ	x	y
-1	1	1
$-\tfrac{1}{2}$	$\tfrac{9}{8}$	$\tfrac{21}{16}$
0	$\tfrac{6}{4}$	$\tfrac{7}{4}$
$\tfrac{1}{2}$	$\tfrac{17}{8}$	$\tfrac{37}{16}$
1	3	3

A less well-behaved 4-node element is shown in Figure 9.16. The element in x–y

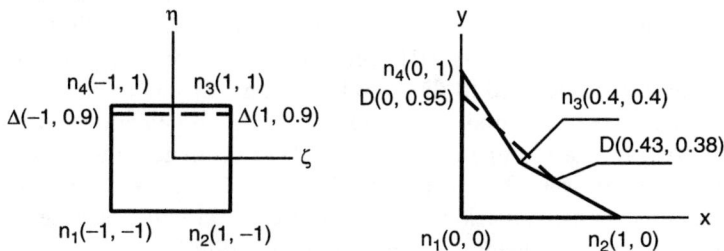

Figure 9.16. Example transformation.

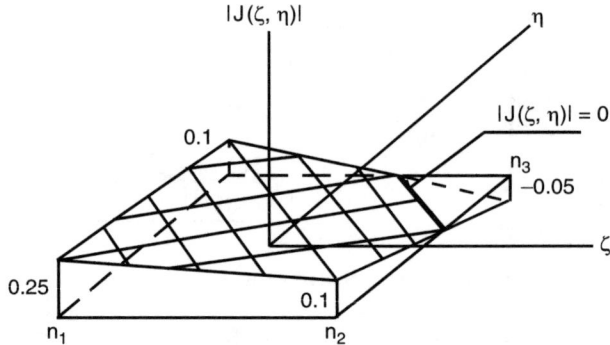

Figure 9.17. Example Jacobian.

space and its master element in ζ–η space are shown. The transformation and Jacobian are developed as above and are

$$x = \tfrac{1}{4}(1 + \zeta)(1.4 - 0.6\eta), \qquad y = \tfrac{1}{4}(1 + \eta)(1.4 - 0.6\zeta),$$

$$|J(\zeta, \ \eta)| = \tfrac{1}{8}(0.8 - 0.6\zeta - 0.6\eta).$$

The Jacobian is a linear polynomial in ζ–η space and is shown plotted in Figure 9.17. It is positive in most regions of the element, negative near node 3, and zero along a line near node 3. One should anticipate that the transformation may give trouble near node 3. The negative values of the Jacobian there are an indicator, and so is the interior angle at node 3, which exceeds 180°.

The line Δ–Δ, whose equation is $\eta = 0.9$ on the master element, maps to the x–y element as the straight line D–D. Its parametric equations in x–y space and equation with the parameter ζ eliminated may be written as was done above and are

$$x = 0.215(1 + \zeta), \qquad y = 0.95(0.7 - 0.3\zeta), \qquad y = -\tfrac{57}{43}x + 0.95.$$

Line D–D plots outside the x–y element in the vicinity of node 3.

9.5.2. Master Element Stiffness Calculations

Much has been said already about the necessity for calculating element stiffness matrices in ζ–η space for the plane stress 4-node quadrilateral finite element. Equations 9.16, 9.17, 9.18, 9.20, and 9.21 were developed for this purpose. These equations, however, are not limited to the 4-node element. Element stiffness matrices for the 8-node serendipity and 9-node Lagrange plane stress elements may be calculated using them. Appropriate shape functions for these elements in terms of ζ and η would be required. There would be eight shape functions for the 8-node element and nine for the 9-node element. Equations 9.16 would contain eight or nine terms, as appropriate on their right-hand side, and would incorporate the eight or nine shape functions and the eight or nine sets of nodal coordinates, as appropriate.

Since the integrands of equations 9.20 and 9.21 are not simple polynomials in ζ and η and are instead polynomial fractions, Gaussian quadrature will not produce exact results. The 2×2 rule integrates polynomials exactly whose terms contain the variables ζ or η or both to a power of 3 or lower. The 3×3 rule is exact for polynomials whose terms contain these variables to a power of 5 or lower. The 4-node element uses bilinear polynomials for its shape functions which result in integrands (equation 9.21b), which are polynomial fractions with quadratic polynomials in the numerator and linear polynomials in the denominator, as suggested earlier. They may be represented as

$$I(\zeta, \eta) = \frac{A + B\zeta + C\eta + D\zeta^2 + E\zeta\eta + F\eta^2 + G\zeta^2\eta + H\zeta\eta^2 + I\zeta^2\eta^2}{a + b\zeta + c\eta}. \qquad (9.33)$$

The 2×2 Gauss quadrature rule will integrate this expression accurately.

The 8- and 9-node plane stress–plane strain elements will be discussed later in this chapter. However, they use quadratic polynomials in ζ and η for shape functions. The numerator polynomials of the integrands for these elements will contain terms with the variables ζ or η or both to the fourth power, and the denominator will contain them to the second power. The 2×2 Gauss quadrature rule clearly will not be as accurate for them. As it happens, the 8-node element, which employs slightly lower degree polynomials than the 9-node element, can usually be integrated with adequate accuracy using the 2×2 Gauss rule. The 9-node element, on the other hand, usually needs the 3×3 Gauss rule to obtain its stiffness matrix.

The element stiffness matrix for the 4-node plane stress finite element is efficiently obtained by implementing the 2×2 Gauss quadrature rule on equations 9.20 and 9.21 as follows:

$$[\mathbf{K}] = t \int_{-1}^{1} \int_{-1}^{1} [\mathbf{B}^{\mathrm{T}}(\zeta, \eta)][\mathbf{E}][\mathbf{B}(\zeta, \eta)]|J(\zeta, \eta)| \, d\zeta \, d\eta \qquad (9.20b)$$

$$= t \int_{-1}^{1} \int_{-1}^{1} [\mathbf{I}(\zeta, \eta)] \, d\zeta \, d\eta \qquad (9.20c)$$

$$= t \left[\sum_{k=1}^{4} \mathbf{I}(\zeta_k, \eta_k) \cdot H_k \right] \qquad (9.34a)$$

$$= t \left[\mathbf{I}\left(-\frac{1}{\sqrt{3}}, -\frac{1}{\sqrt{3}} \right) \cdot 1 \right] + t \left[\mathbf{I}\left(\frac{1}{\sqrt{3}}, -\frac{1}{\sqrt{3}} \right) \cdot 1 \right]$$

$$+ t \left[\mathbf{I}\left(-\frac{1}{\sqrt{3}}, \frac{1}{\sqrt{3}} \right) \cdot 1 \right] + t \left[\mathbf{I}\left(\frac{1}{\sqrt{3}}, \frac{1}{\sqrt{3}} \right) \cdot 1 \right] \qquad (9.34b)$$

The matrix integrand $\mathbf{I}(\zeta_k, \eta_k)$ is evaluated four times, once at each integration point. It is constructed from equations 9.21 as

$$[\mathbf{I}(\zeta_k, \eta_k)] = [\mathbf{B}^{\mathrm{T}}(\zeta_k, \eta_k)][\mathbf{E}][\mathbf{B}(\zeta_k, \eta_k)]|J(\zeta_k, \eta_k)| \qquad (9.35a)$$

$$= \frac{[\mathbf{B}'^{\mathrm{T}}(\zeta_k, \eta_k)][\mathbf{E}][\mathbf{B}'(\zeta_k, \eta_k)]}{|J(\zeta_k, \eta_k)|}, \qquad (9.35b)$$

where $\mathbf{B}'$ is the matrix $\mathbf{B}$ without division by the Jacobian. The Jacobian $|J(\zeta_k, \eta_k)|$ is calculated using equation 9.17c as

$$|J(\zeta_k, \eta_k)| = \frac{\partial x(\zeta_k, \eta_k)}{\partial \zeta} \cdot \frac{\partial y(\zeta_k, \eta_k)}{\partial \eta} - \frac{\partial x(\zeta_k, \eta_k)}{\partial \eta} \cdot \frac{\partial y(\zeta_k, \eta_k)}{\partial \zeta}. \tag{9.36}$$

The derivatives in equation 9.36 may be evaluated using equations 9.15 and 9.16 as

$$\frac{\partial x(\zeta_k, \eta_k)}{\partial \zeta} = -\frac{x_1}{4}(1 - \eta_k) + \frac{x_2}{4}(1 - \eta_k) + \frac{x_3}{4}(1 + \eta_k) - \frac{x_4}{4}(1 + \eta_k), \tag{9.37a}$$

$$\frac{\partial y(\zeta_k, \eta_k)}{\partial \zeta} = -\frac{y_1}{4}(1 - \eta_k) + \frac{y_2}{4}(1 - \eta_k) + \frac{y_3}{4}(1 + \eta_k) - \frac{y_4}{4}(1 + \eta_k), \tag{9.37b}$$

$$\frac{\partial x(\zeta_k, \eta_k)}{\partial \eta} = -\frac{x_1}{4}(1 - \zeta_k) - \frac{x_2}{4}(1 + \zeta_k) + \frac{x_3}{4}(1 + \zeta_k) + \frac{x_4}{4}(1 - \zeta_k), \tag{9.37c}$$

$$\frac{\partial y(\zeta_k, \eta_k)}{\partial \eta} = -\frac{y_1}{4}(1 - \zeta_k) - \frac{y_2}{4}(1 + \zeta_k) + \frac{y_3}{4}(1 + \zeta_k) + \frac{y_4}{4}(1 - \zeta_k). \tag{9.37d}$$

The coordinates (x_i, y_i) are the x–y coordinates of the nodes.
Equations 9.7 apply for the matrix $\mathbf{B}'$:

$$\left[\mathbf{B}'(\zeta_k, \eta_k)\right] = \left[\mathbf{B}'_1(\zeta_k, \eta_k) \quad \mathbf{B}'_2(\zeta_k, \eta_k) \quad \mathbf{B}'_3(\zeta_k, \eta_k) \quad \mathbf{B}'_4(\zeta_k, \eta_k)\right], \tag{9.38a}$$

where

$$\left[\mathbf{B}'_i(\zeta_k, \eta_k)\right] = \begin{bmatrix} \frac{\partial N'_i(\zeta_k, \eta_k)}{\partial x} & 0 \\ 0 & \frac{\partial N'_i(\zeta_k, \eta_k)}{\partial y} \\ \frac{\partial N'_i(\zeta_k, \eta_k)}{\partial y} & \frac{\partial N'_i(\zeta_k, \eta_k)}{\partial x} \end{bmatrix}. \tag{9.38b}$$

The primes on the derivatives with respect to x and y of the shape functions N_i in equation 9.38b indicate that they are not divided by the Jacobian $|J(\zeta_k, \eta_k)|$. These derivatives may be calculated using equations 9.18, 9.37, and 9.15 as

$$\frac{\partial N'_i(\zeta_k, \eta_k)}{\partial x} = \frac{\partial y(\zeta_k, \eta_k)}{\partial \eta}\frac{\zeta_i}{4}(1 + \eta_i\eta_k) - \frac{\partial y(\zeta_k, \eta_k)}{\partial \zeta}\frac{\eta_i}{4}(1 + \zeta_i\zeta_k), \tag{9.39a}$$

$$\frac{\partial N'_i(\zeta_k, \eta_k)}{\partial y} = -\frac{\partial x(\zeta_k, \eta_k)}{\partial \eta}\frac{\zeta_i}{4}(1 + \eta_i\eta_k) + \frac{\partial x(\zeta_k, \eta_k)}{\partial \zeta}\frac{\eta_i}{4}(1 + \zeta_i\zeta_k). \tag{9.39b}$$

The subscript k identifies the coordinates of an integration point; the subscript i identifies the coordinates of a node.

The matrix $\mathbf{E}$ is the plane stress matrix of elastic constants for the isotropic or orthotropic condition (equations 6.5b or 7.20). Its most general form is equation 7.34b:

$$[\mathbf{E}] = \begin{bmatrix} E_{11} & E_{12} & 0 \\ E_{21} & E_{22} & 0 \\ 0 & 0 & E_{33} \end{bmatrix}.$$

The values for the entries of matrix $\mathbf{E}$ depend on which condition applies. They are the entries from equations 6.5b or 7.20, as appropriate, multiplied by their respective scalar.

The numerical values of the derivatives of equation 9.39 at each integration point are assembled to form four sets of matrices $\mathbf{B}'^{\mathrm{T}}$ and $\mathbf{B}'$. These then are arranged with matrix $\mathbf{E}$ in four sets in the order required by equation 9.35b. The matrix multiplications are performed and the entries of each resulting product matrix is divided by its value of the Jacobian $|J(\zeta_k, \eta_k)|$ to produce the four numerical integrand matrices $\mathbf{I}(\zeta_k, \eta_k)$. These are each multiplied by their respective weight factors H_k, and the results are added. The sum is multiplied by the plate thickness t, which yields the numerical 8×8 element stiffness matrix.

Example Calculation.   The 4-node quadrilateral finite element shown in Figure 9.18 is isotropic and in a state of plane stress. Its element stiffness matrix will be calculated as an illustration of the isoparametric transformation scheme. The nodal coordinates for the element are shown in inches, and the element is 0.50 in. thick. Its modulus of elasticity is 30,000 ksi, and its Poisson ratio is 0.25.

The 2×2 Gauss quadrature rule will be used.

Tabulated below are the coordinates of the nodes and integration points on the ζ–η master element, and the values for the derivatives of shape functions with respect to ζ and η at each integration point, for the derivatives of x and y with respect to ζ and η at each integration point, for the Jacobian at each integration point, and for the primed derivatives of shape functions with respect to x and y at each integration point:

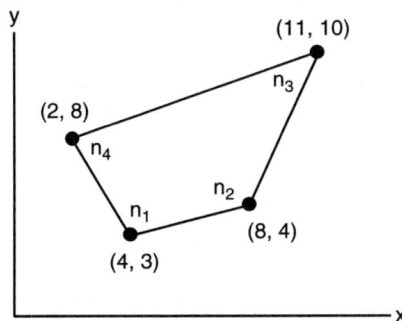

Figure 9.18. Example stiffness matrix.

	Node 1	Node 2	Node 3	Node 4
x_i	4	8	11	2
y_i	3	4	10	8
	Integration Point 1	Integration Point 2	Integration Point 3	Integration Point 4
ζ_k	−0.577350	0.577350	−0.577350	0.577350
η_k	−0.577350	−0.577350	0.577350	0.577350
$\partial N_1/\partial\zeta$	−0.394338	−0.394338	−0.105663	−0.105663
$\partial N_1/\partial\eta$	−0.394338	−0.105663	−0.394338	−0.105663
$\partial N_2/\partial\zeta$	0.394338	0.394338	0.105663	0.105663
$\partial N_2/\partial\eta$	−0.105663	−0.394338	−0.105663	−0.394338
$\partial N_3/\partial\zeta$	0.105663	0.105663	0.394338	0.394338
$\partial N_3/\partial\eta$	0.105663	0.394338	0.105663	0.394338
$\partial N_4/\partial\zeta$	−0.105663	−0.105663	−0.394338	−0.394338
$\partial N_4/\partial\eta$	0.394338	0.105663	0.394338	0.105663
$\partial x/\partial\zeta$	2.528313	2.528313	3.971688	3.971688
$\partial x/\partial\eta$	−0.471688	0.971687	−0.471688	0.971687
$\partial y/\partial\zeta$	0.605663	0.605663	0.894338	0.894338
$\partial y/\partial\eta$	2.605663	2.894338	2.605663	2.894338
$|J|$	6.873613	6.729275	10.770725	10.626388
$\partial N_1'/\partial x$	−0.788675	−1.077350	0.077350	−0.211325
$\partial N_1'/\partial y$	−1.183013	0.116025	−1.616025	−0.316988
$\partial N_2'/\partial x$	1.091506	1.380181	0.369819	0.658494
$\partial N_2'/\partial y$	−0.081143	−1.380181	−0.369819	−1.668856
$\partial N_3'/\partial x$	0.211325	0.066988	0.933013	0.788675
$\partial N_3'/\partial y$	0.316988	0.894338	0.605663	1.183013
$\partial N_4'/\partial x$	−0.514156	−0.369819	−1.380181	−1.235844
$\partial N_4'/\partial y$	0.947169	0.369819	1.380181	0.802831

The numerical integrand matrix at the first integration point is formed from the primed values for the derivatives of the shape functions with respect to x and y at the first integration point, the plane stress matrix of elastic constants, and the Jacobian value at the first integration point (equation 9.35b). Recall that the primed derivatives have not been divided by their Jacobian:

$$\left[\mathbf{I}\!\left(\frac{-1}{\sqrt{3}}, \frac{-1}{\sqrt{3}}\right)\right] = \frac{32,000}{6.873613}\begin{bmatrix} -0.788675 & 0 & -1.183013 \\ 0 & -1.183013 & -0.788675 \\ 1.091506 & 0 & -0.081143 \\ 0 & -0.081143 & 1.091506 \\ 0.211325 & 0 & 0.316988 \\ 0 & 0.316988 & 0.211325 \\ -0.514156 & 0 & 0.947169 \\ 0 & 0.947169 & -0.514156 \end{bmatrix}\begin{bmatrix} 1 & 0.25 & 0 \\ 0.25 & 1 & 0 \\ 0 & 0 & 0.375 \end{bmatrix}\times$$

(continued)

$$
\times \begin{bmatrix}
-0.788675 & 0 & 1.091506 & 0 & 0.211325 & 0 & -0.514156 & 0 \\
0 & -1.183013 & 0 & -0.081143 & 0 & 0.316988 & 0 & 0.947169 \\
-1.183013 & -0.788675 & -0.081143 & 1.091506 & 0.316988 & 0.211325 & 0.947169 & -0.514156
\end{bmatrix}
$$

$$
= \begin{bmatrix}
5339.04 & 2714.77 & -3840.06 & -2179.823 & -1430.59 & -727.42 & -68.39 & 192.47 \\
2714.77 & 7601.35 & -1391.14 & -1055.97 & -727.42 & -2036.78 & -596.20 & -4508.60 \\
-3840.06 & -1391.14 & 5557.97 & -257.70 & 1028.94 & 372.76 & -2746.85 & 1276.09 \\
-2179.82 & -1055.97 & -257.70 & 2110.58 & 584.08 & 282.95 & 1853.44 & -1337.57 \\
-1430.59 & -727.42 & 1028.94 & 584.08 & 383.33 & 194.91 & 18.33 & -51.57 \\
-727.42 & -2036.78 & 372.76 & 282.95 & 194.91 & 545.75 & 159.75 & 1208.08 \\
-68.39 & -596.20 & -2746.85 & 1853.44 & 18.33 & 159.75 & 2796.92 & -1416.99 \\
192.47 & -4508.60 & 1276.09 & -1337.56 & -51.57 & 1208.08 & -1416.99 & 4638.09
\end{bmatrix} \text{ kips/in.}
$$

The process is repeated for the remaining integration points to form the numerical integrand matrices for each. The four numerical integrand matrices are each multiplied by their weight factors, added, and the sum is multiplied by the plate thickness. The result is the element stiffness matrix

$$
[\mathbf{K}] = \begin{bmatrix}
7028.91 & & & & & & & \\
1118.61 & 8926.88 & & & & & & \\
-5133.67 & 608.08 & 11,511.46 & & \text{(symmetric)} & & & \\
-391.92 & -612.85 & -4120.75 & 12,000.62 & & & & \\
-1695.08 & -1987.30 & -311.27 & 1278.71 & 4139.75 & & & \\
-1987.30 & -2908.52 & 278.71 & -5530.89 & 1589.15 & 5666.93 & & \\
-200.15 & 260.61 & -6066.52 & 3233.96 & -2133.40 & 119.45 & 8400.96 & \\
1260.61 & -5405.51 & 3233.96 & -5856.88 & -880.55 & 2772.48 & -3614.02 & 8489.91
\end{bmatrix} \text{ kips/in.}
$$

Further insights into the matrix **K** and the integrand $\mathbf{I}(\zeta, \eta)$ may be gained by treating them in terms of submatrices, as was done with the rectangular finite element (equations 9.11):

$$
[\mathbf{I}(\zeta, \eta)] = \frac{[\mathbf{B}'^{T}(\zeta, \eta)][\mathbf{E}][\mathbf{B}'(\zeta, \eta)]}{|J(\zeta, \eta)|} \tag{9.40a}
$$

$$
= \frac{1}{|J|} \begin{bmatrix} \mathbf{B}'^{T}_{1} \\ \mathbf{B}'^{T}_{2} \\ \mathbf{B}'^{T}_{3} \\ \mathbf{B}'^{T}_{4} \end{bmatrix} [\mathbf{E}][\mathbf{B}'_{1} \quad \mathbf{B}'_{2} \quad \mathbf{B}'_{3} \quad \mathbf{B}'_{4}] \tag{9.40b}
$$

$$
= \frac{1}{|J|} \begin{bmatrix}
\mathbf{B}'^{T}_{1}\mathbf{EB}'_{1} & \mathbf{B}'^{T}_{1}\mathbf{EB}'_{2} & \mathbf{B}'^{T}_{1}\mathbf{EB}'_{3} & \mathbf{B}'^{T}_{1}\mathbf{EB}'_{4} \\
\mathbf{B}'^{T}_{2}\mathbf{EB}'_{1} & \mathbf{B}'^{T}_{2}\mathbf{EB}'_{2} & \mathbf{B}'^{T}_{2}\mathbf{EB}'_{3} & \mathbf{B}'^{T}_{2}\mathbf{EB}'_{4} \\
\mathbf{B}'^{T}_{3}\mathbf{EB}'_{1} & \mathbf{B}'^{T}_{3}\mathbf{EB}'_{2} & \mathbf{B}'^{T}_{3}\mathbf{EB}'_{3} & \mathbf{B}'^{T}_{3}\mathbf{EB}'_{4} \\
\mathbf{B}'^{T}_{4}\mathbf{EB}'_{1} & \mathbf{B}'^{T}_{4}\mathbf{EB}'_{2} & \mathbf{B}'^{T}_{4}\mathbf{EB}'_{3} & \mathbf{B}'^{T}_{4}\mathbf{EB}'_{4}
\end{bmatrix}. \tag{9.40c}
$$

The entries $\mathbf{B}'^{\mathrm{T}}_i\mathbf{E}\mathbf{B}'_j$ are 2×2 submatrices of functions of ζ and η, consistent with the fact that the integrand $[\mathbf{I}(\zeta,\ \eta)]$ is an 8×8 array of functions of ζ and η. The submatrices may be written as

$$[\mathbf{B}'^{\mathrm{T}}_i\mathbf{E}\mathbf{B}'_j] = \begin{bmatrix} \frac{\partial N'_i}{\partial x} & 0 & \frac{\partial N'_i}{\partial y} \\ 0 & \frac{\partial N'_i}{\partial y} & \frac{\partial N'_i}{\partial x} \end{bmatrix} \begin{bmatrix} E_{11} & E_{12} & 0 \\ E_{21} & E_{22} & 0 \\ 0 & 0 & E_{33} \end{bmatrix} \begin{bmatrix} \frac{\partial N'_j}{\partial x} & 0 \\ 0 & \frac{\partial N'_j}{\partial y} \\ \frac{\partial N'_j}{\partial y} & \frac{\partial N'_j}{\partial x} \end{bmatrix} \qquad (9.41a)$$

$$= \begin{bmatrix} E_{11} \frac{\partial N'_i}{\partial x} \frac{\partial N'_j}{\partial x} + E_{33} \frac{\partial N'_i}{\partial y} \frac{\partial N'_j}{\partial y} & E_{12} \frac{\partial N'_i}{\partial x} \frac{\partial N'_j}{\partial y} + E_{33} \frac{\partial N'_i}{\partial y} \frac{\partial N'_j}{\partial x} \\ E_{21} \frac{\partial N'_i}{\partial y} \frac{\partial N'_j}{\partial x} + E_{33} \frac{\partial N'_i}{\partial x} \frac{\partial N'_j}{\partial y} & E_{22} \frac{\partial N'_i}{\partial y} \frac{\partial N'_j}{\partial y} + E_{33} \frac{\partial N'_i}{\partial x} \frac{\partial N'_j}{\partial x} \end{bmatrix}. \qquad (9.41b)$$

To obtain a specific element stiffness coefficient, the correct submatrix $\mathbf{B}'^{\mathrm{T}}_i\mathbf{E}\mathbf{B}'_j$ is identified from the element stiffness integrand matrix. Then the position within this submatrix is identified for the integrand of the stiffness coefficient desired. To obtain element stiffness coefficient k_{36}, submatrix $\mathbf{B}'^{\mathrm{T}}_2\mathbf{E}\mathbf{B}'_3$ is so identified. Within this submatrix, integrand I_{k36} is in the first row and second column.

$$(9.42a)$$

Thus from equation 9.41b,

$$I_{k36}(\zeta,\ \eta) = \frac{E_{12} \frac{\partial N'_2(\zeta,\ \eta)}{\partial x} \frac{\partial N'_3(\zeta,\ \eta)}{\partial y} + E_{33} \frac{\partial N'_2(\zeta,\ \eta)}{\partial y} \frac{\partial N'_3(\zeta,\ \eta)}{\partial x}}{|J(\zeta,\ \eta)|}. \qquad (9.42b)$$

The stiffness coefficient k_{36} is then calculated as

$$k_{36} = t \int_{-1}^{1} \int_{-1}^{1} I_{k36}(\zeta,\ \eta)\ d\zeta\ d\eta \qquad (9.42c)$$

$$= t \sum_{k=1}^{4} I_{k36}(\zeta_k,\ \eta_k) \cdot H_k \qquad (9.42d)$$

$$= t\left(I_{k36}\left(-\frac{1}{\sqrt{3}}, -\frac{1}{\sqrt{3}}\right) \cdot 1 + I_{k36}\left(\frac{1}{\sqrt{3}}, -\frac{1}{\sqrt{3}}\right) \cdot 1 \right.$$

$$\left. + I_{k36}\left(-\frac{1}{\sqrt{3}}, \frac{1}{\sqrt{3}}\right) \cdot 1 + I_{k36}\left(\frac{1}{\sqrt{3}}, \frac{1}{\sqrt{3}}\right) \cdot 1 \right). \tag{9.42e}$$

The numerical values of the primed derivatives of shape functions 2 and 3 with respect to x and y in equation 9.42b are calculated for each integration point from equations 9.39. The numerical values of the Jacobian at each integration point are calculated using equations 9.36 and 9.37. These values and the numerical values of the appropriate elastic constants are substituted into equation 9.42b to obtain the four values of the integrand I_{k36} at the four integration points. These four integrand values along with the plate thickness are then substituted into equation 9.42e to obtain the numerical value for the stiffness coefficient k_{36}.

For the 4-node bilinear quadrilateral plane stress finite element described in the example calculation above, the values for the Jacobian and the primed derivatives at each integration point were tabulated for the Gauss 2×2 rule. Identifying the values needed to calculate the stiffness coefficient k_{36} and using equations 9.42,

$$\int_A I_{k36}\, dA = \frac{0.25(1.091506)(0.316988) + 0.375(-0.081143)(0.211325)}{6.873613} (1.0)$$

$$+ \frac{0.25(1.380181)(0.894338) + 0.375(-1.380181)(0.066988)}{6.729275} (1.0)$$

$$+ \frac{0.25(0.369819)(0.605663) + 0.375(-0.369819)(0.933013)}{10.770725} (1.0)$$

$$+ \frac{0.25(0.658494)(1.183013) + 0.375(-1.668856)(0.788675)}{10.626388} (1.0)$$

$$= 0.0174191.$$

The stiffness coefficient k_{36} then is calculated as

$$k_{36} = \frac{30,000\,(0.5)}{1 - 0.25^2}(0.0174191) = 278.71 \text{ kips/in.}$$

The result compares well to the stiffness coefficient k_{63}, the symmetric counterpart coefficient of k_{36}, obtained from the element stiffness matrix in the example calculation done earlier.

9.5.3. Master Element Equivalent Nodal Loads

In-plane applied loads on a plane stress structure may be concentrated at a point, distributed throughout portions of the structure, or distributed along a line on the structure. The treatment for quadrilateral finite elements is similar to what was discussed for triangular plane stress finite elements (Section 8.2.3), except that calculations to obtain equivalent nodal loads, in most cases, must be done in ζ–η space on the master element.

An *in-plane concentrated load at a point* is best treated by ensuring that a node is located at that point. The load is reduced to its components in the global x–y

directions. The x–y components are then entered directly into the global load vector at the positions corresponding to the gdof of the point.

If a concentrated load cannot be located at a node, equation 6.22f may be applied to obtain its equivalent nodal loads:

$$\{\mathbf{F}_k\} = [\mathbf{N}^T(\zeta_k, \ \eta_k)]\{\mathbf{P}_k\}. \tag{6.22f}$$

The subscript k indicates the point where the concentrated load is located. The matrix $\mathbf{N}(\zeta_k, \ \eta_k)$ is the matrix of shape functions for the element, shown in equation 9.2c and evaluated at the point of the concentrated load. The vector $\mathbf{P}_k$ consists of the two x and y components of the concentrated load. The vector $\mathbf{F}_k$ is the vector of equivalent nodal loads. There will be 8 equivalent nodal loads for the 4-node bilinear quadrilateral element, 16 for the 8-node serendipity element, and 18 for the 9-node Lagrange element.

The shape functions are functions of ζ and η. However, the point where the load acts is usually known only in x–y space as $(x_k, \ y_k)$. The isoparametric transformation will explicitly calculate x–y coordinates from ζ–η coordinates. Since there is no explicit inverse transformation, obtaining the coordinates $(\zeta_k, \ \eta_k)$ of the point of the load in ζ–η space on the master element from the coordinates $(x_k, \ y_k)$ in x–y space must be done by trial and error. An estimate of $(\zeta_k, \ \eta_k)$ may be made and these values used to obtain calculated coordinates $(x_k, \ y_k)$:

$$x_k = \lfloor \mathbf{N}_i(\zeta_k, \ \eta_k) \rfloor \{\mathbf{x}_i\} \quad \text{and} \quad y_k = \lfloor \mathbf{N}_i(\zeta_k, \ \eta_k) \rfloor \{\mathbf{y}_i\}.$$

The terms x_i and y_i are, of course, the coordinates of the element's nodes. If the calculated $(x_k, \ y_k)$ compare favorably to the known $(x_k, \ y_k)$, the estimated $(\zeta_k, \ \eta_k)$ may be used in equation 6.22f to calculate the equivalent nodal loads. For the 4-node bilinear quadrilateral element, equation 6.22f becomes

$$\begin{Bmatrix} F_{1k} \\ F_{2k} \\ F_{3k} \\ F_{4k} \\ F_{5k} \\ F_{6k} \\ F_{7k} \\ F_{8k} \end{Bmatrix} = \begin{bmatrix} N_1(\zeta_k, \ \eta_k) & 0 \\ 0 & N_1(\zeta_k, \ \eta_k) \\ N_2(\zeta_k, \ \eta_k) & 0 \\ 0 & N_2(\zeta_k, \ \eta_k) \\ N_3(\zeta_k, \ \eta_k) & 0 \\ 0 & N_3(\zeta_k, \ \eta_k) \\ N_4(\zeta_k, \ \eta_k) & 0 \\ 0 & N_4(\zeta_k, \ \eta_k) \end{bmatrix} \begin{Bmatrix} P_{kx} \\ P_{ky} \end{Bmatrix} = \begin{Bmatrix} N_1(\zeta_k, \ \eta_k)P_{kx} \\ N_1(\zeta_k, \ \eta_k)P_{ky} \\ N_2(\zeta_k, \ \eta_k)P_{kx} \\ N_2(\zeta_k, \ \eta_k)P_{ky} \\ N)_3(\zeta_k, \ \eta_k)P_{kx} \\ N_3(\zeta_k, \ \eta_k)P_{ky} \\ N_4(\zeta_k, \ \eta_k)P_{kx} \\ N_4(\zeta_k, \ \eta_k)P_{ky} \end{Bmatrix}. \tag{9.43}$$

An *in-plane load distributed throughout the quadrilateral element* is treated with equation 6.22e:

$$\{\mathbf{F}_b\} = \int_V [\mathbf{N}^T(\mathbf{x}, \ \mathbf{y})]\{\mathbf{b}(\mathbf{x}, \ \mathbf{y})\} \, dV. \tag{6.22e}$$

Equation 6.22e adapted for a plane stress quadrilateral finite element is given as

$$\{\mathbf{F}_b\} = t \int_A [\mathbf{N}^T(\mathbf{x}, \ \mathbf{y})]\{\mathbf{b}(\mathbf{x}, \ \mathbf{y})\} \, dx \, dy \tag{9.44a}$$

$$= t \int_{-1}^{1} \int_{-1}^{1} [\mathbf{N}^{T}(\zeta, \eta)] \{\mathbf{b}(\zeta, \eta)\} |J(\zeta, \eta)| \, d\zeta \, d\eta. \tag{9.44b}$$

Applying the Gaussian quadrature rule to equation 9.44b,

$$\{\mathbf{F}_b\} = t \sum_{k=1}^{N} [\mathbf{N}^{T}(\zeta_k, \eta_k)] \{\mathbf{b}(\zeta_k, \eta_k)\} |J(\zeta_k, \eta_k)| (H_k). \tag{9.44c}$$

The matrix $\mathbf{N}(\zeta, \eta)$ of equations 9.44 was formed to express the displacement variation within the quadrilateral element (see equations 9.2c). The vector $\mathbf{b}(\zeta, \eta)$ is the body force per unit of volume. It contains its x and y components, $b_x(\zeta, \eta)$ and $b_y(\zeta, \eta)$, as entries, expressed as functions of ζ and η over the master element. The coordinates (ζ_k, η_k) in equations 9.44 are those of the Gaussian integration points. For the 4-node bilinear quadrilateral element, equation 9.44c is written as

$$
\begin{Bmatrix} F_{b1} \\ F_{b2} \\ F_{b3} \\ F_{b4} \\ F_{b5} \\ F_{b6} \\ F_{b7} \\ F_{b8} \end{Bmatrix} = t \begin{Bmatrix} \sum_{k=1}^{4} N_1(\zeta_k, \eta_k) \cdot b_x(\zeta_k, \eta_k) \cdot |J(\zeta_k, \eta_k)| \cdot (H_k) \\ \sum_{k=1}^{4} N_1(\zeta_k, \eta_k) \cdot b_y(\zeta_k, \eta_k) \cdot |J(\zeta_k, \eta_k)| \cdot (H_k) \\ \sum_{k=1}^{4} N_2(\zeta_k, \eta_k) \cdot b_x(\zeta_k, \eta_k) \cdot |J(\zeta_k, \eta_k)| \cdot (H_k) \\ \sum_{k=1}^{4} N_2(\zeta_k, \eta_k) \cdot b_y(\zeta_k, \eta_k) \cdot |J(\zeta_k, \eta_k)| \cdot (H_k) \\ \sum_{k=1}^{4} N_3(\zeta_k, \eta_k) \cdot b_x(\zeta_k, \eta_k) \cdot |J(\zeta_k, \eta_k)| \cdot (H_k) \\ \sum_{k=1}^{4} N_3(\zeta_k, \eta_k) \cdot b_y(\zeta_k, \eta_k) \cdot |J(\zeta_k, \eta_k)| \cdot (H_k) \\ \sum_{k=1}^{4} N_4(\zeta_k, \eta_k) \cdot b_x(\zeta_k, \eta_k) \cdot |J(\zeta_k, \eta_k)| \cdot (H_k) \\ \sum_{k=1}^{4} N_4(\zeta_k, \eta_k) \cdot b_y(\zeta_k, \eta_k) \cdot |J(\zeta_k, \eta_k)| \cdot (H_k) \end{Bmatrix}. \tag{9.45}
$$

In the example calculation done earlier, the 4-node plane stress bilinear quadrilateral element has a body force of $500 \, \mathrm{lb/ft^3}$ acting in the negative y direction. The body force is constant, so

$$\{\mathbf{b}\} = \begin{Bmatrix} b_x(\zeta, \eta) \\ b_y(\zeta, \eta) \end{Bmatrix} = \begin{Bmatrix} 0 \\ -500 \end{Bmatrix} \, (\mathrm{lb/ft^3}) = \begin{Bmatrix} 0 \\ -0.289352 \end{Bmatrix} \, (\mathrm{lb/in.^3}).$$

To implement equation 9.45, the four shape functions are evaluated at the four integration points on the master element:

	Integration Point 1	Integration Point 2	Integration Point 3	Integration Point 4
$N_1(\zeta, \eta)$	0.622008	0.166667	0.166667	0.044658
$N_2(\zeta, \eta)$	0.166667	0.622008	0.044658	0.166667
$N_3(\zeta, \eta)$	0.044658	0.166667	0.166667	0.622008
$N_4(\zeta, \eta)$	0.166667	0.044658	0.622008	0.166667
$|J(\zeta, \eta)|$	6.873613	6.729275	10.770725	10.626388

The values of the Jacobian at the four integration points for this element were determined earlier and are tabulated again above. The plate thickness is 0.5 in. and the four weight factors H_k for the 2×2 Gauss rule are each 1.0.

The first entry in equation 9.45 is zero, since $b_x = 0$. The second is

$$F_{b2} = 0.5 \Big((0.622008)(-0.289352)(6.873613)\,(1.0)$$

$$+ (0.166667)(-0.289352)(6.729275)\,(1.0)$$
$$+ (0.166667)(-0.289352)(10.770725)\,(1.0)$$

$$+ (0.044658)(-0.289352)(10.626388)\,(1.0) \Big)$$

$$= -1.109182 \, \text{lb.}$$

The remaining entries are calculated similarly to produce

$$\{\mathbf{F}_b\} = \begin{Bmatrix} 0 \\ -1.109182 \\ 0 \\ -1.097126 \\ 0 \\ -1.422647 \\ 0 \\ -1.434703 \end{Bmatrix} \text{lb.}$$

The area of the element is the sum of the products of the values for the Jacobian and the weight factor at each integration point:

$$A = \int_A dA = \int_{-1}^{1}\int_{-1}^{1} |J(\zeta,\ \eta)| \ d\zeta \ d\eta = \sum_{k=1}^{4} |J(\zeta_k,\ \eta_k)|(H_k)$$

$$= 6.873613\,(1.0) + 6.729275\,(1.0) + 10.770725\,(1.0) + 10.626388\,(1.0)$$

$$= 35.00000 \, \text{in}^3.$$

The volume of the element is

$$V = At = (35.00000)\,(0.5) = 17.500 \, \text{in}^3.$$

The total amount of force on the element due to the body force is

$$F_{b,\text{tot}} = bV = (-0.28935)\,(17.500) = -5.0637 \, \text{lb.}$$

This equals the sum of the four numerical entries corresponding to the y direction in the vector $\mathbf{F}_b$ above.

An *in-plane load distributed along a line* in the structure must occur on the edge of one or more of the finite elements used to model the structure. The structure is discretized to ensure this. The magnitude of x and y components of the line load may vary independently along the line. However, they must be constant across the thickness t of the element.

Equation 6.22e applies, but the integrals must be formulated along the edge of the element where the line load acts. This is because the line load does the virtual work there, which is the basis for equation 6.22e. Similarly, the shape functions which express the displacements participating in the virtual work are relevant only along the edge of the element where the line load is applied. If the line along which the load is applied is straight in x–y space, the calculation of the equivalent nodal loads may be done as it was for the straight-sided 3- or 6-node plane stress triangular elements, discussed in Sections 8.2.3, 8.4.3, and 8.5.4. The only difference is that the shape functions for the quadrilateral elements must be used. One notes, however, that both the 3-node triangular and the 4-node quadrilateral elements have shape functions which vary linearly in an identical manner along the edges of their respective elements.

The calculations of the equivalent nodal loads along the line may be done in ζ–η space, and should be if the line along which the load acts is curved. If the line is curved, the 8- or 9-node quadratic quadrilateral elements should be used.

Since the line load is usually expressed in terms of force per unit of length of the line, equation 6.22e may be adapted for it as

$$\{\mathbf{F}_b\} = \int_L [\mathbf{N}^T]\{\mathbf{w}\}\ dL. \tag{9.46}$$

The shape functions $\mathbf{N}$ in ζ–η space are linear variations between the two nodes on the edge of a 4-node quadrilateral element or quadratic variations through the three nodes on the edge of an 8- or 9-node quadrilateral element. They will be functions of either ζ or η, depending on the edge of the element where the load acts.

The line load $\mathbf{w}$ should be considered in ζ–η space also. It will be a function of either ζ or η, depending on the edge of the element where it acts. It may be formulated as a function of ζ or η directly. Alternatively it may be transformed from $\mathbf{w}(\mathbf{x},\ \mathbf{y})$ to $\mathbf{p}(\zeta,\ \eta)$ using the isoparametric transformation equations 9.13 and then to a function of either ζ or η, as follows. If the line load acts along the edge of the element from node 1 to node 2, the equation of the line in ζ–η space, where $\eta = -1$, is used to transform it to a function of ζ as $\mathbf{p}(\zeta,\ -1)$. Similarly, the equations of the other edges of the master element may be used to formulate the load function appropriately when it acts on those other edges.

The line variable L tracks along the line of the load application. The line is an edge of the master element in ζ–η space, where its equation is $\zeta = \pm 1$ or $\eta = \pm 1$. The limits of the integral are from -1 to $+1$. The variable of integration is ζ or η, so that

$$dL = J(\zeta)\ d\zeta \quad \text{or} \quad dL = J(\eta)\ d\eta,$$

where $J(\zeta)$ or $J(\eta)$ is the Jacobian of the transformation of the master element edge to the line where the load acts in x–y space.

The Jacobian for a two-variable transformation is

$$dx\ dy = |J(\zeta,\ \eta)|\ d\zeta\ d\eta. \tag{9.19j}$$

It may be regarded as an area scaling factor at the differential level between area in ζ–η and x–y space. Similarly the Jacobian of a one-variable transformation may be regarded as a length scaling factor at the differential level between ζ–η and x–y space.

The length of the differential line in x–y space is

$$(dL)^2 = (dx)^2 + (dy)^2$$

$$= \left(\frac{\partial x(\zeta,\ \eta)}{d\zeta} d\zeta + \frac{\partial x(\zeta,\ \eta)}{d\eta} d\eta \right)^2 + \left(\frac{\partial y(\zeta,\ \eta)}{d\zeta} d\zeta + \frac{\partial y(\zeta,\ \eta)}{d\eta} d\eta \right)^2 . \qquad (9.47a)$$

When the line is transformed from an edge of the master element parallel to the ζ axis, x and y are functions of ζ only and the derivatives of x and y with respect to η are zero, so that

$$dL = \sqrt{\left(\frac{dx(\zeta)}{d\zeta} \right)^2 + \left(\frac{dy(\zeta)}{d\zeta} \right)^2}\, d\zeta \quad \text{and} \quad J(\zeta) = \sqrt{\left(\frac{dx(\zeta)}{d\zeta} \right)^2 + \left(\frac{dy(\zeta)}{d\zeta} \right)^2}. \qquad (9.47b)$$

When the line is transformed from an edge of the master element parallel to the η axis, x and y are functions of η only and the derivatives of x and y with respect to ζ are zero, so that

$$dL = \sqrt{\left(\frac{dx(\eta)}{d\eta} \right)^2 + \left(\frac{dy(\eta)}{d\eta} \right)^2}\, d\eta \quad \text{and} \quad J(\eta) = \sqrt{\left(\frac{dx(\eta)}{d\eta} \right)^2 + \left(\frac{dy(\eta)}{d\eta} \right)^2}. \qquad (9.47c)$$

Thus the expression for the equivalent nodal loads of the line load (equation 9.46) may be written as

$$\{\mathbf{F}_b\} = \int_{-1}^{1} \left[\mathbf{N}^{\mathrm{T}}(\zeta,\ \pm 1) \right] \{ \mathbf{p}(\zeta,\ \pm 1) \} J(\zeta)\ d\zeta \qquad (9.48a)$$

or

$$\{\mathbf{F}_b\} = \int_{-1}^{1} \left[\mathbf{N}^{\mathrm{T}}(\pm 1,\ \eta) \right] \{ \mathbf{p}(\pm 1,\ \eta) \} J(\eta)\ d\eta. \qquad (9.48b)$$

Bear in mind that in establishing the matrix $\mathbf{N}^{\mathrm{T}}$, the entries are zero at the positions in $\mathbf{N}^{\mathrm{T}}$ which correspond to edof not located on the line where the load acts.

As an example, consider the in-plane line load acting on the 4-node quadrilateral element shown in Figure 9.19. The load is directed down and to the right at an angle of 30° from the x axis.

This line load is identical to the one applied to the triangular element in the example of Section 8.4.3. Since the shape functions of a 4-node quadrilateral element are linear on its edges, they are identical to the shape functions on the edges of a 3-node triangular element. The equivalent nodal loads in this example thus may be calculated in x–y space using local line coordinates, as was done in Section 8.4.3.

The equivalent nodal loads may also be calculated using equation 9.48b. The line load acts on the edge of the element from node 1 to node 4, where the equation of

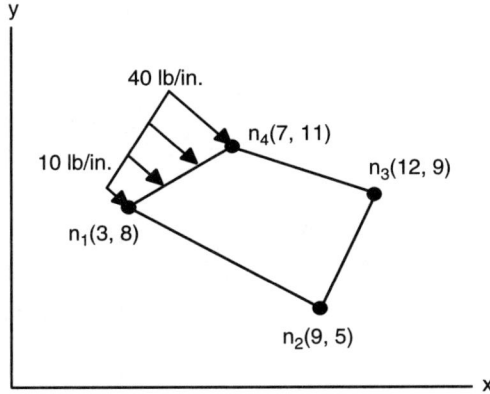

Figure 9.19. Example line load.

that edge in ζ–η space is $\zeta = -1$. The two nonzero shape functions along edge 1–4 may be written as

$$N_1(\eta) = \tfrac{1}{4}[1 - (-1)](1 - \eta) = \tfrac{1}{2}(1 - \eta),$$
$$N_4(\eta) = \tfrac{1}{4}[1 - (-1)](1 + \eta) = \tfrac{1}{2}(1 + \eta),$$
$$N_2(\eta) = N_3(\eta) = 0.$$

The line load function will be formulated directly so that on edge 1–4 its magnitude equals 10 lb in. at $\eta = -1$, 40 lb/in. at $n = +1$, and varies linearly in between. Thus,

$$p(\eta) = 15\eta + 25$$

or

$$\{\mathbf{p}(\eta)\} = \begin{Bmatrix} p_x(\eta) \\ p_y(\eta) \end{Bmatrix} = \begin{Bmatrix} (15\eta + 25)\ \cos(-30°) \\ (15\eta + 25)\ \sin(-30°) \end{Bmatrix} = \begin{Bmatrix} (15\eta + 25)(\tfrac{1}{2}\sqrt{3}) \\ (15\eta + 25)(-\tfrac{1}{2}) \end{Bmatrix}.$$

Since the line is straight where the load acts in x–y space, the Jacobian of the line transformation is constant and equals the ratio of the line lengths in x–y and ζ–η space, or

$$J(\eta) = \frac{\sqrt{(7 - 3)^2 + (11 - 8)^2}}{2} = \frac{5}{2}.$$

The Jacobian may also be calculated more formally from the transformation for the line from ζ–η space to x–y space, a process employed when the line where the load acts is curved and the Jacobian is not constant. The transformation for this line is

$$x = 3 \times \tfrac{1}{2}(1 - \eta) + 7 \times \tfrac{1}{2}(1 + \eta) = 5 + 2\eta,$$
$$y = 8 \times \tfrac{1}{2}(1 - \eta) + 11 \times \tfrac{1}{2}(1 + \eta) = \tfrac{1}{2}(19 + 3\eta).$$

The transformation is differentiated and the results substituted into equation 9.47c:

$$\frac{dx}{d\eta} = 2, \qquad \frac{dy}{d\eta} = \frac{3}{2}, \qquad J(\eta) = \sqrt{(2)^2 + \left(\frac{3}{2}\right)^2} = \frac{5}{2}.$$

Substituting into equation 9.48b,

$$\{\mathbf{F}_b\} = \begin{Bmatrix} \int_{-1}^{1} \frac{1}{2}(1-\eta)(15\eta+25)\left(\frac{\sqrt{3}}{2}\right)\left(\frac{5}{2}\right) d\eta \\ -\int_{-1}^{1} \frac{1}{2}(1-\eta)(15\eta)+25)\left(\frac{1}{2}\right)\left(\frac{5}{2}\right) d\eta \\ 0 \\ 0 \\ 0 \\ 0 \\ \int_{-1}^{1} \frac{1}{2}(1+\eta)(15\eta+25)\left(\frac{\sqrt{3}}{2}\right)\left(\frac{5}{2}\right) d\eta \\ -\int_{-1}^{1} \frac{1}{2}(1+\eta)(15\eta+25)\left(\frac{1}{2}\right)\left(\frac{5}{2}\right) d\eta \end{Bmatrix} = \begin{Bmatrix} 43.30 \\ -25.00 \\ 0 \\ 0 \\ 0 \\ 0 \\ 64.95 \\ -37.50 \end{Bmatrix} \text{ lb.}$$

Exact values for the integrals may be obtained using a two-point Gaussian quadrature rule for a line or by integrating explicitly.

9.6. QUADRATIC QUADRILATERAL ELEMENTS

The equations developed earlier for the 4-node quadrilateral element apply directly for the 8-node or 9-node elements except that they must be modified for the eight or nine shape functions employed. Equations 9.15 must be revised to show the shape functions as eight- or nine-term polynomials for the 8- or 9-node element. Equations 9.16 must be revised to reflect the eight or nine shape functions and eight or nine sets of nodal coordinates required. Equations 9.40 and 9.41 must be modified so that the number of submatrices in matrix $\mathbf{B}$ is eight or nine and in matrix $\mathbf{I}$ is 64×64 or 81×81, as appropriate.

9.6.1. Serendipity Element (8-Node, 16-dof)

Figure 9.20 shows the 8-node serendipity quadrilateral finite element in ζ–η space. From equations 9.3f and 9.4c, its shape functions may be expressed as

$$\{N_i(\zeta, \eta)\} = [\mathbf{h}^{-T}]\{\mathbf{g}\}, \tag{9.49a}$$

where $i = 1, 2, 3, \ldots 8$.

The individual shape functions $N_i(\zeta, \eta)$ may be written following the procedure used earlier. Each shape function contains eight polynomial terms in ζ and η. Its vector $\mathbf{g}$ of these terms is

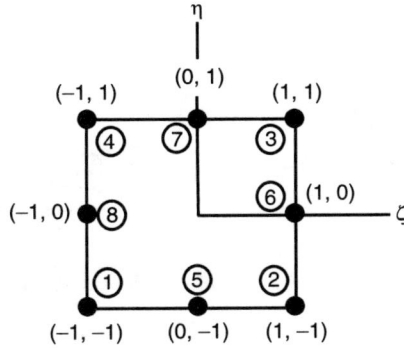

Figure 9.20. Serendipity element.

$$\lfloor \mathbf{g} \rfloor = \lfloor 1 \quad \zeta \quad \eta \quad \zeta^2 \quad \zeta\eta \quad \eta^2 \quad \zeta^2\eta \quad \zeta\eta^2 \rfloor. \tag{9.49b}$$

Its matrix **h** is obtained by evaluating its vector **g** successively at each edof. The eight values for generic polynomial terms obtained in each evaluation of vector **g** are successively the rows of matrix **h**. The result is

$$[\mathbf{h}] = \begin{bmatrix} 1 & -1 & -1 & 1 & 1 & 1 & -1 & -1 \\ 1 & 1 & -1 & 1 & -1 & 1 & -1 & 1 \\ 1 & 1 & 1 & 1 & 1 & 1 & 1 & 1 \\ 1 & -1 & 1 & 1 & -1 & 1 & 1 & -1 \\ 1 & 0 & -1 & 0 & 0 & 1 & 0 & 0 \\ 1 & 1 & 0 & 1 & 0 & 0 & 0 & 0 \\ 1 & 0 & 1 & 0 & 0 & 1 & 0 & 0 \\ 1 & -1 & 0 & 1 & 0 & 0 & 0 & 0 \end{bmatrix}. \tag{9.49c}$$

Matrix **h** is readily inverted and transposed as a numerical matrix. Doing so and substituting the result into equation 9.49a,

$$\begin{Bmatrix} N_1(\zeta, \eta) \\ N_2(\zeta, \eta) \\ N_3(\zeta, \eta) \\ N_4(\zeta, \eta) \\ N_5(\zeta, \eta) \\ N_6(\zeta, \eta) \\ N_7(\zeta, \eta) \\ N_8(\zeta, \eta) \end{Bmatrix} = \frac{1}{4} \begin{bmatrix} -1 & 0 & 0 & 1 & 1 & 1 & -1 & -1 \\ -1 & 0 & 0 & 1 & -1 & 1 & -1 & 1 \\ -1 & 0 & 0 & 1 & 1 & 1 & 1 & 1 \\ -1 & 0 & 0 & 1 & -1 & 1 & 1 & -1 \\ 2 & 0 & -2 & -2 & 0 & 0 & 2 & 0 \\ 2 & 2 & 0 & 0 & 0 & -2 & 0 & -2 \\ 2 & 0 & 2 & -2 & 0 & 0 & -2 & 0 \\ 2 & -2 & 0 & 0 & 0 & -2 & 0 & 2 \end{bmatrix} \begin{Bmatrix} 1 \\ \zeta \\ \eta \\ \zeta^2 \\ \zeta\eta \\ \eta^2 \\ \zeta^2\eta \\ \zeta\eta^2 \end{Bmatrix}. \tag{9.49d}$$

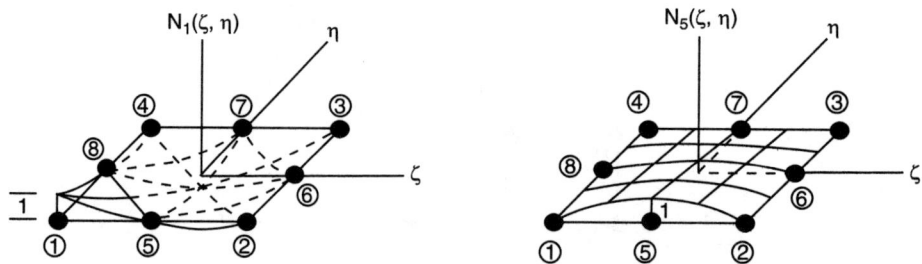

Figure 9.21. Serendipity shape function surfaces.

The individual shape functions may be written directly from equation 9.49d. Regrouping terms as factors yields

$$N_1(\zeta, \; \eta) = \tfrac{1}{4}(1 - \zeta)(1 - \eta)(-1 - \zeta - \eta), \tag{9.50a}$$

$$N_2(\zeta, \; \eta) = \tfrac{1}{4}(1 + \zeta)(1 - \eta)(-1 + \zeta - \eta), \tag{9.50b}$$

$$N_3(\zeta, \; \eta) = \tfrac{1}{4}(1 + \zeta)(1 + \eta)(-1 + \zeta + \eta), \tag{9.50c}$$

$$N_4(\zeta, \; \eta) = \tfrac{1}{4}(1 - \zeta)(1 + \eta)(-1 - \zeta + \eta), \tag{9.50d}$$

$$N_5(\zeta, \; \eta) = \tfrac{1}{2}(1 - \eta)(1 - \zeta^2), \tag{9.50e}$$

$$N_6(\zeta, \; \eta) = \tfrac{1}{2}(1 + \zeta)(1 - \eta^2), \tag{9.50f}$$

$$N_7(\zeta, \; \eta) = \tfrac{1}{2}(1 + \eta)(1 - \zeta^2), \tag{9.50g}$$

$$N_8(\zeta, \; \eta) = \tfrac{1}{2}(1 - \zeta)(1 - \eta^2). \tag{9.50h}$$

Figure 9.21 shows the shape functions $N_1(\zeta, \; \eta)$ and $N_5(\zeta, \; \eta)$ plotted over the $\zeta-\eta$ master element. Each plots as a quadratic surface, equal to one at its node and zero at the other seven nodes. The surfaces for $N_2(\zeta, \; \eta)$, $N_3(\zeta, \; \eta)$, and $N_4(\zeta, \; \eta)$ are similar to the surface for $N_1(\zeta, \; \eta)$, except that they are equal to one at their respective nodes and zero at the other seven nodes. Similarly, the surfaces for $N_6(\zeta, \; \eta)$, $N_7(\zeta, \; \eta)$, and $N_8(\zeta, \; \eta)$ are similar to the surface for $N_5(\zeta, \; \eta)$, except that they are equal to one at their respective nodes and zero at the other seven nodes.

9.6.2. Lagrange Element (9-Node, 18-dof)

Figure 9.22 illustrates the 9-node Lagrange quadrilateral finite element in $\zeta-\eta$ space. Its shape functions are expressed by equation 9.49a, where in this instance the index $i = 1, 2, 3, \ldots, 9$. The individual shape functions may be written following the procedure used earlier. Each shape function will include nine polynomial terms in ζ and η. Its vector $\mathbf{g}$ of generic polynomial terms is

$$\lfloor \mathbf{g} \rfloor = \lfloor 1 \quad \zeta \quad \eta \quad \zeta^2 \quad \zeta\eta \quad \eta^2 \quad \zeta^2\eta \quad \zeta\eta^2 \quad \zeta^2\eta^2 \rfloor. \tag{9.51a}$$

Its matrix $\mathbf{h}$ is obtained, as described for the 8-node serendipity element, by evaluating its vector $\mathbf{g}$ successively at each node. The nine values for generic polynomial

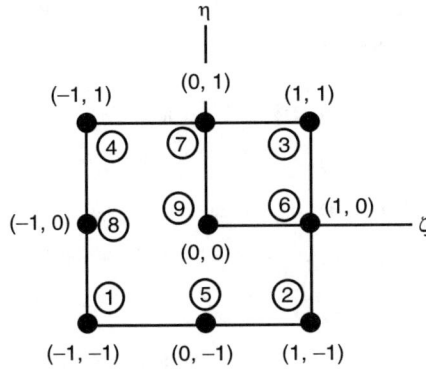

Figure 9.22. Lagrange element.

terms obtained in each evaluation of vector **g** are successively the rows of matrix **h**. The result is

$$
[\mathbf{h}] =
\begin{bmatrix}
1 & -1 & -1 & 1 & 1 & 1 & -1 & -1 & 1 \\
1 & 1 & -1 & 1 & -1 & 1 & -1 & 1 & 1 \\
1 & 1 & 1 & 1 & 1 & 1 & 1 & 1 & 1 \\
1 & -1 & 1 & 1 & -1 & 1 & 1 & -1 & 1 \\
1 & 0 & -1 & 0 & 0 & 1 & 0 & 0 & 0 \\
1 & 1 & 0 & 1 & 0 & 0 & 0 & 0 & 0 \\
1 & 0 & 1 & 0 & 0 & 1 & 0 & 0 & 0 \\
1 & -1 & 0 & 1 & 0 & 0 & 0 & 0 & 0 \\
1 & 0 & 0 & 0 & 0 & 0 & 0 & 0 & 0
\end{bmatrix}.
\tag{9.51b}
$$

Matrix **h** is readily inverted and transposed as a numerical matrix. Doing so and substituting the result into equation 9.49a,

$$
\begin{Bmatrix}
N_1(\zeta, \eta) \\
N_2(\zeta, \eta) \\
N_3(\zeta, \eta) \\
N_4(\zeta, \eta) \\
N_5(\zeta, \eta) \\
N_6(\zeta, \eta) \\
N_7(\zeta, \eta) \\
N_8(\zeta, \eta) \\
N_9(\zeta, \eta)
\end{Bmatrix}
= \frac{1}{4}
\begin{bmatrix}
0 & 0 & 0 & 0 & 1 & 0 & -1 & -1 & 1 \\
0 & 0 & 0 & 0 & -1 & 0 & -1 & 1 & 1 \\
0 & 0 & 0 & 0 & 1 & 0 & 1 & 1 & 1 \\
0 & 0 & 0 & 0 & -1 & 0 & 1 & -1 & 1 \\
0 & 0 & -2 & 0 & 0 & 2 & 2 & 0 & -2 \\
0 & 2 & 0 & 2 & 0 & 0 & 0 & -2 & -2 \\
0 & 0 & 2 & 0 & 0 & 2 & -2 & 0 & -2 \\
0 & -2 & 0 & 2 & 0 & 0 & 0 & 2 & -2 \\
4 & 0 & 0 & -4 & 0 & -4 & 0 & 0 & 4
\end{bmatrix}
\begin{Bmatrix}
1 \\
\zeta \\
\eta \\
\zeta^2 \\
\zeta\eta \\
\eta^2 \\
\zeta^2\eta \\
\zeta\eta^2 \\
\zeta^2\eta^2
\end{Bmatrix}.
$$

$$
\tag{9.51c}
$$

The individual shape functions may be written directly from equation 9.51c, and the terms regrouped as factors, to yield

$$N_1(\zeta,\ \eta) = \tfrac{1}{4}\zeta\eta(1-\zeta)(1-\eta), \tag{9.52a}$$

$$N_2(\zeta,\ \eta) = -\tfrac{1}{4}\zeta\eta(1+\zeta)(1-\eta), \tag{9.52b}$$

$$N_3(\zeta,\ \eta) = \tfrac{1}{4}\zeta\eta(1+\zeta)(1+\eta), \tag{9.52c}$$

$$N_4(\zeta,\ \eta) = -\tfrac{1}{4}\zeta\eta(1-\zeta)(1+\eta), \tag{9.52d}$$

$$N_5(\zeta,\ \eta) = -\tfrac{1}{2}\eta(1-\eta)(1-\zeta^2), \tag{9.52e}$$

$$N_6(\zeta,\ \eta) = \tfrac{1}{2}\zeta(1+\zeta)(1-\eta^2), \tag{9.52f}$$

$$N_7(\zeta,\ \eta) = \tfrac{1}{2}\eta(1+\eta)(1-\zeta^2), \tag{9.52g}$$

$$N_8(\zeta,\ \eta) = -\tfrac{1}{2}\zeta(1-\zeta)(1-\eta^2), \tag{9.52h}$$

$$N_9(\zeta,\ \eta) = (1-\zeta^2)(1-\eta^2). \tag{9.52i}$$

Figure 9.23 shows the shape functions $N_1(\zeta,\ \eta)$, $N_5(\zeta,\ \eta)$, and $N_9(\zeta,\ \eta)$ plotted over the ζ–η master element. Each plots as a quadratic surface, equal to one at its node and zero at the other eight nodes. The surfaces for $N_2(\zeta,\ \eta)$, $N_3(\zeta,\ \eta)$, and $N_4(\zeta,\ \eta)$ are similar to the surface for $N_1(\zeta,\ \eta)$, except that they equal one at their respective nodes and zero at the other eight nodes. The surfaces for $N_6(\zeta,\ \eta)$,

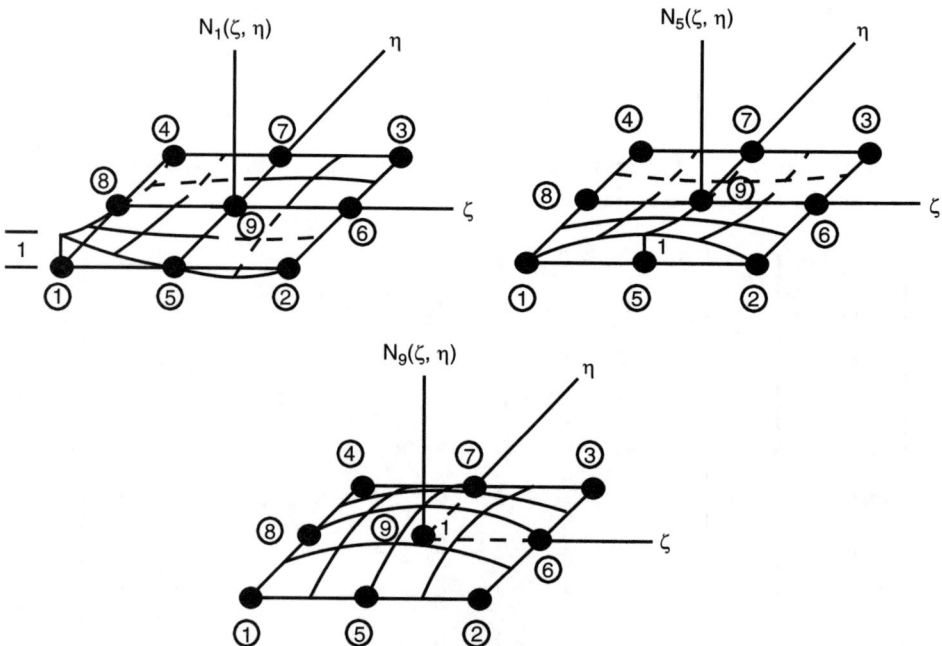

Figure 9.23. Lagrange shape function surfaces.

$N_7(\zeta,\ \eta)$, and $N_8(\zeta,\ \eta)$ are similar to the surface for $N_5(\zeta,\ \eta)$, except that they equal one at their respective nodes and zero at the other eight nodes.

9.7. CLOSING REMARKS

Plane stress quadrilateral finite elements formulated with the isoparametric transformation scheme provide an efficient and accurate way to model and analyze plane stress structures. The 4-node quadrilateral element is more accurate than the 3-node triangular element because its shape functions are higher degree, permitting variation of strain and stress within the element. Both elements are versatile at modeling irregularly shaped plane stress structures, though each models the boundaries of such structures with piecewise straight lines.

The 8- and 9-node quadrilateral elements use second-degree polynomials in ζ and η for shape functions, permitting more variation of strain and stress within the element with the consequent greater accuracy for the analysis. In addition, the isoparametric transformation used to achieve the required calculations for element stiffness matrices results in parabolic curved-sided elements in x–y space. This latter characteristic permits them to model the boundaries of irregularly shaped structures with piecewise parabolic curves, improving the accuracy of the analysis.

Elements with more nodes require more calculation effort to obtain their stiffness matrices than those with fewer nodes. The added effort comes about because of more shape functions, more and higher degree terms in the shape functions, more nodal coordinates to combine with these shape functions in the calculation algorithm, and often the necessity to use a higher order quadrature rule to achieve the accuracy desired. There is a trade-off, and the analyst must consider whether to achieve the level of accuracy needed for the overall analysis with a large number of elements which possess only three or four nodes or with fewer elements which possess more nodes.

The isoparametric transformation scheme is so-called because it employs the identical shape functions for both the coordinate transformation from ζ–η space to x–y space and the displacement variation within the element. In most plane stress applications the mathematical symmetry of isoparametric transformation yields the most efficient approach for the accuracy desired. For other conditions, however, a subparametric transformation scheme may be desirable, one for which the shape functions used for the coordinate transformation are lower degree polynomials than those used for displacement variation. Subparametric transformations are not discussed in this book.

BIBLIOGRAPHY

Bathe, K. J., *Finite Element Procedures in Engineering Analysis* (Chapter 5), Prentice-Hall, Englewood Cliffs, NJ, 1982.

Cook, R. D., Malkus, D. S., and Plesha, M. E., *Concepts and Applications of Finite Element Analysis* (Chapter 6), Wiley, New York, 1989.

Logan, D. L., *A First Course in the Finite Element Method* (Chapter 11), PWS-Kent, Boston, 1992.

MacNeal, R. H., *Finite Elements: Their Design and Performance* (Chapter 4), Dekker, New York, 1994.

Prathap, G., *The Finite Element Method in Structural Mechanics* (Chapter 7), Kluwer Academic, Dordrecht, The Netherlands, 1993.

Shames, I. H., and Dym, C. L., *Energy and Finite Elements in Structural Mechanics* (Chapter 11), Hemisphere (McGraw-Hill), New York, 1985.

Weaver, W. Jr., and Johnston, P. R., *Finite Elements for Structural Analysis* (Chapter 3, Appendix B), Prentice-Hall, Englewood Cliffs, NJ, 1984.

Yang, T. Y., *Finite Element Structural Analysis* (Chapter 11), Prentice-Hall, Englewood Cliffs, NJ, 1986.

Zienkiewcz, O. C., and Taylor, R. L., *The Finite Element Method* (Chapter 8), 4th edn., McGraw-Hill, Berkshire, England, 1989.

PROBLEMS

The use of computer software, such as MATLAB or a spreadsheet, which manipulates individual matrices and evaluates functions at a point in their space will facilitate solution of many of these problems.

9.1. A thin flat rectangular plate element is subjected to in-plane loads. The plate's thickness $t = 0.5$ in., its Poisson ratio $v = 0.25$, and modulus $E = 30,000$ ksi.

  a. Write the element's shape functions.

  b. Write the matrix **B** for the element.

  c. Using the procedure suggested by equations 9.9–9.11, determine the element stiffness coefficient k_{42}. Evaluate the integral analytically.

  d. Repeat part c for the element stiffness coefficient k_{55}.

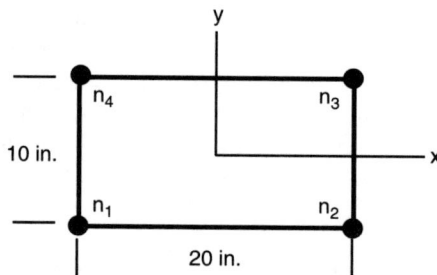

Problem 9.1

9.2. Repeat Problems 9.1a–c for the plate element shown, except determine stiffness coefficient k_{14}. The plate's thickness $t = 1.25$ in., its Poisson ratio $v = 0.25$, and modulus $E = 30,000$ ksi.

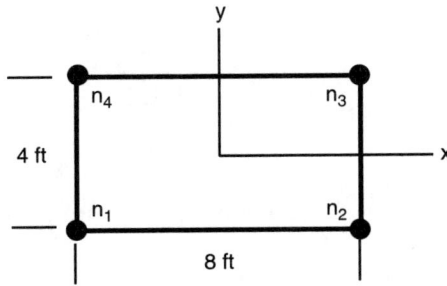

Problem 9.2

9.3. Evaluate the integral

$$I = \int_1^7 \frac{x^2 \, dx}{x+1}.$$

a. Analytically.

b. Numerically using a two-point Gauss quadrature rule and integrating with respect to x.

c. Numerically using a two-point Gauss quadrature rule and integrating with respect to ζ, a transformed coordinate whose origin is at the center of the integration interval.

d. Numerically using a three-point Gauss quadrature rule.

9.4. Evaluate the integral

$$I = \int_0^6 \frac{(x^4 + 2x + 5) \, dx}{x^2 + 2x + 1}$$

a. Analytically. (*Hint*: Translate the variable x.)

b. Numerically using a three-point Gauss quadrature rule and integrating with respect to x.

c. Numerically using a three-point Gauss quadrature rule and integrating with respect to ζ, a transformed coordinate whose origin is at the center of the integration interval.

d. Numerically using a two-point Gauss quadrature rule.

e. Compare the results from parts a–d.

9.5. Evaluate the integral I over the rectangular area shown:

$$I = \int_A \frac{x^2 y}{x+1} \, dA$$

a. Analytically.

b. Numerically using the 2×2 Gauss quadrature rule and integrating with respect to x and y.

c. Numerically using the 2×2 Gauss quadrature rule and integrating with respect to coordinates ζ and η with their origin at the center of a transformed 4-node 2×2 square. The ζ axis is parallel to the x axis; the η axis is parallel to the y axis.

d. Repeat part b using the 3×3 Gauss quadrature rule.

e. Repeat part c using the 3×3 Gauss quadrature rule.

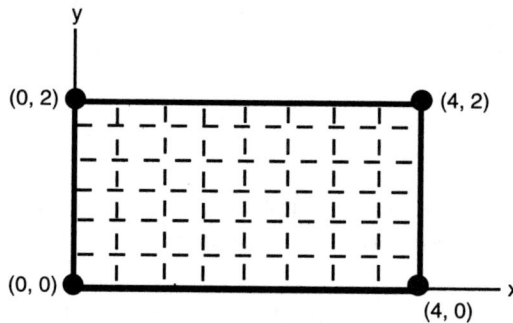

Problem 9.5

9.6. Repeat Problem 9.1c but evaluate the integral numerically:

a. Using the 2×2 Gauss quadrature rule.

b. Using the 3×3 Gauss quadrature rule.

9.7. Repeat problem 9.1d but evaluate the integral numerically:

a. Using the 2×2 Gauss quadrature rule.

b. Using the 3×3 Gauss quadrature rule.

9.8. Determine the complete element stiffness matrix **K** for the element described in Problem 9.1. Use the numerical procedure described briefly at the end of Section 9.2 and use the 2×2 Gauss quadrature rule.

9.9. For the quadrilateral and x–y coordinates shown, use the transformation to a 4-node 2×2 square with ζ–η coordinates which have their origin at the center of the square.

a. Map the n_1–n_3 diagonal of the 2×2 square of ζ–η space onto x–y space and sketch it on the quadrilateral in x–y space.

b. Calculate the area of the quadrilateral by:
  (1) Geometric or algebraic methods.
  (2) Integrating the Jacobian numerically with respect to ζ–η over the area of the quadrilateral using the Gauss 2×2 quadrature rule.

c. Evaluate the integral of the function $x^2 y$ over the area of the quadrilateral using the 2×2 Gauss quadrature rule.

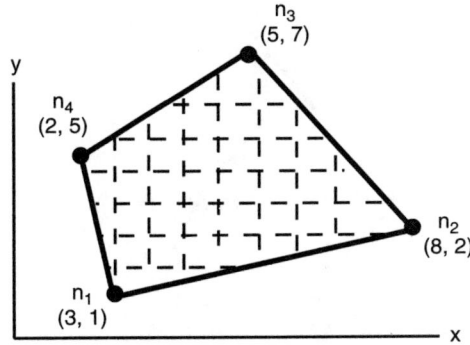

Problem 9.9

9.10. For the quadrilateral element of Problem 9.14:

a. Find its area by geometric or algebraic means.

b. Find its area by integrating the Jacobian of the transformation of the 2×2 square master element to the quadrilateral element. Use the 2×2 Gauss quadrature rule.

c. Evaluate the integral of the function $x^2 y$ over the area of the quadrilateral using the 2×2 Gauss quadrature rule.

9.11. The integral $I = \int_A (x^2 + xy + y^2)\, dA$ is defined in x–y space and over the 4-node area shown in the sketch. Sides n_1–n_4 and n_2–n_3 are circular arcs with their center at the x–y origin. The following transformation to polar coordinates applies:

$$x = r \cos\ \theta, \qquad y = r \sin\ \theta.$$

a. Map and sketch the area of x–y space shown as a corresponding area of r–θ space using the transformation given above. The r–θ axes are parallel to the x–y axes, respectively.

b. Evaluate the integral I, analytically, over the area, integrating with respect to r–θ so that the result is the correct value for the integral evaluated over x–y space.

c. Map and sketch the area of the r–θ space to a 4-node 2×2 square with coordinates ζ–η whose origin is at the center of the square. The ζ–η axes are parallel to the r–θ axes.

d. Evaluate the integral I, numerically, over the area, integrating with respect to ζ–η so that the result is the correct value for the integral evaluated over x–y space.
  (1) Use the 2×2 Gauss quadrature rule.
  (2) Use the 3×3 Gauss quadrature rule.

e. Why do the results in parts b and d differ?

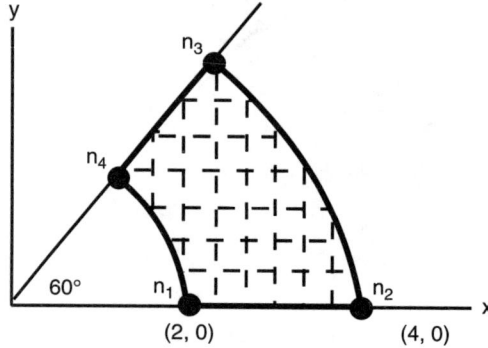

(2, 0)   (4, 0)

Problem 9.11

9.12. The thin, flat quadrilateral element shown is subjected only to in-plane loads. The plate's thickness $t = 0.75$ in., its modulus $E = 30,000$ ksi, and Poisson ratio $\nu = 0.25$. The coordinates shown are in feet. Using the 2×2 Gauss quadrature rule:

a. Determine matrix **B** for the element, evaluated numerically at the first integration point.

b. Using the procedure suggested by equations 9.40–9.42, determine the element stiffness coefficient k_{42}.

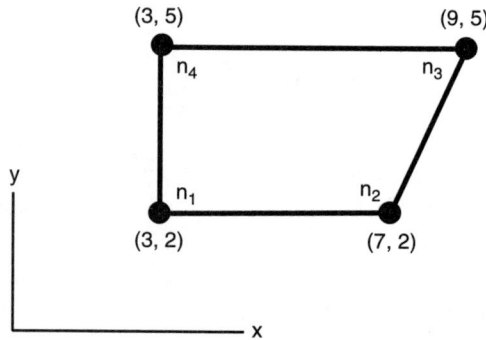

(3, 2)   (7, 2)

Problem 9.12

9.13. Determine the complete element stiffness matrix **K** for the element described in Problem 9.12. Use the numerical procedure described briefly at the end of Section 9.2 and illustrated in Section 9.5.2. Use the 2×2 Gauss quadrature rule.

9.14. Repeat Problem 9.12, but use the quadrilateral element shown in the accompanying sketch. Coordinates are in feet. Properties of the element are stated in Problem 9.12.

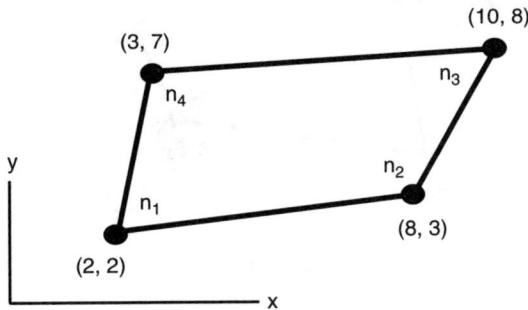

Problem 9.14

9.15. Repeat Problem 9.13, but use the quadrilateral element of Problem 9.14.

9.16. The in-plane concentrated load and linearly distributed line loads shown on the quadrilateral element in the sketch each act normal to their respective element sides. The unit weight for the plate is $500 \, \text{lb/ft}^3$. The plate thickness $t = 0.75 \, \text{in}$. Coordinates shown are in feet. Determine the vector of equivalent nodal loads for the element.

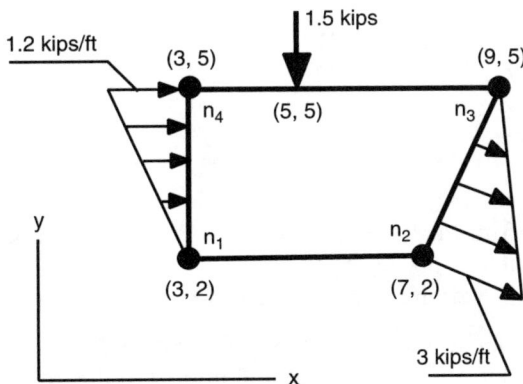

Problem 9.16

9.17. The in-plane uniform and linearly distributed line loads shown on the quadrilateral element in the sketch each act normal to their respective element sides. The unit weight for the plate is $500 \, lb/ft^3$. The plate thickness $t = 0.75 \, in$. Coordinates shown are in feet. Determine the vector of equivalent nodal loads for the element.

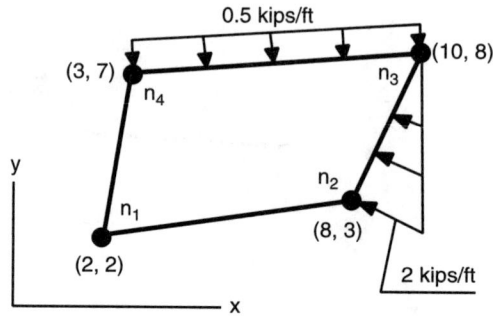

Problem 9.17

9.18. A finite element mesh employing 4-node quadrilateral elements is defined on a flat plate by the coordinates of the global nodes as tabulated below:

Global Node	x (ft)	y (ft)	Element	Global Nodes
1	5	5	a	1, 2, 6, 5
2	11	3	b	2, 3, 7, 6
3	18	6	c	3, 4, 8, 7
4	24	13	d	5, 6, 10, 9
5	5	12	e	6, 7, 11, 10
6	10	13	f	7, 8, 12, 11
7	15	15		
8	23	18		
9	2	21		
10	9	22		
11	14	22		
12	19	23		

The plate is 3.0 in. thick, its modulus $E = 30,000 \, ksi$, Poisson ratio $\nu = 0.25$, and unit weight $\gamma = 500 \, lb/ft^3$.

Three linearly varying line loads act in-plane on the plate as:

(i) Perpendicular to the edge from global nodes 3 to 4 and directed up and left; the value is $300 \, lb/ft$ at global node 3 and zero at global node 4.

(ii) Horizontal and directed right from global nodes 4–8; the value is zero at global node 4 and $500 \, lb/ft$ at global node 8.

(iii) Horizontal and directed right from global nodes 8–12; the value is $500 \, lb/ft$ at global node 8 and zero at global node 12.

a. Sketch the finite element mesh, showing its nodes, their coordinates, its elements, and the applied loads.

b. For element c, where element node 1 is coincident with global node 3, determine:

(1) The complete load vector.

(2) The element stiffness coefficient k_{63}. The indices 6 and 3 are element dof numbers using the usual element dof numbering scheme.

(3) The complete element stiffness matrix.

c. Repeat part b above but for element f, and determine k_{44} instead of k_{63}. Element node 1 is coincident with global node 7.

9.19. The structure shown is a thin plate with a single in-plane concentrated load at point D. It is supported with a pin support at A, a roller at C, and lateral support along its entire length. Determine the support reactions at points A and C and the displacement of point B using the four 4-node, 8-dof plane stress rectangular finite elements indicated. Neglect the weight of the plate. The plate's modulus of elasticity E, Poisson ratio v, and thickness t are given as

$$E = 30,000 \text{ ksi}, \qquad v = 0.25, \qquad t = 3/4 \text{ in.}$$

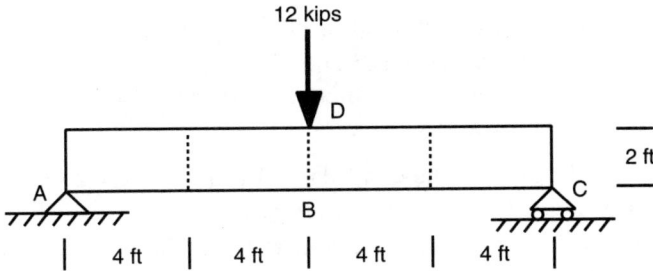

Problem 9.19

9.20. Repeat Problem 9.19, changing the roller support at C to a pin support.

9.21. Repeat Problem 9.19 using the four 4-node, 8-dof plane stress quadrilateral finite elements oriented as shown. Why do these results differ from those of Problem 9.19?

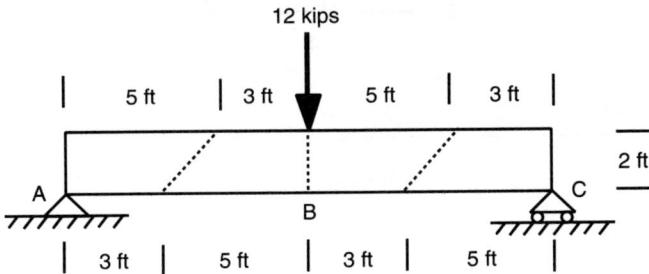

Problem 9.21

9.22. Repeat Problem 9.21, changing the roller support at C to a pin support. Why do these results differ from those of Problem 9.20?

9.23. Repeat Problem 9.19 using the four 4-node, 8-dof plane stress quadrilateral finite elements oriented as shown. Why do these results differ from those of Problems 9.19 and 9.21?

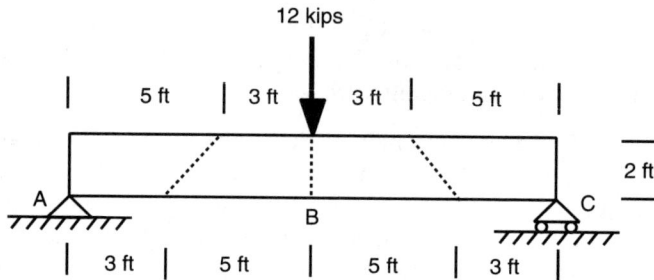

Problem 9.23

9.24. Repeat Problem 9.23 changing the roller support at C to a pin support. Why do these results differ from those of Problems 9.20 and 9.22?

9.25. Extend the equations and illustrations developed for the quadrilateral 4-node plane stress element to the quadrilateral 8-node serendipity plane stress element shown on the sketch. Coordinates are in feet. Use appropriate software, such as MATLAB or a spreadsheet:

a. Determine the area of the quadrilateral using the 2×2 Gauss quadrature rule and the 3×3 Gauss quadrature rule.

b. Evaluate the integral of the function $x^2 y$ over the area of the quadrilateral using the 2×2 and the 3×3 Gauss quadrature rule.

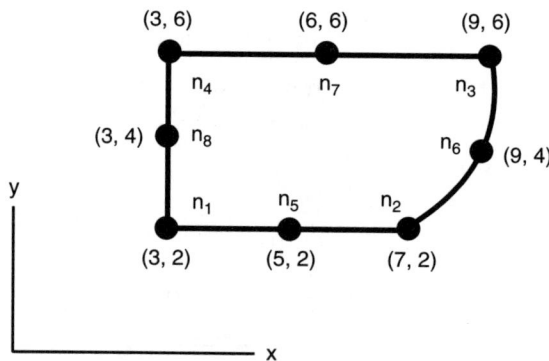

Problem 9.25

9.26. Repeat Problem 9.25 for the quadrilateral element shown on the sketch. Coordinates are in feet.

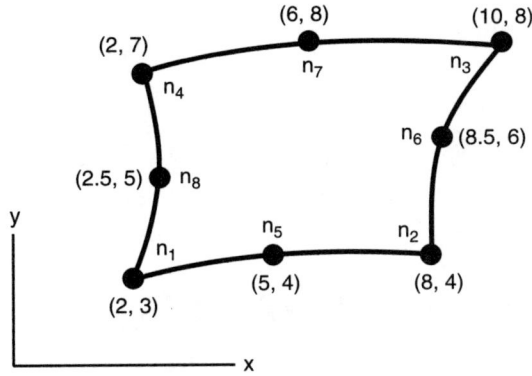

Problem 9.26

9.27. Find the vector of equivalent nodal loads for the quadrilateral element of Problem 9.26 for a 2-kip/ft of edge uniformly distributed line load acting horizontally to the right on the side defined by nodes n_2-n_6-n_3.

9.28. Find the vector of equivalent nodal loads for the quadrilateral element of Problem 9.26 for a linearly distributed line load acting horizontally to the right on the side defined by nodes n_2-n_6-n_3. The intensity of the load is 3 kips/ft of edge at node 2 and zero at node 3.

9.29. Find the vector of equivalent nodal loads for the quadrilateral element of Problem 9.26 if its downward-acting unit weight $\gamma = 500\,\text{lb/ft}^3$ and its thickness $t = 1.5\,\text{in}$.

9.30 Prove that the Jacobian for the 4-node isoparametric quadrilateral element (equation 9.17c) is a linear function of ζ and η.

CHAPTER 10

FLAT PLATE FLEXURAL FINITE ELEMENTS

10.1. INTRODUCTION

Thin, flat plate elements in flexure are the 2D counterparts to thin, straight beam elements in flexure. Some of the phenomena studied in previous chapters on beam elements occur also in plate elements. There will be new phenomena to address in plates, however, brought about by the second dimension of the structural element and the interplay of phenomena between the two dimensions.

In the plane stress condition, loads are applied in-plane to a flat plate, resulting in stress and strain variation in the plane of the plate only. Stress and strain are uniform in the transverse direction in plane stress and transverse displacements are insignificant. In plate flexure, loads are applied transverse to the plane of the plate. The resulting stress and strain variation is both in-plane and transverse, and the dominant displacements are in the transverse direction. Though the in-plane strains will still be the first derivatives of the in-plane displacements with respect to the in-plane coordinates as they were in the plane stress condition, treating the transverse displacements will bring about second derivatives of the transverse displacements with respect to the in-plane coordinates.

A plate is thin when its thickness is much less than either of its other two dimensions. Moreover, when the transverse deflections of a thin, flat plate are small relative to the plate thickness, three important simplifying assumptions may be made:

- The midplane of the plate is assumed to displace only in the transverse direction during bending, much as the neutral surface of a thin beam does during its bending.

- Straight lines normal to the midplane of the plate are assumed to remain straight and normal to the midplane during bending, with a result that the effects of transverse shear deformation are eliminated. The same is assumed for thin beams.

- Stresses normal to the midplane are negligible.

When the transverse deflections cannot be regarded as small with respect to the plate thickness, the midplane of the plate deforms and displaces laterally so that the first assumption cited above is no longer valid. The other two assumptions remain valid as long as the transverse deflections are still small with respect to the plate width and length.

For a plate undergoing deflections whose magnitudes are comparable to the plate thickness but are still much smaller than the other plate dimensions, the in-plane strains may be regarded as having two sources. The first is the same as for thin, flat plate flexure when the deflections are small with respect to plate thickness. It causes strains that are the first derivatives of the in-plane displacements. The second results in a membrane effect. It involves nonlinear terms of derivatives of the transverse displacement. Thin plates undergoing such transverse displacements, comparable to or larger than the plate thickness, will not be treated in this book.

10.2. THIN, FLAT PLATE SMALL DEFLECTION FLEXURE THEORY

The thin, flat plate theory described below conforms to the three assumptions cited above and thus is classical thin plate theory. It ignores the effect of transverse shear deformations during bending. The classical analysis of thin beams does the same. As long as the plates and beams are "thin," the effects of transverse shear deformation are unimportant. For thick plates or beams, however, the effects of shear deformations should be considered.

Figure 10.1 illustrates a thin, flat plate in pure flexure. The in-plane directions are x and y and in-plane displacements are u and v. The transverse direction is z and transverse displacement is w. The stresses of interest are in-plane and are σ_x, σ_y, and τ_{xy}. The second assumption cited above eliminates the transverse shear strains on transverse cross sections. Without shear strain in the transverse direction on the vertical faces of the differential element, there will be no shear strains on horizontal surfaces of the element either.

The strains of interest are in-plane and are ϵ_x, ϵ_y, and γ_{xy}. When plate deflections are small with respect to the plate thickness, these strains are defined as they were for plane stress:

$$\epsilon_x = \frac{\partial u}{\partial x}, \qquad \epsilon_y = \frac{\partial v}{\partial y}, \qquad \gamma_{xy} = \frac{\partial u}{\partial y} + \frac{\partial v}{\partial x}$$

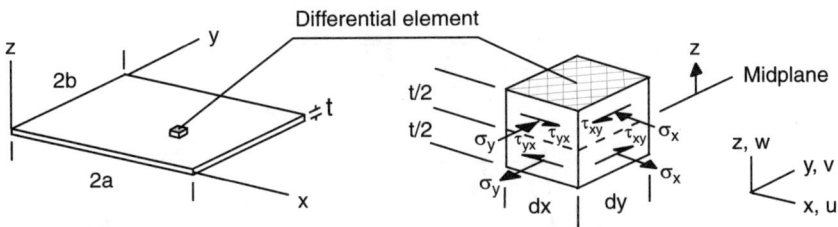

Figure 10.1. Thin plate flexure, small deflection.

The second assumption cited above that straight lines normal to the midplane remain straight and normal during bending is illustrated in Figure 10.2. It requires linear variations of the in-plane displacements u and v in the transverse direction. The displaced position of the cross section normal to the x direction portrays the variation of the displacement u. Since it is normal to the midplane, it must also be inclined at the slope of the displaced midplane with respect to the x–coordinate direction. A similar observation may be made for the cross section normal to the y-coordinate direction. Thus,

$$u = -z\frac{\partial w}{\partial x} \quad \text{and} \quad v = -z\frac{\partial w}{\partial y}.$$

The negative signs are needed since the displacement u and v are negative when the coordinate z is positive. The origin of coordinates is on the midplane.

These relationships and the strain definitions allow the in-plane strains to be related to the transverse displacement w:

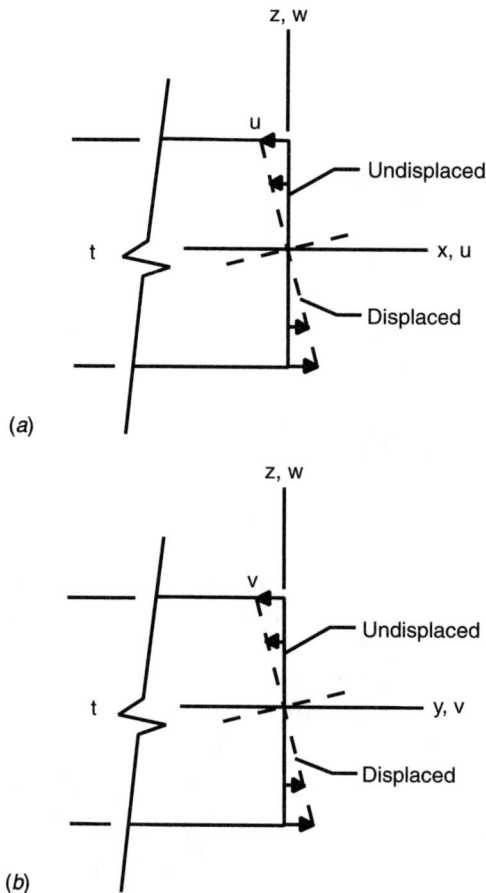

Figure 10.2. Displacement variation: (*a*) *x* cross section; (*b*) *y* cross section.

$$\epsilon_x = \frac{\partial u}{\partial x} = -z\frac{\partial^2 w}{\partial x^2}, \tag{10.1a}$$

$$\epsilon_y = \frac{\partial v}{\partial y} = -z\frac{\partial^2 w}{\partial y^2}, \tag{10.1b}$$

$$\gamma_{xy} = \frac{\partial u}{\partial y} + \frac{\partial v}{\partial x} = -z\frac{\partial^2 w}{\partial x\,\partial y} - z\frac{\partial^2 w}{\partial y\,\partial x} = -2z\frac{\partial^2 w}{\partial x\,\partial y}. \tag{10.1c}$$

The second derivative of the transverse displacement w with respect to x is the curvature of the midplane in the x direction, ϕ_{xx}. The second derivative of w with respect to y is the curvature of the midplane in the y direction, ϕ_{yy}. Twice the mixed derivative of w with respect to x and y is the twist of the midplane, or the cross rate of slope change, ϕ_{xy}. These strain–displacement–curvature relationships in matrix form are

$$\left\{\begin{array}{c} \epsilon_x \\ \epsilon_y \\ \gamma_{xy} \end{array}\right\} = -z\left\{\begin{array}{c} \frac{\partial^2 w}{\partial x^2} \\ \frac{\partial^2 w}{\partial y^2} \\ 2\frac{\partial^2 w}{\partial x\,\partial y} \end{array}\right\} = -z\left\{\begin{array}{c} \phi_{xx} \\ \phi_{yy} \\ \phi_{xy} \end{array}\right\}, \tag{10.1d}$$

$$\{\epsilon\} = -z\{\phi\}. \tag{10.1e}$$

The in-plane stress and strain vectors are the same as they were for the plane stress condition. The stress–strain–curvature relationships are

$$\{\sigma\} = [\mathbf{E}]\{\epsilon\} = -z[\mathbf{E}]\{\phi\}, \tag{10.2a}$$

$$\left\{\begin{array}{c} \sigma_x \\ \sigma_y \\ \tau_{xy} \end{array}\right\} = \frac{-zE}{1-v^2}\begin{bmatrix} 1 & v & 0 \\ v & 1 & 0 \\ 0 & 0 & \frac{1-v}{2} \end{bmatrix}\left\{\begin{array}{c} \phi_{xx} \\ \phi_{yy} \\ \phi_{xy} \end{array}\right\}. \tag{10.2b}$$

As in the theory for thin, straight beams in pure flexure, the linear variation of in-plane displacements in the transverse direction in a thin, flat plate in pure flexure results in linear normal strain and stress variations in the transverse direction. These normal stress distributions give rise to bending moments in the plate. The normal stress and bending moments are illustrated in Figure 10.3.

The moment M_{xx} is a distributed bending moment per unit of width of plate in the y direction. It is calculated from the distribution of normal stress σ_x on a unit width of the cross section perpendicular to the x direction as

$$M_{xx} = \int_{-t/2}^{t/2} (-z\sigma_x)\,dz. \tag{10.3a}$$

The negative sign is present because σ_x is compression when the transverse coordinate z is positive. From equation 10.2b, the expression for σ_x is written as

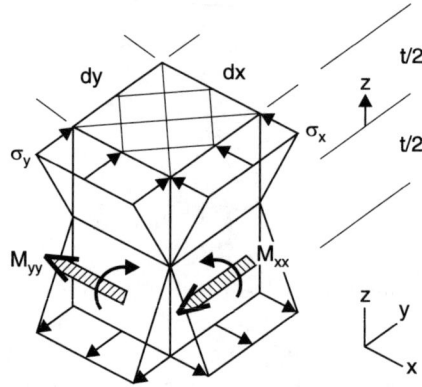

Figure 10.3. Distributed bending moment.

$$\sigma_x = \frac{-zE}{1 - v^2}(\phi_{xx} + v\phi_{yy}).$$

Substituting this expression into equation 10.3a,

$$M_{xx} = \frac{E}{1 - v^2}\int_{-t/2}^{t/2}(\phi_{xx} + v\phi_{yy})z^2\ dz. \tag{10.3b}$$

The curvatures ϕ_{xx} and ϕ_{yy} are not functions of the transverse coordinate z so they may be moved from under the integral. The remaining integral is evaluated to give

$$M_{xx} = \frac{Et^3}{12(1 - v^2)}(\phi_{xx} + v\phi_{yy}). \tag{10.3c}$$

Note that $\frac{1}{12}t^3$ is the moment of inertia about the midplane of a unit width of the plate.

The expression for the distributed bending moment M_{yy} per unit of width in the x direction may be obtained similarly. It is calculated from the distribution of normal stress σ_y on a unit width of cross section perpendicular to the y direction:

$$M_{yy} = \frac{Et^3}{12(1 - v^2)}(v\phi_{xx} + \phi_{yy}). \tag{10.3d}$$

The in-plane shear stresses τ_{xy} and τ_{yx} on a differential element whose height is finite and equal to the plate thickness t are illustrated in Figure 10.4. With τ_{xy} distributed as shown on the vertical face of the element normal to the x axis, τ_{yx} must be distributed identically and oriented as shown on the vertical face of the element normal to the y axis. The requirements for moment equilibrium about the vertical axes of differential elements whose heights are the infinitesimal dz dictate this.

These shear stress distributions give rise to the distributed twisting moments M_{xy} and M_{yx} per unit of plate width in the x and y directions. To produce the distributed twisting moment M_{xy} in the positive x direction, τ_{xy} must be in the negative y

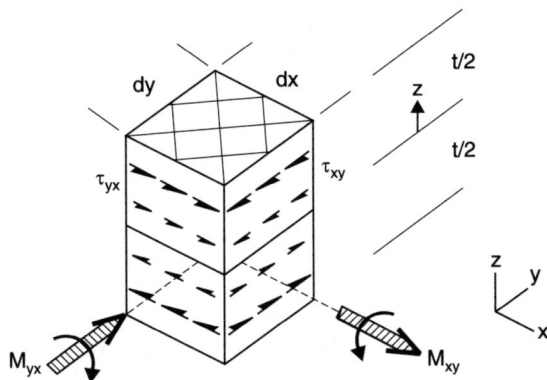

Figure 10.4. Distributed twisting moment.

direction above the midplane, in the positive y direction below it, and zero at the midplane. The corresponding distribution of τ_{yx} produces the positive distributed twisting moment M_{yx}. Because the two shear stress distributions are equal in magnitude, so are the two distributed twisting moments.

The thin plate may be viewed as layers of membranes in a plane stress condition, each strained in ways consistent with the transverse distributions of displacement in the plate discussed above. If the membranes are displaced relative to one another so that the normal strains in the plate are linearly distributed in the transverse direction, the horizontal in-plane shear strains on the transverse faces will be linearly distributed in the transverse direction also. The distributed twisting moment M_{xy} per unit of width in the x direction may then be calculated from the distribution of shear stress τ_{xy} on a unit width of cross section perpendicular to the x axis:

$$M_{xy} = \int_{-t/2}^{t/2} (-z\tau_{xy}) \, dz. \tag{10.4a}$$

The negative sign is present because τ_{xy} is oriented in the negative y direction when the transverse coordinate z is positive. From equation 10.2b, the expression for τ_{xy} is written as

$$\tau_{xy} = \frac{-zE}{1 - v^2} \left(\frac{1 - v}{2} \right) (\phi_{xy}).$$

Substituting this expression for τ_{xy} into equation 10.4a,

$$M_{xy} = \frac{E}{1 - v^2} \left(\frac{1 - v}{2} \right) \int_{-t/2}^{t/2} (\phi_{xy}) z^2 \, dz. \tag{10.4b}$$

The twist ϕ_{xy} is not a function of the transverse coordinate z so it may be moved from under the integral. Evaluating the integral,

$$M_{xy} = \frac{Et^3}{12(1 - v^2)} \left(\frac{1 - v}{2} \right) (\phi_{xy}). \tag{10.4c}$$

The distributed moment–curvature equations in matrix form are

$$\left\{ \begin{array}{c} M_{xx} \\ M_{yy} \\ M_{xy} \end{array} \right\} = \frac{Et^3}{12(1 - v^2)} \begin{bmatrix} 1 & v & 0 \\ v & 1 & 0 \\ 0 & 0 & \frac{1-v}{2} \end{bmatrix} \left\{ \begin{array}{c} \phi_{xx} \\ \phi_{yy} \\ \phi_{xy} \end{array} \right\}, \qquad (10.5a)$$

$$\{\mathbf{M}\} = [\bar{\mathbf{E}}]\{\boldsymbol{\phi}\}. \qquad (10.5b)$$

Equations 10.5 are the generalized stress–strain equations for thin plate flexure. They identify the generalized stress vector as the moment vector $\mathbf{M}$, the generalized strain vector as the curvature vector $\boldsymbol{\phi}$, and the generalized matrix of elastic constants as the plane stress matrix $\mathbf{E}$ of elastic constants multiplied by the scalar $\frac{1}{12}t^3$. Compare equations 10.5 to equations 6.24 in the earlier discussion of generalized stress and strain.

The stress–curvature relationship (equation 10.2a) may be rewritten using the moment curvature relationship (equation 10.5b):

$$\begin{aligned} \{\boldsymbol{\sigma}\} &= -\frac{12z}{t^3}[\bar{\mathbf{E}}]\{\boldsymbol{\phi}\} \\ &= -\frac{12z}{t^3}\{\mathbf{M}\} = -\frac{\{\mathbf{M}\}z}{t^3/12} = -\frac{\{\mathbf{M}\}z}{I}. \end{aligned} \qquad (10.6)$$

Equation 10.6 is the "flexure formula" for thin plate flexure, where the $\boldsymbol{\sigma}$ and $\mathbf{M}$ are vectors rather than scalars. The term I is the moment of inertia of a unit width of plate about the midplane. If the distributed moment vector $\mathbf{M}$ is available on a plate element, it is easy to calculate the stress vector $\boldsymbol{\sigma}$. Moreover, if the in-plane principal stresses and maximum shear stresses are desired, they may be readily calculated from the vector $\boldsymbol{\sigma}$ employing the biaxial stress transformation relationships (equations 7.3 and 7.5) discussed earlier. The largest in-plane stresses occur when the coordinate z is set equal to $\pm\frac{1}{2}t$.

Because the stress vector $\boldsymbol{\sigma}$ is the distributed moment vector $\mathbf{M}$ multiplied by a scalar quantity z/I, "principal distributed bending moments" and "maximum distributed twisting moments" may be envisioned. They may be calculated directly from the moment vector $\mathbf{M}$, employing the biaxial stress transformation relationships (equations 7.3 and 7.5), by replacing the stresses with the corresponding distributed moments.

Although the assumptions made for the thin plate in flexure preclude the presence of transverse shear strains γ_{zx} and γ_{zy}, it must be apparent that transverse shear stresses τ_{zx} and τ_{zy} are present when transverse loads are applied to the plate.

The development of transverse shear in thin plates is analogous to its development in elementary thin beam theory. The expressions for the thin beam transverse shear force and stress are based entirely on equilibrium considerations of a beam differential element subjected to a distributed transverse line load $w(x)$:

$$\frac{dV(x)}{dx} = w(x) \quad \text{and} \quad \frac{dM(x)}{dx} = V(x). \qquad (10.7)$$

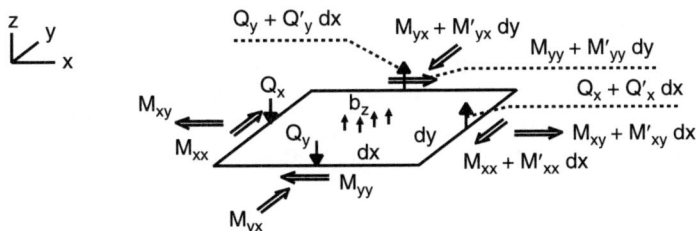

Figure 10.5. Distributed moments and shear force.

The beam's transverse shear force is $V(x)$, and its bending moment is $M(x)$. These expressions lead to reasonable predictions of shear phenomena in beams provided that the beams are thin.

Accounting for shear force in thin flexure plates may be done in a similar manner. Figure 10.5 shows the midplane of a differential element of a thin flexure plate in equilibrium, subjected to a distributed transverse load $b_z(x, y)$. The shear forces $Q_x(x, y)$ and $Q_y(x, y)$ are distributed shear force per unit of length in the y and x directions, respectively. The moments $M_{xx}(x, y)$, $M_{yy}(x, y)$, and $M_{xy}(x, y)$ are distributed also and were described earlier. The primed symbols are partial derivatives of the quantity indicated with respect to its first subscript.

The equilibrium of forces in the z direction requires

$$b_z \, dx \, dy - Q_x \, dy + \left(Q_x + \frac{\partial Q_x}{\partial x} \right) dy - Q_y \, dx + \left(Q_y + \frac{\partial Q_y}{\partial y} \right) dx = 0,$$

or,

$$\frac{\partial Q_x}{\partial x} + \frac{\partial Q_y}{\partial y} + b_z = 0. \tag{10.8a}$$

Equilibrium of moments about the right edge of the element which is parallel to the y axis requires

$$b_z \, dx \, dy \frac{dx}{2} + M_{xx} \, dy - \left(M_{xx} + \frac{\partial M_{xx}}{\partial x} \, dx \right) dy + M_{yx} \, dy - \left(M_{yx} + \frac{\partial M_{yx}}{\partial y} \, dy \right) dx$$

$$- Q_x \, dy \, dx - Q_y \, dx \frac{dx}{2} + \left(Q_y + \frac{\partial Q_y}{\partial y} \, dy \right) dx \frac{dx}{2} = 0.$$

Neglecting the second-order terms,

$$\frac{\partial M_{xx}}{\partial x} + \frac{\partial M_{yx}}{\partial y} + Q_x = 0. \tag{10.8b}$$

Summing moments in a similar way about the bottom edge of the element which is parallel to the x axis results in

$$\frac{\partial M_{xy}}{\partial x} + \frac{\partial M_{yy}}{\partial y} + Q_y = 0. \tag{10.8c}$$

Equations 10.8 are thin flexure plate counterparts to the thin beam shear and moment equations 10.7. Equations 10.8b and 10.8c provide a way to calculate the distributed shear force in thin flexure plates once the variation of the distributed moments has been determined.

In elementary thin beam theory, when the transverse shear force V is determined, the transverse shear stress τ_z is calculated as

$$\tau_z = \frac{VS}{It}. \tag{10.9a}$$

The term S is the first moment, with respect to the neutral axis of the beam cross-sectional area A, of that portion of the area A which extends from the point of the transverse shear stress τ_z to the outer edge of the area A. Here, I is the centroidal moment of inertia of area A and t is the lateral thickness of the area A at the point of the transverse shear stress τ_z. Because the transverse shear stress is assumed to not vary laterally across the width t of the beam, equation 10.9a is accurate only when the ratio between the depth and width of the beam is large, at least 5 or more. When equation 10.9a is applied to a thin, rectangular cross-sectional area, the resulting distribution of transverse shear stress τ_z is uniform in the lateral direction and parabolic in the transverse direction.

The distribution is characterized by

$$\tau_z = \begin{cases} 0 & \text{at } z = \pm\frac{t}{2}, \\ \frac{3V}{2A} & \text{at } z = 0. \end{cases} \tag{10.9b}$$

The transverse cross-sectional area of a differential element of a thin plate in flexure is an infinitesimally thin rectangle so that the distributions of transverse shear stresses τ_{zx} and τ_{zy} are also parabolic. Recalling that the transverse shear forces Q_x and Q_y are per unit width of plate,

$$\tau_{yz} = \tau_{zx} = 0 \quad \text{at } z = \pm\frac{t}{2}$$

$$\tau_{yz} = \frac{3Q_y}{2t} \quad \text{at } z = 0 \tag{10.10}$$

$$\tau_{zx} = \frac{3Q_x}{2t} \quad \text{at } z = 0.$$

10.3. RECTANGULAR THIN PLATE ELEMENT (4-NODE, 12-DOF)

The rectangular, thin flat plate finite element, based on classical thin plate theory, is the simplest one which deals with the complexities of plate flexure. It is called the Kirchhoff element. Studying it is useful to gain an understanding of the application of the finite element method to plates in flexure.

Equation 10.6 allows the stress vector $\boldsymbol{\sigma}$ in the plate to be calculated from the distributed moment vector $\mathbf{M}$. Equation 10.5a allows the distributed moment vector $\mathbf{M}$ to be calculated from the curvature and twist vector $\boldsymbol{\phi}$. The curvatures and twist are second derivatives with respect to the in-plane coordinates of the transverse displacement $w(x, y)$. Thus the solution of the plate flexure problem requires first that the transverse displacement function $w(x, y)$ be determined. From it, the desired moments and stresses are calculated directly.

In thin flat thin plates in flexure, the two in-plane displacements u and v and the transverse displacement w are important to the development of the analysis. The in-plane displacements u and v, which are functions of x, y, z, may be replaced by the product of the partial derivative of the transverse displacement w, which is a function of x and y only, and the transverse coordinate z. These relationships were developed earlier as

$$u(x, \ y, \ z) = -z\frac{\partial w(x, \ y)}{\partial x} \quad \text{and} \quad v(x, \ y, \ z) = -z\frac{\partial w(x, \ y)}{\partial y}.$$

This suggests that the three edof at each node on the element may be chosen as the values of the nodal transverse displacement w, its first partial derivative with respect to the x coordinate, and its first partial derivative with respect to the y coordinate. These partial derivatives are slopes of the midplane in the x and y directions. This is a situation analogous to the beam finite element where the two edof at each node are the transverse displacement and its first derivative with respect to the x coordinate along the axis of the beam.

Figure 10.6 illustrates a rectangular thin plate finite element in flexure with these dof. As with the plane stress rectangular element, the length of the rectangle in the x direction is $2a$, and in the y direction it is $2b$. The thickness of the plate is t. The origin of coordinates is at the center of the rectangle, and local dimensionless coordinates are taken as $\zeta = x/a$ and $\eta = y/b$.

The edof are numbered at each node. The first subscript identifies the node where the dof acts. The second identifies the dof. At node 1, the first dof q_{11} is the transverse displacement, a translation in the positive z direction. The second dof, q_{12}, is a rotation in the positive x direction. Notice that it is physically a positive or counterclockwise slope of edge of the midplane in the y direction. The third dof, q_{13}, is a rotation in the positive y direction. It is a clockwise slope of the edge of the midplane in the x direction. The submatrix of dof at node i is written as

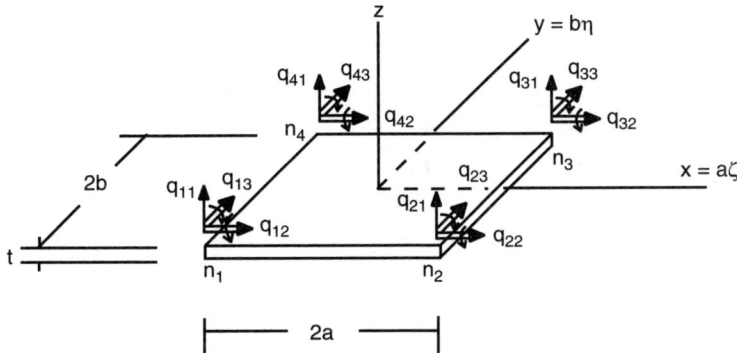

Figure 10.6. Rectangular plate element.

$$\{\mathbf{q}_i\} = \left\{ \begin{array}{c} q_{i1} \\ q_{i2} \\ q_{i3} \end{array} \right\} = \left\{ \begin{array}{c} w_i \\ \frac{\partial w}{\partial y}_i \\ -\frac{\partial w}{\partial x}_i \end{array} \right\}. \tag{10.11}$$

Specifying, at each end of an element edge, one transverse translation and one slope along the edge allows the enforcement of a cubic polynomial variation of transverse displacement along that edge. This is the identical circumstance for the 1D beam element. With the same cubic polynomial representing the transverse displacements for each pair of edges of adjoining plate elements, it is clear that transverse displacements and slopes along edges will be continuous from element to element.

However, only one slope is specified normal to the edge at each end of an element edge. For slopes normal to edges of two adjoining elements, a linear variation of slope along the edge is the highest degree polynomial that can be enforced. Actual slopes normal to the edges of plate elements in flexure vary more severely. The consequence is that slopes normal to the edges from element to element are discontinuous, making this 4-node, 12-dof rectangular element nonconforming. The nonconforming characteristic of the element, however, need not significantly affect its accuracy.

The internal forces and moments at nodes which correspond to the nodal displacement are shown in Figure 10.7. The transverse action is a shear force in the positive z direction. The two in-plane actions are distributed moment vectors in the x and y directions.

As with the edof described above, the actions at nodes may also be identified by node. The submatrix of actions at node i is

$$\{\mathbf{F}_i\} = \left\{ \begin{array}{c} F_{zi} \\ M_{xi} \\ M_{yi} \end{array} \right\}. \tag{10.12}$$

10.3.1. Shape and Displacement Functions

The displacement function $w(\zeta, \eta)$ is a 12-term polynomial, a complete cubic polynomial in ζ and η with two additional higher degree terms. The function is

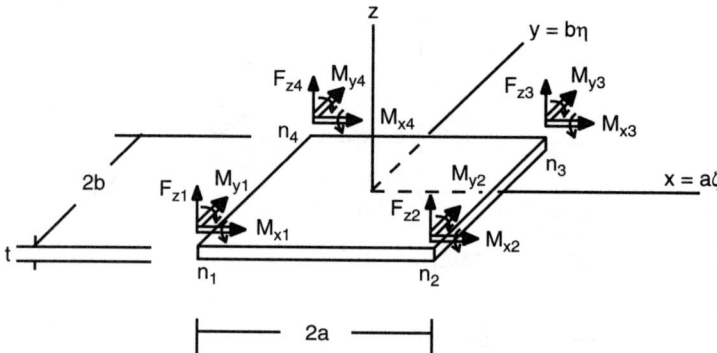

Figure 10.7. Nodal actions.

$$w = D_1 + D_2\zeta + D_3\eta + D_4\zeta^2 + D_5\zeta\eta + D_6\eta^2 + D_7\zeta^3 + D_8\zeta^2\eta + D_9\zeta\eta^2 + D_{10}\eta^3$$
$$+ D_{11}\zeta^3\eta + D_{12}\zeta\eta^3$$

$$= \lfloor 1 \quad \zeta \quad \eta \quad \zeta^2 \quad \zeta\eta \quad \eta^2 \quad \zeta^3 \quad \zeta^2\eta \quad \zeta\eta^2 \quad \eta^3 \quad \zeta^3\eta \quad \zeta\eta^3 \rfloor \begin{Bmatrix} D_1 \\ D_2 \\ D_3 \\ D_4 \\ D_5 \\ D_6 \\ D_7 \\ D_8 \\ D_9 \\ D_{10} \\ D_{11} \\ D_{12} \end{Bmatrix}$$

$$= \lfloor \mathbf{g} \rfloor \{\mathbf{D}\}. \tag{10.13}$$

The displacement function may also be represented in terms of the nodal displacements q_{ij} and shape functions $N_{ij}(\zeta, \eta)$:

$$w = q_{11}N_{11} + q_{12}N_{12} + q_{13}N_{13} + q_{21}N_{21} + q_{22}N_{22} + q_{23}N_{23} + q_{31}N_{31} + q_{32}N_{32}$$
$$+ q_{33}N_{33} + q_{41}N_{41} + q_{42}N_{42} + q_{43}N_{43}$$

$$= \lfloor N_{11} \quad N_{12} \quad N_{13} \quad N_{21} \quad N_{22} \quad N_{23} \quad N_{31} \quad N_{32} \quad N_{33} \quad N_{41} \quad N_{42} \quad N_{43} \rfloor \begin{Bmatrix} q_{11} \\ q_{12} \\ q_{13} \\ q_{21} \\ q_{22} \\ q_{23} \\ q_{31} \\ q_{32} \\ q_{33} \\ q_{41} \\ q_{42} \\ q_{43} \end{Bmatrix}$$

$$= \lfloor \mathbf{N} \rfloor \{\mathbf{q}\}. \tag{10.14}$$

For equation 10.14 to be equivalent to equation 10.13, the shape functions $\mathbf{N}(\zeta, \eta)$ must each contain the same 12 generic polynomial terms that the displacement function $w(\zeta, \eta)$ does.

Having grouped the three displacements at node i as a submatrix $\mathbf{q}_i$, it will be convenient also to group the three corresponding shape functions at node i and treat them as a submatrix:

$$\lfloor \mathbf{N}_i \rfloor = \lfloor N_{i1}(\zeta,\ \eta)\quad N_{i2}(\zeta,\ \eta)\quad N_{i3}(\zeta,\ \eta) \rfloor. \tag{10.15}$$

There are a total of 12 shape functions, 3 at each node and 1 for each edof. They may be obtained by following the procedure used earlier. The vector $\mathbf{g}$ of generic polynomial terms is shown above in equation 10.13. The 12 shape functions take the form of equation 9.49a, which is rewritten as

$$\{\mathbf{N}(\zeta,\ \eta)\} = [\mathbf{h}^{-T}]\{\mathbf{g}\}. \tag{10.16}$$

The matrix $\mathbf{h}$ is obtained by evaluating the vector $\mathbf{g}$ of generic polynomial terms successively at each edof. The second and third edof at each node are slopes of the midplane along the edges of the element in the y and x directions. To evaluate them, the polynomial terms in vector $\mathbf{g}$ must first be differentiated with respect to the y or x coordinate. The half-width a of the element will occur when $\zeta = x/a$ is differentiated with respect to x, and the half-length b will occur when $\eta = y/b$ is differentiated with respect to y.

The matrix of generic functions and their derivatives with respect to y and x are

$$\begin{bmatrix} \mathbf{g} \\ \frac{\partial \mathbf{g}}{\partial y} \\ -\frac{\partial \mathbf{g}}{\partial x} \end{bmatrix} = \begin{bmatrix} 1 & \zeta & \eta & \zeta^2 & \zeta\eta & \eta^2 & \zeta^3 & \zeta^2\eta & \zeta\eta^2 & \eta^3 & \zeta^3\eta & \zeta\eta^3 \\ 0 & 0 & \frac{1}{b} & 0 & \frac{\zeta}{b} & \frac{2\eta}{b} & 0 & \frac{\zeta^2}{b} & \frac{2\zeta\eta}{b} & \frac{3\eta^2}{b} & \frac{\zeta^3}{b} & \frac{3\zeta\eta^2}{b} \\ 0 & -\frac{1}{a} & 0 & -\frac{2\zeta}{a} & -\frac{\eta}{a} & 0 & -\frac{3\zeta^2}{a} & -\frac{2\zeta\eta}{a} & -\frac{\eta^2}{a} & 0 & -\frac{3\zeta^2\eta}{a} & -\frac{\eta^3}{a} \end{bmatrix}. \tag{10.17a}$$

Substituting the nodal coordinates $(\zeta_i,\ \eta_i)$, where $i = 1,\ 2,\ 3,\ 4$, successively into equation 10.17a, the matrix $\mathbf{h}$ may be written as

$$[\mathbf{h}] = \begin{bmatrix}
1 & -1 & -1 & 1 & 1 & 1 & -1 & -1 & -1 & -1 & 1 & 1 \\
0 & 0 & \frac{1}{b} & 0 & -\frac{1}{b} & -\frac{2}{b} & 0 & \frac{1}{b} & \frac{2}{b} & \frac{3}{b} & -\frac{1}{b} & -\frac{3}{b} \\
0 & -\frac{1}{a} & 0 & \frac{2}{a} & \frac{1}{a} & 0 & -\frac{3}{a} & -\frac{2}{a} & -\frac{1}{a} & 0 & \frac{3}{a} & \frac{1}{a} \\
1 & 1 & -1 & 1 & -1 & 1 & 1 & -1 & 1 & -1 & -1 & -1 \\
0 & 0 & \frac{1}{b} & 0 & \frac{1}{b} & -\frac{2}{b} & 0 & \frac{1}{b} & -\frac{2}{b} & \frac{3}{b} & \frac{1}{b} & \frac{3}{b} \\
0 & -\frac{1}{a} & 0 & -\frac{2}{a} & \frac{1}{a} & 0 & -\frac{3}{a} & \frac{2}{a} & -\frac{1}{a} & 0 & \frac{3}{a} & \frac{1}{a} \\
1 & 1 & 1 & 1 & 1 & 1 & 1 & 1 & 1 & 1 & 1 & 1 \\
0 & 0 & \frac{1}{b} & 0 & \frac{1}{b} & \frac{2}{b} & 0 & \frac{1}{b} & \frac{2}{b} & \frac{3}{b} & \frac{1}{b} & \frac{3}{b} \\
0 & -\frac{1}{a} & 0 & -\frac{2}{a} & -\frac{1}{a} & 0 & -\frac{3}{a} & -\frac{2}{a} & -\frac{1}{a} & 0 & -\frac{3}{a} & -\frac{1}{a} \\
1 & -1 & 1 & 1 & -1 & 1 & -1 & 1 & -1 & 1 & -1 & -1 \\
0 & 0 & \frac{1}{b} & 0 & -\frac{1}{b} & \frac{2}{b} & 0 & \frac{1}{b} & -\frac{2}{b} & \frac{3}{b} & -\frac{1}{b} & -\frac{3}{b} \\
0 & -\frac{1}{a} & 0 & \frac{2}{a} & -\frac{1}{a} & 0 & -\frac{3}{a} & \frac{2}{a} & -\frac{1}{a} & 0 & -\frac{3}{a} & -\frac{1}{a}
\end{bmatrix}. \tag{10.17b}$$

$N_{11}(\zeta, \eta)$ $N_{12}(\zeta, \eta)$ $N_{13}(\zeta, \eta)$

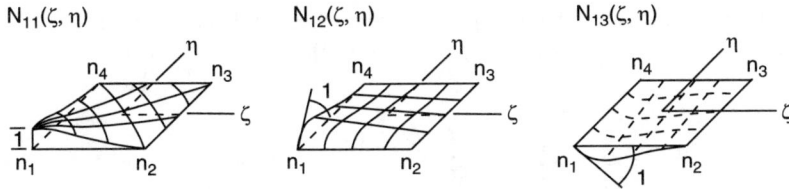

Figure 10.8. Shape functions at node 1.

Equations 10.16 and 10.17 may be manipulated so that the shape functions are expressed explicitly in terms of the plate dimensions a and b and the nodal coordinates (ζ_i, η_i):

$$N_{i1} = \tfrac{1}{8}(1 + \zeta_i\zeta)(1 + \eta_i\eta)(2 + \zeta_i\zeta + \eta_i\eta - \zeta^2 - \eta^2), \tag{10.18a}$$

$$N_{i2} = -\tfrac{1}{8}b\eta_i(1 + \zeta_i\zeta)(1 - \eta_i\eta)(1 + \eta_i\eta)^2, \tag{10.18b}$$

$$N_{i3} = \tfrac{1}{8}a\zeta_i(1 - \zeta_i\zeta)(1 + \eta_i\eta)(1 + \zeta_i\zeta)^2. \tag{10.18c}$$

Figure 10.8 illustrates the three displacement shape functions at node 1. They conform to the requirement for shape functions which is that they equal one at their own edof and zero at the other edof on the element. Shape functions at the other three nodes are similar, except that they are equal to one at their respective edof and zero at the other edof on the element.

Alternatively, the matrix **h** may also be inverted and transposed as a numerical matrix when numerical values for a and b for specific plate elements are used. The shape functions for these elements may then be written using equation 10.16.

10.3.2. Element Stiffness Matrix: Physical Displacements

The element stiffness matrix will be formulated in terms of generalized stress and strain and physical nodal displacements. The generalized stresses are the distributed moments, and the generalized strains are the curvatures and twist.

The formulation begins with the expression for the matrix **K**, written from the one developed for a beam finite element in terms of its generalized stress and strain (equation 6.27c). For the plate element, the integration is over the area of the element:

$$[\mathbf{K}] = \int_A [\bar{\mathbf{B}}^T][\bar{\mathbf{E}}][\bar{\mathbf{B}}] \, dA \tag{10.19a}$$

$$= \int_{-a}^{a} \int_{-b}^{b} [\bar{\mathbf{B}}^T][\bar{\mathbf{E}}][\bar{\mathbf{B}}] \, dx \, dy \tag{10.19b}$$

$$= \int_{-1}^{1} \int_{-1}^{1} [\bar{\mathbf{B}}^T][\bar{\mathbf{E}}][\bar{\mathbf{B}}](ab) \, d\zeta \, d\eta. \tag{10.19c}$$

The strain–displacement matrix $\bar{\mathbf{B}}$ for generalized stress and strain was defined earlier as

$$\lfloor \bar{\mathbf{B}} \rfloor = \lfloor \bar{\mathbf{d}} \rfloor [\mathbf{N}]. \tag{6.26b}$$

Equations 10.1, in describing the strain–displacement relationships for the rectangular plate, suggest that the vector of differential operators $\bar{\mathbf{d}}$ for the generalized stress is

$$
\left\{ \begin{array}{c} \epsilon_x \\ \epsilon_y \\ \gamma_{xy} \end{array} \right\} = -z \left\{ \begin{array}{c} \frac{\partial^2 w}{\partial x^2} \\ \frac{\partial^2 w}{\partial y^2} \\ 2 \frac{\partial^2 w}{\partial x\, \partial y} \end{array} \right\} = -z \left\{ \begin{array}{c} \frac{\partial^2}{\partial x^2} \\ \frac{\partial^2}{\partial y^2} \\ 2 \frac{\partial^2}{\partial x\, \partial y} \end{array} \right\} w(\zeta, \eta), \qquad (10.20a)
$$

$$
\{\epsilon\} = -z\{\bar{\mathbf{d}}\} w(\zeta, \eta). \tag{10.20b}
$$

Thus,

$$
\{\bar{\mathbf{d}}\} = \left\{ \begin{array}{c} \frac{\partial^2}{\partial x^2} \\ \frac{\partial^2}{\partial y^2} \\ 2 \frac{\partial^2}{\partial x\, \partial y} \end{array} \right\}. \tag{10.20c}
$$

The strain–displacement matrix $\bar{\mathbf{B}}$ may now be written for the 4-node rectangular thin plate flexure element as

$$
\{\bar{\mathbf{B}}\} = \left\{ \begin{array}{c} \frac{\partial^2}{\partial x^2} \\ \frac{\partial^2}{\partial y^2} \\ 2 \frac{\partial^2}{\partial x\, \partial y} \end{array} \right\} \lfloor \mathbf{N}_1(\zeta, \eta) \quad \mathbf{N}_2(\zeta, \eta) \quad \mathbf{N}_3(\zeta, \eta) \quad \mathbf{N}_4(\zeta, \eta) \rfloor. \tag{10.21}
$$

The submatrix of shape functions $\mathbf{N}_i(\zeta, \eta)$, identified for node i, is defined by equations 10.15 and 10.18. It will be convenient to express the matrix $\bar{\mathbf{B}}$ in terms of submatrices also, identified at node i:

$$
[\bar{\mathbf{B}}_i] = \{\bar{\mathbf{d}}\} \lfloor \mathbf{N}_i \rfloor = \left\{ \begin{array}{c} \frac{\partial^2}{\partial x^2} \\ \frac{\partial^2}{\partial y^2} \\ 2 \frac{\partial^2}{\partial x\, \partial y} \end{array} \right\} \lfloor N_{i1} \quad N_{i2} \quad N_{i3} \rfloor = \begin{bmatrix} \frac{\partial^2 N_{i1}}{\partial x^2} & \frac{\partial^2 N_{i2}}{\partial x^2} & \frac{\partial^2 N_{i3}}{\partial x^2} \\ \frac{\partial^2 N_{i1}}{\partial y^2} & \frac{\partial^2 N_{i2}}{\partial y^2} & \frac{\partial^2 N_{i3}}{\partial y^2} \\ 2 \frac{\partial^2 N_{i1}}{\partial x\, \partial y} & 2 \frac{\partial^2 N_{i2}}{\partial x\, \partial y} & 2 \frac{\partial^2 N_{i3}}{\partial x\, \partial y} \end{bmatrix}. \tag{10.22a}
$$

The nine second partial derivatives of the shape functions indicated in the strain–displacement matrix $\bar{\mathbf{B}}_i$ for node i may be written explicitly from equations 10.18, so that

$$
[\bar{\mathbf{B}}_i] = \begin{bmatrix} -\frac{3\zeta_i}{4a^2} \zeta(1 + \eta_i \eta) & 0 & -\frac{\zeta_i}{4a}(1 + 3\zeta_i \zeta)(1 + \eta_i \eta) \\ -\frac{3\eta_i}{4b^2} \eta(1 + \zeta_i \zeta) & \frac{\eta_i}{4b}(1 + \zeta_i \zeta)(1 + 3\eta_i \eta) & 0 \\ \frac{\zeta_i \eta_i}{4ab}(4 - 3\zeta^2 - 3\eta^2) & -\frac{\zeta_i}{4a}(1 - 3\eta_i \eta)(1 + \eta_i \eta) & \frac{\eta_i}{4b}(1 - 3\zeta_i \zeta)(1 + \zeta_i \zeta) \end{bmatrix}. \tag{10.22b}
$$

The terms $(\zeta_i, \; \eta_i)$ are the coordinates of the nodes.

The integrand of equation 10.19c is a 12×12 matrix of functions $\mathbf{I}(\zeta, \; \eta)$:

$$[\mathbf{I}(\zeta, \eta)] = [\bar{\mathbf{B}}^{\mathrm{T}}][\bar{\mathbf{E}}][\bar{\mathbf{B}}](ab). \tag{10.23a}$$

It may be written in terms of the generalized strain–displacement submatrix $\bar{\mathbf{B}}_i$ at nodes 1, 2, 3, 4 and the generalized matrix of elastic constants $\bar{\mathbf{E}}$:

$$[\mathbf{I}(\zeta, \eta)] = \left\{ \begin{array}{c} \bar{B}_1^{\mathrm{T}} \\ \bar{B}_2^{\mathrm{T}} \\ \bar{B}_3^{\mathrm{T}} \\ \bar{B}_4^{\mathrm{T}} \end{array} \right\} [\bar{E}] \lfloor \bar{B}_1 \quad \bar{B}_2 \quad \bar{B}_3 \quad \bar{B}_4 \rfloor (ab). \tag{10.23b}$$

Each submatrix shown is a 3×3 array so that when the submatrix multiplications indicated are performed, the result is a 4×4 array of 3×3 submatrices which yields the 12×12 matrix $\mathbf{I}(\zeta, \eta)$:

$$[\mathbf{I}(\zeta, \eta)] = \begin{bmatrix} I_{11} & I_{12} & I_{13} & I_{14} \\ I_{21} & I_{22} & I_{23} & I_{24} \\ I_{31} & I_{32} & I_{33} & I_{34} \\ I_{41} & I_{42} & I_{43} & I_{44} \end{bmatrix} = \begin{bmatrix} \bar{B}_1^{\mathrm{T}} \bar{E} B_1 & \bar{B}_1^{\mathrm{T}} \bar{E} B_2 & \bar{B}_1^{\mathrm{T}} \bar{E} B_3 & \bar{B}_1^{\mathrm{T}} \bar{E} B_4 \\ \bar{B}_2^{\mathrm{T}} \bar{E} B_1 & \bar{B}_2^{\mathrm{T}} \bar{E} B_2 & \bar{B}_2^{\mathrm{T}} \bar{E} B_3 & \bar{B}_2^{\mathrm{T}} \bar{E} B_4 \\ \bar{B}_3^{\mathrm{T}} \bar{E} B_1 & \bar{B}_3^{\mathrm{T}} \bar{E} B_2 & \bar{B}_3^{\mathrm{T}} \bar{E} B_3 & \bar{B}_3^{\mathrm{T}} \bar{E} B_4 \\ \bar{B}_4^{\mathrm{T}} \bar{E} B_1 & \bar{B}_4^{\mathrm{T}} \bar{E} B_2 & \bar{B}_4^{\mathrm{T}} \bar{E} B_3 & \bar{B}_4^{\mathrm{T}} \bar{E} B_4 \end{bmatrix} (ab). \tag{10.23c}$$

The indices $i = 1, 2, 3, 4$ and $j = 1, 2, 3, 4$ identify the position of submatrix $\mathbf{I}_{ij}$ in the array of submatrices $\mathbf{I}(\zeta, \eta)$. They also identify the submatrix $\bar{\mathbf{B}}_i^{\mathrm{T}}$ and the submatrix $\bar{\mathbf{B}}_j$, which must be used to form submatrix $\mathbf{I}_{ij}$:

$$[\mathbf{I}_{ij}] = [\bar{\mathbf{B}}_i^{\mathrm{T}}][\bar{\mathbf{E}}][\bar{\mathbf{B}}_j](ab).$$

If the integrand for a particular element stiffness coefficient is desired, it may be obtained from among the nine entries of the appropriate 3×3 nodal submatrix $\mathbf{I}_{ij}$.

For example, the integrand for stiffness coefficient k_{74} occurs as the entry in the first row and first column of submatrix $\mathbf{I}_{32}$, as illustrated below in equation 10.24. To obtain k_{74}, the 3×3 submatrix $\mathbf{I}_{32}$ must be formed, and its entry I_{k74} obtained, which is at row 1 and column 1 of this submatrix:

Node number   1   2   3   4
edof number   1 2 3 4   5 6 7   8 9   10 11 12

$$[\mathbf{I}(\zeta, \eta)] = \begin{array}{c} \\ 1 \\ 2 \\ 3 \\ 4 \end{array} \begin{array}{c} 1 \\ 2 \\ 3 \\ 4 \\ 5 \\ 6 \\ 7 \\ 8 \\ 9 \\ 10 \\ 11 \\ 12 \end{array} \left[\begin{array}{cccccccccccc} & & & & & & & & & & & \\ & & & & & & & & & & & \\ & & & & & & & & & & & \\ & & & & & & & & & & & \\ & & & & & \times & & & & & & \\ & & & & & & & & & & & \\ & & & & & & & & & & & \\ & & & & & & & & & & & \\ & & & & & & & & & & & \end{array} \right], \tag{10.24}$$

$$[\mathbf{I}_{32}] = [\bar{\mathbf{B}}_3^\mathsf{T}][\bar{\mathbf{E}}][\bar{\mathbf{B}}_2](ab) = \begin{bmatrix} I_{k74} & I_{k75} & I_{k76} \\ I_{k84} & I_{k85} & I_{k86} \\ I_{k94} & I_{k95} & I_{k96} \end{bmatrix}, \tag{10.25a}$$

$$[\mathbf{I}_{32}] = \frac{Et^3(ab)}{12(1-v^2)} \begin{bmatrix} \frac{\partial^2 N_{31}}{\partial x^2} & \frac{\partial^2 N_{31}}{\partial y^2} & 2\frac{\partial^2 N_{31}}{\partial x\, \partial y} \\ \frac{\partial^2 N_{32}}{\partial x^2} & \frac{\partial^2 N_{32}}{\partial y^2} & 2\frac{\partial^2 N_{32}}{\partial x\, \partial y} \\ \frac{\partial^2 N_{33}}{\partial x^2} & \frac{\partial^2 N_{33}}{\partial y^2} & 2\frac{\partial^2 N_{33}}{\partial x\, \partial y} \end{bmatrix} \begin{bmatrix} 1 & v & 0 \\ v & 1 & 0 \\ 0 & 0 & \frac{1}{2}(1-v) \end{bmatrix} \begin{bmatrix} \frac{\partial^2 N_{21}}{\partial x^2} & \frac{\partial^2 N_{22}}{\partial x^2} & \frac{\partial^2 N_{23}}{\partial x^2} \\ \frac{\partial^2 N_{21}}{\partial y^2} & \frac{\partial^2 N_{22}}{\partial y^2} & \frac{\partial^2 N_{23}}{\partial y^2} \\ 2\frac{\partial^2 N_{21}}{\partial x\, \partial y} & 2\frac{\partial^2 N_{22}}{\partial x\, \partial y} & 2\frac{\partial^2 N_{23}}{\partial x\, \partial y} \end{bmatrix}. \tag{10.25b}$$

To obtain the entry in the first row and first column of submatrix $\mathbf{I}_{32}$, the partial derivatives in the first row of submatrix $\bar{\mathbf{B}}_3^\mathsf{T}$ and those in the first column of submatrix $\bar{\mathbf{B}}_2$ will be required. They may be written from the partial derivative expressions of the shape functions given in equations 10.22b. Performing the matrix manipulations indicated in equation 10.25b with these differentiated expressions, the resulting integrand is

$$I_{74} = \frac{Et^3}{384a^3b^3(1-v^2)}\big[18b^4\zeta^2(1-\eta^2) - 36va^2b^2\zeta\eta^2(1+\zeta) - 18a^4\eta^2(1+\zeta)^2$$
$$- (1-v)a^2b^2(4 - 3\zeta^2 - 3\eta^2)^2\big].$$

The stiffness coefficient k_{74} then is calculated from

$$k_{74} = \int_{-1}^{1} \int_{-1}^{1} I_{k74} \, d\zeta \, d\eta.$$

The integral may be evaluated analytically or numerically for known element dimensions a and b and elastic constants E and v. Because the integrand contains terms with ζ^4 and η^4, the 2×2 Gauss quadrature rule will produce an approximate numerical result. The 3×3 Gauss quadrature rule will evaluate the integral exactly.

As an example, the stiffness coefficient k_{74} for a rectangular steel plate $5\,\text{ft} \times 3\,\text{ft} \times 0.75\,\text{in.}$ will be computed. The modulus of elasticity E is $30{,}000\,\text{ksi}$ and Poisson's ratio ν is 0.25.

The scalar to the equation is given as

$$\frac{Et^3}{384a^3b^3(1-\nu^2)} = \frac{(30{,}000)\,(0.75^3)}{384\,(30^3)\,(18^3)\,(1-0.25^2)}$$
$$= 2.23265 \times 10^{-7}\ \text{kips/in.}^5$$

Employing the Gauss 2×2 rule,

$$K_{74} = \frac{Et^3}{384a^3b^3(1-\nu^2)}[I_{k74}(\zeta_1,\ \eta_1)H_1 + I_{k74}(\zeta_2,\ \eta_2)H_2 + I_{k74}(\zeta_3,\ \eta_3)H_3 + I_{k74}(\zeta_4,\ \eta_4)H_4]$$
$$= 2.23265 \times 10^{-7}(-1{,}109{,}580 - 13{,}343{,}410 - 1{,}109{,}580 - 13{,}343{,}410)$$
$$= -6.4537\ \textbf{kips/in.}$$

Employing the Gauss 3×3 rule,

$$k_{74} = \frac{Et^3}{384a^3b^3(1-\nu^2)}[I_{k74}(\zeta_1,\ \eta_1)H_1 + I_{k74}(\zeta_2,\ \eta_2)H_2 + I_{k74}(\zeta_3,\ \eta_3)H_3$$
$$+ I_{k74}(\zeta_4,\ \eta_4)H_4 + I_{k74}(\zeta_5,\ \eta_5)H_5 + I_{k74}(\zeta_6,\ \eta_6)H_6$$
$$+ I_{k74}(\zeta_7,\ \eta_7)H_7 + I_{k74}(\zeta_8,\ \eta_8)H_8 + I_{k74}(\zeta_9,\ \eta_9)H_9]$$
$$= 2.23265 \times 10^{-7}\Big(248{,}975 \times \tfrac{25}{81} - 9{,}806{,}510\,\tfrac{40}{81} - 29{,}295{,}100 \times \tfrac{25}{81}$$
$$+ 75{,}232.8 \times \tfrac{40}{81} - 3{,}499{,}200 \times \tfrac{64}{81} + 75{,}232.8 \times \tfrac{40}{81}$$
$$+ 248{,}975 \times \tfrac{25}{81} - 9{,}806{,}510 \times \tfrac{40}{81} - 29{,}295{,}100 \times \tfrac{25}{81}\Big)$$
$$= -6.7662\ \textbf{kips/in.}$$

The latter value of k_{74} is correct to the significant figures shown. It may also be obtained by integrating I_{74} analytically between the limits $-1 < \zeta < +1$ and $-1 < \eta < +1$. The magnitude of k_{74} obtained by the Gauss 2×2 rule is too small by 4.6%.

The integral equation 10.19c for the entire stiffness matrix is efficiently evaluated numerically using a Gauss quadrature rule when values for the element dimensions a and b and elastic constants E and ν are known. The 2×2 rule is approximate for most of the stiffness coefficients, since most contain terms with ζ^4 or η^4. The 3×3 rule, on the other hand, is exact. Equation 10.19c rewritten to reflect the application of a quadrature rule is

$$[\mathbf{K}] = (ab)\sum_{k=1}^{n}[\bar{\mathbf{B}}^{\mathrm{T}}(\zeta_k,\ \eta_k)][\bar{\mathbf{E}}][\bar{\mathbf{B}}(\zeta_k,\ \eta_k)]H_k. \tag{10.26}$$

To calculate the stiffness matrix $\mathbf{K}$, the 3×12 generalized strain–displacement matrix $\bar{\mathbf{B}}$ of functions of ζ and η is developed from equation 10.22b, where $(\zeta_i,\ \eta_i) = (-1, -1),\ (1, -1),\ (1, 1),$ and $(-1, 1)$. Each functional entry is evaluated at the first integration point (ζ_{k1}, η_{k1}). The resulting 3×12 numerical matrix $\bar{\mathbf{B}}(\zeta_{k1}, \eta_{k1})$ is manipulated with the 3×3 numerical generalized elastic matrix $\bar{\mathbf{E}}$, as indicated by equation 10.26, to produce a 12×12 numerical product matrix. This product matrix is multiplied by the weight factor H_{k1} associated with the first integration point, resulting in a 12×12 numerical matrix for the first integration point. The process is repeated for the remaining integration points, and the resulting 12×12 numerical matrices for each integration point are summed. The sum is a 12×12 numerical matrix which is multiplied by the plate half-lengths a and b to produce the element stiffness matrix.

For the 5 ft $\times$ 3 ft $\times$ 0.75 in. steel plate example used to compute k_{74} above, the 3×12 matrix $\bar{\mathbf{B}}$, evaluated at the first integration point $(-0.57735, -0.57735)$ for the Gauss 2×2 rule, is

$$[\bar{\mathbf{B}}_{k1}] = \begin{bmatrix} -0.75890 & 0.0000 & 35.9117 & 0.75890 & 0.0000 & 9.6225 & 0.20335 & 0.0000 & 2.5783 & -0.20335 & 0.0000 & 9.6225 \\ -2.1081 & -59.8528 & 0.0000 & -0.56485 & -16.0375 & 0.0000 & 0.56485 & -4.2972 & 0.0000 & 2.1081 & -16.0375 & 0.0000 \\ 0.92593 & -9.6225 & 16.0375 & -0.92593 & 9.6225 & -16.0375 & 0.92593 & -9.6225 & 16.0375 & -0.92593 & 9.6225 & -16.0375 \end{bmatrix} \times 10^{-3}.$$

Its units are reciprocal inches squared in the first, fourth, seventh, and tenth columns, and reciprocal inches in the remaining columns.

The 3×3 generalized elastic matrix $\bar{\mathbf{E}}$ is given as

$$[\bar{\mathbf{E}}] = \frac{(30,000)(0.75)^3}{12(1-0.25^2)} \begin{bmatrix} 1 & 0.25 & 0 \\ 0.25 & 1 & 0 \\ 0 & 0 & 0.375 \end{bmatrix} = \begin{bmatrix} 1125 & 281.25 & 0 \\ 281.25 & 1125 & 0 \\ 0 & 0 & 421.875 \end{bmatrix} \quad \text{(kip-in.)}$$

The 12×12 product matrix $\bar{\mathbf{B}}_{k1}^T \bar{\mathbf{E}} \bar{\mathbf{B}}_{k1}$ for the first integration point multiplied by the weight factor $H_{k1} = 1$ for the Gauss 2×2 rule is

$$[\bar{\mathbf{B}}_{k1}^T \bar{\mathbf{E}} \bar{\mathbf{B}}_{k1}] = \begin{bmatrix}
0.00691 & 0.15096 & -0.04569 & 0.00000 & 0.04522 & -0.02019 & -0.00139 & 0.00735 & 0.00253 & -0.00552 & 0.04522 & -0.02019 \\
0.15096 & 4.06921 & -0.66963 & 0.02902 & 1.04081 & -0.09688 & -0.04522 & 0.32841 & -0.10851 & -0.13476 & 1.04081 & -0.09688 \\
-0.04569 & -0.66963 & 1.55936 & 0.01869 & -0.09688 & 0.28025 & 0.02019 & -0.10851 & 0.21267 & 0.00681 & -0.09688 & 0.28025 \\
0.00000 & 0.02902 & 0.01869 & 0.00113 & 0.00301 & 0.01295 & -0.00046 & 0.00557 & -0.00447 & -0.00067 & 0.00301 & 0.01295 \\
0.04522 & 1.04081 & -0.09688 & 0.00301 & 0.32841 & -0.10851 & -0.00735 & 0.03847 & 0.05347 & -0.04088 & 0.32841 & -0.10851 \\
-0.02019 & -0.09688 & 0.28025 & 0.01295 & -0.10851 & 0.21267 & -0.00253 & 0.05347 & -0.08060 & 0.00977 & -0.10851 & 0.21267 \\
-0.00139 & -0.04522 & 0.02019 & -0.00046 & -0.00735 & -0.00253 & 0.00083 & -0.00674 & 0.00726 & 0.00102 & -0.00735 & -0.00253 \\
0.00735 & 0.32841 & -0.10851 & 0.00557 & 0.03847 & 0.05347 & -0.00674 & 0.05984 & -0.06822 & -0.00619 & 0.03847 & 0.05347 \\
0.00253 & -0.10851 & 0.21267 & -0.00447 & 0.05347 & -0.08060 & 0.00726 & -0.06822 & 0.11599 & -0.00533 & 0.05347 & -0.08060 \\
-0.00552 & -0.13476 & 0.00681 & -0.00067 & -0.04088 & 0.00977 & 0.00102 & -0.00619 & -0.00533 & 0.00517 & -0.04088 & 0.00977 \\
0.04522 & 1.04081 & -0.09688 & 0.00301 & 0.32841 & -0.10851 & -0.00735 & 0.03847 & 0.05347 & -0.04088 & 0.32841 & -0.10851 \\
-0.02019 & -0.09688 & 0.28025 & 0.01295 & -0.10851 & 0.21267 & -0.00253 & 0.05347 & -0.08060 & 0.00977 & -0.10851 & 0.21267
\end{bmatrix}.$$

The units for the product matrix may be seen in terms of 3×3 product submatrices (equation 10.23c), whose units are

$$[\bar{\mathbf{B}}_i^T \bar{\mathbf{E}} \bar{\mathbf{B}}_j] = \begin{bmatrix} \text{kips/in.}^3 & \text{kips/in.}^2 & \text{kips/in.}^2 \\ \text{kips/in.}^2 & \text{kips/in.} & \text{kips/in.} \\ \text{kips.in.}^2 & \text{kips/in.} & \text{kips/in.} \end{bmatrix}.$$

When the 12×12 product matrices are formed for the remaining three integration points, each multiplied by its weight factor, all four added, and the matrix sum multiplied by the plate half-lengths a and b, the result is the plate element stiffness matrix $\mathbf{K}$:

$$[\mathbf{K}] = \begin{bmatrix}
7.579 & 108.854 & -30.313 & 1.102 & 47.396 & -22.500 & -2.227 & 52.083 & -11.250 & -6.454 & 104.167 & -3.438 \\
108.854 & 2584.375 & -281.250 & 47.396 & 1165.625 & 0.000 & -52.083 & 709.375 & 0.000 & -104.167 & 1165.625 & 0.000 \\
-30.313 & -281.250 & 1134.375 & 22.500 & 0.000 & 215.625 & 11.250 & 0.000 & 459.375 & -3.438 & 0.000 & 215.625 \\
1.102 & 47.396 & 22.500 & 7.579 & 108.854 & 30.313 & -6.454 & 104.167 & 3.438 & -2.227 & 52.083 & 11.250 \\
47.396 & 1165.625 & 0.000 & 108.854 & 2584.375 & 281.250 & -104.167 & 1165.625 & 0.000 & -52.083 & 709.375 & 0.000 \\
-22.500 & 0.000 & 215.625 & 30.313 & 281.250 & 1134.375 & 3.438 & 0.000 & 215.625 & -11.250 & 0.000 & 459.375 \\
-2.227 & -52.083 & 11.250 & -6.454 & -104.167 & 3.438 & 7.579 & -108.854 & 30.313 & 1.102 & -47.396 & 22.500 \\
52.083 & 709.375 & 0.000 & 104.167 & 1165.625 & 0.000 & -108.854 & 2584.375 & -281.250 & -47.396 & 1165.625 & 0.000 \\
-11.250 & 0.000 & 459.375 & 3.438 & 0.000 & 215.625 & 30.313 & -281.250 & 1134.375 & -22.500 & 0.000 & 215.625 \\
-6.454 & -104.167 & -3.438 & -2.227 & -52.083 & -11.250 & 1.102 & -47.396 & -22.500 & 7.579 & -108.854 & -30.313 \\
104.167 & 1165.625 & 0.000 & 52.083 & 709.375 & 0.000 & -47.396 & 1165.625 & 0.000 & -108.854 & 2584.375 & 281.250 \\
-3.438 & 0.000 & 215.625 & 11.250 & 0.000 & 459.375 & 22.500 & 0.000 & 215.625 & -30.313 & 281.250 & 1134.375
\end{bmatrix}.$$

Units for matrix $\mathbf{K}$ in terms of the nodal submatrix are

$$[\mathbf{K}_{ij}] = \begin{bmatrix}
\text{kips/in.} & \text{kips} & \text{kips} \\
\text{kips} & \text{kip-in.} & \text{kip-in.} \\
\text{kips} & \text{kip-in.} & \text{kip-in.}
\end{bmatrix}.$$

When the Gauss 3×3 rule is applied, the result is

$$[\mathbf{K}] = \begin{bmatrix}
7.891 & 111.667 & -35.000 & 0.789 & 44.583 & -27.188 & -1.914 & 49.271 & -6.563 & -6.766 & 106.979 & 1.250 \\
111.667 & 2635.000 & -281.250 & 44.583 & 1115.000 & 0.000 & -49.271 & 658.750 & 0.000 & -106.979 & 1216.250 & 0.000 \\
-35.000 & -281.250 & 1275.000 & 27.188 & 0.000 & 356.250 & 6.563 & 0.000 & 318.750 & 1.250 & 0.000 & 75.000 \\
0.789 & 44.583 & 27.188 & 7.891 & 111.667 & 35.000 & -6.766 & 106.979 & -1.250 & -1.914 & 49.271 & 6.563 \\
44.583 & 1115.000 & 0.000 & 111.667 & 2635.000 & 281.250 & -106.979 & 1216.250 & 0.000 & -49.271 & 658.750 & 0.000 \\
-27.188 & 0.000 & 356.250 & 35.000 & 281.250 & 1275.000 & -1.250 & 0.000 & 75.000 & -6.563 & 0.000 & 318.750 \\
-1.914 & -49.271 & 6.563 & -6.766 & -106.979 & -1.250 & 7.891 & -111.667 & 35.000 & 0.789 & -44.583 & 27.188 \\
49.271 & 658.750 & 0.000 & 106.979 & 1216.250 & 0.000 & -111.667 & 2635.000 & -281.250 & -44.583 & 1115.000 & 0.000 \\
-6.563 & 0.000 & 318.750 & -1.250 & 0.000 & 75.000 & 35.000 & -281.250 & 1275.000 & -27.188 & 0.000 & 356.250 \\
-6.766 & -106.979 & 1.250 & -1.914 & -49.271 & -6.563 & 0.789 & -44.583 & -27.188 & 7.891 & -111.667 & -35.000 \\
106.979 & 1216.250 & 0.000 & 49.271 & 658.750 & 0.000 & -44.583 & 1115.000 & 0.000 & -111.667 & 2635.000 & 281.250 \\
1.250 & 0.000 & 75.000 & 6.563 & 0.000 & 318.750 & 27.188 & 0.000 & 356.250 & -35.000 & 281.250 & 1275.000
\end{bmatrix}.$$

There are important differences between the exact values for stiffness coefficients obtained using the Gauss 3×3 rule and those obtained using the Gauss 2×2 rule. The error in the magnitudes of stiffness coefficients from the Gauss 2×2 rule range from zero to 188%, for an average error of 32%. An average error weighted on the magnitudes of the stiffness coefficients is 9.1%. The overall accuracy of the plate element will depend on more than the accuracy of the quadrature rule employed; however, one should anticipate that the Gauss 2×2 quadrature rule may cause important inaccuracy.

10.3.3. Element Stiffness Matrix: Generalized dof

If the element stiffness matrix is developed with the vector $\mathbf{D}$ of generalized edof (equation 10.13) instead of the vector $\mathbf{q}$ of nodal displacements (equation 10.14), the single-term generic functions contained in the vector $\mathbf{g}$ will be used instead of the 12-term shape functions contained in the vector $\mathbf{N}$. The use of single-term rather than 12-term polynomial functions provides a significant simplification. The single-term functions are readily determined and more easily differentiated and their products make much simpler integrands. Generalized dof and the relationships among them and the displacement dof were discussed in Chapter 6.

The stiffness equations for generalized edof were shown to be

$$[\mathbf{K}_D]\{\mathbf{D}\} = \{\mathbf{F}_D\} + \{\mathbf{F}_{Db}\} + \{\mathbf{F}_{Dk}\}, \tag{6.39a}$$

where

$$\{\mathbf{F}_D\} = [\mathbf{h}^T]\{\mathbf{F}\}, \tag{6.39b}$$

$$\{\mathbf{F}_{Db}\} = [\mathbf{h}^T]\{\mathbf{F}_b\}, \tag{6.39c}$$

$$\{\mathbf{F}_{Dk}\} = [\mathbf{h}^T]\{\mathbf{F}_k\}, \tag{6.39d}$$

$$\{\mathbf{D}\} = [\mathbf{h}^{-1}]\{\mathbf{q}\}. \tag{6.33}$$

The generalized element stiffness matrix is matrix $\mathbf{K}_D$ and may be computed from

$$[\mathbf{K}_D] = \int_V [\mathbf{B}_D^T][\mathbf{E}][\mathbf{B}_D] \, dV. \tag{6.37f}$$

It may then be transformed to the element stiffness matrix referred to the physical displacement edof vector $\mathbf{q}$ using

$$[\mathbf{K}] = [\mathbf{h}^{-T}][\mathbf{K}_D][\mathbf{h}^{-1}]. \tag{6.37e}$$

The matrix $\mathbf{h}$ was developed earlier for the rectangular plate in flexure as equation 10.17b.

Equation 6.37f may be adapted to the rectangular plate element as

$$[\mathbf{K}_D] = \int_{-1}^{1} \int_{-1}^{1} [\bar{\mathbf{B}}_D^T][\bar{\mathbf{E}}][\bar{\mathbf{B}}_D](ab) \, d\zeta \, d\eta \tag{10.27a}$$

$$= \int_{-1}^{1} \int_{-1}^{1} [\mathbf{I}_D] \, d\zeta \, d\eta. \tag{10.27b}$$

The strain–displacement matrix $\bar{\mathbf{B}}_D$ for generalized stress and strain and generalized edof is

$$[\bar{\mathbf{B}}_D] = \{\bar{\mathbf{d}}\}\lfloor \mathbf{g}\rfloor, \tag{10.28a}$$

$$= \left\{ \begin{array}{c} \frac{\partial^2}{\partial x^2} \\ \frac{\partial^2}{\partial y^2} \\ 2\frac{\partial^2}{\partial x\, \partial y} \end{array} \right\} \lfloor 1 \quad \zeta \quad \eta \quad \zeta^2 \quad \zeta\eta \quad \eta^2 \quad \zeta^3 \quad \zeta^2\eta \quad \zeta\eta^2 \quad \eta^3 \quad \zeta^3\eta \quad \zeta\eta^3 \rfloor. $$

$$\tag{10.28b}$$

Performing the differentiations indicated with respect to x to y leads to

$$[\bar{\mathbf{B}}_D] = \frac{1}{(ab)^2} \begin{bmatrix} 0 & 0 & 0 & 2b^2 & 0 & 0 & 6b^2\zeta & 2b^2\eta & 0 & 0 & 6b^2\zeta\eta & 0 \\ 0 & 0 & 0 & 0 & 0 & 2a^2 & 0 & 0 & 2a^2\zeta & 6a^2\eta & 0 & 6a^2\zeta\eta \\ 0 & 0 & 0 & 0 & 2ab & 0 & 0 & 4ab\zeta & 4ab\eta & 0 & 6ab\zeta^2 & 6ab\eta^2 \end{bmatrix}.$$

$$\tag{10.28c}$$

The integrand matrix $\mathbf{I}_D$ is formed as the product of matrices $\bar{\mathbf{B}}_D^T$, $\bar{\mathbf{E}}$ and $\bar{\mathbf{B}}_D$ and the scalar plate half-lengths a and b, as indicated in equation 10.27. It is a 12×12 matrix of 71 zeros and 73 very simple functions of ζ and η. All but 8 are single-term polynomial functions and all but 2 are of degree less than 4. Those terms that contain an odd degree in ζ or η, that is, ζ, η, ζ^3, or η^3, are zero when integrated analytically between $-1 < \zeta < +1$ and $-1 < \eta < +1$. As a consequence, there are only 21 non-zero entries for the generalized stiffness matrix $\mathbf{K}_D$:

$$[\mathbf{K}_D] = \begin{bmatrix} 0 & 0 & 0 & 0 & 0 & 0 & 0 & 0 & 0 & 0 & 0 & 0 \\ 0 & 0 & 0 & 0 & 0 & 0 & 0 & 0 & 0 & 0 & 0 & 0 \\ 0 & 0 & 0 & 0 & 0 & 0 & 0 & 0 & 0 & 0 & 0 & 0 \\ 0 & 0 & 0 & k_{D44} & 0 & k_{D46} & 0 & 0 & 0 & 0 & 0 & 0 \\ 0 & 0 & 0 & 0 & k_{D55} & 0 & 0 & 0 & 0 & 0 & k_{D5,11} & k_{D5,12} \\ 0 & 0 & 0 & k_{D64} & 0 & k_{D66} & 0 & 0 & 0 & 0 & 0 & 0 \\ 0 & 0 & 0 & 0 & 0 & 0 & k_{D77} & 0 & k_{D79} & 0 & 0 & 0 \\ 0 & 0 & 0 & 0 & 0 & 0 & 0 & k_{D88} & 0 & k_{D8,10} & 0 & 0 \\ 0 & 0 & 0 & 0 & 0 & 0 & k_{D97} & 0 & k_{D99} & 0 & 0 & 0 \\ 0 & 0 & 0 & 0 & 0 & 0 & 0 & k_{D10,8} & 0 & k_{D10,10} & 0 & 0 \\ 0 & 0 & 0 & 0 & k_{D11,5} & 0 & 0 & 0 & 0 & 0 & k_{D11,11} & k_{D11,12} \\ 0 & 0 & 0 & 0 & k_{D12,5} & 0 & 0 & 0 & 0 & 0 & k_{D12,11} & k_{D12,12} \end{bmatrix}.$$

$$\tag{10.29a}$$

The 21 expressions for the coefficients are obtained by integrating analytically:

$$k'_{D44} = 16b^4,$$

$$k'_{D46} = k'_{D64} = k'_{D79} = k'_{D97} = k'_{D8,10} = k'_{D10,8} = 16a^2b^2\nu,$$

$$k'_{D55} = k'_{D5,11} = k'_{D11,5} = k'_{D5,12} = k'_{D12,5} = 8a^2b^2(1 - \nu),$$

$$k'_{D66} = 16a^4,$$

$$k'_{D77} = 48b^4,$$

$$k'_{D88} = \tfrac{16}{3}\left(b^4 + 2a^2b^2(1 - \nu)\right),$$

$$k'_{D99} = \tfrac{16}{3}\left(a^4 + 2a^2b^2(1 - \nu)\right),$$

$$k'_{D10,10} = 48a^4,$$

$$k'_{D11,11} = \tfrac{8}{5}\left(10b^4 + 9a^2b^2(1 - \nu)\right),$$

$$k'_{D11,12} = k'_{D12,11} = 8a^2b^2(1 + \nu),$$

$$k'_{D12,12} = \tfrac{8}{5}\left(10a^4 + 9a^2b^2(1 - \nu)\right).$$

(10.29b)

The final value of each generalized stiffness coefficient is obtained by multiplying each expression in equations 10.29b by the scalar constant, so that

$$k_{Dij} = \frac{Et^3}{12(1 - \nu^2)(ab)^3} k'_{Dij}. \tag{10.29c}$$

Although a Gauss quadrature rule can be applied to obtain numerical values for the generalized stiffness coefficients, there is little reason to do so. The generalized stiffness matrix, or any of its coefficients singly, may be readily, efficiently, and exactly determined from equations 10.29.

The element stiffness matrix referred to physical displacements as dof is calculated using

$$[\mathbf{K}] = [\mathbf{h}^{-T}][\mathbf{K}_D][\mathbf{h}^{-1}]. \tag{6.37e}$$

If the plate structure in flexure being analyzed is rectangular, or if it can be represented as several rectangular major sections, each section may be discretized into rectangular elements which are the same throughout the section. Thus the matrices [**h**], [**h**$^{-1}$], [**h**$^{-T}$], [**K**$_D$], and [**K**] will be the same for each element within a section and need be determined only once for each section.

The 5 ft × 3 ft × 0.75 in. steel plate example used earlier will be employed to illustrate the computation of a rectangular plate element stiffness matrix using equations 10.29 and 10.17b. The modulus of elasticity $E = 30,000$ ksi and Poisson's ratio $\nu = 0.25$. The element half-lengths $a = 30$ in. and $b = 18$ in. For this element, the matrices **h** and **K**$_D$ may be computed directly from equations 10.17b and 10.29 as

$$[\mathbf{h}] = \frac{1}{90} \begin{bmatrix} 90 & -90 & -90 & 90 & 90 & 90 & -90 & -90 & -90 & -90 & 90 & 90 \\ 0 & 0 & 5 & 0 & -5 & -10 & 0 & 5 & 10 & 15 & -5 & -15 \\ 0 & -3 & 0 & 6 & 3 & 0 & -9 & -6 & -3 & 0 & 9 & 3 \\ 90 & 90 & -90 & 90 & -90 & 90 & 90 & -90 & 90 & -90 & -90 & -90 \\ 0 & 0 & 5 & 0 & 5 & -10 & 0 & 5 & -10 & 15 & 5 & 15 \\ 0 & -3 & 0 & -6 & 3 & 0 & -9 & 6 & -3 & 0 & 9 & 3 \\ 90 & 90 & 90 & 90 & 90 & 90 & 90 & 90 & 90 & 90 & 90 & 90 \\ 0 & 0 & 5 & 0 & 5 & 10 & 0 & 5 & 10 & 15 & 5 & 15 \\ 0 & -3 & 0 & -6 & -3 & 0 & -9 & -6 & -3 & 0 & -9 & -3 \\ 90 & -90 & 90 & 90 & -90 & 90 & -90 & 90 & -90 & 90 & -90 & -90 \\ 0 & 0 & 5 & 0 & -5 & 10 & 0 & 5 & -10 & 15 & -5 & -15 \\ 0 & -3 & 0 & 6 & -3 & 0 & -9 & 6 & -3 & 0 & -9 & -3 \end{bmatrix},$$

$$[\mathbf{K}_D] = \frac{1}{324} \begin{bmatrix} 0 & 0 & 0 & 0 & 0 & 0 & 0 & 0 & 0 & 0 & 0 & 0 \\ 0 & 0 & 0 & 0 & 0 & 0 & 0 & 0 & 0 & 0 & 0 & 0 \\ 0 & 0 & 0 & 0 & 0 & 0 & 0 & 0 & 0 & 0 & 0 & 0 \\ 0 & 0 & 0 & 3{,}888 & 0 & 2{,}700 & 0 & 0 & 0 & 0 & 0 & 0 \\ 0 & 0 & 0 & 0 & 4{,}050 & 0 & 0 & 0 & 0 & 0 & 4{,}050 & 4{,}050 \\ 0 & 0 & 0 & 2{,}700 & 0 & 30{,}000 & 0 & 0 & 0 & 0 & 0 & 0 \\ 0 & 0 & 0 & 0 & 0 & 0 & 11{,}664 & 0 & 2{,}700 & 0 & 0 & 0 \\ 0 & 0 & 0 & 0 & 0 & 0 & 0 & 6{,}696 & 0 & 2{,}700 & 0 & 0 \\ 0 & 0 & 0 & 0 & 0 & 0 & 2{,}700 & 0 & 15{,}400 & 0 & 0 & 0 \\ 0 & 0 & 0 & 0 & 0 & 0 & 0 & 2{,}700 & 0 & 90{,}000 & 0 & 0 \\ 0 & 0 & 0 & 0 & 4{,}050 & 0 & 0 & 0 & 0 & 0 & 11{,}178 & 6{,}750 \\ 0 & 0 & 0 & 0 & 4{,}050 & 0 & 0 & 0 & 0 & 0 & 6{,}750 & 37{,}290 \end{bmatrix}.$$

The numerical matrix $\mathbf{h}^{-1}$ is readily calculated with the aid of a computer by inverting the matrix $\mathbf{h}$. The matrix $\mathbf{h}^{-T}$ is determined by transposing matrix $\mathbf{h}^{-1}$. These latter two matrices and the matrix $\mathbf{K}_D$ are then manipulated as indicated in equation 6.37e. The result is the element stiffness matrix $\mathbf{K}$ referred to nodal displacements. It is identical to the one shown as the matrix $\mathbf{K}$ obtained in the example done earlier using the Gauss 3×3 quadrature rule.

10.3.4. Nodal Actions

The internal forces and moments at nodes for the 4-node, 12-dof rectangular thin plate flexure element are illustrated in Figure 10.7 and in equation 10.12. When external transverse loads are applied to a plate element so that flexure occurs, their equivalent forces and moments must be computed, applied at the element's nodes to replace the applied transverse loads, and incorporated into the global load vector for the structure.

A concentrated transverse load on a plate may be dealt with in either of two ways. If the plate is discretized so that a node occurs at the location of the concentrated load, its global dof is identified and the load is entered into the corresponding position of the global load vector. If the concentrated load occurs on the plate element between nodes, then equivalent forces and moments at the four nodes must be computed.

Equation 6.22f applies. It may be written for the plate element as

$$\{F_k\} = \begin{Bmatrix} F_{z1k} \\ M_{x1k} \\ M_{y1k} \\ F_{z2k} \\ M_{x2k} \\ M_{y2k} \\ F_{z3k} \\ M_{x3k} \\ M_{y3k} \\ F_{z4k} \\ M_{x4k} \\ M_{y4k} \end{Bmatrix} = \{N(\zeta_k, \ \eta_k)P\} = \begin{Bmatrix} N_{11}(\zeta_k, \ \eta_k)P \\ N_{12}(\zeta_k, \ \eta_k)P \\ N_{13}(\zeta_k, \ \eta_k)P \\ N_{21}(\zeta_k, \ \eta_k)P \\ N_{22}(\zeta_k, \ \eta_k)P \\ N_{23}(\zeta_k, \ \eta_k)P \\ N_{31}(\zeta_k, \ \eta_k)P \\ N_{32}(\zeta_k, \ \eta_k)P \\ N_{33}(\zeta_k, \ \eta_k)P \\ N_{41}(\zeta_k, \ \eta_k)P \\ N_{42}(\zeta_k, \ \eta_k)P \\ N_{43}(\zeta_k, \ \eta_k)P \end{Bmatrix}. \tag{10.30}$$

The coordinates $(\zeta_k, \ \eta_k)$ are the coordinates of the point of application of the concentrated transverse load P. One needs only to evaluate the 12 shape functions $N_{ij}(\zeta, \eta)$ at the location of the load and substitute these values and the magnitude of the load P into equation 10.30. The result is a vector of 12 equivalent forces and moments at the nodes of the plate element. Their global dof are then identified and their values inserted into the global load vector at the corresponding gdof positions.

If the transverse load applied to the plate element is a distributed load $b_z(x, \ y)$ whose dimensions are force per unit of area, equation 6.22e applies. The equation may be written for the plate element as

$$\begin{aligned} \{F_b\} &= \int_A \{N(x, \ y)\} b_z(x, \ y) \ dx \ dy \\ &= \int_{-1}^{1} \int_{-1}^{1} \{N(\zeta, \ \eta)\} b_z(\zeta, \ \eta) |J(\zeta, \ \eta)| \ d\zeta \ d\eta. \end{aligned} \tag{10.31a}$$

The shape functions are expressed in terms of ζ and η so the integration is done with ζ and η as the variables of integration. The transverse distributed load b_z must be expressed as a function of ζ and η also. The transformation for the rectangular plate is a simple one, $x = a\zeta$ and $y = b\eta$, allowing the Jacobian of the transformation to be written as

$$|J(\zeta, \ \eta)| = \frac{\partial x}{\partial \zeta}\frac{\partial y}{\partial \eta} - \frac{\partial x}{\partial \eta}\frac{\partial y}{\partial \zeta} = ab. \tag{10.31b}$$

The vector of equivalent nodal forces and moments for the distributed transverse load may be written as

$$\{\mathbf{F}_b\} = \begin{Bmatrix} F_{z1b} \\ M_{x1b} \\ M_{y1b} \\ F_{z2b} \\ M_{x2b} \\ M_{y2b} \\ F_{z3b} \\ M_{x3b} \\ M_{y3b} \\ F_{z4b} \\ M_{x4b} \\ M_{y4b} \end{Bmatrix} = \begin{Bmatrix} ab \int_{-1}^{1} \int_{-1}^{1} N_{11}(\zeta, \eta) b_z(\zeta, \eta) \, d\zeta \, d\eta \\ ab \int_{-1}^{1} \int_{-1}^{1} N_{12}(\zeta, \eta) b_z(\zeta, \eta) \, d\zeta \, d\eta \\ ab \int_{-1}^{1} \int_{-1}^{1} N_{13}(\zeta, \eta) b_z(\zeta, \eta) \, d\zeta \, d\eta \\ ab \int_{-1}^{1} \int_{-1}^{1} N_{21}(\zeta, \eta) b_z(\zeta, \eta) \, d\zeta \, d\eta \\ ab \int_{-1}^{1} \int_{-1}^{1} N_{22}(\zeta, \eta) b_z(\zeta, \eta) \, d\zeta \, d\eta \\ ab \int_{-1}^{1} \int_{-1}^{1} N_{23}(\zeta, \eta) b_z(\zeta, \eta) \, d\zeta \, d\eta \\ ab \int_{-1}^{1} \int_{-1}^{1} N_{31}(\zeta, \eta) b_z(\zeta, \eta) \, d\zeta \, d\eta \\ ab \int_{-1}^{1} \int_{-1}^{1} N_{32}(\zeta, \eta) b_z(\zeta, \eta) \, d\zeta \, d\eta \\ ab \int_{-1}^{1} \int_{-1}^{1} N_{33}(\zeta, \eta) b_z(\zeta, \eta) \, d\zeta \, d\eta \\ ab \int_{-1}^{1} \int_{-1}^{1} N_{41}(\zeta, \eta) b_z(\zeta, \eta) \, d\zeta \, d\eta \\ ab \int_{-1}^{1} \int_{-1}^{1} N_{42}(\zeta, \eta) b_z(\zeta, \eta) \, d\zeta \, d\eta \\ ab \int_{-1}^{1} \int_{-1}^{1} N_{43}(\zeta, \eta) b_z(\zeta, \eta) \, d\zeta \, d\eta \end{Bmatrix}. \qquad (10.31c)$$

The individual integrals are usually evaluated numerically, except for uniform distributed loads. For a distributed load which is linear or quadratic in ζ or η, the integrands will possess ζ or η terms of fourth or fifth degree. For these, exact equivalent forces and moments may be obtained with the Gauss 3×3 rule. Higher degree distributed loads will require higher order Gauss quadrature rules.

Consider a uniformly distributed transverse load and its equivalent forces and moments. Figure 10.9 illustrates the applied distributed load b_z, force per unit of area, as a positive upward load. Its equivalent nodal forces and moments are shown also.

Since b_z is constant, the integrals of equation 10.31 are of the shape functions. They may be evaluated analytically in terms of the nodal coordinates (ζ_i, η_i):

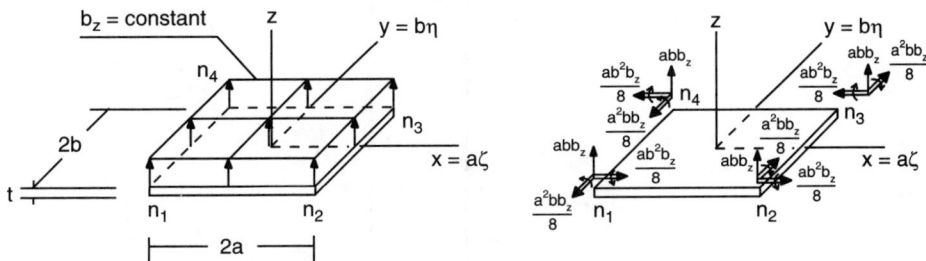

Figure 10.9. Equivalent nodal forces and moments.

$$F_{zib} = abb_z \int_{-1}^{1} \int_{-1}^{1} N_{i1}(\zeta, \eta) \, d\zeta \, d\eta$$

$$= \frac{abb_z}{8} \int_{-1}^{1} \int_{-1}^{1} (1 + \zeta_i\zeta)(1 + \eta_i\eta)(2 + \zeta_i\zeta + \eta_i\eta - \zeta^2 - \eta^2) \, d\zeta \, d\eta = abb_z,$$

$$M_{xib} = abb_z \int_{-1}^{1} \int_{-1}^{1} N_{i2}(\zeta, \eta)) \, d\zeta \, d\eta$$

$$= -\frac{ab^2 b_z \eta_i}{8} \int_{-1}^{1} \int_{-1}^{1} (1 + \zeta_i\zeta)(1 - \eta_i\eta)(1 + \eta_i\eta)^2 \, d\zeta d\eta = -\frac{ab^2 b_z \eta_i}{3},$$

$$M_{yib} = abb_z \int_{-1}^{1} \int_{-1}^{1} N_{i3}(\zeta, \eta) \, d\zeta \, d\eta$$

$$= \frac{a^2 bb_z \zeta_i}{8} \int_{-1}^{1} \int_{-1}^{1} (1 - \zeta_i\zeta)(1 + \eta_i\eta)(1 + \zeta_i\zeta)^2 \, d\zeta \, d\eta = \frac{a^2 bb_z \zeta_i}{3}.$$

Clearly if the distributed transverse load $b_z(\zeta, \eta)$ were not uniform, all 12 integrals would be evaluated numerically using the Gauss 3×3 or higher quadrature rule.

When the stiffness matrix for a thin plate flexure element is calculated using generalized dof, the numerical matrices $\mathbf{h}$, $\mathbf{h}^{-1}$, and $\mathbf{h}^{-T}$ must be developed for the element. With the matrix $\mathbf{h}^{-T}$ already available, it is then simpler to compute the equivalent forces and moments for transverse loads on the element using the generalized dof form also.

The relationship between the vectors $\mathbf{g}$ and $\mathbf{N}(\zeta, \eta)$ was developed earlier in Chapter 6 and cited again above as equation 10.16.

For a concentrated transverse load P applied between nodes, the vector of equivalent forces and moments $\mathbf{F}_k$ is obtained by substituting equation 10.16 into equation 10.30:

$$\{\mathbf{F}_k\} = [\mathbf{h}^{-T}]\{\mathbf{g}_k P\} = [\mathbf{h}^{-T}] \begin{Bmatrix} 1P \\ \zeta_k P \\ \eta_k P \\ \zeta_k^2 P \\ \zeta_k \eta_k P \\ \eta_k^2 P \\ \zeta_k^3 P \\ \zeta_k^2 \eta_k P \\ \zeta_k \eta_k^2 P \\ \eta_k^3 P \\ \zeta_k^3 \eta_k P \\ \zeta_k \eta_k^3 P \end{Bmatrix} = [\mathbf{h}^{-T}]\{\mathbf{F}_{Dk}\}. \tag{10.32}$$

Equation 10.32 with single-term generic functions is more easily evaluated than equation 10.30 with 12-term shape functions, though the numerical matrix $\mathbf{h}^{-T}$ for the element will be needed.

When the distributed transverse load $b_z(\zeta,\ \eta)$ varies over the plate element, equation 10.16 is substituted into equation 10.31a to produce

$$\{\mathbf{F}_b\} = [\mathbf{h}^{-T}] \int_{-1}^{1} \int_{-1}^{1} \{\mathbf{g}\} b_z(\zeta,\ \eta) |J(\zeta,\ \eta)|\ d\zeta\ d\eta$$

$$= [\mathbf{h}^{-T}](ab) \int_{-1}^{1} \int_{-1}^{1} \{\mathbf{g}\} b_z(\zeta,\ \eta)\ d\zeta\ d\eta \qquad (10.33a)$$

$$= [\mathbf{h}^{-T}]\{\mathbf{F}_{Db}\}.$$

The complete vector of equivalent nodal forces and moments $\mathbf{F}_{Db}$ referred to generalized dof may be written as

$$\{\mathbf{F}_{Db}\} = \begin{Bmatrix} ab \int_{-1}^{1} \int_{-1}^{1} 1 b_z(\zeta,\ \eta)\ d\zeta\ d\eta \\ ab \int_{-1}^{1} \int_{-1}^{1} \zeta b_z(\zeta,\ \eta)\ d\zeta\ d\eta \\ ab \int_{-1}^{1} \int_{-1}^{1} \eta b_z(\zeta,\ \eta)\ d\zeta\ d\eta \\ ab \int_{-1}^{1} \int_{-1}^{1} \zeta^2 b_z(\zeta,\ \eta)\ d\zeta\ d\eta \\ ab \int_{-1}^{1} \int_{-1}^{1} \zeta\eta b_z(\zeta,\ \eta)\ d\zeta\ d\eta \\ ab \int_{-1}^{1} \int_{-1}^{1} \eta^2 b_z(\zeta,\ \eta)\ d\zeta\ d\eta \\ ab \int_{-1}^{1} \int_{-1}^{1} \zeta^3 b_z(\zeta,\ \eta)\ d\zeta\ d\eta \\ ab \int_{-1}^{1} \int_{-1}^{1} \zeta^2\eta b_z(\zeta,\ \eta)\ d\zeta\ d\eta \\ ab \int_{-1}^{1} \int_{-1}^{1} \zeta\eta^2 b_z(\zeta,\ \eta)\ d\zeta\ d\eta \\ ab \int_{-1}^{1} \int_{-1}^{1} \eta^3 b_z(\zeta,\ \eta)\ d\zeta\ d\eta \\ ab \int_{-1}^{1} \int_{-1}^{1} \zeta^3\eta b_z(\zeta,\ \eta)\ d\zeta\ d\eta \\ ab \int_{-1}^{1} \int_{-1}^{1} \zeta\eta^3 b_z(\zeta,\ \eta)\ d\zeta\ d\eta \end{Bmatrix}. \qquad (10.33b)$$

The integrands of equation 10.33b are much simpler to integrate than those of equation 10.31c, either analytically or numerically. With the matrix $\mathbf{h}^{-T}$ for the element already evaluated to compute the element stiffness matrix in the generalized dof form, the computation of equivalent nodal forces and moments for a variable distributed transverse load is also simpler and more efficiently done in the generalized form using equations 10.33.

Some plate structures in flexure do not have transverse distributed loads applied across their entire surface. An efficient way to deal with this circumstance is to discretize the plate so that an element either has no distributed transverse load on it or has it across its entire surface.

10.3.5. Effect of an Elastic Foundation

A thin plate in flexure which derives some or all of its support from an elastic foundation is similar in concept to the thin beam finite element on an elastic foundation (discussed in Chapter 5). The plate, however, is a 2D structure, whereas the beam is 1D. The subgrade reaction $r_z(x, y)$ to the plate element is a transverse distributed force per unit of area which is proportional to the transverse displacement $w(x, y)$ of the plate element. Thus the subgrade reaction is dependent on and proportional to the element's displaced shape. The constant of proportionality between the element transverse displacements and the distributed subgrade reaction is the subgrade modulus k_s. Figure 10.10 shows a thin plate flexure element on an elastic foundation.

The general expressions for element stiffness equations were developed in Chapter 6 employing the principle of virtual work. The virtual strain energy δU of the finite element in its virtual displaced state was equated to the virtual work δW done by the element's external actions as it achieved its virtual displaced state. The form of the virtual strain energy expression δU is unaffected by the presence of the elastic foundation. However, the virtual work δW of the external actions must include the virtual work δW_{sg} done by the subgrade reaction. The virtual work of the subgrade reaction will be negative since the reaction must oppose the displacement. It may be expressed as

$$\delta W_{sq} = \int_A \delta w(x, y) r_z(x, y) \, dx \, dy$$
$$= -\int_A \delta w(x, y) k_s w(x, y) \, dx \, dy. \tag{10.34a}$$

The real and virtual transverse displacements are represented by

$$w(x, y) = \lfloor N(x, y) \rfloor \{q\}, \tag{10.34b}$$
$$\delta w(x, y) = \lfloor \delta q \rfloor \{N(x, y)\}. \tag{10.34c}$$

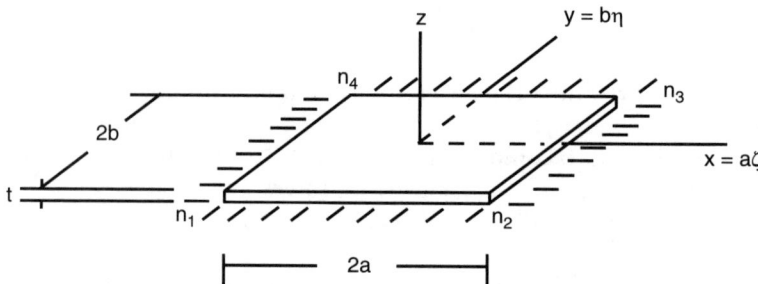

Figure 10.10. Plate element on an elastic foundation.

Substituting these expressions into equation 10.34a and noting that the vectors $\mathbf{q}$ and $\delta\mathbf{q}$ are constant and may be moved from under the integral as long as the order of matrix multiplication is not changed,

$$\begin{aligned}\delta W_{sq} &= -\lfloor\delta\mathbf{q}\rfloor\left[\int_A \{\mathbf{N}(\mathbf{x},\ \mathbf{y})\}k_s\lfloor\mathbf{N}(\mathbf{x},\ \mathbf{y})\rfloor\ dx\ dy\right]\{\mathbf{q}\} \\ &= -\lfloor\delta\mathbf{q}\rfloor[\mathbf{K}_r]\{\mathbf{q}\}.\end{aligned}$$

(10.34d)

Equations 6.22 were written earlier as an expression of the principle of virtual work, showing the general form of the element stiffness equations, but without the contribution to the virtual work done by the subgrade reaction of an elastic foundation. Rewriting equations 6.22 to include the contribution δW_{sg},

$$\lfloor\delta\mathbf{q}\rfloor[\mathbf{K}]\{\mathbf{q}\} = \lfloor\delta\mathbf{q}\rfloor\{\mathbf{F}\} + \lfloor\delta\mathbf{q}\rfloor\{\mathbf{F}_b\} + \lfloor\delta\mathbf{q}\rfloor\{\mathbf{F}_k\} - \lfloor\delta\mathbf{q}\rfloor[\mathbf{K}_r]\{\mathbf{q}\}.$$

(10.35)

Since equation 10.35 is a valid matrix equation, so also is

$$\begin{aligned}[\mathbf{K}]\{\mathbf{q}\} &= \{\mathbf{F}\} + \{\mathbf{F}_b\} + \{\mathbf{F}_k\} - [\mathbf{K}_r]\{\mathbf{q}\}, \\ [\mathbf{K} + \mathbf{K}_r]\{\mathbf{q}\} &= \{\mathbf{F}\} + \{\mathbf{F}_b\} + \{\mathbf{F}_k\}.\end{aligned}$$

(10.36)

Equation 10.36 gives the element stiffness equations for the thin plate in flexure supported by an elastic foundation. All the terms in the equation, except the subgrade stiffness matrix $\mathbf{K}_r$, are the same ones discussed earlier. Matrix $\mathbf{K}_r$ is the effect of the elastic foundation on the plate element stiffness equations. One must compute it and add it to the plate element stiffness matrix $\mathbf{K}$. Matrix $\mathbf{K}_r$ is a 12×12 matrix of integrals of products of shape functions, given as

$$\begin{aligned}[\mathbf{K}_r] &= \left[\int_A k_s\{\mathbf{N}(\mathbf{x},\ \mathbf{y})\}\lfloor\mathbf{N}(\mathbf{x},\ \mathbf{y})\rfloor\ dx\ dy\right] \\ &= \left[\int_{-1}^{1}\int_{-1}^{1} k_s\{\mathbf{N}(\zeta,\ \eta)\}\lfloor\mathbf{N}(\zeta,\ \eta)\rfloor|J(\zeta,\ \eta)\ d\zeta\ d\eta\right] \\ &= \left[ab\int_{-1}^{1}\int_{-1}^{1} k_s\{\mathbf{N}(\zeta,\ \eta)\}\lfloor\mathbf{N}(\zeta,\ \eta)\rfloor\ d\zeta\ d\eta\right].\end{aligned}$$

(10.37)

The usual application for a thin rectangular flexure plate on an elastic foundation is for a constant subgrade modulus k_s on individual plate elements, though the modulus may be different on different elements. For this simplest of circumstances, the 144 integrands in equation 10.37 are each products of the 12-term cubic polynomial shape functions in ζ and η. Some will contain up to 23 terms, and many will contain terms of sixth degree in ζ and η. Analytical integration would be tedious at best. The integrals may be evaluated numerically for numerical values of the plate half-lengths a and b. A Gauss 4×4 quadrature rule will be necessary for exact results.

The subgrade stiffness matrix $\mathbf{K}_r$ is more easily obtained from the generalized dof form. This development is similar to the physical dof form except single-term generic functions replace the 12-term shape functions, and the resulting generalized subgrade stiffness matrix $\mathbf{K}_{Dr}$ must be transformed to be referred to physical displacements as matrix $\mathbf{K}_r$.

Substituting equation 10.16 into the expression for the matrix $\mathbf{K}_r$ (equation 10.37) yields

$$
\begin{aligned}
[\mathbf{K}_r] &= [\mathbf{h}^{-T}]\left[ab\int_{-1}^{1}\int_{-1}^{1} k_s\{\mathbf{g}\}\lfloor \mathbf{g}\rfloor \, d\zeta \, d\eta\right][\mathbf{h}^{-1}] \\
&= [\mathbf{h}^{-T}][\mathbf{K}_{Dr}][\mathbf{h}^{-1}].
\end{aligned}
\tag{10.38}
$$

The single-term generic functions in equation 10.38 result in 144 single-term polynomial integrands, a few of which contain sixth-degree terms in ζ or η. They are easily integrated analytically. As discussed earlier, those terms of odd degree in ζ and η are zero when integrated between the limits $-1 < \zeta < +1$ and $-1 < \eta < +1$. The result of this integration is

$$
[\mathbf{K}_{Dr}] = \frac{4abk_s}{1575}
\begin{bmatrix}
1575 & 0 & 0 & 525 & 0 & 525 & 0 & 0 & 0 & 0 & 0 & 0 \\
0 & 525 & 0 & 0 & 0 & 0 & 315 & 0 & 175 & 0 & 0 & 0 \\
0 & 0 & 525 & 0 & 0 & 0 & 0 & 175 & 0 & 315 & 0 & 0 \\
525 & 0 & 0 & 315 & 0 & 175 & 0 & 0 & 0 & 0 & 0 & 0 \\
0 & 0 & 0 & 0 & 175 & 0 & 0 & 0 & 0 & 0 & 105 & 105 \\
525 & 0 & 0 & 175 & 0 & 315 & 0 & 0 & 0 & 0 & 0 & 0 \\
0 & 315 & 0 & 0 & 0 & 0 & 225 & 0 & 105 & 0 & 0 & 0 \\
0 & 0 & 175 & 0 & 0 & 0 & 0 & 105 & 0 & 105 & 0 & 0 \\
0 & 715 & 0 & 0 & 0 & 0 & 105 & 0 & 105 & 0 & 0 & 0 \\
0 & 0 & 315 & 0 & 0 & 0 & 0 & 105 & 0 & 225 & 0 & 0 \\
0 & 0 & 0 & 0 & 105 & 0 & 0 & 0 & 0 & 0 & 75 & 63 \\
0 & 0 & 0 & 0 & 105 & 0 & 0 & 0 & 0 & 0 & 63 & 75
\end{bmatrix}.
\tag{10.39}
$$

To obtain the subgrade stiffness matrix $\mathbf{K}_r$ referred to physical displacements, the matrices $\mathbf{h}^{-T}$ and $\mathbf{h}^{-1}$ are calculated from matrix $\mathbf{h}$ (equation 10.17b), and the matrix manipulations indicated in equation 10.38 are performed.

The $5\,\text{ft} \times 3\,\text{ft} \times 0.75\,\text{in.}$ plate element used in the previous examples will illustrate the subgrade stiffness matrix determined from the generalized dof form. For a subgrade modulus of $25\,\text{psi/in.}$, the plate's generalized stiffness matrix is written directly from equation 10.39 as

$$[K_{Dr}] = \frac{1}{175}\begin{bmatrix} 9450 & 0 & 0 & 3150 & 0 & 3150 & 0 & 0 & 0 & 0 & 0 & 0 \\ 0 & 3150 & 0 & 0 & 0 & 0 & 1890 & 0 & 1050 & 0 & 0 & 0 \\ 0 & 0 & 3150 & 0 & 0 & 0 & 0 & 1050 & 0 & 1890 & 0 & 0 \\ 3150 & 0 & 0 & 1890 & 0 & 1050 & 0 & 0 & 0 & 0 & 0 & 0 \\ 0 & 0 & 0 & 0 & 1050 & 0 & 0 & 0 & 0 & 0 & 630 & 630 \\ 3150 & 0 & 0 & 1050 & 0 & 1890 & 0 & 0 & 0 & 0 & 0 & 0 \\ 0 & 1890 & 0 & 0 & 0 & 0 & 1350 & 0 & 630 & 0 & 0 & 0 \\ 0 & 0 & 1050 & 0 & 0 & 0 & 0 & 630 & 0 & 630 & 0 & 0 \\ 0 & 1050 & 0 & 0 & 0 & 0 & 630 & 0 & 630 & 0 & 0 & 0 \\ 0 & 0 & 1890 & 0 & 0 & 0 & 0 & 630 & 0 & 1350 & 0 & 0 \\ 0 & 0 & 0 & 0 & 630 & 0 & 0 & 0 & 0 & 0 & 450 & 378 \\ 0 & 0 & 0 & 0 & 630 & 0 & 0 & 0 & 0 & 0 & 378 & 450 \end{bmatrix}.$$

The plate element's matrix **h** was written earlier when its stiffness matrix for physical displacements was calculated from its generalized stiffness matrix. The numerical matrices $\mathbf{h}^{-1}$ and $\mathbf{h}^{-T}$ are readily calculated from the numerical matrix **h** with the aid of a computer. The matrices $\mathbf{h}^{-1}$, $\mathbf{h}^{-T}$ and $\mathbf{K}_{Dr}$ are then manipulated as indicated in equation 10.38 to produce

$$[K_r] = \begin{bmatrix} 7.401 & 35.563 & -59.271 & 2.627 & 15.351 & 35.229 & 0.844 & -8.949 & 14.914 & 2.627 & -21.137 & -25.586 \\ 35.563 & 222.171 & -291.600 & 15.351 & 111.086 & 194.400 & 8.949 & -83.314 & 129.600 & 21.137 & -166.629 & -194.400 \\ -59.271 & -291.600 & 617.143 & -35.229 & -194.400 & -462.857 & -14.914 & 129.600 & -231.429 & -25.586 & 194.400 & 308.571 \\ 2.627 & 15.351 & -35.229 & 7.401 & 35.563 & 59.271 & 2.627 & -21.137 & 25.586 & 0.844 & -8.949 & -14.914 \\ 15.351 & 111.086 & -194.400 & 35.563 & 222.171 & 291.600 & 21.137 & -166.629 & 194.400 & 8.949 & -83.314 & -129.600 \\ 35.229 & 194.400 & -462.857 & 59.271 & 291.600 & 617.143 & 25.586 & -194.400 & 308.571 & 14.914 & -129.600 & -231.429 \\ 0.844 & 8.949 & -14.914 & 2.627 & 21.137 & 25.586 & 7.401 & -35.563 & 59.271 & 2.627 & -15.351 & -35.229 \\ -8.949 & -83.314 & 129.600 & -21.137 & -166.629 & -194.400 & -35.563 & 222.171 & -291.600 & -15.351 & 111.086 & 194.400 \\ 14.914 & 129.600 & -231.429 & 25.586 & 194.400 & 308.571 & 59.271 & -291.600 & 617.143 & 35.229 & -194.400 & -462.857 \\ 2.627 & 21.137 & -25.586 & 0.844 & 8.949 & 14.914 & 2.627 & -15.351 & 35.229 & 7.401 & -35.563 & -59.271 \\ -21.137 & -166.629 & 194.400 & -8.949 & -83.314 & -129.600 & -15.351 & 111.086 & -194.400 & -35.563 & 222.171 & 291.600 \\ -25.586 & -194.400 & 308.571 & -14.914 & -129.600 & -231.429 & -35.229 & 194.400 & -462.857 & -59.271 & 291.600 & 617.143 \end{bmatrix}.$$

The units for the plate element subgrade matrix $\mathbf{K}_r$ are the same as for the plate element stiffness matrix **K**.

An effective way to deal with the thin rectangular flexure plate element on an elastic foundation is to use generalized dof to determine its stiffness equations and then transform them to their physical displacement form. The matrices $\mathbf{K}_D$ and $\mathbf{K}_{Dr}$ referred to generalized dof are calculated using equations 10.29 and 10.39 and added to obtain the total generalized stiffness matrix $\mathbf{K}_{DE}$. Matrix $\mathbf{K}_{DE}$ is transformed to the total element stiffness matrix $\mathbf{K}_E$ referred to physical displacements using the matrices $\mathbf{h}^{-T}$ and $\mathbf{h}^{-1}$ and equation 10.38. The vectors $\mathbf{F}_{Dk}$ and $\mathbf{F}_{Db}$ referred to generalized dof are calculated using equations 10.32 and 10.33 and transformed to the vectors $\mathbf{F}_k$ and $\mathbf{F}_b$ referred to physical displacements. Once the element's stiffness matrix and equivalent force and moment vectors are obtained in the physical displacement form, the finite element method proceeds as described earlier.

The rectangular plate in flexure on an elastic foundation may be used to model many kinds of structural foundations and bearing plates. However, a word of caution is in order. The model assumes that the subgrade reaction $r_z(x, y)$ equals $-k_s \cdot w(x, y)$. For most foundations and bearing plates this assumption is reasonable as long as the entire transverse displacement distribution $w(x, y)$ of the plate penetrates into the elastic foundation, causing a compressive distributed subgrade reaction $r_z(x, y)$ across the entire area of the plate. If the loading on the plate and the relative stiffnesses of the plate and subgrade cause the plate to curve out of the elastic foundation in some areas of the plate, the model will develop a tensile contact subgrade reaction in those areas. In these circumstances, foundations and bearing plates which cannot develop this tension must be analyzed by other models.

Plates whose flexural stiffness coefficients are small compared to the subgrade stiffness coefficients and whose transverse loads are concentrated in a small region of the plate area are likely to bend out of the elastic foundation at the edges. If the plate element stiffness coefficients are larger than the subgrade stiffness coefficients, or if the transverse loading on the plate is not concentrated in a single small area, the plate should adequately model the structural foundation or bearing plate.

10.3.6. Restraints and Boundary Conditions

Plate structures must be subjected to the physical restraints needed to hold the structure in equilibrium. Restraint may take the form of an elastic foundation, which was discussed in the last section. It modifies the element stiffness equations by adding a full element subgrade stiffness matrix to the plate element stiffness matrix and manifests itself in the global stiffness equations when the total element stiffness matrices are assembled to form the global stiffness matrix.

Restraints may take the form of applied transverse loads. These were discussed earlier also and modify the element stiffness equations as vectors of equivalent forces and moments. They manifest themselves on the global stiffness equations when they are assembled into the global load vector.

Restraints also take the form of directly specified displacement values at points, along lines, or over areas of the plate structure. The usual displacement restraints are along edges of the plate and are called its boundary conditions. Plate edge boundary conditions are free, clamped, or simply supported.

A free edge is one in which no restraint whatever is imposed anywhere along the edge. If nothing is specified anywhere in the global stiffness equations, all the plate's edges will be modeled as free edges. For such a plate, sufficient restraint must be provided in other ways, such as with an elastic foundation. An elastic foundation is reactive in nature and will develop whatever restraint is necessary to satisfy equilibrium requirements.

A clamped edge has its transverse translation, slope tangent to its edge, and slope normal to its edge fully restrained—usually set equal to zero. In a finite element analysis of a plate structure, its displacements can be specified only at nodes. Thus a clamped edge must be modeled by specifying all three displacements at each node along the edge and allowing the edge distance between nodes to be free. This is accomplished by specifying the transverse translation and rotation displacement values at the appropriate gdof positions in the global displacement vector before solution of the global stiffness equations.

A simply supported edge has its transverse translation and slope tangent to the edge specified—usually zero. The slope normal to its edge is not specified, leading to the distributed moment about its edge being zero. In the finite element model of the plate structure, the transverse translation and slope tangent to the edge are specified at nodes along the simply supported edge, in the appropriate positions in the global displacement vector before solution of the global stiffness equations. The slope normal to the edge is not specified, and the distance between nodes is left free and unspecified.

The boundary conditions described are for the classical thin plate theory, adapted to finite element analyses. They work well for thin rectangular plate elements when the mesh is not too coarse, despite the necessity to impose the classical conditions only at nodes and the inability to impose them continuously along the edges.

Displacement restraints need not be zero. They may be specified as any appropriate value. Moreover, in finite element analyses, they may be specified at any appropriate gdof, not just those occurring on the edges of plates. The most common interior nodal displacement restraints are specified transverse translations, though rotations might be specified also if it suits the finite element model of the plate structure.

A plate may be restrained by elastically yielding supports at points which develop concentrated reactions normal to the plate. These reactions are proportional to the plate transverse translation displacement at the point of the support. They may be treated by ensuring that each support point is at a node and then adding the constants of proportionality between reactions and displacements to the corresponding diagonal entries in the plate global stiffness matrix and specifying that the corresponding entries in the global load vector be zero. This is identical to the mixed boundary condition for a thin beam discussed in Chapter 2.

A mixture of restraints and boundary conditions may be imposed in a finite element analysis of a plate structure. These may include a mixture of specified transverse translation displacements and slopes at nodes, applied transverse loads, elastic foundations for parts of the plate structure, and elastically yielding point supports at nodes.

10.3.7. Shape Functions and the Generalized dof Form

It is apparent that in using generalized dof for the development of element stiffness equations and transforming them to the physical displacement form, element shape functions are not required directly. Neither are they needed directly for the assembly and solution of the global stiffness equations or the subsequent identification of the element displacement vectors $\mathbf{q}$. Transverse displacements between nodes are computed by combining equations 6.33 and 10.13:

$$w(\zeta, \ \eta) = \lfloor \mathbf{g} \rfloor \{\mathbf{D}\} = \lfloor \mathbf{g} \rfloor [\mathbf{h}^{-1}]\{\mathbf{q}\}. \tag{10.40a}$$

The curvature vector $\phi(\zeta, \ \eta)$ is calculated by applying equations 10.1e and 10.20b to equation 10.40a:

$$\{\phi\} = [\bar{\mathbf{d}}]w(\zeta, \ \eta) = [\bar{\mathbf{d}}]\lfloor \mathbf{g} \rfloor [\mathbf{h}^{-1}]\{\mathbf{q}\}. \tag{10.40b}$$

Distributed moments are then calculated from equation 10.5 and in-plate stresses from equation 10.6.

Removing the early determination and direct use of shape functions in the finite element analysis of plate structures, and using instead generalized dof and single-term generic functions, is an important simplification and savings of computation time and effort.

10.4. OTHER FLAT PLATE FLEXURE FINITE ELEMENTS

The Kirchhoff 4-node, 12-dof thin plate rectangular flexure element is one of the earliest and perhaps the simplest used. It performs acceptably despite its noncon-forming nature, provided that the important assumptions underlying it are observed. That is, the plate structure must be capable of being modeled by rectangular elements, plate transverse deflections must be small compared to its thickness, and the plate length and width must be large compared to its thickness. Acceptable results with this element do require that a sufficient number of elements be used.

The detailed development of the Kirchhoff element earlier in this chapter was intended to provide the reader with an in-depth understanding of the finite element method applied to flat plates in flexure. It should facilitate understanding the many other plate flexure elements developed and available. Many of these give improved performance, and some attempt to overcome the assumptions restricting the Kirchhoff element. Three are described below.

10.4.1. Rectangular Thin Plate Element (4-Node, 16-dof)

A 4-node, 16-dof thin plate rectangular flexure element is similar to the Kirchhoff element except that it employs a fourth dof at each node. The added dof is the twist or $\partial^2 w/\partial x \partial y$, complementing the Kirchhoff element transverse translation w and slopes $\partial w/\partial y$ and $-\partial w/\partial x$.

This element is conforming and performs somewhat better than the Kirchhoff element, due mainly to its higher degree 16-term polynomial displacement function. Its conforming characteristics do not affect its accuracy significantly. It requires more computation than the Kirchhoff element to produce its results, however.

This element is easily formulated in terms of generalized dof in the manner that the Kirchhoff element was. Additional terms must be added to the matrices $\bar{\mathbf{B}}_D$, $\mathbf{K}_D$, $\mathbf{h}$, $\mathbf{F}_{Dk}$, and $\mathbf{F}_{Db}$ to incorporate the effects of the four additional dof. The element displacement function in terms of its generic function vector $\mathbf{g}$ and generalized edof vector $\mathbf{D}$ is given as

$$
\begin{aligned}
w(\zeta,\ \eta) = {}& D_1 + D_2\zeta + D_3\eta + D_4\zeta^2 + D_5\zeta\eta + D_6\eta^2 + D_7\zeta^3 + D_8\zeta^2\eta + D_9\zeta\eta^2 \\
& + D_{10}\eta^3 + D_{11}\zeta^3\eta + D_{12}\zeta^2\eta^2 + D_{13}\zeta\eta^3 + D_{14}\zeta^3\eta^2 + D_{15}\zeta^2\eta^3 + D_{16}\zeta^3\eta^3 \\
= {}& \lfloor \mathbf{g} \rfloor \{\mathbf{D}\}.
\end{aligned}
$$

$$(10.41a)$$

The matrix $\mathbf{h}$ for this element is found as it was for the Kirchhoff element (equations 10.17). It will reflect the four dof at each node and the 16 generic functions of its vector **g**. The matrix of generic functions and their derivatives with respect to x and y is

$$\begin{bmatrix} \mathbf{g} \\ \frac{\partial \mathbf{g}}{\partial y} \\ -\frac{\partial \mathbf{g}}{\partial x} \\ \frac{\partial^2 \mathbf{g}}{\partial x \, \partial y} \end{bmatrix} = \begin{bmatrix} 1 & \zeta & \eta & \zeta^2 & \zeta\eta & \eta^2 & \zeta^3 & \zeta^2\eta & \zeta\eta^2 & \eta^3 & \zeta^3\eta & \zeta^2\eta^2 & \zeta\eta^3 & \zeta^3\eta^2 & \zeta^2\eta^3 & \zeta^3\eta^3 \\ 0 & 0 & \frac{1}{b} & 0 & \frac{\zeta}{b} & \frac{2\eta}{b} & 0 & \frac{\zeta^2}{b} & \frac{2\zeta\eta}{b} & \frac{3\eta^2}{b} & \frac{\zeta^3}{b} & \frac{2\zeta^2\eta}{b} & \frac{3\zeta\eta^2}{b} & \frac{2\zeta^3\eta}{b} & \frac{3\zeta^2\eta^2}{b} & \frac{3\zeta^3\eta^2}{b} \\ 0 & -\frac{1}{a} & 0 & -\frac{2\zeta}{a} & -\frac{\eta}{a} & 0 & -\frac{3\zeta^2}{a} & -\frac{2\zeta\eta}{a} & -\frac{\eta^2}{a} & 0 & -\frac{3\zeta^2\eta}{a} & -\frac{2\zeta\eta^2}{a} & -\frac{\eta^3}{a} & -\frac{3\zeta^2\eta^2}{a} & -\frac{2\zeta\eta^3}{a} & -\frac{3\zeta^2\eta^3}{a} \\ 0 & 0 & 0 & 0 & \frac{1}{ab} & 0 & 0 & \frac{2\zeta}{ab} & \frac{2\eta}{ab} & 0 & \frac{3\zeta^2}{ab} & \frac{4\zeta\eta}{ab} & \frac{3\eta^2}{ab} & \frac{6\zeta^2\eta}{ab} & \frac{6\zeta\eta^2}{ab} & \frac{9\zeta^2\eta^2}{ab} \end{bmatrix}.$$

$$(10.41b)$$

Substituting the nodal coordinates (ζ_i, η_i) successively into equation 10.41b, the matrix **h** may be written as

$$[\mathbf{h}] = \begin{bmatrix} 1 & -1 & -1 & 1 & 1 & 1 & -1 & -1 & -1 & -1 & 1 & 1 & 1 & -1 & -1 & 1 \\ 0 & 0 & \frac{1}{b} & 0 & -\frac{1}{b} & -\frac{2}{b} & 0 & \frac{1}{b} & \frac{2}{b} & \frac{3}{b} & -\frac{1}{b} & -\frac{2}{b} & -\frac{3}{b} & \frac{2}{b} & \frac{3}{b} & -\frac{3}{b} \\ 0 & -\frac{1}{a} & 0 & \frac{2}{a} & \frac{1}{a} & 0 & -\frac{3}{a} & -\frac{2}{a} & -\frac{1}{a} & 0 & \frac{3}{a} & \frac{2}{a} & \frac{1}{a} & -\frac{3}{a} & -\frac{2}{a} & \frac{3}{a} \\ 0 & 0 & 0 & 0 & \frac{1}{ab} & 0 & 0 & -\frac{2}{ab} & -\frac{2}{ab} & 0 & \frac{3}{ab} & \frac{4}{ab} & \frac{3}{ab} & -\frac{6}{ab} & -\frac{6}{ab} & \frac{9}{ab} \\ 1 & 1 & -1 & 1 & -1 & 1 & 1 & -1 & 1 & -1 & -1 & 1 & -1 & 1 & -1 & -1 \\ 0 & 0 & \frac{1}{b} & 0 & \frac{1}{b} & -\frac{2}{b} & 0 & \frac{1}{b} & -\frac{2}{b} & \frac{3}{b} & \frac{1}{b} & -\frac{2}{b} & \frac{3}{b} & -\frac{2}{b} & \frac{3}{b} & \frac{3}{b} \\ 0 & -\frac{1}{a} & 0 & -\frac{2}{a} & \frac{1}{a} & 0 & -\frac{3}{a} & \frac{2}{a} & -\frac{1}{a} & 0 & \frac{3}{a} & -\frac{2}{a} & \frac{1}{a} & -\frac{3}{a} & \frac{2}{a} & \frac{3}{a} \\ 0 & 0 & 0 & 0 & \frac{1}{ab} & 0 & 0 & \frac{2}{ab} & -\frac{2}{ab} & 0 & \frac{3}{ab} & -\frac{4}{ab} & \frac{3}{ab} & -\frac{6}{ab} & \frac{6}{ab} & \frac{9}{ab} \\ 1 & 1 & 1 & 1 & 1 & 1 & 1 & 1 & 1 & 1 & 1 & 1 & 1 & 1 & 1 & 1 \\ 0 & 0 & \frac{1}{b} & 0 & \frac{1}{b} & \frac{2}{b} & 0 & \frac{1}{b} & \frac{2}{b} & \frac{3}{b} & \frac{1}{b} & \frac{2}{b} & \frac{3}{b} & \frac{2}{b} & \frac{3}{b} & \frac{3}{b} \\ 0 & -\frac{1}{a} & 0 & -\frac{2}{a} & -\frac{1}{a} & 0 & -\frac{3}{a} & -\frac{2}{a} & -\frac{1}{a} & 0 & -\frac{3}{a} & -\frac{2}{a} & -\frac{1}{a} & -\frac{3}{a} & -\frac{2}{a} & -\frac{3}{a} \\ 0 & 0 & 0 & 0 & \frac{1}{ab} & 0 & 0 & \frac{2}{ab} & \frac{2}{ab} & 0 & \frac{3}{ab} & \frac{4}{ab} & \frac{3}{ab} & \frac{6}{ab} & \frac{6}{ab} & \frac{9}{ab} \\ 1 & -1 & 1 & 1 & -1 & 1 & -1 & 1 & -1 & 1 & -1 & 1 & -1 & -1 & 1 & -1 \\ 0 & 0 & \frac{1}{b} & 0 & -\frac{1}{b} & \frac{2}{b} & 0 & \frac{1}{b} & -\frac{2}{b} & \frac{3}{b} & -\frac{1}{b} & \frac{2}{b} & -\frac{3}{b} & -\frac{2}{b} & \frac{3}{b} & -\frac{3}{b} \\ 0 & -\frac{1}{a} & 0 & \frac{2}{a} & -\frac{1}{a} & 0 & -\frac{3}{a} & 2 & -\frac{1}{a} & 0 & -\frac{3}{a} & 2 & -\frac{1}{a} & -\frac{3}{a} & 2 & -\frac{3}{a} \\ 0 & 0 & 0 & 0 & \frac{1}{ab} & 0 & 0 & -\frac{2}{ab} & \frac{2}{ab} & 0 & \frac{3}{ab} & -\frac{4}{ab} & \frac{3}{ab} & \frac{6}{ab} & -\frac{6}{ab} & \frac{9}{ab} \end{bmatrix}.$$

$$(10.41c)$$

The strain displacement matrix $\bar{\mathbf{B}}_D$ for generalized stress and dof is as expressed in equation 10.28 for the Kirchhoff element, except that the 16-term vector **g** must be used. The vector $\bar{\mathbf{d}}$ of generalized strain differential operators is unchanged, since the generalized strain in this element is the same as for the Kirchhoff element. Thus the matrix $\bar{\mathbf{B}}_D$ for this element is written

$$[\bar{\mathbf{B}}_D] = \frac{1}{(ab)^2} \begin{bmatrix} 0 & 0 & 0 & 2b^2 & 0 & 0 & 6b^2\zeta & 2b^2\eta & 0 & 0 & 6b^2\zeta\eta & 2b^2\eta^2 & 0 & 6b^2\zeta\eta^2 & 2b^2\eta^3 & 6b^2\zeta\eta^3 \\ 0 & 0 & 0 & 0 & 0 & 2a^2 & 0 & 0 & 2a^2\zeta & 6a^2\eta & 0 & 2a^2\zeta^2 & 6a^2\zeta\eta & 2a^2\zeta^3 & 6a^2\zeta^2\eta & 6a^2\zeta^3\eta \\ 0 & 0 & 0 & 0 & 2ab & 0 & 0 & 4ab\zeta & 4ab\eta & 0 & 6ab\zeta^2 & 8ab\zeta\eta & 6ab\eta^2 & 12ab\zeta^2\eta & 12ab\zeta\eta^2 & 18ab\zeta^2\eta^2 \end{bmatrix}$$

$$(10.41d)$$

The stiffness matrix $\mathbf{K}_D$ for generalized dof for this element may be obtained explicitly. The 16×16 integrand matrix $\mathbf{I}_D$ of simple functions of ζ and η, indicated in equation 10.27, is formed and integrated between the limits $-1 < \zeta < +1$ and

$-1 < \eta < +1$. Many of the integrand terms will be zero, and many others will contain an odd degree in ζ or η. The result is a generalized element stiffness matrix with 213 zero entries and only 43 nonzero entries:

$$[\mathbf{K}_D] = \begin{bmatrix}
0 & 0 & 0 & 0 & 0 & 0 & 0 & 0 & 0 & 0 & 0 & 0 & 0 & 0 & 0 & 0 \\
0 & 0 & 0 & 0 & 0 & 0 & 0 & 0 & 0 & 0 & 0 & 0 & 0 & 0 & 0 & 0 \\
0 & 0 & 0 & 0 & 0 & 0 & 0 & 0 & 0 & 0 & 0 & 0 & 0 & 0 & 0 & 0 \\
0 & 0 & 0 & k_{D44} & 0 & k_{D46} & 0 & 0 & 0 & 0 & 0 & k_{D4,12} & 0 & 0 & 0 & 0 \\
0 & 0 & 0 & 0 & k_{D55} & 0 & 0 & 0 & 0 & 0 & k_{D5,11} & 0 & k_{D5,13} & 0 & 0 & k_{D5,16} \\
0 & 0 & 0 & k_{D64} & 0 & k_{D66} & 0 & 0 & 0 & 0 & 0 & k_{D6,12} & 0 & 0 & 0 & 0 \\
0 & 0 & 0 & 0 & 0 & 0 & k_{D77} & 0 & k_{D79} & 0 & 0 & 0 & 0 & k_{D7,14} & 0 & 0 \\
0 & 0 & 0 & 0 & 0 & 0 & 0 & k_{D88} & 0 & k_{D8,10} & 0 & 0 & 0 & 0 & k_{D8,15} & 0 \\
0 & 0 & 0 & 0 & 0 & 0 & k_{D97} & 0 & k_{D99} & 0 & 0 & 0 & 0 & k_{D9,14} & 0 & 0 \\
0 & 0 & 0 & 0 & 0 & 0 & 0 & k_{D10,8} & 0 & k_{D10,10} & 0 & 0 & 0 & 0 & k_{D10,15} & 0 \\
0 & 0 & 0 & 0 & k_{D11,5} & 0 & 0 & 0 & 0 & 0 & k_{D11,11} & 0 & k_{D11,13} & 0 & 0 & k_{D11,16} \\
0 & 0 & 0 & k_{D12,4} & 0 & k_{D12,6} & 0 & 0 & 0 & 0 & 0 & k_{D12,12} & 0 & 0 & 0 & 0 \\
0 & 0 & 0 & 0 & k_{D13,5} & 0 & 0 & 0 & 0 & 0 & k_{D13,11} & 0 & k_{D13,13} & 0 & 0 & k_{D13,16} \\
0 & 0 & 0 & 0 & 0 & 0 & k_{D14,7} & 0 & k_{D14,9} & 0 & 0 & 0 & 0 & k_{D14,14} & 0 & 0 \\
0 & 0 & 0 & 0 & 0 & 0 & 0 & k_{D15,8} & 0 & k_{D15,10} & 0 & 0 & 0 & 0 & k_{D15,15} & 0 \\
0 & 0 & 0 & 0 & k_{D16,5} & 0 & 0 & 0 & 0 & 0 & k_{D16,11} & 0 & k_{D16,13} & 0 & 0 & k_{D16,16}
\end{bmatrix}. \quad (10.41e)$$

The first 11 rows and columns for the generalized stiffness coefficients for both this element and the Kirchhoff element are identical. The expressions for these stiffness coefficients for both elements are given in equations 10.29b.

The expressions for the remaining five rows and columns for this element (equation 10.41e) are

$$k'_{D4,12} = k'_{D12,4} = \tfrac{16}{3}(b^4 + a^2 b^2 v),$$
$$k'_{D5,13} = k'_{D13,5} = k'_{D5,16} = k'_{D16,5} = 8a^2 b^2 (1 - v),$$
$$k'_{D6,12} = k'_{D12,6} = \tfrac{16}{3}(a^4 + a^2 b^2 v),$$
$$k'_{D7,14} = k'_{D14,7} = \tfrac{16}{5}(5b^4 + 3a^2 b^2 v),$$
$$k'_{D8,15} = k'_{D15,8} = \tfrac{16}{15}(3b^4 + 5a^2 b^2 (2 - v)),$$
$$k'_{D9,14} = k'_{D14,9} = \tfrac{16}{15}(3a^4 + 5a^2 b^2 (2 - v)),$$
$$k'_{D10,15} = k'_{D15,10} = \tfrac{16}{5}(5a^4 + 3a^2 b^2 v)),$$
$$k'_{D11,13} = k'_{D13,11} = 8a^2 b^2 (1 + v), \qquad (10.41f)$$
$$k'_{D11,16} = k'_{D16,11} = \tfrac{24}{5}(2b^4 + a^2 b^2 (3 - v)),$$
$$k'_{D12,12} = \tfrac{16}{45}(9a^4 + 9b^4 + 10a^2 b^2 (4 - 3v)),$$
$$k'_{D13,13} = \tfrac{8}{5}(10a^4 + 9a^2 b^2 (1 - v)),$$
$$k'_{D13,16} = k'_{D16,13} = \tfrac{24}{5}(2a^4 + a^2 b^2 (3 - v)),$$
$$k'_{D14,14} = \tfrac{16}{35}(5a^4 + 21b^4 + 14a^2 b^2 (3 - 2v)),$$
$$k'_{D15,15} = \tfrac{16}{35}(21a^4 + 5b^4 + 14a^2 b^2 (3 - 2v)),$$
$$k'_{D16,16} = \tfrac{24}{175}(50a^4 + 50b^4 + 21a^2 b^2 (9 - 5v)).$$

As with equations 10.29b, each stiffness coefficient of equations 10.41f must be multiplied by the scalar shown in equation 10.29c.

The $60 \times 36 \times 0.75$ in. rectangular plate used earlier as an example for the Kirchhoff element stiffness matrix will be used again as an example for this 4-node, 16-dof conforming rectangular element. Its modulus of elasticity $E = 30,000$ ksi and its Poisson ratio $v = 0.25$. Equation 10.41c is used to compute its matrix $\mathbf{h}$, which is then inverted numerically to matrix $\mathbf{h}^{-1}$ and transposed to matrix $\mathbf{h}^{-T}$. Equations 10.41e, 10.41f, and 10.29c are used to establish its generalized stiffness matrix $\mathbf{K}_D$. Equation 6.36e is used to transform matrix $\mathbf{K}_D$ to the element stiffness matrix $\mathbf{K}$ in the physical displacement form. The resulting matrix $\mathbf{K}$ is given as

$$[\mathbf{K}] = \begin{bmatrix}
8.784 & 130.439 & -96.510 & 1,188.179 & -0.104 & 25.811 & -0.329 & -402.446 & -1.021 & 38.186 & -33.421 & -633.071 & -7.659 & 118.064 & 62.760 & 957.554 \\
130.439 & 3,029.486 & -1,469.429 & 26,834.571 & 25.811 & 720.514 & 402.446 & -12,015.429 & -38.186 & 556.971 & -633.071 & -7,859.143 & -118.064 & 1,318.029 & 957.554 & 10,553.357 \\
-96.510 & -1,469.429 & 2,198.095 & -19,855.714 & 0.329 & -402.446 & -95.000 & 6,502.500 & 33.421 & -633.071 & 770.000 & 9,990.000 & 62.760 & -957.554 & -848.095 & -9,505.714 \\
1,188.179 & 26,834.571 & -19,855.714 & 376,868.571 & 402.446 & 12,015.429 & 6,502.500 & -197,794.286 & -633.071 & 7,859.143 & -9,990.000 & -103,525.714 & -957.554 & 10,553.357 & 9,505.714 & 81,977.143 \\
-0.104 & 25.811 & 0.329 & 402.446 & 8.784 & 130.439 & 96.510 & -1,188.179 & -7.659 & 118.064 & -62.760 & -957.554 & -1.021 & 38.186 & 33.421 & 633.071 \\
25.811 & 720.514 & -402.446 & 12,015.429 & 130.439 & 3,029.486 & 1,469.429 & -26,834.571 & -118.064 & 1,318.029 & -957.554 & -10,553.357 & -38.186 & 556.971 & 633.071 & 7,859.143 \\
-0.329 & 402.466 & -95.000 & 6502.500 & 96.510 & 1,469.429 & 2,198.095 & -19,855.714 & -62.760 & 957.554 & -848.095 & -9,505.714 & -33.421 & 633.071 & 770.000 & 9,990.000 \\
-402.446 & -12,015.429 & 6,502.500 & -197,794.286 & -1,188.179 & -26,834.571 & -19,855.714 & 376,868.571 & 957.554 & -10,553.357 & 9,505.714 & 81,977.143 & 633.071 & -7,859.143 & -9,990.000 & -103,525.714 \\
-1.021 & -38.186 & 33.421 & -633.071 & -7.659 & -118.064 & -62.760 & 957.554 & 8.784 & -130.439 & 96.510 & 1,188.179 & -0.104 & -25.811 & 0.329 & -402.446 \\
38.186 & 556.971 & -633.071 & 7,859.143 & 118.064 & 1,318.029 & 957.554 & -10,553.357 & -130.439 & 3,029.486 & -1,469.429 & -26,834.571 & -25.811 & 720.514 & 402.446 & 12,015.429 \\
-33.421 & -633.071 & 770.000 & -9,990.000 & -62.760 & -957.554 & -848.095 & 9,505.714 & 96.510 & -1,469.429 & 2,198.095 & 19,855.714 & -0.329 & -402.446 & -95.000 & -6,502.500 \\
-633.071 & -7,859.143 & 9,990.000 & -103,525.714 & -957.554 & -10,553.357 & -9,505.714 & 81,977.143 & 1,188.179 & -26,834.571 & 19,855.714 & 376,868.571 & 402.446 & -12,015.429 & -6,502.500 & -197,794.286 \\
-7.659 & 118.064 & 62.760 & -957.554 & -1.021 & -38.186 & -33.421 & 633.071 & -0.104 & -25.811 & -0.329 & 402.446 & 8.784 & -130.439 & -96.510 & -1,188.179 \\
118.064 & 1318.029 & -957.554 & 10,553.357 & 38.186 & 556.971 & 633.071 & -7,859.143 & -25.811 & 720.514 & -402.446 & -12,015.429 & -130.439 & 3,029.486 & 1,469.429 & 26,834.571 \\
62.760 & 957.554 & -848.095 & 9,505.714 & 33.421 & 633.071 & 770.000 & -9,990.000 & 0.329 & 402.446 & -95.000 & -6,502.500 & -96.510 & 1,469.429 & 2,198.095 & 19,855.714 \\
957.554 & 10,553.357 & -9,505.714 & 81,977.143 & 633.071 & 7,859.143 & 9,990.000 & -103,525.714 & -402.446 & 12,015.429 & -6,502.500 & -197,794.286 & -1,188.179 & 26,834.571 & 19,855.714 & 376,868.571
\end{bmatrix}$$

Note that these results reflect the dof order which is comparable to the Kirchhoff element but with the twist added as a fourth dof. Moreover, the development is similar to that of the Kirchhoff element and is in terms of centroidal natural coordinates ζ and η with the element dimensions a and b as the half-length and half-width. The dof order and units for the stiffness matrix as a 4×4 nodal submatrix are

$$\{\mathbf{q}_i\} = \begin{Bmatrix} w_i \\ \dfrac{\partial w}{\partial y}\big|_i \\ -\dfrac{\partial w}{\partial x}\big|_i \\ \dfrac{\partial^2 w}{\partial x\,\partial y}\big|_i \end{Bmatrix}, \quad [\mathbf{K}_{ij}] = \begin{bmatrix} \text{kips/in.} & \text{kips} & \text{kips} & \text{kip-in.} \\ \text{kips} & \text{kip-in.} & \text{kip-in.} & \text{kip-in.}^2 \\ \text{kips} & \text{kip-in.} & \text{kip-in.} & \text{kip-in.}^2 \\ \text{kip-in.} & \text{kip-in.}^2 & \text{kip-in.}^2 & \text{kip-in.}^3 \end{bmatrix}.$$

The sequence of nodes $i = 1, 2, 3, 4$ is read counterclockwise around the element.

When a distributed transverse load $b_z(\zeta, \eta)$ varies over the plate element, equations 10.33 for the 4-node, 12-dof thin plate rectangular element may be extended and applied. Equations 10.33a will be the same, and equation 10.33b is extended to use its first 12 entries for this 4-node, 16-dof rectangular element. The remaining 4 entries are formed by using as integrands the last four generic polynomial terms of the element's vector $\mathbf{g}$ as products with the element's load $b_z(\zeta, \eta)$.

10.4.2. Triangular Thin Plate Element (3-Node, 9-dof)

Triangular finite elements possess an inherent advantage over rectangular elements in discretizing irregular plate structures. Triangular elements can model almost any irregular shaped plate structure and hence are a useful addition to the family of plate flexure elements. Though the complexity of their mathematics is comparable to the rectangular flexure element, the results they produce are usually less accurate. The accuracy of structural finite elements is better for higher degree displacement functions. Triangular elements usually have fewer nodes than comparable rectangular elements, fewer dof, and thus lower degree displacement functions.

One of the simplest triangular thin plate flexure element employs three nodes and nine dof. It is comparable to the 4-node, 12-dof Kirchhoff element and is referred to as the Kirchhoff triangular plate element. It exhibits the same nonconforming characteristics. The only differences between the Kirchhoff triangular and rectangular elements are their shape and number of dof. For analytical integration over the element, however, the triangular shape will require transformation of integrands to area coordinates and the integration formulas discussed in Chapter 8. Study of this element is instructive for the treatment of triangular plate flexure elements.

Figure 10.11 illustrates the Kirchhoff triangular flexure element and its dof. It is convenient to use the plate global x–y axes to describe it. The node numbering is the same as for the plane stress triangular element, and the dof numbering scheme is the same as for the Kirchhoff rectangular element.

This element is easily formulated in terms of generalized dof. Its displacement function, expressed in terms of a generic function vector $\mathbf{g}$ and generalized edof vector $\mathbf{D}$, is

$$w(x,\ y) = D_1 + D_2 x + D_3 y + D_4 x^2 + D_5 xy + D_6 y^2 + D_7 x^3 + D_8(x^2 y + xy^2) + D_9 y^3$$
$$= \lfloor \mathbf{g} \rfloor \{D\}.$$

$$(10.42a)$$

The function $w(x,\ y)$ has 9 terms since there are 9 dof on the element. However, a complete cubic polynomial in x and y includes 10 terms. Thus the element does not represent the displacement variation with a cubic polynomial in its fullest sense and loses some accuracy and flexibility as a result. A number of combinations of polynomial terms are possible. The one shown is geometrically isotropic and works acceptably in most cases. However, if the element is an isosceles right triangle

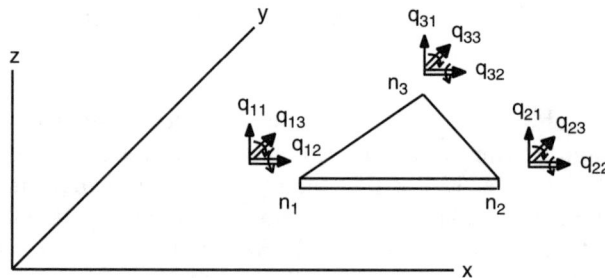

Figure 10.11. Triangular plate element.

whose equal legs are parallel to the x and y axes and whose hypotenuse slopes negatively with respect to the x and y axes, its matrix $\mathbf{h}$ will be singular. Accordingly, this circumstance, and conditions close to it, must be avoided when discretizing the plate structure.

The matrix $\mathbf{h}$ for this element is written as was done for the rectangular elements earlier. It will reflect three dof at each node and the nine generic functions of its vector $\mathbf{g}$. The matrix of generic functions and their derivatives with respect to x and y is

$$
\begin{bmatrix} g \\ \frac{\partial g}{\partial y} \\ -\frac{\partial g}{\partial x} \end{bmatrix} = \begin{bmatrix} 1 & x & y & x^2 & xy & y^2 & x^3 & x^2y + xy^2 & y^3 \\ 0 & 0 & 1 & 0 & x & 2y & 0 & x^2 + 2xy & 3y^2 \\ 0 & -1 & 0 & -2x & -y & 0 & -3x^2 & -(2xy + y^2) & 0 \end{bmatrix}. \quad (10.42b)
$$

Substituting the nodal coordinates (x_i, y_i) successively into equation 10.42b, the matrix $\mathbf{h}$ may be written as

$$
[\mathbf{h}] = \begin{bmatrix}
1 & x_1 & y_1 & x_1^2 & x_1y_1 & y_1^2 & x_1^3 & x_1^2y_1 + x_1y_1^2 & y_1^3 \\
0 & 0 & 1 & 0 & x_1 & 2y_1 & 0 & x_1^2 + 2x_1y_1 & 3y_1^2 \\
0 & -1 & 0 & -2x_1 & -y_1 & 0 & -3x_1^2 & -(2x_1y_1 + y_1^2) & 0 \\
1 & x_2 & y_2 & x_2^2 & x_2y_2 & y_2^2 & x_2^3 & x_2^2y_2 + x_2y_2^2 & y_2^3 \\
0 & 0 & 1 & 0 & x_2 & 2y_2 & 0 & x_2^2 + 2x_2y_2 & 3y_2^2 \\
0 & -1 & 0 & -2x_2 & -y_2 & 0 & -3x_2^2 & -(2x_2y_2 + y_2^2) & 0 \\
1 & x_3 & y_3 & x_3^2 & x_3y_3 & y_3^2 & x_3^3 & x_3^2y_3 + x_3y_3^2 & y_3^3 \\
0 & 0 & 1 & 0 & x_3 & 2y_3 & 0 & x_3^2 + 2x_3y_3 & 3y_3^2 \\
0 & -1 & 0 & -2x_3 & -y_3 & 0 & -3x_3^2 & -(2x_3y_3 + y_3^2) & 0
\end{bmatrix}. \quad (10.42c)
$$

The generalized strain–displacement matrix $\bar{\mathbf{B}}_D$ is the same as for the Kirchhoff rectangular element (equation 10.28), except that the nine-term vector $\mathbf{g}$ must be used. The vector $\bar{\mathbf{d}}$ of generalized strain differential operators is unchanged since the strains for this element are the same as for the Kirchhoff rectangular element. Thus the matrix $\bar{\mathbf{B}}_D$ for this element is written as

$$
[\bar{\mathbf{B}}_D] = \begin{bmatrix} 0 & 0 & 0 & 2 & 0 & 0 & 6x & 2y & 0 \\ 0 & 0 & 0 & 0 & 0 & 2 & 0 & 2x & 6y \\ 0 & 0 & 0 & 0 & 2 & 0 & 0 & 4(x+y) & 0 \end{bmatrix}. \quad (10.42d)
$$

The stiffness matrix $\mathbf{K}_D$ for generalized dof is obtained from

$$
[\mathbf{K}_D] = \int_A [\mathbf{I}_D] \, dx \, dy = \int_A [\bar{\mathbf{B}}_D^T][\bar{\mathbf{E}}][\bar{\mathbf{B}}_D] \, dx \, dy. \quad (10.43)
$$

The generalized elastic matrix $\bar{\mathbf{E}}$ is the same as for the Kirchhoff rectangular element. The integrand matrix $\mathbf{I}_D$ of simple functions of the global x and y coordinates is formed from the matrices $\bar{\mathbf{B}}_D$ and $\bar{\mathbf{E}}$, as indicated in equation 10.43. There will be 53 zero entries and 28 will be nonzero.

Analytical integration over the triangular area of the element is easily done as described in Chapter 8. The integrand functions of global x and y coordinates are transformed to functions of triangular area coordinates (ζ_1, ζ_2, ζ_3) using the equations

$$x = \zeta_1 x_1 + \zeta_2 x_2 + \zeta_3 x_3, \qquad (8.17a)$$
$$y = \zeta_1 y_1 + \zeta_2 y_2 + \zeta_3 y_3. \qquad (8.17b)$$

The area of the triangle is calculated using the equation

$$A = \frac{1}{2} \begin{vmatrix} 1 & 1 & 1 \\ x_1 & x_2 & x_3 \\ y_1 & y_2 & y_3 \end{vmatrix}. \qquad (8.4)$$

The integration is then accomplished with the equation

$$\int_A B\zeta_1^a \zeta_2^b \zeta_3^c \, dA = 2A \frac{B a! b! c!}{(a+b+c+2)!} \qquad (8.21a)$$

Applying these equations to integrate the polynomial terms occurring in the matrix $\mathbf{I}_D$ over the area of the triangular element, it may be shown that

$$\int_A (1) \, dA = A,$$

$$\int_A (x) \, dA = \frac{A x_s}{3}, \quad \text{where } x_s = x_1 + x_2 + x_3,$$

$$\int_A (y) \, dA = \frac{A y_s}{3}, \quad \text{where } y_s = y_1 + y_2 + y_3,$$

$$\int_A (x^2) \, dA = \frac{A}{12}(x_{ss} + x_s^2), \quad \text{where } x_{ss} = x_1^2 + x_2^2 + x_3^2,$$

$$\int_A (y^2) \, dA = \frac{A}{12}(y_{ss} + y_s^2), \quad \text{where } y_{ss} = y_1^2 + y_2^2 + y_3^2,$$

$$\int_A (xy) \, dA = \frac{A}{12}(p_{xy} + x_s y_s), \quad \text{where } p_{xy} = x_1 y_1 + x_2 y_2 + x_3 y_3.$$

$$(10.44)$$

The generalized element stiffness matrix $\mathbf{K}_D$ is obtained by applying equations 10.44 to matrix $\mathbf{I}_D$:

$$[\mathbf{K}_d] = \begin{bmatrix} 0 & 0 & 0 & 0 & 0 & 0 & 0 & 0 & 0 \\ 0 & 0 & 0 & 0 & 0 & 0 & 0 & 0 & 0 \\ 0 & 0 & 0 & 0 & 0 & 0 & 0 & 0 & 0 \\ 0 & 0 & 0 & k_{D44} & 0 & k_{D46} & k_{D47} & k_{D48} & k_{D49} \\ 0 & 0 & 0 & 0 & k_{D55} & 0 & 0 & k_{D58} & 0 \\ 0 & 0 & 0 & k_{D64} & 0 & k_{D66} & k_{D67} & k_{D68} & k_{D69} \\ 0 & 0 & 0 & k_{D74} & 0 & k_{D76} & k_{D77} & k_{D78} & k_{D79} \\ 0 & 0 & 0 & k_{D84} & k_{D85} & k_{D86} & k_{D87} & k_{D88} & k_{D89} \\ 0 & 0 & 0 & k_{D94} & 0 & k_{D96} & k_{D97} & k_{D98} & k_{D99} \end{bmatrix}. \tag{10.45a}$$

Expressions for these stiffness coefficients k_{Dij} are derived as

$$
\begin{aligned}
k'_{D44} &= k'_{D66} = 4A, \\
k'_{D46} &= k'_{D64} = 4vA, \\
k'_{D47} &= k'_{D74} = 4Ax_s, \\
k'_{D48} &= k'_{D84} = \frac{4A}{3}(vx_s + y_s), \\
k'_{D49} &= k'_{D94} = 4vAy_s, \\
k'_{D55} &= 2(1-v)A, \\
k'_{D58} &= k'_{D85} = \frac{4(1-v)A}{3}(x_s + y_s), \\
k'_{D67} &= k'_{D76} = 4vAx_s, \\
k'_{D68} &= k'_{D86} = \frac{4A}{3}(x_s + vy_s), \\
k'_{D69} &= k'_{D96} = 4Ay_s, \\
k'_{D77} &= 3A(x_{ss} + x_s^2), \\
k'_{D78} &= k'_{D87} = A(p_{xy} + x_s y_s + vx_{ss} + vx_s^2), \\
k'_{D79} &= k'_{D97} = 3vA(p_{xy} + x_s y_s), \\
k'_{D88} &= \frac{A}{3}\left((3-2v)(x_{ss} + x_s^2 + y_{ss} + y_s^2) + 2(2-v)(p_{xy} + x_s y_s)\right), \\
k'_{D89} &= k'_{D98} = A(p_{xy} + x_s y_s + vy_{ss} + vy_s^2), \\
k'_{D99} &= 3A(y_{ss} + y_s^2).
\end{aligned}
\tag{10.45b}
$$

The values of each generalized stiffness coefficient of equations 10.45b must be multiplied by the appropriate scalar constant,

$$ k_{Dij} = \left(\frac{Et^3}{12(1-v^2)}\right)k'_{Dij}. \tag{10.45c}$$

The element stiffness matrix **K** in the physical displacement form is calculated from matrix $\mathbf{K}_D$ using equation 6.37e.

To illustrate the Kirchhoff triangular thin plate flexure element, the stiffness matrix will be calculated for a triangular element from the lower left half of the $60 \times 36 \times \frac{3}{4}$ in. steel plate used earlier. The coordinates of nodes 1–3 are (0, 0), (60, 0), (0, 36). The elastic modulus $E = 30,000$ ksi and Poisson's ratio $\nu = 0.25$. The matrix **h** is written directly from equation 10.42c, inverted numerically to matrix $\mathbf{h}^{-1}$, and transposed to matrix $\mathbf{h}^{-T}$. The generalized stiffness matrix $\mathbf{K}_D$ is written from equation 10.45, and the element stiffness matrix **K** in the physical displacement form is calculated using equation 6.37e:

$$
[\mathbf{K}] = \begin{bmatrix}
9.806 & 308.021 & 105.521 & -1.125 & -95.000 & -14.688 & -8.681 & 99.479 & -158.333 \\
308.021 & 13,940.625 & 7,328.125 & -21.562 & -5,831.250 & 112.500 & -286.458 & 2,203.125 & -8,734.375 \\
105.521 & 7,328.125 & 6,165.625 & 16.875 & -3,609.375 & 628.125 & -122.396 & 687.500 & -5,781.250 \\
-1.125 & -21.563 & 16.875 & 1.125 & 16.875 & 22.500 & 0.000 & 4.688 & 28.125 \\
-95.000 & -5,831.250 & -3,609.375 & 16.875 & 3,346.875 & 28.125 & 78.125 & -328.125 & 4,593.750 \\
-14,688 & 112.500 & 628.125 & 22.500 & 28.125 & 675.000 & -7.813 & 140.625 & 46.875 \\
-8.681 & -286.458 & -122.396 & 0.000 & 78.125 & -7.813 & 8.681 & -104.167 & 130.208 \\
99.479 & 2,203.125 & 687.500 & 4.688 & -328.125 & 140.625 & -104.167 & 1,875.000 & -546.875 \\
-158.333 & -8,734.375 & -5,781.250 & 28.125 & 4,593.750 & 46.875 & 130.208 & -546.875 & 7,421.875
\end{bmatrix}.
$$

Units for matrix **K** use the same nodal submatrix form as for the rectangular Kirchhoff element. Note that despite the orientation of this right triangle, its matrix **h** is not singular because the triangle is not isosceles.

When a distributed transverse load $b_z(x, y)$, whose dimensions are force per unit area, is applied to a triangular flexural plate element, equivalent forces and moments at the element's nodes must be computed from the load. Equation 6.22e applies and may be written for the element in terms of generalized displacements as

$$
\{\mathbf{F}_b\} = [\mathbf{h}^{-T}] \int_A \{\mathbf{g}\} b_z(x, y)\, dA. \tag{10.46}
$$

The generalized displacement form is especially efficient when the matrix **h** is available from a generalized displacement development of the element stiffness matrix. The matrix **h** and vector **g** were shown earlier in equations 10.42.

To integrate equation 10.46 over the area of the triangular element, the approach used to develop equations 10.44 and 10.45 for the generalized stiffness matrix for the triangular flexural element may be followed. In addition to the expressions shown in equations 10.44 for the integrals of generic polynomial terms over the triangular area of the element, the following may also be readily developed. The term A is the area of the triangular element:

$$\int_A (x^3)\, dA = \frac{A}{60}(5x_{sc} + 3x_{pss} + x_s^3),$$

$$\int_A (y^3)\, dA = \frac{A}{60}(5y_{sc} + 3y_{pss} + y_s^3), \tag{10.47}$$

$$\int_A (x^2 y + xy^2)\, dA = \frac{A}{60}\left(5(x_{spy} + y_{spx}) + x_{ssy} + y_{ssx} + x_s^2 y_s + x_s y_s^2\right).$$

where

$$x_s = x_1 + x_2 + x_3,$$
$$x_{sc} = x_1^3 + x_2^3 + x_3^3,$$
$$x_{pss} = x_1 x_2(x_1 + x_2) + x_2 x_3(x_2 + x_3) + x_3 x_1(x_3 + x_1),$$
$$y_s = y_1 + y_2 + y_3,$$
$$y_{sc} = y_1^3 + y_2^3 + y_3^3,$$
$$y_{pss} = y_1 y_2(y_1 + y_2) + y_2 y_3(y_2 + y_3) + y_3 y_1(y_3 + y_1),$$
$$x_{spy} = x_1^2 y_1 + x_2^2 y_2 + x_3^2 y_3,$$
$$y_{spx} = x_1 y_1^2 + x_2 y_2^2 + x_3 y_3^2,$$
$$x_{ssy} = (x_1^2 + 2x_2 x_3)(y_2 + y_3) + (x_2^2 + 2x_3 x_1)(y_3 + y_1)$$
$$\quad + (x_3^2 + 2x_1 x_2)(y_1 + y_2),$$
$$y_{ssx} = (y_1^2 + 2y_2 y_3)(x_2 + x_3) + (y_2^2 + 2y_3 y_1)(x_3 + x_1)$$
$$\quad + (y_3^2 + 2y_1 y_2)(x_1 + x_2).$$

With a uniform distributed transverse load over the element, b_z is constant so that equation 10.46 may be written explicitly as

$$\{\mathbf{F}_b\} = A b_z [\mathbf{h}^{-T}] \left\{ \begin{array}{c} 1 \\ \frac{x_s}{3} \\ \frac{y_s}{3} \\ \frac{x_{ss} + x_s^2}{12} \\ \frac{p_{xy} + x_s y_s}{12} \\ \frac{y_{ss} + y_s^2}{12} \\ \frac{5x_{sc} + 3x_{pss} + x_s^2}{60} \\ \frac{5(x_{spy} + y_{spx}) + x_{ssy} + y_{ssx} + x_s^2 y_s + x_s y_s^2}{60} \\ \frac{5y_{sc} + 3y_{pss} + y_s^2}{60} \end{array} \right\}. \tag{10.48}$$

When the distributed transverse load is variable, equation 10.46 may be implemented with a numerical quadrature rule.

10.4.3. Rectangular Thick Plate Element (4-Node, 12-dof)

For thick flat plate structures, the thin plate flexure theory is modified by changing its second assumption: Straight lines initially normal to the midplane will remain straight, but they will not remain normal to the midplane during bending. Instead, they will rotate through angular displacements θ_x and θ_y, independent of the transverse translation w or its derivatives $\partial w/\partial x$ and $\partial w/\partial y$. This allows transverse shear strains and stresses to develop during bending, a matter of importance for thick plates.

Figure 10.12 illustrates a deformed differential element of a plate with the transverse shear deformations present. The line OA was normal to the midplane prior to deformation. The point A displaced through translations u, v, and w. For small rotations, the in-plane displacements are

$$u(x, \ y) = -z\theta_x(x, \ y), \tag{10.49a}$$

$$v(x, \ y) = -z\theta_y(x, \ y). \tag{10.49b}$$

The in-plane strains are

$$\epsilon_x = \frac{\partial u}{\partial x} = -z\frac{\partial \theta_x}{\partial x}, \tag{10.50a}$$

$$\epsilon_y = \frac{\partial v}{\partial y} = -z\frac{\partial \theta_y}{\partial y}, \tag{10.50b}$$

$$\gamma_{xy} = \frac{\partial u}{\partial y} + \frac{\partial v}{\partial x} = -z\left(\frac{\partial \theta_x}{\partial y} + \frac{\partial \theta_y}{\partial x}\right). \tag{10.50c}$$

The transverse shear strains are the angular changes in the right-angle corners of the differential element as it deforms:

$$\gamma_{yz} = \theta_y - \frac{\partial w}{\partial y}, \tag{10.50d}$$

$$\gamma_{zx} = \theta_x - \frac{\partial w}{\partial x}. \tag{10.50e}$$

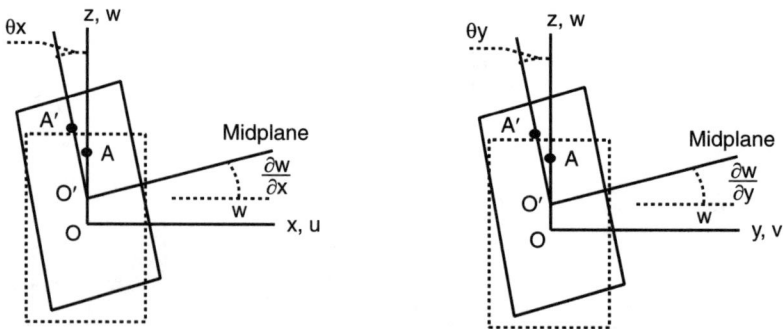

Figure 10.12. Thick plate differential element.

The stress–strain relationships for plane stress must be extended to include the transverse shear stress and strain:

$$\begin{Bmatrix} \sigma_x \\ \sigma_y \\ \tau_{xy} \\ \tau_{yz} \\ \tau_{zx} \end{Bmatrix} = \frac{E}{1-\nu^2} \begin{bmatrix} 1 & \nu & 0 & 0 & 0 \\ \nu & 1 & 0 & 0 & 0 \\ 0 & 0 & \frac{1-\nu}{2} & 0 & 0 \\ 0 & 0 & 0 & \frac{1-\nu}{2} & 0 \\ 0 & 0 & 0 & 0 & \frac{1-\nu}{2} \end{bmatrix} \begin{Bmatrix} \epsilon_x \\ \epsilon_y \\ \gamma_{xy} \\ \gamma_{yz} \\ \gamma_{zx} \end{Bmatrix}, \tag{10.51a}$$

$$\{\boldsymbol{\sigma}\} = [\mathbf{E}]\{\boldsymbol{\epsilon}\}. \tag{10.51b}$$

The moment–curvature relationships for thick plate flexure are developed following the procedures used earlier for thin plate flexure (Section 10.2), but using equations 10.50 and 10.51 in place of equations 10.1 and 10.2:

$$M_{xx} = \int_{-t/2}^{t/2} (-z\sigma_x)\, dz = \frac{Et^3}{12(1-\nu^2)} \left(\frac{\partial \theta_x}{\partial x} + \nu \frac{2\theta_y}{\partial y} \right), \tag{10.52a}$$

$$M_{yy} = \int_{-t/2}^{t/2} (-z\sigma_y)\, dz = \frac{Et^3}{12(1-\nu^2)} \left(\nu \frac{\partial \theta_x}{\partial x} + \frac{\partial \theta_y}{\partial y} \right), \tag{10.52b}$$

$$M_{xy} = \int_{-t/2}^{t/2} (-z\tau_{xy})\, dz = \frac{Et^3}{12(1-\nu^2)} \left(\frac{1-\nu}{2} \right) \left(\frac{\partial \theta_x}{\partial y} + \frac{\partial \theta_y}{\partial x} \right), \tag{10.52c}$$

$$Q_y = \int_{-t/2}^{t/2} (-z\tau_{yz})\, dz = \frac{Et^3}{12(1-\nu^2)} \left(\frac{1-\nu}{2} \right) \left(\frac{12}{t^2} \right) \left(\theta_y - \frac{\partial w}{\partial y} \right), \tag{10.52d}$$

$$Q_x = \int_{-t/2}^{t/2} (-z\tau_{zx})\, dz = \frac{Et^3}{12(1-\nu^2)} \left(\frac{1-\nu}{2} \right) \left(\frac{12}{t^2} \right) \left(\theta_x - \frac{\partial w}{\partial x} \right). \tag{10.52e}$$

In matrix form these relationships are given as

$$\begin{Bmatrix} M_{xx} \\ M_{yy} \\ M_{xy} \\ Q_y \\ Q_x \end{Bmatrix} = \frac{Et^3}{12(1-\nu^2)} \begin{bmatrix} 1 & \nu & 0 & 0 & 0 \\ \nu & 1 & 0 & 0 & 0 \\ 0 & 0 & \frac{1-\nu}{2} & 0 & 0 \\ 0 & 0 & 0 & \frac{6(1-\nu)}{t^2} & 0 \\ 0 & 0 & 0 & 0 & \frac{6(1-\nu)}{t^2} \end{bmatrix} \begin{Bmatrix} \frac{\partial \theta_x}{\partial x} \\ \frac{\partial \theta_y}{\partial y} \\ \frac{\partial \theta_x}{\partial y} + \frac{\partial \theta_y}{\partial x} \\ \theta_y - \frac{\partial w}{\partial y} \\ \theta_x - \frac{\partial w}{\partial x} \end{Bmatrix}, \tag{10.52f}$$

$$\{\mathbf{M}\} = [\bar{\mathbf{E}}]\{\boldsymbol{\phi}\}.$$

The moment vector $\mathbf{M}$ is the generalized stress vector $\bar{\sigma}$ and the curvature vector ϕ is the generalized strain vector $\bar{\epsilon}$.

The Mindlin element is a 4-node, 12-dof rectangular thick plate element developed from this situation. The element possesses three independent displacement functions: the transverse translation $w(x, y)$, the rotation $\theta_x(x, y)$, and the rotation $\theta_y(x, y)$. These will give rise to 3 physical edof of w_i, θ_{xi}, and θ_{yi} at each node and a total of 12 dof for a 4-node element. The displacement functions for the element are expressed as a vector of 3 functions:

$$\{\mathbf{w(x,\ y)}\} = \left\{ \begin{array}{c} w(x,\ y) \\ \theta_x(x,\ y) \\ \theta_y(x,\ y) \end{array} \right\}. \tag{10.53a}$$

Since the 3 displacement functions are independent of one another, the same 4 shape functions may be used to express them. These shape functions are the same as the ones used for the 4-node, 8-dof quadrilateral plane stress element (equations 9.6). The 3 displacement functions of the Mindlin element are expressed in terms of the 4 shape functions and its 12 nodal displacements as

$$\left\{ \begin{array}{c} w(\zeta,\ \eta) \\ \theta_x(\zeta,\ \eta) \\ \theta_y(\zeta,\ \eta) \end{array} \right\} = \begin{bmatrix} N_1 & 0 & 0 & N_2 & 0 & 0 & N_3 & 0 & 0 & N_4 & 0 & 0 \\ 0 & N_1 & 0 & 0 & N_2 & 0 & 0 & N_3 & 0 & 0 & N_4 & 0 \\ 0 & 0 & N_1 & 0 & 0 & N_2 & 0 & 0 & N_3 & 0 & 0 & N_4 \end{bmatrix} \left\{ \begin{array}{c} w_1 \\ \theta_{x1} \\ \theta_{y1} \\ w_2 \\ \theta_{x2} \\ \theta_{y2} \\ w_3 \\ \theta_{x3} \\ \theta_{y3} \\ w_4 \\ \theta_{x4} \\ \theta_{y4} \end{array} \right\},$$

$$\tag{10.53b}$$

$$\{\mathbf{w}(\zeta,\ \eta)\} = [\mathbf{N}(\zeta,\ \eta)]\{\mathbf{g}\}. \tag{10.53c}$$

Using the strain displacement relationships (equations 10.50) and the curvatures shown in equations 10.52,

$$\{\bar{\epsilon}\} = \{\boldsymbol{\phi}\} = \left\{ \begin{array}{c} \frac{\partial \theta_x}{\partial x} \\ \frac{\partial \theta_y}{\partial y} \\ \frac{\partial \theta_x}{\partial y} + \frac{\partial \theta_y}{\partial x} \\ \theta_y - \frac{\partial w}{\partial y} \\ \theta_x - \frac{\partial w}{\partial x} \end{array} \right\} = \begin{bmatrix} 0 & \frac{\partial}{\partial x} & 0 \\ 0 & 0 & \frac{\partial}{\partial y} \\ 0 & \frac{\partial}{\partial y} & \frac{\partial}{\partial x} \\ -\frac{\partial}{\partial y} & 0 & 1 \\ -\frac{\partial}{\partial x} & 1 & 0 \end{bmatrix} \left\{ \begin{array}{c} w \\ \theta_x \\ \theta_y \end{array} \right\}, \tag{10.54a}$$

$$\{\bar{\epsilon}\} = [\bar{\mathbf{d}}]\{\mathbf{w}\}. \tag{10.54b}$$

Substituting equation 10.53c, the generalized strain–displacement relationships are given as

$$\{\bar{\epsilon}\} = \lfloor \bar{\mathbf{d}} \rfloor [\mathbf{N}]\{\mathbf{q}\} = \lfloor \bar{\mathbf{B}} \rfloor \{\mathbf{q}\}. \tag{10.54c}$$

The generalized strain–displacement submatrix $\bar{\mathbf{B}}_i$ at node i is given as

$$[\bar{\mathbf{B}}_i] = [\bar{\mathbf{d}}][\mathbf{N}_i] = \begin{bmatrix} 0 & \frac{\partial}{\partial x} & 0 \\ 0 & 0 & \frac{\partial}{\partial y} \\ 0 & \frac{\partial}{\partial y} & \frac{\partial}{\partial x} \\ -\frac{\partial}{\partial y} & 0 & 1 \\ -\frac{\partial}{\partial x} & 1 & 0 \end{bmatrix} \begin{bmatrix} N_i & 0 & 0 \\ 0 & N_i & 0 \\ 0 & 0 & N_i \end{bmatrix} = \begin{bmatrix} 0 & \frac{\partial N_i}{\partial x} & 0 \\ 0 & 0 & \frac{\partial N_i}{\partial y} \\ 0 & \frac{\partial N_i}{\partial y} & \frac{\partial N_i}{\partial x} \\ -\frac{\partial N_i}{\partial y} & 0 & N_i \\ -\frac{\partial N_i}{\partial x} & N_i & 0 \end{bmatrix}. \tag{10.55}$$

The shape functions $N_i(\zeta, \eta)$ are the four bilinear shape functions developed for the plane stress plate element (equations 9.6). The complete strain displacement matrix $\bar{\mathbf{B}}$ is obtained by evaluating equation 10.55 at each of the four nodes to produce a 5×12 matrix of functions of ζ and η.

From this point, the element stiffness matrix **K** for the 4-node, 12-dof Mindlin plate element is developed by following the process used for the thin plate flexure element stiffness matrix:

$$\underset{12 \times 12}{[\mathbf{K}]} = \int_{-1}^{1} \int_{-1}^{1} \underset{12 \times 5}{[\bar{\mathbf{B}}^{\mathrm{T}}(\zeta, \eta)]} \underset{5 \times 5}{[\bar{\mathbf{E}}]} \underset{5 \times 12}{[\bar{\mathbf{B}}(\zeta, \eta)](ab)} \, d\zeta \, d\eta. \tag{10.56a}$$

Though equation 10.56a might be evaluated analytically, it is more conveniently done with numerical quadrature. Because the shape functions are bilinear, the terms of the integrand of equation 10.56a may be evaluated exactly with the Gauss 2×2 rule. Equation 10.56a expressed for evaluation with a quadrature rule is

$$\underset{12 \times 12}{[\mathbf{K}]} = (ab) \sum_{k=1}^{n} \underset{12 \times 5}{[\bar{\mathbf{B}}^{\mathrm{T}}(\zeta_k, \eta_k)]} \underset{5 \times 5}{[\bar{\mathbf{E}}]} \underset{5 \times 12}{[\bar{\mathbf{B}}(\zeta_i, \eta_k)]} H_k. \tag{10.56b}$$

The Mindlin element works well as a rectangular element on thick plates. It does poorly on thin rectangular plates, however, due to a phenomenon known as "shear locking." What occurs is that as the element thickness becomes small compared to its length and width, the element transverse shear phenomena become increasingly dominant, overshadowing the flexural phenomena. However, in the physical behavior of thin plates, the reverse is true. Flexural behavior dominates and transverse shear behavior is insignificant.

To gain insight into this aspect of the Mindlin element, the integrand matrices $\bar{\mathbf{B}}$ and $\bar{\mathbf{E}}$ may be partitioned to separate the effects of the in-plane strains ϵ_x, ϵ_y, and γ_{xy} from the transverse shear strains γ_{yz} and γ_{zx}:

$$[\mathbf{I}(\zeta, \eta)] = [\bar{\mathbf{B}}^{\mathrm{T}}] \ [\bar{\mathbf{E}}] \ [\bar{\mathbf{B}}]$$

$$\underset{12\times12}{} \qquad \underset{12\times5 \ \ 5\times5 \ \ 5\times12}{}$$

$$= \begin{bmatrix} \bar{\mathbf{B}}_f^{\mathrm{T}} & | & \bar{\mathbf{B}}_s^{\mathrm{T}} \\ {\scriptstyle 12\times3} & | & {\scriptstyle 12\times2} \end{bmatrix} \begin{bmatrix} \bar{\mathbf{E}}_f & | & \mathbf{0} \\ {\scriptstyle 3\times3} & | & {\scriptstyle 3\times2} \\ \hline \mathbf{0} & | & \bar{\mathbf{E}}_s \\ {\scriptstyle 2\times3} & | & {\scriptstyle 2\times2} \end{bmatrix} \begin{bmatrix} \bar{\mathbf{B}}_f \\ {\scriptstyle 3\times12} \\ \hline \bar{\mathbf{B}}_s \\ {\scriptstyle 2\times12} \end{bmatrix}. \tag{10.56c}$$

The subscript f identifies the submatrices associated with the in-plane strains, or flexure effects. The subscript s identifies those associated with the transverse shear strain, or transverse shear effects. When equation 10.56c is multiplied out in terms of its submatrices, the element stiffness matrix may be written as

$$[\mathbf{K}] = \int_{-1}^{1} \int_{-1}^{1} ([\bar{\mathbf{B}}_f^{\mathrm{T}} \bar{\mathbf{E}}_f \bar{\mathbf{B}}_f] + [\bar{\mathbf{B}}_s^{\mathrm{T}} \bar{\mathbf{E}}_s \bar{\mathbf{B}}_s])(ab) \ d\zeta \ d\eta \tag{10.56d}$$

$$= [\mathbf{K}_f] + [\mathbf{K}_s].$$

For a square plate element whose half-length and half-width are both a, the matrices $\mathbf{K}_f$ and $\mathbf{K}_s$ possess entries whose magnitudes are determined by a scalar D_f or D_s, the half-length a, and products of two shape functions, a shape function and a first partial derivative of a shape function, or two first partial derivatives of shape functions. The scalars are given as

$$D_f = \frac{Et^3}{12(1 - v^2)} \quad \text{and} \quad D_s = \frac{Et^3}{12(1 - v^2)} \frac{6(1 - v)}{t^2} = \frac{Et}{2(1 + v)}.$$

Some entries of the matrix $\mathbf{K}_f$ are multiplied by Poisson's ratio v or $\frac{1}{2}(1 - v)$. Some entries of matrix $\mathbf{K}_s$ are multiplied by the half-length a or its square a^2. Thus the ratios of stiffness coefficients of matrix $\mathbf{K}_s$ to matrix $\mathbf{K}_f$ are approximated by

$$\frac{k_s}{k_f} \approx \frac{6}{t^2} \quad \text{or} \quad \frac{6a}{t^2} \quad \text{or} \quad \frac{6a^2}{t^2}.$$

Clearly, as the plate thickness t becomes small with respect to its dimension a, the effect of the transverse shear matrix $\mathbf{K}_s$ on the element stiffness matrix $\mathbf{K}$ becomes dominant and causes the element to be governed by shear phenomena, a result at odds with the physical behavior of thin plates.

To further illustrate the "shear locking" characteristic of the Mindlin plate element, the matrices $\mathbf{K}_f$, $\mathbf{K}_s$, and $\mathbf{K}$ for the $60 \times 36 \times 0.75$ in. plate used earlier will be calculated. Equations 10.52, 10.55, and 10.56 are used and integrated using the Gauss 2×2 quadrature rule. The Mindlin flexure matrix $\mathbf{K}_f$ is given as

$$[\mathbf{K}_f] = \begin{bmatrix} 0.000 & 0.000 & 0.000 & 0.000 & 0.000 & 0.000 & 0.000 & 0.000 & 0.000 & 0.000 & 0.000 & 0.000 \\ 0.000 & 1837.500 & 703.125 & 0.000 & -431.250 & -140.625 & 0.000 & -918.750 & -703.125 & 0.000 & -487.500 & 140.625 \\ 0.000 & 703.125 & 2837.500 & 0.000 & 140.625 & 912.500 & 0.000 & -703.125 & -1418.750 & 0.000 & -140.625 & -2331.250 \\ 0.000 & 0.000 & 0.000 & 0.000 & 0.000 & 0.000 & 0.000 & 0.000 & 0.000 & 0.000 & 0.000 & 0.000 \\ 0.000 & -431.250 & 140.625 & 0.000 & 1837.500 & -703.125 & 0.000 & -487.500 & -140.625 & 0.000 & -918.750 & 703.125 \\ 0.000 & -140.625 & 912.500 & 0.000 & -703.125 & 2837.500 & 0.000 & 140.625 & -2331.250 & 0.000 & 703.125 & -1418.750 \\ 0.000 & 0.000 & 0.000 & 0.000 & 0.000 & 0.000 & 0.000 & 0.000 & 0.000 & 0.000 & 0.000 & 0.000 \\ 0.000 & -918.750 & -703.125 & 0.000 & -487.500 & 140.625 & 0.000 & 1837.500 & 703.125 & 0.000 & -431.250 & -140.625 \\ 0.000 & -703.125 & -1418.750 & 0.000 & -140.625 & -2331.250 & 0.000 & 703.125 & 2837.500 & 0.000 & 140.625 & 912.500 \\ 0.000 & 0.000 & 0.000 & 0.000 & 0.000 & 0.000 & 0.000 & 0.000 & 0.000 & 0.000 & 0.000 & 0.000 \\ 0.000 & -487.500 & -140.625 & 0.000 & -918.750 & 703.125 & 0.000 & -431.250 & 140.625 & 0.000 & 1837.500 & -703.125 \\ 0.000 & 140.625 & -2331.250 & 0.000 & 703.125 & -1418.750 & 0.000 & -140.625 & 912.500 & 0.000 & -703.125 & 2837.500 \end{bmatrix}.$$

Each entry of rows and columns 1, 4, 7, and 10 are zeros. Thus the coefficients for these rows and columns of the element's stiffness matrix **K** are not influenced by the flexure effects of matrix $\mathbf{K}_f$. The root-mean-square value of the 144 stiffness coefficients for the matrix $\mathbf{K}_f$ of this example is 66.66.

The Mindlin transverse shear matrix $\mathbf{K}_s$ is given as

$$[\mathbf{K}_s] = \begin{bmatrix} 27,200.000 & 400.000 & 666.667 & 2,800.000 & 400.000 & 333.333 & -13,600.000 & 200.000 & 333.333 & -16,400.000 & 200.000 & 666.667 \\ 400.000 & 29.630 & 0.000 & -400.000 & 14.815 & 0.000 & -200.000 & 7.407 & 0.000 & 200.000 & 14.815 & 0.000 \\ 666.667 & 0.000 & 29.630 & 333.333 & 0.000 & 14.815 & -333.333 & 0.000 & 7.407 & -666.667 & 0.000 & 14.815 \\ 2,800.000 & -400.000 & 333.333 & 27,200.000 & -400.000 & 666.667 & -16,400.000 & -200.000 & 666.667 & -13,600.000 & -200.00 & 333.333 \\ 400.000 & 14.815 & 0.000 & -400.000 & 29.630 & 0.000 & -200.000 & 14.815 & 0.000 & 200.000 & 7.407 & 0.000 \\ 333.333 & 0.000 & 14.815 & 666.667 & 0.000 & 29.630 & -666.667 & 0.000 & 14.815 & -333.333 & 0.000 & 7.407 \\ -13,600.000 & -200.000 & -333.333 & -16,400.000 & -200.000 & -666.667 & 27,200.000 & -400.000 & -666.667 & 2,800.000 & -400.000 & -333.333 \\ 200.000 & 7.407 & 0.000 & -200.000 & 14.815 & 0.000 & -400.000 & 29.630 & 0.000 & 400.000 & 14.815 & 0.000 \\ 333.333 & 0.000 & 7.407 & 666.667 & 0.000 & 14.815 & -666.667 & 0.000 & 29.630 & -333.333 & 0.000 & 14.815 \\ -16,400.000 & 200.000 & -666.667 & -13,600.000 & 200.000 & -333.333 & 2,800.000 & 400.000 & -333.333 & 27,200.000 & 400.000 & -666.667 \\ 200.000 & 14.815 & 0.000 & -200.000 & 7.407 & 0.000 & -400.000 & 14.815 & 0.000 & 400.000 & 29.630 & 0.000 \\ 666.667 & 0.000 & 14.815 & 333.333 & 0.000 & 7.407 & -333.333 & 0.000 & 14.815 & -666.667 & 0.000 & 29.630 \end{bmatrix}.$$

The root-mean-square value of the 144 stiffness coefficients of the matrix $\mathbf{K}_s$ for this example is 482.05, over seven times as large as the value for matrix $\mathbf{K}_f$.

Adding matrices $\mathbf{K}_f$ and $\mathbf{K}_s$ produces the matrix **K** for this example:

$$[\mathbf{K}] = \begin{bmatrix} 27,200.000 & 400.000 & 666.667 & 2,800.000 & 400.000 & 333.333 & -13,600.000 & 200.000 & 333.333 & -16,400.000 & 200.000 & 666.667 \\ 400.000 & 1,867.130 & 703.125 & -400.000 & -416.435 & -140.625 & -200.000 & -911.343 & -703.125 & 200.000 & -472.685 & 140.625 \\ 666.667 & 703.125 & 2,867.130 & 333.333 & 140.625 & 927.315 & -333.333 & -703.125 & -1,411.343 & -666.667 & -140.625 & -2,316.435 \\ 2,800.000 & -400.000 & 333.333 & 27,200.000 & -400.000 & 666.667 & -16,400.000 & -200.000 & 666.667 & -13,600.000 & -200.000 & 333.333 \\ 400.000 & -416.435 & 140.625 & -400.000 & 1,867.130 & -703.125 & -200.000 & -472.685 & -140.625 & 200.000 & -911.343 & 703.125 \\ 333.333 & -140.625 & 927.315 & 666.667 & -703.125 & 2,867.130 & -666.667 & 140.625 & -2,316.435 & -333.333 & 703.125 & -1,411.343 \\ -13,600.000 & -200.000 & -333.333 & -16,400.000 & -200.000 & -666.667 & 27,200.000 & -400.000 & -666.667 & 2,800.000 & -400.000 & -333.333 \\ 200.000 & -911.343 & -703.125 & -200.000 & -472.685 & 140.625 & -400.000 & 1,867.130 & 703.125 & 400.000 & -416.435 & -140.625 \\ 333.333 & -703.125 & -1,411.343 & 666.667 & -140.625 & -2,316.435 & -666.667 & 703.125 & 2,867.130 & -333.333 & 140.625 & 927.315 \\ -16,400.000 & 200.000 & -666.667 & -13,600.000 & 200.000 & -333.333 & 2,800.000 & 400.000 & -333.333 & 27,200.000 & 400.000 & -666.667 \\ 200.000 & -472.685 & -140.625 & -200.000 & -911.343 & 703.125 & -400.000 & -416.435 & 140.625 & 400.000 & 1,867.130 & -703.125 \\ 666.667 & 140.625 & -2,316.435 & 333.333 & 703.125 & -1,411.343 & -333.333 & -140.625 & 927.315 & -666.667 & -703.125 & 2,867.130 \end{bmatrix}.$$

The root-mean-square value of the 144 stiffness coefficients of this matrix is 486.67. In terms of root-mean-square values of stiffness coefficients, the Mindlin stiffness matrix **K** for this example exhibits the character of its transverse stiffness matrix $\mathbf{K}_s$, almost exclusively.

Compare the Mindlin stiffness matrix to the exact Kirchhoff stiffness matrix for this example; the latter was shown earlier in Section 10.3.2. Recall that the Kirchhoff element is based entirely on flexure behavior, the dominant phenomenon for thin plates. The Mindlin stiffness coefficients are many times larger, making it a much stiffer element. The root-mean-square value of the 144 Kirchhoff stiffness coefficients for this example is 48.47, about one-tenth of the corresponding value for the Mindlin element. Thus the Mindlin element, as described, does not model this thin rectangular plate well.

10.5. QUADRILATERAL THICK FLAT PLATE FLEXURE ELEMENT

A quadrilateral element possesses an ability to model irregular plate structural shapes better than rectangular plate elements and comparable to what triangular plate elements can do. Because they possess higher degree polynomial displacement functions than comparable triangular elements, they can usually achieve better accuracy than a comparable number of such triangular elements in a plate structure. They have a disadvantage, however. Their mathematics is more complicated than triangular or rectangular elements, with the result that computation effort is greater.

A quadrilateral thick plate flexure element is based on the 4-node rectangular Mindlin element which employs, for its displacement function, the four bilinear polynomial shape functions $N_i(\zeta, \eta)$ of equations 9.6. Moreover, the coordinate transformation to the 4-node quadrilateral element from a 2 × 2 square master element is the same as for the 4-node quadrilateral plane stress element. It is described in Sections 9.3.1, 9.5.1, and 9.5.2 and is defined by equations 9.13 in terms of the four nodal coordinates (x_i, y_i) and the same four bilinear polynomial shape functions $N_i(\zeta, \eta)$. Thus the transformation and calculations are isoparametric. The calculations for the derivatives of shape functions with respect to x and y are the same as for the plane stress quadrilateral element, but their manipulation will be different. It will be based on the requirements for the rectangular Mindlin element, as described in Section 10.4.3.

The expression for the stiffness matrix of the quadrilateral thick plate flexure element is an extension of the expression for the stiffness matrix of the Mindlin element (equation 10.56a):

$$\underset{12 \times 12}{[\mathbf{K}]} = \int_{-1}^{1} \int_{-1}^{1} \underset{12 \times 5}{[\bar{\mathbf{B}}^{\mathsf{T}}(\zeta, \eta)]} \underset{5 \times 5}{[\bar{\mathbf{E}}]} \underset{5 \times 12}{\bar{\mathbf{B}}(\zeta, \eta)} |J(\zeta, \eta)| \, d\zeta \, d\eta. \tag{10.57a}$$

Expressed in a form for evaluation with a quadrature rule, it is

$$\underset{12 \times 12}{[\mathbf{K}]} = \sum_{k=1}^{n} \underset{12 \times 5}{[\bar{\mathbf{B}}^{\mathsf{T}}(\zeta_k, \eta_k)]} \underset{5 \times 5}{[\bar{\mathbf{E}}]} \underset{5 \times 12}{\bar{\mathbf{B}}(\zeta_k, \eta_k)} |J(\zeta_k, \eta_k)| H_k. \tag{10.57b}$$

The matrix $\bar{\mathbf{B}}$ is written from the matrix $\bar{\mathbf{B}}_i$ (equation 10.55 for the Mindlin element) by letting the nodal index $i = 1, 2, 3, 4$. The derivatives $\partial N_i(\zeta, \eta)/\partial x$ and

$\partial N_i(\zeta, \eta)/\partial y$ at node i of the matrix $\bar{\mathbf{B}}$ are obtained through the isoparametric coordinate transformation, as was done for the plane stress quadrilateral element. Equations 9.16–9.18 may be used for this purpose. The matrix $\bar{\mathbf{E}}$ is expressed in equation 10.52f, the same as for the Mindlin element. The Jacobian $|J(\zeta, \eta)|$ is calculated from equations 9.16 and 9.17, the same as for the plane stress quadrilateral element.

10.6. CLOSING REMARKS

There are scores of plate flexure elements which have been developed and published in the literature. The ones discussed in this chapter work adequately—as well as most and better than many. They were chosen to provide the reader with an understanding of the finite element treatment of plate flexure.

All of the elements discussed require that transverse plate deflections be small, much less than the thickness of the plate. Larger transverse deflections bring about in-plane normal strains which are often described as a membrane effect. These strains are in addition to the in-plane flexure strains and cause the plate element to behave with greater stiffness. They are nonlinear in character and are beyond the scope of this text.

Thin plate elements are those where thickness is much less than their lateral dimensions. The 4-node, 12-dof nonconforming Kirchhoff rectangular flexure element, 4-node, 16-dof conforming rectangular flexure element, and 3-node, 9-dof nonconforming Kirchhoff triangular flexure element were presented in detail. They are best developed in terms of generalized stress and strain and generalized dof to keep their mathematics and computation effort as simple as possible. In so doing, their stiffness matrices may be determined explicitly.

The thick plate 4-node, 12-dof Mindlin rectangular flexure element was presented in terms of generalized stress and strain. As presented, it is suitable for thick plates but not for thin plates. Its stiffness matrix was obtained by Gaussian quadrature.

A 4-node, 12-dof straight-sided quadrilateral flexure element was discussed also. It is based on the Mindlin rectangular element. It lends itself readily to an isoparametric formulation which is similar to the isoparametric formulation for the quadrilateral plane stress plate element which was presented in detail in Chapter 9. The thick quadrilateral flexure element was discussed in terms of generalized stress and strain, and its stiffness matrix was obtained by Gaussian quadrature.

BIBLIOGRAPHY

Cook, R. D., Malkus, D. S., and Plesha, M. E., *Concepts and Applications of Finite Element Analysis* (Chapters 6, 11), Wiley, New York, 1989.

MacNeal, R. H., *Finite Elements: Their Design and Performance* (Chapters 4, 6, 8, 9), Dekker, New York, 1994.

Prathap, G., *The Finite Element Method in Structural Mechanics* (Chapter 7), Kluwer Academic, Dordrecht, The Netherlands, 1993.

Shames, I. H., and Dyrm, C. L., *Energy and Finite Elements in Structural Mechanics* (Chapters 6, 13), hemisphere (McGraw-Hill), New York, 1985.

Weaver, W., Jr., and Johnston, P. R., *Finite Elements for Structural Analysis* (Chapters 3, 6; Appendix B), Prentice-Hall, Englewood Cliffs, NJ, 1984.

Yang, T. Y., *Finite Element Structural Analysis* (Chapters 11, 12), Prentice-Hall, Englewood Cliffs, NJ, 1986.

PROBLEMS

The use of computer software, such as MATLAB or a spreadsheet, which manipulates individual matrices and evaluates functions at a point in their space will facilitate solution of some of these problems.

10.1. A thin, flat rectangular plate in flexure has modulus of elasticity $E = 30,000$ ksi, Poisson ratio $v = \frac{1}{4}$ and plate thickness $t = \frac{3}{4}$ in. Its strain vector, referred to x–y axes parallel to its sides and taken at a point $\frac{1}{4}$ in. distant from the neutral surface of bending, is given as

$$\left\{ \begin{array}{c} \epsilon_x \\ \epsilon_y \\ \gamma_{xy} \end{array} \right\} = \left\{ \begin{array}{c} -0.00015 \\ -0.00010 \\ 0.00055 \end{array} \right\}.$$

a. Compute the stress vector at the x–y coordinates of the point and referred to the x–y axes but at the greatest distance possible from the neutral surface of bending.

b. Compute the in-plane principal stresses and maximum shearing stress at the point that is the greatest distance possible from the neutral surface of bending.

c. Compute the distributed bending and twisting moment vector at the x–y coordinates of the point and referred to the x–y axes.

d. Compute the principal distributed bending moments and maximum distributed twisting moment at the point.

10.2. Repeat Problem 10.1 for a plate $\frac{1}{2}$ in. thick whose strain vector $\frac{1}{4}$ in. distant from the neutral surface of bending is

$$\left\{ \begin{array}{c} \epsilon_x \\ \epsilon_y \\ \gamma_{xy} \end{array} \right\} = \left\{ \begin{array}{c} 0.00012 \\ 0.00018 \\ 0.00020 \end{array} \right\}.$$

10.3. Using equations 10.16 and 10.17b, write the shape function N_{11} for a rectangular thin plate flexure element which is 6 units wide (in the ζ direction) and 4 units high (in the η direction).

10.4. Repeat Problem 10.3 for the shape function N_{21}.

10.5. Repeat Problem 10.3 for the shape function N_{31}

10.6. Repeat Problem 10.3 for the shape function N_{12}, verifying your result using equation 10.18b.

10.7. Repeat Problem 10.3 for the shape function N_{13}, verifying your result using equation 10.18c.

10.8. A rectangular thin flat plate flexure element is shown. Coordinates are in feet. The plate's elastic modulus is 30,000 ksi, its Poisson ratio is $\frac{1}{4}$, and its thickness t is $\frac{5}{8}$ in.

    a. Following the approach suggested in equations 10.22–10.25, write the integral expression for the element stiffness coefficient k_{74}.

    b. Evaluate the integral obtained above for k_{74}, numerically using the Gauss 2×2 quadrature rule.

    c. Evaluate the integral obtained above for k_{74}, numerically using the Gauss 3×3 quadrature rule.

    d. Evaluate the integral obtained above for k_{74}, analytically.

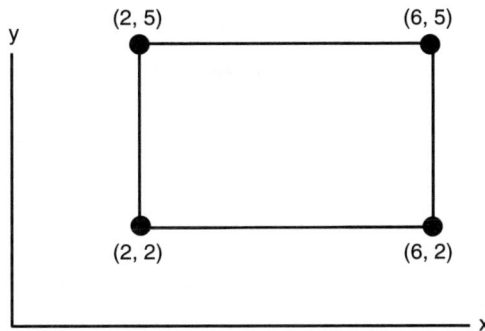

Problem 10.8

10.9. Repeat Problem 10.8 for element stiffness coefficient k_{42}.

10.10. Employing equation 10.26, calculate the stiffness matrix **K** for the rectangular thin plate flexure element illustrated in Problem 10.8. Use the Gauss 3×3 quadrature rule.

10.11. Employing equation 10.26, calculate the stiffness matrix **K** for a rectangular thin plate flexure element which is 5 ft long (ζ direction), 2 ft high (η direction), and 1 in. thick. Its elastic modulus $E = 30,000$ ksi and Poisson ratio $v = \frac{1}{4}$. Use the Gauss 3×3 quadrature rule.

10.12. Repeat Problem 10.10, but use the generalized edof approach (equations 10.29).

10.13. Repeat Problem 10.11, but use the generalized edof approach (equations 10.29).

10.14. Using the procedure suggested in equations 10.29, which includes integrating terms of the matrix $\mathbf{I}_D$ over the rectangular area of the element, derive the expression for k_{D44}, a stiffness coefficient for a 4-node, 12-dof rectangular thin plate flexure element in terms of generalized dof.

10.15. Repeat Problem 10.14 for coefficient k_{D55}.

10.16. Repeat Problem 10.14 for coefficient k_{D64}.

10.17. The rectangular thin plate flexure element of Problem 10.8 is shown with a vertical, downward-acting 20-kip concentrated load applied. Calculate the vector of equivalent forces and moments to act on the element using:

   a. Physical displacements as edof, equation 10.30.
   b. Generalized edof, equation 10.32.

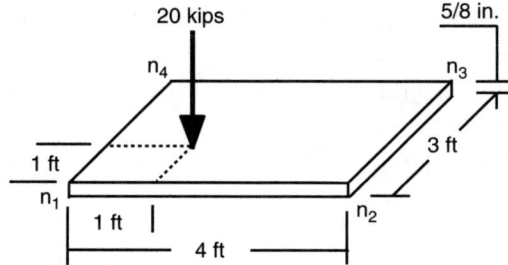

Problem 10.17

10.18. Repeat Problem 10.17, but with the 20-kip load applied at the center of the element.

10.19. The rectangular thin plate flexure element of Problem 10.17 is subjected to a 1.44-kip/ft^2 vertical, downward-acting uniformly distributed load across the entire area of the element. Determine the element's resulting vector of equivalent forces and moments.

10.20. The rectangular thin plate flexure element of Problem 10.17 is shown with a vertical, downward-acting linearly distributed load acting over its entire surface. The intensity of the applied load is zero at edge n_1–n_4 and increases linearly to 2.88 kips/ft^2 at edge n_2–n_3. Determine its vector of equivalent forces and moments.

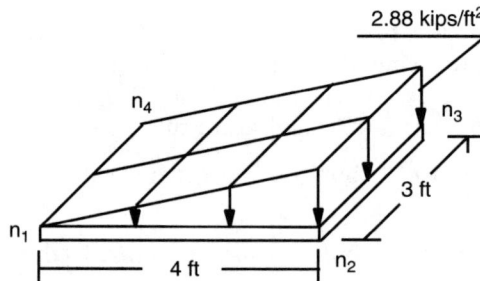

Problem 10.20

10.21. Repeat Problem 10.20 but with a plate that is 5×3 ft rather than 4×3 ft. The intensity of the applied load is zero at edge n_1–n_4 and increases linearly to 4.32 kips/ft^2 at edge n_2–n_3.

10.22. The rectangular thin plate flexure element of Problem 10.17 is shown resting on an elastic foundation. The modulus of subgrade reaction is 18 psi/in. Determine the element's total stiffness matrix $\mathbf{K}_E$.

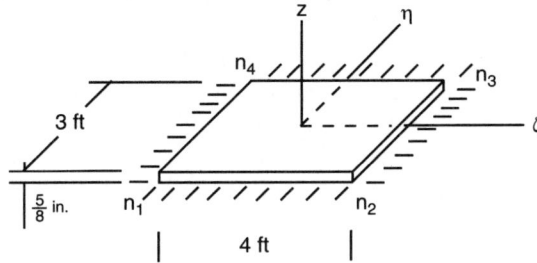

Problem 10.22

10.23. Using equations 10.44, evaluate the integrals over the triangular area shown of:

a. $x + y$.
b. $x^2 + xy + y^2$.
c. $(x + y)^2$.
d. $(x + y + 2)^2$.

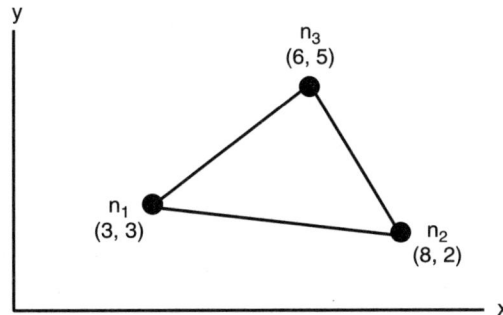

Problem 10.23

10.24. Repeat Problem 10.23 using equations 8.21.

10.25. Repeat Problem 10.23 using equations 8.36.

10.26. A triangular thin rectangular plate flexure element is located as shown on the sketch in Problem 10.23. It has elastic modulus $E = 30,000$ ksi, Poisson ratio $v = \frac{1}{4}$, and thickness $t = \frac{3}{8}$ in. Coordinates shown are inches. Determine the element's stiffness matrix.

10.27. Using the procedure suggested in equations 10.45, which includes integrating terms of the matrix $\mathbf{I}_D$ over the triangular area of the element, derive the expression for k_{D44}, a stiffness coefficient for a 3-node, 9-dof triangular thin plate flexure element in terms of generalized dof.

10.28. Repeat Problem 10.27 for coefficient k_{D55}.

10.29. Repeat Problem 10.27 for coefficient k_{D64}.

10.30. For the flexure plate supported and loaded as shown, determine the reactions at points A, B, C, and D and the transverse deflection at point E. The supports are pin supports at each corner of the plate which provide complete restraint for transverse translation but no rotational restraint. Use two 4-node, 12-dof thin plate flexural rectangular elements. Here, $E = 30,000$ ksi, $v = \frac{1}{4}$, and $t = 2.00$ in. The unit weight of the plate is 500 lb/ft^3. Verify that the applied loads and the support reactions satisfy the conditions for equilibrium.

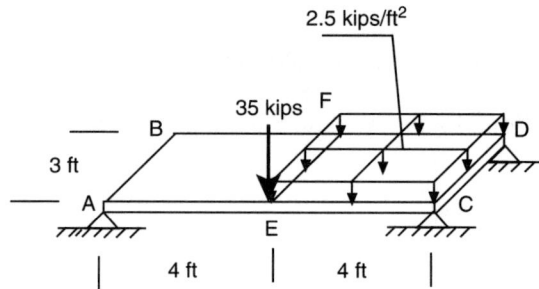

Problem 10.30

10.31. Repeat Problem 10.30 except assume that the four supports provide complete restraint for transverse translation and rotation displacement about both the x and y axes.

10.32. Repeat Problem 10.30 except move the 35-kip concentrated load and point E to the center of the plate and use four 4-node, 12-dof thin plate flexural rectangular elements.

10.33. Repeat Problem 10.32 except assume that the four supports provide complete restraint for transverse translation and rotation displacement about the y axis, but no restraint for rotation displacement about the x axis.

10.34. Repeat Problem 10.30 except use four 3-node, 9-dof thin plate triangular flexure elements. Elements 1–4 are triangles AEF, AFB, ECD, and EDF, respectively.

10.35. Repeat Problem 10.34 except assume that the four supports provide complete restraint for transverse translation and rotation displacement about the x and y axes.

10.36. Repeat Problem 10.34 except move the 35-kip concentrated load and point E to the center of the plate and use eight 4-node, 9-dof thin plate triangular flexural elements which are of course smaller and are oriented as in Problem 10.34.

10.37. Repeat Problem 10.36 except assume that the four supports provide complete restraint for transverse translation and rotation displacement about the y axis but no restraint for rotation displacement about the x axis.

CHAPTER 11

AXISYMMETRIC STRUCTURAL FINITE ELEMENTS

11.1. INTRODUCTION

An axisymmetric solid structure is solid of revolution, generated by rotating a plane shape about an axis (Figure 11.1). The plane shape shown is a rectangle and the z axis is the axis of revolution. The r axis is normal to the z axis and extends radially outward from it. The θ axis is normal to both the z and r axes and traverses circumferentially about the z axis. Orthogonal cylindrical coordinates (r, θ, z) readily describe axisymmetric solid structures. Figure 11.1 also denotes displacements along the cylindrical coordinate r, θ, and z axes as u, v, and w, respectively.

The geometry of an axisymmetric solid structure is symmetric with respect to the axis of rotation. If the loads on the structure, its supports, and its material properties are axisymmetric also, the structure may be analyzed mathematically as a 2D problem. That is, if the body, its loads, supports, and material properties are independent of the circumferential coordinate θ, then so will be its displacements, strains, and stresses. The problem is then treated in terms of r–z coordinates alone.

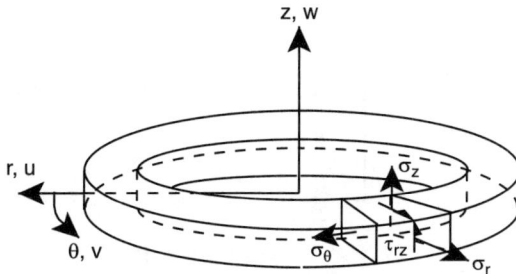

Figure 11.1. Axisymmetric solid.

An axisymmetric solid structure whose loads are not axisymmetric will possess displacements, strains, and stresses which are not distributed axisymmetrically. In this circumstance, superposition may be employed. The loads are expressed as the sum of a series of component loads using Fourier analysis. The structure then is analyzed for each component load separately as a 2D problem. The solutions for each of these problems are summed to produce the structural displacements, strains, and stresses caused by the original nonaxisymmetric load. By employing superposition in this manner, the series of 2D problems undertaken in lieu of a single 3D problem will usually be done at a worthwhile savings of time and computational effort.

Axisymmetric solid structures subjected to nonaxisymmetric loads are not treated further in this book.

11.2. AXISYMMETRIC STRESS, STRAIN, AND ELASTIC PROPERTIES

Figure 11.1 shows an element of an axisymmetric solid structure with the stresses σ_r, σ_θ, σ_z, and τ_{rz} which occur as a result of applied axisymmetric loads. Their corresponding strains ϵ_r, ϵ_θ, ϵ_z, and γ_{rz}, though not shown, are present. The displacements u and v in the radial and axial directions are present also. Because of axial symmetry, the circumferential displacement v, shear stresses $\tau_{r\theta}$ and $\tau_{\theta z}$, and shear strains $\gamma_{r\theta}$ and $\gamma_{\theta z}$ are all zero.

The relationships among the nonzero stresses and strains may be written as

$$\{\sigma\} = [\mathbf{E}]\{\epsilon\}, \tag{11.1a}$$

$$
\begin{Bmatrix} \sigma_r \\ \sigma_\theta \\ \sigma_z \\ \tau_{rz} \end{Bmatrix} = \begin{bmatrix} E_{11} & E_{12} & E_{13} & E_{14} \\ E_{21} & E_{22} & E_{23} & E_{24} \\ E_{31} & E_{32} & E_{33} & E_{34} \\ E_{41} & E_{42} & E_{43} & E_{44} \end{bmatrix} \begin{Bmatrix} \epsilon_r \\ \epsilon_\theta \\ \epsilon_z \\ \gamma_{rz} \end{Bmatrix}. \tag{11.1b}
$$

Examination of the general 3D stress–strain equation 6.4n in Chapter 6 suggests that it may be modified by replacing σ_x with σ_r, σ_y with σ_θ, and τ_{xy} with τ_{rz}, by leaving σ_z as is, and by deleting the fourth and fifth rows and columns of the matrices in the equation system to produce

$$
\begin{Bmatrix} \sigma_r \\ \sigma_\theta \\ \sigma_z \\ \tau_{rz} \end{Bmatrix} = \frac{E}{(1+v)(1-2v)} \begin{bmatrix} 1-v & v & v & 0 \\ v & 1-v & v & 0 \\ v & v & 1-v & 0 \\ 0 & 0 & 0 & \frac{1-2v}{2} \end{bmatrix} \begin{Bmatrix} \epsilon_r \\ \epsilon_\theta \\ \epsilon_z \\ \gamma_{rz} \end{Bmatrix}. \tag{11.1c}
$$

Equation 11.1c shows the stress–strain relationships for an elastic, axisymmetric, isotropic solid structure. One notes their similarity to the plane strain stress–strain relationships (equations 6.5b). The axisymmetric condition is similar in many ways to the plane strain condition. The displacements u, v, and w are similar. The axisymmetric normal stresses and strains σ_r, σ_z, ϵ_r, and ϵ_z compare to the plane strain normal stresses and strains σ_x, σ_y, ϵ_x, and ϵ_y respectively. Only the lateral normal strains and stresses are different. In the plane strain condition lateral strain is zero and lateral stress is constant, whereas in the axisymmetric condition ϵ_θ is nonzero and σ_θ is a function of all of the normal strains.

In the axisymmetric condition the radial and axial displacements u and w determine the strains. For small strains, the form of the strain displacement relationships is the same as it is for plane strain except for the circumferential strain:

$$\epsilon_r = \frac{\partial u}{\partial r}, \tag{11.2a}$$

$$\epsilon_z = \frac{\partial w}{\partial z}, \tag{11.2b}$$

$$\gamma_{rz} = \frac{\partial u}{\partial z} + \frac{\partial w}{\partial r}. \tag{11.2c}$$

Because the circumferential displacement v is zero, the circumferential strain ϵ_θ is not affected directly by it. However, there is circumferential small strain present. If an element, such as the one shown in Figure 11.1, displaces radially outward, it must stretch in the circumferential direction. If it displaces inward, it must compress. This is because even though there can be no displacement in the circumferential direction, circumferential lengths are proportional to the radius. Thus,

$$\epsilon_\theta = \frac{2\pi(r + u) - 2\pi r}{2\pi r} = \frac{u}{r}. \tag{11.2d}$$

In matrix form, these strain–displacement relationships are

$$
\begin{Bmatrix} \epsilon_r \\ \epsilon_\theta \\ \epsilon_z \\ \gamma_{rz} \end{Bmatrix} =
\begin{bmatrix} \frac{\partial}{\partial r} & 0 \\ \frac{1}{r} & 0 \\ 0 & \frac{\partial}{\partial z} \\ \frac{\partial}{\partial z} & \frac{\partial}{\partial r} \end{bmatrix}
\begin{Bmatrix} u \\ w \end{Bmatrix}, \tag{11.2e}
$$

$$\{\epsilon\} = [\mathbf{d}]\{\mathbf{u}\}. \tag{11.2f}$$

Note that the entry in the second row of matrix $\mathbf{d}$ is an algebraic multiplier rather than a differential operator.

11.3. AXISYMMETRIC STRUCTURAL FINITE ELEMENTS

Because the axisymmetric condition is similar to the plane strain condition, and thus to the plane stress condition, axisymmetric finite elements will be very similar to plane strain and plane stress elements. The main differences are that a fourth row and a fourth column are present in the matrices for the axisymmetric elements to treat the circumferential stress and strain interactions with other stresses and strains. In addition, integrations are uniformly around the circular circumference of the

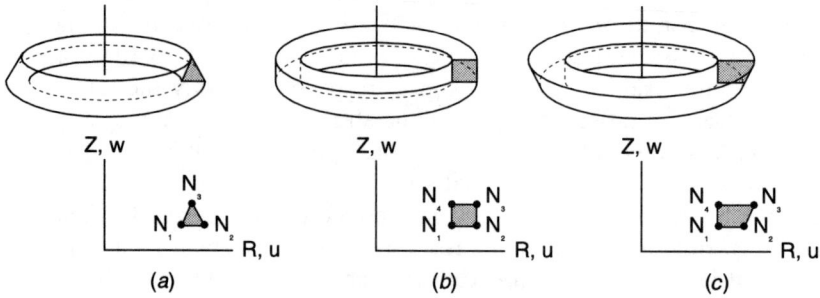

Figure 11.2. Axisymmetric structural finite elements: (*a*) triangular; (*b*) rectangular; (*c*) quadrilateral.

axisymmetric element rather than uniformly across the thickness of the plane strain or plane stress element. This results in a 2π multiplier for the axisymmetric element as opposed to the plate thickness multiplier for the plane strain or plane stress element. The discussion of the triangular and quadrilateral plane stress finite elements in Chapters 8 and 9 will be relied on heavily in the sections that follow.

The typical axisymmetric finite element is a circular ring. The element is characterized by its cross-sectional shape and nodes on that shape. In reality, axisymmetric finite elements have nodal circles rather than nodal points. The nodal points on the element's cross-sectional shape represent the nodal circles when treating the element in the r–z plane, just as the cross-sectional shape represents the circular ring. Figure 11.2 illustrates triangular, rectangular, and quadrilateral axisymmetric finite elements.

11.4. TRIANGULAR FINITE ELEMENT (3-NODE, 6-DOF)

A simple axisymmetric finite element is one whose cross-sectional shape is triangular with nodes at each vertex and two dof at each node. Its development is very similar to the development of the plane stress constant strain triangular finite element described in Section 8.2. Figure 11.3 shows the axisymmetric triangular

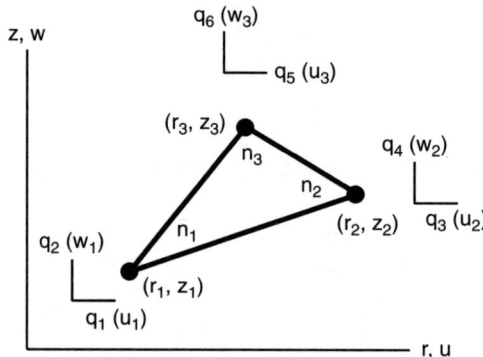

Figure 11.3. Triangular element.

finite element; Figure 8.1 shows the plane stress constant strain triangular finite element.

11.4.1. Shape and Displacement Functions

The displacements u and w in the axisymmetric triangular finite element are independent of one another, as they are in the plane stress constant stress triangular finite element. Both elements also have the same number of nodes and dof. Thus for both, aside from using the coordinates r and z rather than x and y, the displacement and shape functions are the same.

The displacement functions for the axisymmetric element are

$$u(r, \; z) = N_1(r, \; z)q_1 + N_2(r, \; z)q_3 + N_3(r, \; z)q_5, \tag{11.3a}$$
$$w(r, \; z) = N_1(r, \; z)q_2 + N_2(r, \; z)q_4 + N_3(r, \; z)q_6. \tag{11.3b}$$

In matrix form they are

$$\left\{ \begin{array}{c} u \\ v \end{array} \right\} = \left[\begin{array}{cccccc} N_1 & 0 & N_2 & 0 & N_3 & 0 \\ 0 & N_1 & 0 & N_2 & 0 & N_3 \end{array} \right] \left\{ \begin{array}{c} q_1 \\ q_2 \\ q_3 \\ q_4 \\ q_5 \\ q_6 \end{array} \right\}, \tag{11.3c}$$

$$\{\mathbf{u}\} = [\mathbf{N}]\{\mathbf{q}\}. \tag{11.3d}$$

The matrix $\mathbf{N}$ is the array of shape functions used for the axisymmetric triangular element. The symbol N_i indicates individual shape functions.

The equations for the axisymmetric shape functions are identical to those for the plane stress element (equations 8.5). Thus,

$$N_1(r, \; z) = \frac{1}{2A_e}[(r_2z_3 - r_3z_2) - (z_3 - z_2)r + (r_3 - r_2)z], \tag{11.4a}$$

$$N_2(r, \; z) = \frac{1}{2A_e}[(r_3z_1 - r_1z_3) - (z_1 - z_3)r + (r_1 - r_3)z], \tag{11.4b}$$

$$N_3(r, \; z) = \frac{1}{2A_e}[(r_1z_2 - r_2z_1) - (z_2 - z_1)r + (r_2 - r_1)z]. \tag{11.4c}$$

Numerical subscripts refer to element nodes. The term A_e is the area of the element's triangular cross section,

$$A_e = \frac{1}{2} \begin{vmatrix} 1 & r_1 & z_1 \\ 1 & r_2 & z_2 \\ 1 & r_3 & z_3 \end{vmatrix}. \tag{11.5}$$

11.4.2. Element Stiffness Matrix: Physical dof

The element stiffness matrix is developed from the equation

$$[\mathbf{K}] = \int_V [\mathbf{B}^T][\mathbf{E}][\mathbf{B}] \, dV. \qquad (6.22d)$$

The elastic matrix $\mathbf{E}$ for the isotropic axisymmetric condition was developed earlier and is expressed in equation 11.1c.

The strain–displacement matrix $\mathbf{B}$ is given by

$$[\mathbf{B}] = [\mathbf{d}][\mathbf{N}]. \qquad (11.6a)$$

Substituting for operator matrix $\mathbf{d}$ from equation 11.2 and matrix $\mathbf{N}$ from equation 11.3,

$$[\mathbf{B}] = \begin{bmatrix} \frac{\partial}{\partial r} & 0 \\ \frac{1}{r} & 0 \\ 0 & \frac{\partial}{\partial z} \\ \frac{\partial}{\partial z} & \frac{\partial}{\partial r} \end{bmatrix} \begin{bmatrix} N_1 & 0 & N_2 & 0 & N_3 & 0 \\ 0 & N_1 & 0 & N_2 & 0 & N_3 \end{bmatrix}, \qquad (11.6b)$$

$$[\mathbf{B}] = \begin{bmatrix} \frac{\partial N_1}{\partial r} & 0 & \frac{\partial N_2}{\partial r} & 0 & \frac{\partial N_3}{\partial r} & 0 \\ \frac{N_1}{r} & 0 & \frac{N_2}{r} & 0 & \frac{N_3}{r} & 0 \\ 0 & \frac{\partial N_1}{\partial z} & 0 & \frac{\partial N_2}{\partial z} & 0 & \frac{\partial N_3}{\partial z} \\ \frac{\partial N_1}{\partial z} & \frac{\partial N_1}{\partial r} & \frac{\partial N_2}{\partial z} & \frac{\partial N_2}{\partial r} & \frac{\partial N_3}{\partial z} & \frac{\partial N_3}{\partial r} \end{bmatrix}. \qquad (11.6c)$$

The matrix $\mathbf{B}$ for the axisymmetric 3-node, 6-dof triangular element differs from the matrix $\mathbf{B}$ for the plane stress constant strain triangular element. The axisymmetric matrix $\mathbf{B}$ contains an additional row, arising as a result of the algebraic multiplier $1/r$ in matrix $\mathbf{d}$. Moreover, when the operations indicated in equation 11.6c are performed on the shape functions N_i expressed in equations 11.4, the resulting matrix $\mathbf{B}$ is not an array of constants as it is for the plane stress constant strain triangular element:

$$[\mathbf{B}] = \frac{1}{2A_e} \begin{bmatrix} -(z_3 - z_2) & 0 & | & -(z_1 - z_3) & 0 & | & -(z_2 - z_1) & 0 \\ \frac{2A_e N_1(r, z)}{r} & 0 & | & \frac{2A_e N_2(r, z)}{r} & 0 & | & \frac{2A_e N_3(r, z)}{r} & 0 \\ 0 & r_3 - r_2 & | & 0 & r_1 - r_3 & | & 0 & r_2 - r_1 \\ r_3 - r_2 & -(z_3 - z_2) & | & r_1 - r_3 & -(z_1 - z_3) & | & r_2 - r_1 & -(z_2 - z_1) \end{bmatrix}. \qquad (11.6d)$$

Row 2 of the matrix $\mathbf{B}$ shown above contains functions of r and z that range throughout the triangular cross section.

Adapting the volume integral equation 6.22d, cited earlier, to the axisymmetric condition,

$$[\mathbf{K}] = \int_A \int_0^{2\pi} [\mathbf{B}^\mathrm{T}(\mathbf{r}, \; \mathbf{z})][\mathbf{E}][\mathbf{B}(\mathbf{r}, \; \mathbf{z})]r \; d\theta \; dA \tag{11.7a}$$

$$= 2\pi \int_A [\mathbf{B}^\mathrm{T}(\mathbf{r}, \; \mathbf{z})][\mathbf{E}][\mathbf{B}(\mathbf{r}, \; \mathbf{z})]r \; dA. \tag{11.7b}$$

The integrand of equation 11.7b is a 6×6 matrix of functions of r and z. It is easily integrated over the area of the triangular cross section using numerical methods. However, in applying a quadrature rule to equation 11.7b, terms whose denominators are the coordinate r must be evaluated. Consequently, the axisymmetric element being treated cannot have an integration point on the axis of symmetry where $r_k = 0$. If an element must have an edge of the triangular cross section on the axis, then the quadrature rule described in Section 8.5.3 cannot be used because its integration points are at the midpoint of each side of the triangle.

An alternative three-point quadrature rule may be employed. For it, the locations of the three integration points are interior to and near each vertex of the triangle. The natural triangular area coordinates $(\zeta_1, \; \zeta_2, \; \zeta_3)$ for these three integration points are $(\frac{2}{3}, \frac{1}{6}, \frac{1}{6})$, $(\frac{1}{6}, \frac{2}{3}, \frac{1}{6})$, $(\frac{1}{6}, \frac{1}{6}, \frac{2}{3})$. Area coordinates for triangles were discussed in Chapter 8. The cylindrical coordinates for the integration points are obtained from the area coordinate transformation (equations 8.17):

$$r = r_1\zeta_1 + r_2\zeta_2 + r_3\zeta_3 \quad \text{and} \quad z = z_1\zeta_1 + z_2\zeta_2 + z_3\zeta_3. \tag{11.8}$$

The weight factor for each integration point is $\frac{1}{3}A_e$, the same as for the rule described in Section 8.5.3.

Equation 11.7b in terms of either three-point quadrature rule is

$$[\mathbf{K}] = \frac{2\pi A_e}{3} \sum_{k=1}^{3} \left([\mathbf{B}^\mathrm{T}(r_k, \; z_k)][\mathbf{E}][\mathbf{B}(r_k, \; z_k)]r_k \right). \tag{11.9}$$

An approximate procedure can be employed which reduces the integrand of equation 11.7b to an array of constants and avoids problems with the axis of symmetry. The matrix **B** is simply evaluated at the centroid of the triangular cross section of the element. That is, replace the variables r and z in the second row of equation 11.6d with

$$r_m = \frac{r_1 + r_2 + r_3}{3} \quad \text{and} \quad z_m = \frac{z_1 + z_2 + z_3}{3} \tag{11.10}$$

Equation 11.7b then becomes a matrix of constants,

$$[\mathbf{K}] = 2\pi A_e \left[\mathbf{B}^\mathrm{T}(\mathbf{r}_m, \mathbf{z}_m) \right][\mathbf{E}][\mathbf{B}(\mathbf{r}_m, \mathbf{z}_m)]r_m. \tag{11.11}$$

Acceptable results can be achieved using equation 11.11 as long as the axisymmetric structure is discretized into a fine triangular mesh, especially where the stress distribution varies sharply.

To illustrate the determination of the stiffness matrix **K** for an isotropic axisymmetric triangular 3-node, 6-dof finite element, consider the element shown in Figure 11.4. The integration points are shown for the quadrature rule described above. The

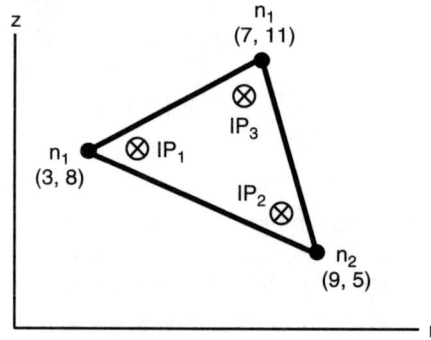

Figure 11.4. Example element.

coordinates are in inches. The elastic modulus $E = 30,000$ ksi and Poisson's ratio $v = \frac{1}{4}$. The area A_e of the element is given as

$$A_e = \frac{1}{2} \begin{vmatrix} 1 & 3 & 8 \\ 1 & 9 & 5 \\ 1 & 7 & 11 \end{vmatrix} = 15.000 \text{ in}^2.$$

The elastic matrix **E** for the element (equation 11.1c) is written as

$$[\mathbf{E}] = 1000 \begin{bmatrix} 36 & 12 & 12 & 0 \\ 12 & 36 & 12 & 0 \\ 12 & 12 & 36 & 0 \\ 0 & 0 & 0 & 12 \end{bmatrix} \text{ kips/in.}$$

The matrix **B** (equation 11.6d) is given as

$$[\mathbf{B}] = \frac{1}{30} \begin{bmatrix} -6 & 0 & 3 & 0 & 3 & 0 \\ \frac{30N_1}{r} & 0 & \frac{30N_2}{r} & 0 & \frac{30N_3}{r} & 0 \\ 0 & -2 & 0 & -4 & 0 & 6 \\ -2 & 6 & -4 & 3 & 6 & 3 \end{bmatrix} \text{ in}^{-1}.$$

The expressions $30N_i/r$ in matrix **B** are written using equations 11.4:

$$\frac{30N_1}{r} = \frac{64}{r} - 6 - \frac{27z}{r}, \qquad \frac{30N_2}{r} = \frac{22}{r} + 3 - \frac{4z}{r}, \qquad \frac{30N_3}{r} = -\frac{57}{r} + 3 + \frac{6z}{r}.$$

The cylindrical coordinates of the three integration points IP_1, IP_2, and IP_3 for the three-point quadrature rule are described above and are obtained with equation 11.8. They are, respectively, $(\frac{14}{3}, 8)$, $(\frac{23}{3}, \frac{13}{2})$, $(\frac{20}{3}, \frac{57}{6})$. Evaluating the matrix **B** at the first integration point $(\frac{14}{3}, 8)$,

$$[\mathbf{B}] = \frac{1}{30} \begin{bmatrix} -6 & 0 & 3 & 0 & 3 & 0 \\ \frac{32}{7} & 0 & \frac{8}{7} & 0 & \frac{8}{7} & 0 \\ 0 & -2 & 0 & -4 & 0 & 6 \\ -2 & 6 & -4 & 3 & 6 & 3 \end{bmatrix} \text{ in}^{-1}.$$

Evaluating the integrand matrix of equation 11.9 at the first integration point and multiplying by 2π and the weight factor $\frac{1}{3}A_e = 5.000$,

$$[\mathbf{K}_1] = 1000 \begin{bmatrix} 226.115 & 30.159 & -50.425 & 1.676 & -89.520 & -31.835 \\ 30.159 & 93.829 & 30.997 & 11.729 & -86.289 & -105.558 \\ -50.425 & 30.997 & 103.353 & -55.292 & 25.163 & 24.295 \\ 1.676 & 11.729 & -55.292 & 111.422 & 3.351 & -123.150 \\ -89.520 & -86.289 & 25.163 & 3.351 & 142.449 & 82.938 \\ -31.835 & -105.558 & 24.295 & -123.150 & 82.938 & 228.708 \end{bmatrix} \text{ kips/in.}$$

Repeating the process for the remaining two integration points produces the other two numerical matrices $\mathbf{K}_2$ and $\mathbf{K}_3$. The element stiffness matrix is the sum of these three matrices:

$$[\mathbf{K}] = [\mathbf{K}_1] + [\mathbf{K}_2] + [\mathbf{K}_3]$$

$$= 1000 \begin{bmatrix} 857.100 & 165.876 & -355.768 & 92.991 & -513.098 & -258.867 \\ 165.876 & 382.018 & 118.124 & 47.752 & -359.398 & -429.770 \\ -355.768 & 118.124 & 494.630 & -241.274 & 141.247 & 123.150 \\ 92.991 & 47.752 & -241.274 & 453.646 & -2.513 & -501.398 \\ -513.098 & -359.398 & 141.247 & -2.513 & 663.024 & 361.911 \\ -258.867 & -429.770 & 123.150 & -501.398 & 361.911 & 931.168 \end{bmatrix} \text{ kips/in.}$$

The element stiffness matrix can be obtained more easily but slightly less accurately using the approximate procedure outlined with equation 11.11. Employing this procedure,

$$[\mathbf{K}_{approx}] = 1000 \begin{bmatrix} 800.103 & 165.876 & -344.274 & 92.991 & -503.448 & -258.867 \\ 165.876 & 382.018 & 118.124 & 47.752 & -359.398 & -429.770 \\ -344.274 & 118.124 & 477.147 & -241.274 & 158.799 & 123.150 \\ 92.991 & 47.752 & -241.274 & 453.646 & -2.513 & -501.398 \\ -503.448 & -359.398 & 158.799 & -2.513 & 636.321 & 361.911 \\ -258.867 & -429.770 & 123.150 & -501.398 & 361.911 & 931.168 \end{bmatrix} \text{ kips/in.}$$

Only 9 of the 36 elements of matrix $\mathbf{K}_{approx}$ differ from the corresponding elements of matrix $\mathbf{K}$. The other 27 elements are exactly the same. The magnitudes of these entries in matrix K_{approx} differ from 1.8 to 12.4% from the corresponding ones in matrix $\mathbf{K}$. The average error among the 9 is 5.5%.

11.4.3. Element Stiffness Matrix: Generalized dof

The axisymmetric 3-node, 6-dof triangular element stiffness matrix may also be developed in terms of generalized dof. Generalized dof were introduced in Chapter 6 and used extensively in Chapter 10 on thin plate flexure elements.

The displacement functions for the axisymmetric 3-node, 6-dof triangular element in terms of generalized dof are

$$u(r,\ z) = D_1 + D_2 r + D_3 z, \tag{11.12a}$$

$$w(r,\ z) = D_4 + D_5 r + D_6 z, \tag{11.12b}$$

$$\left\{ \begin{array}{c} u \\ w \end{array} \right\} = \left[\begin{array}{cccccc} 1 & r & z & 0 & 0 & 0 \\ 0 & 0 & 0 & 1 & r & z \end{array} \right] \left\{ \begin{array}{c} D_1 \\ D_2 \\ D_3 \\ D_4 \\ D_5 \\ D_6 \end{array} \right\}, \tag{11.12c}$$

$$\{\mathbf{u}\} = [\mathbf{g}]\{\mathbf{D}\}. \tag{11.12d}$$

The matrix $\mathbf{g}$ is the appropriate array of single-term generic functions. The vector $\mathbf{D}$ is the array of constants in the displacement functions. These constants are the generalized dof.

The matrix $\mathbf{h}$ is the matrix $\mathbf{g}$ evaluated successively by rows at each node:

$$[\mathbf{h}] = \left[\begin{array}{cccccc} 1 & r_1 & z_1 & 0 & 0 & 0 \\ 0 & 0 & 0 & 1 & r_1 & z_1 \\ 1 & r_2 & z_2 & 0 & 0 & 0 \\ 0 & 0 & 0 & 1 & r_2 & z_2 \\ 1 & r_3 & z_3 & 0 & 0 & 0 \\ 0 & 0 & 0 & 1 & r_3 & z_3 \end{array} \right]. \tag{11.13}$$

Following the procedure introduced in Chapter 6, the element stiffness matrix $\mathbf{K}$ based on physical dof is expressed as a triple product of forms of its matrix $\mathbf{h}$ and stiffness matrix $\mathbf{K}_D$ based on generalized dof:

$$[\mathbf{K}] = [\mathbf{h}^{-T}][\mathbf{K}_D][\mathbf{h}^{-1}]. \tag{11.14}$$

The matrix $\mathbf{K}_D$ as introduced in Chapter 6 (equation 6.37f) is given as

$$[\mathbf{K}_D] = \int_V [\mathbf{B}_D^T][\mathbf{E}][\mathbf{B}_D] \, dV.$$

Adapting the equation 6.37f to the axisymmetric condition,

$$[\mathbf{K}_D] = 2\pi \int_A [\mathbf{B}_D^T][\mathbf{E}][\mathbf{B}_D] r \, dA. \tag{11.15}$$

The elastic matrix **E** is the same as before (equation 11.1c).

The strain–displacement matrix $\mathbf{B}_D$ based on generalized dof is the product of the operator matrix **d** (equation 11.2) and the generic function matrix **g** (equation 11.12):

$$[\mathbf{B}_D] = [\mathbf{d}][\mathbf{g}]$$

$$= \begin{bmatrix} \frac{\partial}{\partial r} & 0 \\ \frac{1}{r} & 0 \\ 0 & \frac{\partial}{\partial z} \\ \frac{\partial}{\partial z} & \frac{\partial}{\partial r} \end{bmatrix} \begin{bmatrix} 1 & r & z & 0 & 0 & 0 \\ 0 & 0 & 0 & 1 & r & z \end{bmatrix} = \begin{bmatrix} 0 & 1 & 0 & 0 & 0 & 0 \\ \frac{1}{r} & 1 & \frac{z}{r} & 0 & 0 & 0 \\ 0 & 0 & 0 & 0 & 0 & 1 \\ 0 & 0 & 1 & 0 & 1 & 0 \end{bmatrix}. \tag{11.16}$$

Substituting for matrices **E** and $\mathbf{B}_D$ (equations 11.1c and 11.16) in equation 11.15 and carrying out the multiplications indicated,

$$[\mathbf{K}_D] = \frac{2\pi E}{(1+v)(1-2v)} \int_A \begin{bmatrix} (1-v)\frac{1}{r} & 1 & (1-v)\frac{z}{r} & 0 & 0 & v \\ 1 & 2r & z & 0 & 0 & (2v)r \\ (1-v)\frac{z}{r} & z & (1-v)\frac{z^2}{r} + \left(\frac{1-2v}{2}\right)r & 0 & \left(\frac{1-2v}{2}\right)r & (v)r \\ 0 & 0 & 0 & 0 & 0 & 0 \\ 0 & 0 & \left(\frac{1-2v}{2}\right)r & 0 & \left(\frac{1-2v}{2}\right)r & 0 \\ v & (2v)r & (v)z & 0 & 0 & (1-v)r \end{bmatrix} dA. \tag{11.17}$$

Although equation 11.17 may be integrated explicitly over the triangular area of the element, the process is lengthy and the resulting expressions are cumbersome. Some contain the functions $\ln(r_i/r_j)$ and $1/(r_i - r_j)$, where r_i and r_j are nodal coordinates on the triangular cross section. With these expressions, if a node falls on the axis of symmetry where $r_j = 0$ or if a side of the element is parallel to the axis of symmetry where $r_i - r_j = 0$, special techniques must be used to evaluate the matrix $\mathbf{K}_D$.

It is simpler and more efficient to employ the quadrature rule described earlier in this chapter. Equation 11.15 expressed in terms of the quadrature rule is

$$[\mathbf{K}_D] = \frac{2\pi A_e}{3} \sum_{k=1}^{3} \left([\mathbf{B}_D^T(\mathbf{r}_k, \ \mathbf{z}_k)][\mathbf{E}][\mathbf{B}_D(\mathbf{r}_k, \ \mathbf{z}_k)]r_k\right). \tag{11.18}$$

Because the integration points for the quadrature rule are interior to the element's triangular cross section, elements whose nodes fall on the axis of symmetry do not require special handling.

The quadrature rule may be applied directly to equation 11.17. However, the numerical process will be more simply carried out by evaluating the matrices $\mathbf{B}_D^T$ and $\mathbf{B}_D$ with the quadrature rule and then carrying out the multiplications indicated in equation 11.18.

The approximate determination of matrix $\mathbf{K}$, described earlier in equation 11.11, may be adapted to the generalized dof procedure by evaluating the matrices $\mathbf{B}_D^T$ and $\mathbf{B}_D$ at the centroid of the element's triangular area. The result, as before, is a matrix of constants, not requiring a quadrature rule and reducing equation 11.15 to

$$[\mathbf{K}_D] = 2\pi A_e \big[\mathbf{B}_D^T(\mathbf{r}_m,\ \mathbf{z}_m)\big][\mathbf{E}][\mathbf{B}_D(\mathbf{r}_m,\ \mathbf{z}_m)]r_m. \tag{11.19}$$

The terms r_m and z_m are the coordinates of the triangle's centroid (equation 11.10).

The element's stiffness matrix based on physical dof is obtained by executing the triple product of matrices $\mathbf{h}^{-T}$, $\mathbf{K}_D$, and $\mathbf{h}^{-1}$, as indicated by equation 11.14.

To illustrate the determination of the axisymmetric triangular element's stiffness matrix $\mathbf{K}$ based on physical dof but using generalized dof, consider the example in the previous section which used physical dof. The element's cylindrical coordinates of nodes and its area, elastic matrix $\mathbf{E}$, and cylindrical coordinates of integration points are as given in the example above.

The matrix $\mathbf{B}_D$ (equation 11.16) evaluated at the first integration point is given as

$$[\mathbf{B}_{D1}] = \begin{bmatrix} 0 & 1 & 0 & 0 & 0 & 0 \\ \frac{3}{14} & 1 & \frac{12}{7} & 0 & 0 & 0 \\ 0 & 0 & 0 & 0 & 0 & 1 \\ 0 & 0 & 1 & 0 & 1 & 0 \end{bmatrix}.$$

Evaluating the integrand matrix of equation 11.18 at the first integration point by multiplying out the triple product of numerical matrices $\mathbf{B}_{D1}^T$, $\mathbf{E}$, and $\mathbf{B}_{D1}$ and multiplying the result of this triple product by the scalars 2π and weight factor $\frac{1}{3}A_e = 5.000$ produces

$$[\mathbf{K}_{D1}] = 1000 \begin{bmatrix} 242.351 & 1,507.964 & 1,938.811 & 0.000 & 0.000 & 376.991 \\ 1,507.964 & 14,074.335 & 12,063.716 & 0.000 & 0.000 & 3,518.584 \\ 1,938.811 & 12,063.716 & 17,269.784 & 0.000 & 1,759.292 & 3,015.929 \\ 0.000 & 0.000 & 0.000 & 0.000 & 0.000 & 0.000 \\ 0.000 & 0.000 & 1,759.292 & 0.000 & 1,759.292 & 0.000 \\ 376.991 & 3,518.584 & 3,015.929 & 0.000 & 0.000 & 5,277.876 \end{bmatrix}.$$

Repeating the process for the remaining two integration points produces the other two numerical matrices $\mathbf{K}_{D2}$ and $\mathbf{K}_{D3}$. The element stiffness matrix based on generalized dof is

$$[\mathbf{K}_D] = [\mathbf{K}_{D1}] + [\mathbf{K}_{D2}] + [\mathbf{K}_{D3}]$$

$$= 1000 \begin{bmatrix} 559.516 & 4{,}523.893 & 4{,}509.317 & 0.000 & 0.000 & 1{,}130.973 \\ 4{,}523.893 & 57{,}302.650 & 36{,}191.147 & 0.000 & 0.000 & 14{,}325.663 \\ 4{,}509.317 & 36{,}191.147 & 44{,}216.521 & 0.000 & 7{,}162.831 & 9{,}047.787 \\ 0.000 & 0.000 & 0.000 & 0.000 & 0.000 & 0.000 \\ 0.000 & 0.000 & 7{,}162.831 & 0.000 & 7{,}162.831 & 0.000 \\ 1{,}130.973 & 14{,}325.663 & 9{,}047.787 & 0.000 & 0.000 & 21{,}488.494 \end{bmatrix}.$$

The matrix $\mathbf{h}$ (equation 11.13) is given as

$$[\mathbf{h}] = \begin{bmatrix} 1 & 3 & 8 & 0 & 0 & 0 \\ 0 & 0 & 0 & 1 & 3 & 8 \\ 1 & 9 & 5 & 0 & 0 & 0 \\ 0 & 0 & 0 & 1 & 9 & 5 \\ 1 & 7 & 11 & 0 & 0 & 0 \\ 0 & 0 & 0 & 1 & 7 & 11 \end{bmatrix}.$$

Next this numerical matrix $\mathbf{h}$ is inverted to produce matrix $\mathbf{h}^{-1}$. Then matrix $\mathbf{h}^{-1}$ is transposed to produce matrix $\mathbf{h}^{-T}$. Finally, the triple product of numerical matrices $\mathbf{h}^{-T}$, $\mathbf{K}_D$, and $\mathbf{h}^{-1}$ is carried out to produce the element's stiffness matrix $\mathbf{K}$ based on physical dof (equation 11.14). It is left as an exercise for the reader to carry out these remaining calculations.

The element stiffness matrix $\mathbf{K}$ can be obtained more easily but somewhat less accurately using the approximate procedure suggested with equations 11.19 and 11.14. It is left as an exercise for the reader to carry out these calculations.

The use of generalized dof to determine the element stiffness matrix of the axisymmetric 3-node, 6-dof triangular finite element is simpler and somewhat faster than using physical dof.

11.5. QUADRILATERAL FINITE ELEMENTS (4-NODE, 8-DOF)

Quadrilateral axisymmetric finite elements are an alternative formulation to comparable triangular elements subjected to axisymmetric loads. The quadrilateral shape is as versatile as the triangle in modeling structural shapes. Because quadrilateral elements possess more nodes than triangular elements, they require displacement and shape functions of higher polynomial degree. Thus there will be opportunities for better accuracy with them.

As with the plane stress condition, quadrilateral elements in the axisymmetric condition involve more complicated mathematics than do triangular elements. Numerical quadrature will almost always be the preferred procedure to obtain their stiffness coefficients and equivalent nodal forces.

11.5.1. Shape and Displacement Functions

The development of the 4-node, 8-dof quadrilateral element for the axisymmetric condition is similar to its development for plane stress, as described in Section 9.3.

Figure 11.5 illustrates an axisymmetric 4-node, 8-dof quadrilateral element. Comparing it to the plane stress quadrilateral element shown in Figure 9.4, the only difference is that the global rectilinear axes are r–z for the axisymmetric element. The local natural ζ–η axes are the same. These natural axes are "skewed" coordinate lines, which bisect the angle between their corresponding sides of the quadrilateral.

The natural coordinate values for the element's nodes are shown in Figure 11.5. The origin of the natural coordinate axes is defined to be at

$$r_g = \tfrac{1}{4}(r_1 + r_2 + r_3 + r_4), \tag{11.20a}$$

$$z_g = \tfrac{1}{4}(z_1 + z_2 + z_3 + z_4). \tag{11.20b}$$

As with the plane stress quadrilateral element, the axisymmetric element is best treated in terms of the natural coordinates ζ–η and thought of in terms of r–z and ζ–η spaces. The calculations are done entirely in terms of these natural coordinates, in ζ–η space. Figure 11.6 illustrates the element in r–z and ζ–η spaces.

The equations for the shape and displacement functions for the axisymmetric element are identical to those for the plane stress element:

$$\left\{ \begin{array}{c} u(\zeta,\ \eta) \\ v(\zeta,\ \eta) \end{array} \right\} = \begin{bmatrix} N_1(\zeta,\ \eta) & 0 & N_2(\zeta,\ \eta) & 0 & N_3(\zeta,\ \eta) & 0 & N_4(\zeta,\ \eta) & 0 \\ 0 & N_1(\zeta,\ \eta) & 0 & N_2(\zeta,\ \eta) & 0 & N_3(\zeta,\ \eta) & 0 & N_4(\zeta,\ \eta) \end{bmatrix} \left\{ \begin{array}{c} q_1 \\ q_2 \\ q_3 \\ q_4 \\ q_5 \\ q_6 \\ q_7 \\ q_8 \end{array} \right\},$$

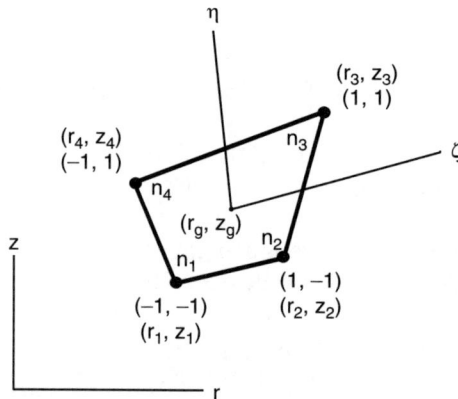

$$\tag{11.21}$$

Figure 11.5. Quadrilateral finite element r–z space.

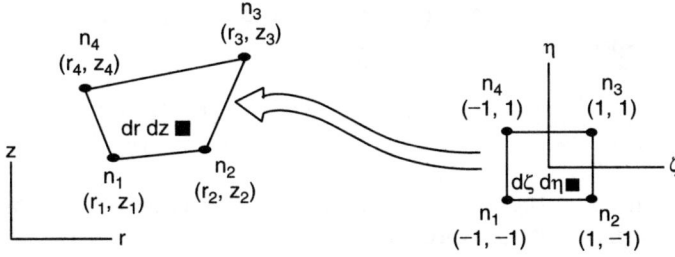

Figure 11.6. Element spaces.

where

$$N_i(\zeta, \eta) = \tfrac{1}{4}(1 + \zeta_i\zeta)(1 + \eta_i\eta). \tag{11.22}$$

11.5.2. Isoparametric Transformation

The discussion of the transformation from ζ–η to r–z space is identical to what is given for the plane stress quadrilateral element in Chapter 9. The reader is referred to Section 9.3.1 and equations 9.13–9.19. To relate this discussion to the axisymmetric condition, simply replace the plane stress x and y coordinates with the axisymmetric r and z coordinates wherever they occur in the text and equations in Section 9.3.1.

The basic coordinate transformation from ζ–η to r–z is

$$r = r_1 N_1(\zeta, \eta) + r_2 N_2(\zeta, \eta) + r_3 N_3(\zeta, \eta) + r_4 N_4(\zeta, \eta), \tag{11.23a}$$

$$z = z_1 N_1(\zeta, \eta) + z_2 N_2(\zeta, \eta) + z_3 N_3(\zeta, \eta) + z_4 N_4(\zeta, \eta). \tag{11.23b}$$

11.5.3. Element Stiffness Matrix

To calculate the axisymmetric element stiffness matrix in ζ–η space, the plane stress integral equations for the matrix $\mathbf{K}$ (equations 9.20) are adjusted to reflect the axisymmetric condition:

$$
\begin{aligned}
[\mathbf{K}] &= 2\pi \int_A [\mathbf{B}^{\mathrm{T}}(\zeta, \eta)][\mathbf{E}][\mathbf{B}(\zeta, \eta)]r(\zeta, \eta)\, dA \\
&= 2\pi \int_{-1}^{1} \int_{-1}^{1} [\mathbf{B}^{\mathrm{T}}(\zeta, \eta)][\mathbf{E}][\mathbf{B}(\zeta, \eta)]|J(\zeta, \eta)|r(\zeta, \eta)\, d\zeta\, d\eta.
\end{aligned}
\tag{11.24a}
$$

Expressed as a Gauss quadrature rule,

$$[\mathbf{K}] = 2\pi \sum_{k=1}^{n} [\mathbf{B}^{\mathrm{T}}(\zeta_k, \eta_k)][\mathbf{E}][\mathbf{B}(\zeta_k, \eta_k)]|J(\zeta_k, \eta_k)|r(\zeta_k, \eta_k)H_k. \tag{11.24b}$$

The matrix $\mathbf{K}$ is efficiently calculated using equation 11.24b.

The quantity $r(\zeta_k, \eta_k)$ is calculated using equation 11.23a. The elastic matrix $\mathbf{E}$ is the same as for the triangular axisymmetric element and is expressed in equation 11.1c.

The Jacobian $|J(\zeta_k,\ \eta_k)|$ is obtained using equation 9.17c, replacing the x and y in it with r and z, respectively:

$$|J(\zeta_k,\ \eta_k)| = \frac{\partial r(\zeta_k,\ \eta_k)}{\partial \zeta} \cdot \frac{\partial z(\zeta_k,\ \eta_k)}{\partial \eta} - \frac{\partial r(\zeta_k,\ \eta_k)}{\partial \eta} \cdot \frac{\partial z(\zeta_k,\ \eta_k)}{\partial \zeta}. \tag{11.25}$$

The derivatives in equation 11.25 are evaluated using equations 9.37a–d and replacing x and y with r and z, respectively:

$$\frac{\partial r(\zeta_k,\ \eta_k)}{\partial \zeta} = -\frac{r_1}{4}(1 - \eta_k) + \frac{r_2}{4}(1 - \eta_k) + \frac{r_3}{4}(1 + \eta_k) - \frac{r_4}{4}(1 + \eta_k), \tag{11.26a}$$

$$\frac{\partial z(\zeta_k,\ \eta_k)}{\partial \zeta} = -\frac{z_1}{4}(1 - \eta_k) + \frac{z_2}{4}(1 - \eta_k) + \frac{z_3}{4}(1 + \eta_k) - \frac{z_4}{4}(1 + \eta_k), \tag{11.26b}$$

$$\frac{\partial r(\zeta_k,\ \eta_k)}{\partial \eta} = -\frac{r_1}{4}(1 - \zeta_k) - \frac{r_2}{4}(1 + \zeta_k) + \frac{r_3}{4}(1 + \zeta_k) + \frac{r_4}{4}(1 - \zeta_k), \tag{11.26c}$$

$$\frac{\partial z(\zeta_k,\ \eta_k)}{\partial \eta} = -\frac{z_1}{4}(1 - \zeta_k) - \frac{z_2}{4}(1 + \zeta_k) + \frac{z_3}{4}(1 + \zeta_k) + \frac{z_4}{4}(1 - \zeta_k). \tag{11.26d}$$

The r–z coordinates with numerical subscripts are the coordinates of the nodes of the element. Those with the subscript k are the coordinates of the integration points.

The strain–displacement matrix $\mathbf{B}$ is given by

$$[\mathbf{B}] = [\mathbf{d}][\mathbf{N}]. \tag{11.27a}$$

For the 4-node, 8-dof quadrilateral axisymmetric element under development here, the operator matrix $\mathbf{d}$ is identical to the operator matrix $\mathbf{d}$ for the axisymmetric triangular element developed earlier (equation 11.2). It must account for the presence of the lateral strain ϵ_θ.

The element's array of shape functions, that is, matrix $\mathbf{N}$, is shown in equation 11.21. It is identical to the matrix $\mathbf{N}$ for the plane stress rectangular element developed in Chapter 9 (equation 9.2). The matrix $\mathbf{N}$ must account for the presence of four nodes on the element.

Thus the matrix $\mathbf{B}$ for the 4-node, 8-dof quadrilateral axisymmetric element is written as

$$[\mathbf{B}] = \begin{bmatrix} \frac{\partial}{\partial r} & 0 \\ \frac{1}{r} & 0 \\ 0 & \frac{\partial}{\partial z} \\ \frac{\partial}{\partial z} & \frac{\partial}{\partial r} \end{bmatrix} \begin{bmatrix} N_1 & 0 & N_2 & 0 & N_3 & 0 & N_4 & 0 \\ 0 & N_1 & 0 & N_2 & 0 & N_3 & 0 & N_4 \end{bmatrix} \tag{11.27b}$$

$$= \begin{bmatrix} \frac{\partial N_1}{\partial r} & 0 & \frac{\partial N_2}{\partial r} & 0 & \frac{\partial N_3}{\partial r} & 0 & \frac{\partial N_4}{\partial r} & 0 \\ \frac{N_1}{r} & 0 & \frac{N_2}{r} & 0 & \frac{N_3}{r} & 0 & \frac{N_4}{r} & 0 \\ 0 & \frac{\partial N_1}{\partial z} & 0 & \frac{\partial N_2}{\partial z} & 0 & \frac{\partial N_3}{\partial z} & 0 & \frac{\partial N_4}{\partial z} \\ \frac{\partial N_1}{\partial z} & \frac{\partial N_1}{\partial r} & \frac{\partial N_2}{\partial z} & \frac{\partial N_2}{\partial r} & \frac{\partial N_3}{\partial z} & \frac{\partial N_3}{\partial r} & \frac{\partial N_4}{\partial z} & \frac{\partial N_4}{\partial r} \end{bmatrix}. \tag{11.27c}$$

The derivatives of shape functions with respect to r and z in equation 11.27c are obtained using equations 9.39a–b. The variables x and y must be replaced with r and z in the plane stress equations from Chapter 9. The result is

$$\frac{\partial N_i(\zeta_k, \eta_k)}{\partial r} = \frac{1}{|J(\zeta_k, \eta_k)|}\left(\frac{\partial z(\zeta_k, \eta_k)}{\partial \eta}\frac{\zeta_i}{4}(1 + \eta_i\eta_k) - \frac{\partial z(\zeta_k, \eta_k)}{\partial \zeta}\frac{\eta_i}{4}(1 + \zeta_i\zeta_k)\right),$$

$$\frac{\partial N_i(\zeta_k, \eta_k)}{\partial z} = \frac{1}{|J(\zeta_k, \eta_k)|}\left(-\frac{\partial r(\zeta_k, \eta_k)}{\partial \eta}\frac{\zeta_i}{4}(1 + \eta_i\eta_k) + \frac{\partial r(\zeta_k, \eta_k)}{\partial \zeta}\frac{\eta_i}{4}(1 + \zeta_i\zeta_k)\right).$$

$$(11.28)$$

The expression N_i/r in matrix **B** of equation 11.27c may be written from the shape function equations 11.22 and coordinate transformation equations 11.23:

$$\frac{N_i(\zeta_k, \eta_k)}{r(\zeta_k, \eta_k)} = \frac{(1 + \zeta_i\zeta_k)(1 + \eta_i\eta_k)/4}{r_1(1 - \zeta_k)(1 - \eta_k)/4 + r_2(1 + \zeta_k)(1 - \eta_k)/4 + r_3(1 + \zeta_k)(1 + \eta_k)/4 + r_4(1 - \zeta_k)(1 + \eta_k)}.$$

$$(11.29)$$

As before, for equations 11.29, the subscript i, or numerical subscript, identifies a node; the subscript k identifies an integration point.

The expression for $r(\zeta_k, \eta_k)$ in the denominator of equation 11.29 is the same term in the integrand of equation 11.24b.

Using the 2×2 Gauss rule, the element stiffness matrix **K** is given as

$$[\mathbf{K}] = 2\pi \sum_{k=1}^{4}[\mathbf{I}(\zeta_k, \eta_k) \cdot H_k]$$

$$= 2\pi\left(\mathbf{I}\left(-\frac{1}{\sqrt{3}}, -\frac{1}{\sqrt{3}}\right)\cdot\mathbf{1} + \mathbf{I}\left(\frac{1}{\sqrt{3}}, -\frac{1}{\sqrt{3}}\right)\cdot\mathbf{1} + \mathbf{I}\left(-\frac{1}{\sqrt{3}}, \frac{1}{\sqrt{3}}\right)\cdot\mathbf{1} + \mathbf{I}\left(\frac{1}{\sqrt{3}}, \frac{1}{\sqrt{3}}\right)\cdot\mathbf{1}\right),$$

$$(11.30a)$$

where

$$[\mathbf{I}(\zeta_k, \eta_k)] = [\mathbf{B}^{\mathrm{T}}(\zeta_k, \eta_k)][\mathbf{E}][\mathbf{B}(\zeta_k, \eta_k)]|J(\zeta_k, \eta_k)|r(\zeta_k, \eta_k). \qquad (11.30b)$$

If the element is rectangular, a special case of quadrilateral, equations 11.23–11.30 may be followed directly to determine its stiffness matrix. However, as was seen in Section 9.2, the quantities needed to implement equation 11.30b for a rectangular element may be simplified.

The quantity $r(\zeta_k, \eta_k)$ may be written from the geometry of the rectangle and equation 11.20a as

$$r(\zeta_k, \eta_k) = r_g + a\zeta_k. \qquad (11.31a)$$

The Jacobian for a rectangle is constant and may be written from equations 9.17 and 9.19 as

$$J = ab = \tfrac{1}{4}A_e. \qquad (11.31b)$$

The derivatives in the matrix **B** for a rectangle may be written from equations 9.8 and 9.9 as

$$\frac{\partial N_i(\zeta_k, \eta_k)}{\partial r} = \frac{1}{a}\frac{\partial N_i(\zeta_k, \eta_k)}{\partial \zeta} = \frac{1}{A_e}(b\zeta_i)(1 + \eta_i\eta), \qquad (11.31c)$$

$$\frac{\partial N_i(\zeta_k, \eta_k)}{\partial z} = \frac{1}{b}\frac{\partial N_i(\zeta_k, \eta_k)}{\partial \eta} = \frac{1}{A_3}(a\eta_i)(1 + \zeta_i\zeta). \qquad (11.31d)$$

To illustrate the use of equations 11.23–11.30 to calculate the stiffness matrix of a 4-node, 8-dof axisymmetric quadrilateral element, consider the axisymmetric element shown in Figure 11.7. The coordinates are in inches, the modulus $E = 30,000\,\text{ksi}$, and Poisson's ratio $\nu = \frac{1}{4}$.

The element shown has the same cross section as the example depicted in Figure 9.18. The Chapter 9 example illustrates the calculations for the 4-node, 8-dof plane stress quadrilateral. The data tabulated in Section 9.5.2 for nodal coordinates, coordinates of integration points, derivatives of shape functions with respect to ζ and η, and derivatives of shape functions with respect to x and y are identical to the corresponding values for the axisymmetric element if the x and y coordinates are replaced with r and z.

The matrix **B** for the axisymmetric element is expressed in equation 11.27c. To evaluate it at the first integration point, the derivatives of the four shape functions $N_i(\zeta_k, \eta_k)$ with respect to r and z, expressed in equations 11.28, may be obtained from the tabulated data in Section 9.5.2. Note that the values of derivatives shown there are not divided by the corresponding values of the Jacobian:

$$\frac{\partial N_1}{\partial r} = \frac{-0.788675}{6.873618} = -0.114740, \qquad \frac{\partial N_1}{\partial z} = -\frac{1.183013}{6.873618} = -0.172109,$$

$$\frac{\partial N_2}{\partial r} = \frac{1.091506}{6.873618} = 0.158797, \qquad \frac{\partial N_2}{\partial z} = -\frac{0.081143}{6.873618} = -0.011805,$$

$$\frac{\partial N_3}{\partial r} = \frac{0.211325}{6.873618} = 0.030744, \qquad \frac{\partial N_3}{\partial z} = \frac{0.316988}{6.873618} = 0.046117,$$

$$\frac{\partial N_4}{\partial r} = -\frac{0.514156}{6.873618} = -0.074801, \qquad \frac{\partial N_4}{\partial z} = \frac{0.947169}{6.873618} = 0.137798.$$

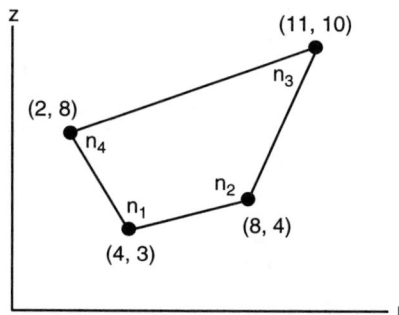

Figure 11.7. Example stiffness matrix.

The second row of the matrix **B** is calculated from the values of the shape functions at the integration point divided by the values of the coordinate r at the integration point (equation 11.29). The values of the shape functions are obtained from equation 11.22 or the expression in the numerator of equation 11.29. The value of the coordinate r is obtained from equation 11.23a or the expression in the denominator of equation 11.29. In each case, the expressions are evaluated at the first integration point:

$$\frac{N_1}{r} = \frac{0.622008}{4.645941} = 0.133882, \qquad \frac{N_2}{r} = \frac{0.166667}{4.645941} = 0.035874,$$

$$\frac{N_3}{r} = \frac{0.044658}{4.645941} = 0.009612, \qquad \frac{N_4}{r} = \frac{0.166667}{4.645941} = 0.035874.$$

Assembling these values of derivatives of shape functions with respect to r and z and of shape functions divided by the coordinate r, the matrix **B**, evaluated at the first integration point, is written as indicated from equation 11.27c:

$$[\mathbf{B}] = \begin{bmatrix} -0.114740 & 0 & 0.158797 & 0 & 0.030744 & 0 & -0.074801 & 0 \\ 0.133882 & 0 & 0.035874 & 0 & 0.009612 & 0 & 0.035874 & 0 \\ 0 & -0.172109 & 0 & -0.011805 & 0 & 0.046117 & 0 & 0.137798 \\ -0.172109 & -0.114740 & -0.011805 & 0.158797 & 0.046117 & 0.030744 & 0.137798 & -0.074801 \end{bmatrix} \text{ in}^{-1}.$$

The matrix **E** (equation 11.1c) is written as

$$[\mathbf{E}] = 1000 \begin{bmatrix} 36 & 12 & 12 & 0 \\ 12 & 36 & 12 & 0 \\ 12 & 12 & 36 & 0 \\ 0 & 0 & 0 & 12 \end{bmatrix} \text{ kips/in.}$$

The value of the Jacobian (equation 11.25) at the first integration point may be taken from the data listed in Section 9.5.2:

$$J = 6.873613.$$

The coordinate r (denominator in equation 11.29) at the first integration point was determined above as

$$r = 4.64594 \text{ in.}$$

The integrand matrix of equations 11.24 or 11.30, evaluated at the first integration point, is obtained by carrying out the operations indicated in the equations. The matrices $\mathbf{B}^{\mathrm{T}}$, **E**, and **B**, the Jacobian J, and the radius r, all evaluated at the first integration point, are multiplied and then multiplied by the scalar 2π. The result is

$$[\mathbf{K}_1] = 1000 \begin{bmatrix} 221.92 & 39.62 & -50.75 & -66.35 & -28.04 & -10.62 & 5.56 & 37.35 \\ 39.62 & 245.67 & -77.41 & -29.19 & -29.46 & -65.83 & -21.94 & -150.65 \\ -50.75 & -77.41 & 219.21 & -10.05 & 42.78 & 20.74 & -73.17 & 66.72 \\ -66.35 & -29.19 & -10.05 & 61.72 & 16.49 & 7.82 & 53.79 & -40.35 \\ -28.04 & -29.46 & 42.78 & 16.49 & 14.04 & 7.90 & 2.10 & 5.08 \\ -10.62 & -65.83 & 20.74 & 7.82 & 7.90 & 17.64 & 5.88 & 40.37 \\ 5.56 & -21.94 & -73.17 & 53.79 & 2.10 & 5.88 & 82.51 & -37.73 \\ 37.35 & -150.65 & 66.72 & -40.35 & 5.08 & 40.37 & -37.73 & 150.63 \end{bmatrix} \text{ kips/in.}$$

Repeating the process for the remaining three integration points produces the other three numerical matrices $\mathbf{K}_2$, $\mathbf{K}_3$, and $\mathbf{K}_4$. The element stiffness matrix is the sum of these four matrices:

$$[\mathbf{K}] = [\mathbf{K}_1] + [\mathbf{K}_2] + [\mathbf{K}_4] + [\mathbf{K}_4]$$

$$= 1000 \begin{bmatrix} 604.25 & -0.41 & -417.33 & 33.55 & -90.90 & -141.56 & 106.23 & 105.42 \\ -0.41 & 593.92 & 49.12 & -54.42 & -216.96 & -207.40 & -57.94 & -332.10 \\ -417.33 & 49.12 & 1367.87 & -563.27 & 82.32 & 139.80 & -515.40 & 374.35 \\ 36.55 & -54.42 & -563.27 & 1265.06 & -23.57 & -640.69 & 286.39 & -569.95 \\ -90.90 & -216.96 & 82.32 & -23.57 & 598.02 & 239.85 & -119.06 & 0.67 \\ -141.56 & -207.40 & 139.80 & -640.69 & 239.85 & 575.15 & -11.89 & 272.95 \\ 106.23 & -57.94 & -515.40 & 286.39 & -119.06 & -11.89 & 744.71 & -216.55 \\ 105.42 & -332.10 & 374.35 & -569.95 & 0.67 & 272.95 & -216.55 & 629.10 \end{bmatrix} \text{ kips/in.}$$

11.6. NODAL ACTIONS

When axisymmetric loads are applied to an axisymmetric structure, they produce axisymmetric patterns of displacements, strains, and stresses. These loads are specific cases of a concentrated force acting at a point, distributed line forces along a circular line, distributed surface forces acting over circular and axisymmetric areas, and body forces acting throughout a volume of the structure. The first three are illustrated in Figure 11.8.

11.6.1. Concentrated Axial Force

A concentrated force may act only along the axis of the structure, usually at its ends. The force may have only an axial component. Located elsewhere or with a radial component, it would not be axisymmetric. Such a force is inserted directly into the structure's global force vector at the position in the vector corresponding to its gdof.

11.6.2. Uniform Line Force

A uniformly distributed line force may act along a circular line which is perpendicular to the axis of the structure with its center on the axis, as shown in Figure 11.8. The force may have both axial and radial components, but not a circumferential

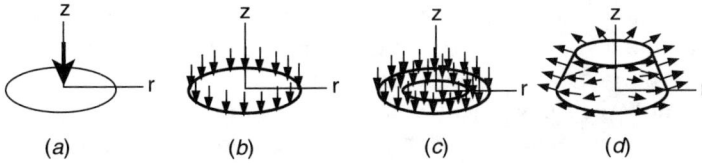

Figure 11.8. axisymmetric applied loads: (*a*) concentrated axial force; (*b*) uniform line force; (*c*) distributed surface force; (*d*) distributed surface force.

component. The magnitudes of the force and its components must be constant. Each of these restrictions must be met if the line force is axisymmetric.

Usually the analyst will discretize the structure so that a uniform line force acts on a circular ring node i of the structure. In so doing, the line load p_i (force per unit length of circular line) becomes a concentrated force P_i acting at node i on the element's cross section. The radial and axial components of P_i are inserted directly into the structure's global load vector at the positions corresponding to their gdof:

$$P_i = 2\pi r_i p_i. \tag{11.32a}$$

If the structure is not discretized so that the uniform line force occurs at a node, then equivalent forces must be computed and applied at the element's nodes to replace the original line force. For a uniform line force p_k acting along a circular ring at coordinates (r_k, z_k), the concentrated force P_k on the element's cross section is given as

$$P_k = 2\pi r_k p_k. \tag{11.32b}$$

The equivalent nodal forces are computed from equation 6.22f, adapted to the axisymmetric condition. Thus,

$$\{F_k\} = [\mathbf{N}^{\mathrm{T}}(\mathbf{r}_k, \ \mathbf{z}_k)]\{\mathbf{P}_k\}. \tag{11.32c}$$

For the triangular axisymmetric element,

$$\{F_b\} = \begin{bmatrix} N_1(r_k, z_k) & 0 \\ 0 & N_1(r_k, z_k) \\ N_2(r_k, z_k) & 0 \\ 0 & N_2(r_k, z_k) \\ N_3(r_k, z_k) & 0 \\ 0 & N_3(r_k, z_k) \end{bmatrix} \begin{Bmatrix} P_{kr} \\ P_{kz} \end{Bmatrix} = \begin{Bmatrix} N_1(r_k, z_k)P_{kr} \\ N_1(r_k, z_k)P_{kz} \\ N_2(r_k, z_k)P_{kr} \\ N_2(r_k, z_k)P_{kz} \\ N_3(r_k, z_k)P_{kr} \\ N_3(r_k, z_k)P_{kz} \end{Bmatrix}. \tag{11.32d}$$

The shape functions $N_i(r, z)$ for the triangular axisymmetric element are expressed in equations 11.4.

For the quadrilateral axisymmetric element, the matrix of shape functions $\mathbf{N}$ is expressed in equations 11.21 and its shape function entries are in equations 11.22. Note that since there are four shape functions, its matrix $\mathbf{N}^{\mathrm{T}}(\zeta_k, \eta_k)$ is 8×2. Note also that its shape functions $N_i(\zeta, \eta)$ are functions of the element's natural coordinates ζ and η. To write this matrix $\mathbf{N}^\mathrm{T}(\zeta_k, \eta_k)$, the coordinates (ζ_k, η_k) must be obtained by trial and error using the isoparametric transformation equations 11.23.

If the stiffness matrices for triangular axisymmetric elements are being computed with generalized dof, their equivalent nodal forces should be also. Equations 6.38d and 10.32, for a concentrated force not acting at a node, may be adapted for this purpose:

$$\{\mathbf{F}_k\} = [\mathbf{h}^{-\mathrm{T}}][\mathbf{g}_k^\mathrm{T}]\{\mathbf{P}_k\} = [\mathbf{h}^{-\mathrm{T}}] \begin{bmatrix} 1 & 0 \\ r_k & 0 \\ z_k & 0 \\ 0 & 1 \\ 0 & r_k \\ 0 & z_k \end{bmatrix} \begin{Bmatrix} P_{kr} \\ P_{kz} \end{Bmatrix} = [\mathbf{h}^{-\mathrm{T}}] \begin{Bmatrix} P_{kr} \\ r_k P_{kr} \\ z_k P_{kr} \\ P_{kz} \\ r_k P_{kz} \\ z_k P_{kz} \end{Bmatrix}. \qquad (11.32e)$$

The matrix $\mathbf{h}^{-\mathrm{T}}$ is the inverse of the transpose of the matrix $\mathbf{h}$ for the triangular axisymmetric element given by equation 11.13.

11.6.3. Distributed Surface Force

A distributed surface force may act on a circular area or a band of a circular area which is perpendicular to the axis of the structure with its center on the axis. It may also act on the axisymmetric outer or inner surface of the structure. Both situations are shown in Figure 11.8. The force may have both axial and radial components, but not a circumferential one. The magnitude of the force and its components may be functions of the radial and axial coordinates if the surface on which they act is a function of these coordinates. They may not, however, be functions of the circumferential coordinate θ. The structure must be discretized so that its axisymmetric distributed surface forces act on axisymmetric faces of the element. This results in a distributed line force $b(L)$ acting along an edge of the cross section of the element. The line force is most easily treated as a function of a line variable L which varies along the side of the element's cross section where the line force acts. Usually the variable L equals zero at one end of the side and the side's length L_s at the other end.

Equation 6.22e, adapted to the axisymmetric condition, is used for the computation of equivalent nodal forces of a distributed surface force. The integrals, however, must be along the edge of the element's cross section where the line force acts. Thus the equation becomes

$$\{\mathbf{F}_b\} = 2\pi \int_0^{L_s} [\mathbf{N}^\mathrm{T}(L)]\{\mathbf{b}(L)\}r(L)\ dL. \qquad (11.33)$$

Equation 11.33 may be used directly on the distributed line force $b(L)$ acting on side n_1–n_3 of the 3-node, 6-dof triangular axisymmetric element in Figure 11.9. The length of side n_1–n_3 is L_{13}.

The matrix $\mathbf{N}$ of shape functions for the triangular element is expressed in equation 11.3. Shape function N_2 is zero along side n_1–n_3, so zeros must be entered in its positions in matrix $\mathbf{N}^\mathrm{T}$. Shape functions N_1 and N_3 are linear functions along side

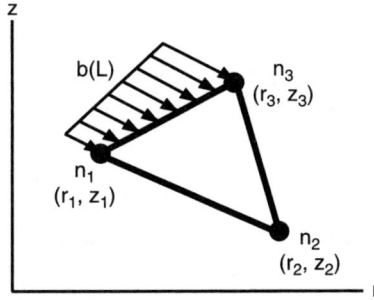

Figure 11.9. Surface force.

n_1–n_3. Setting the line variable at zero at node 1 and at L_{13} at node 3, these shape functions may be expressed as

$$N_1(L) = 1 - \frac{L}{L_{13}} \quad \text{and} \quad N_3(L) = \frac{L}{L_{13}}. \tag{11.34a}$$

The line force $b(L)$ is expressed as components $b_r(L)$ and $b_z(L)$ in the r and z coordinate directions.

The radial position $r(L)$ on side n_1–n_3 corresponding to the line value L must also be expressed explicitly; it is part of the integrand of equation 11.33. It may be written from the geometry of the element, shown in Figure 11.9:

$$r(L) = r_1 + \frac{r_2 - r_1}{L_{13}} L. \tag{11.34b}$$

The vector of equivalent forces for the surface force represented in Figure 11.9, written from equation 11.33, is

$$\{F_b\} = 2\pi \int_0^{L_{13}} \begin{bmatrix} N_1(L) & 0 \\ 0 & N_1(L) \\ 0 & 0 \\ 0 & 0 \\ N_3(L) & 0 \\ 0 & N_3(L) \end{bmatrix} \begin{Bmatrix} b_r(L) \\ b_z(L) \end{Bmatrix} r(L)\, dL = \begin{Bmatrix} 2\pi \int_0^{L_{13}} N_1(L) b_r(L) r(L)\, dL \\ 2\pi \int_0^{L_{13}} N_1(L) b_z(L) r(L)\, dL \\ 0 \\ 0 \\ 2\pi \int_0^{L_{13}} N_3(L) b_r(L) r(L)\, dL \\ 2\pi \int_0^{L_{13}} N_3(L) b_z(L) r(L)\, dL \end{Bmatrix}$$

$$\tag{11.34c}$$

Once the nature of the variation of the force $b(L)$ is known, the integrals of equation 11.34c may be evaluated explicitly, or numerically with a Gauss quadrature rule.

Equations 11.34c are for a line force applied along edge 1–3 of the element's cross section where shape function N_2 equals zero. For a force along edge 1–2, shape function N_3 would be zero along edge 1–2 and N_1 and N_2 would be nonzero.

Similarly for a force along edge 2–3, shape function N_1 would be zero and N_2 and N_3 would be nonzero.

Equations 11.34 may be employed to compute equivalent forces for surface forces acting on 4-node, 8-dof quadrilateral axisymmetric elements. In r–z space the sides of the quadrilateral element's cross section are straight lines, just as they are for the 3-node, 6-dof triangular axisymmetric element. Although explicit shape functions for the quadrilateral elements are not available in r–z space, it should be clear that they are bilinear in character. Consequently evaluated on the straight-line edge of a quadrilateral element, they will be linear in exactly the same form as for the triangular element. Thus equations 11.34a and 11.34b may be used directly for either element. Equation 11.34c will differ between the two elements only to accommodate the different numbers of nodes and dof. For the 4-node, 8-dof quadrilateral element, the matrix $\mathbf{N}^T$ is 8×2 and contains the two shape functions which are nonzero on the side of the element's cross section where the force acts.

Equation 11.33 may be adapted for the quadrilateral axisymmetric element to compute its equivalent forces in ζ–η space. For a distributed line force acting on a side of the element's cross section parallel to the ζ axis in ζ–η space, it is

$$\{\mathbf{F}_b\} = 2\pi \int_{-1}^{1} \left[\mathbf{N}^T(\zeta)\right]\{\mathbf{b}(\zeta)\}r(\zeta)|J(\zeta)| \, d\zeta. \tag{11.35}$$

For a force on a side parallel to the η axis, ζ is replaced by η.

An axisymmetric quadrilateral element's cross section with a distributed surface force acting on side n_1–n_2 is shown in Figure 11.10. The nonzero shape functions on the side are

$$N_1(\zeta) = \tfrac{1}{2}(1 - \zeta), \qquad N_2(\zeta) = \tfrac{1}{2}(1 + \zeta).$$

The radial position along side n_1–n_2 is

$$r(\zeta) = r_1 N_1(\zeta) + r_2 N_2(\zeta).$$

The Jacobian $|J(\zeta)|$ for the transformation of lines is a length scaling factor at the differential level, as discussed in Section 9.5.3. For straight lines, it is the ratio of the length of the lines in the two spaces, or $L_{12}/2$.

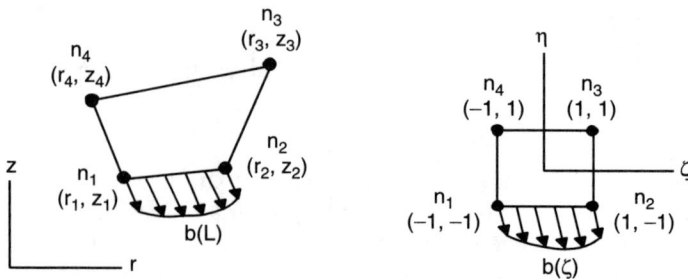

Figure 11.10. Element line force.

Using equation 11.35, the equivalent forces for the element are

$$
\{\mathbf{F}_b\} = \pi L_{12} \int_{-1}^{1}
\begin{bmatrix}
N_1(\zeta) & 0 \\
0 & N_1(\zeta) \\
N_2(\zeta) & 0 \\
0 & N_2(\zeta) \\
0 & 0 \\
0 & 0 \\
0 & 0 \\
0 & 0
\end{bmatrix}
\left\{
\begin{array}{c}
b_r(\zeta) \\
b_z(\zeta)
\end{array}
\right\} r(\zeta)\, d\zeta =
\left\{
\begin{array}{c}
\pi L_{12} \int_{-1}^{1} N_1(\zeta) b_r(\zeta) r(\zeta)\, d\zeta \\
\pi L_{12} \int_{-1}^{1} N_1(\zeta) b_z(\zeta) r(\zeta)\, d\zeta \\
\pi L_{12} \int_{-1}^{1} N_2(\zeta) b_r(\zeta) r(\zeta)\, d\zeta \\
\pi L_{12} \int_{-1}^{1} N_2(\zeta) b_z(\zeta) r(\zeta)\, d\zeta \\
0 \\
0 \\
0 \\
0
\end{array}
\right\}.
$$

$$(11.36)$$

Once the nature of the variation of the force $b(\zeta)$ is known, the integrals of equation 11.36 may be evaluated explicitly, or numerically with a Gauss quadrature rule.

A similar development may be made when the distributed surface force acts on other sides of the quadrilateral element's cross section.

11.6.4. Distributed Body Force

A body force acts throughout an axisymmetric structure. The force may have both axial and radial components, but not a circumferential one. The magnitude of the force may be a function of both the radial and axial coordinates r and z but not of the circumferential coordinate θ.

Since an axisymmetric body force acts throughout each axisymmetric element, it results in a distributed area force $b(r, z)$ in the r–z plane acting throughout the cross sections of each element.

Equation 6.22e adapted for axisymmetric body forces is

$$
\{\mathbf{F}_b\} = 2\pi \int_A [\mathbf{N}^{\mathrm{T}}(\mathbf{r},\ \mathbf{z})]\{\mathbf{b}(\mathbf{r},\ \mathbf{z})\} r\, dA. \tag{11.37a}
$$

For the 3-node, 6-dof triangular axisymmetric element it is

$$
\{\mathbf{F}_b\} = 2\pi \int_A
\begin{bmatrix}
N_1(r,\ z) & 0 \\
0 & N_1(r,\ z) \\
N_2(r,\ z) & 0 \\
0 & N_2(r,\ z) \\
N_3(r,\ z) & 0 \\
0 & N_3(r,\ z)
\end{bmatrix}
\left\{
\begin{array}{c}
b_r(r,\ z) \\
b_z(r,\ z)
\end{array}
\right\} r\, dA = 2\pi
\left\{
\begin{array}{c}
\int_A N_1(r,\ z) b_r(r,\ z) r\, dA \\
\int_A N_1(r,\ z) b_z(r,\ z) r\, dA \\
\int_A N_2(r,\ z) b_r(r,\ z) r\, dA \\
\int_A N_2(r,\ z) b_z(r,\ z) r\, dA \\
\int_A N_3(r,\ z) b_r(r,\ z) r\, dA \\
\int_A N_3(r,\ z) b_z(r,\ z) r\, dA
\end{array}
\right\}.
$$

$$(11.37b)$$

The shape functions $N_i(r,\ z)$ are expressed in equations 11.4.

Though equation 11.37b may be evaluated explicitly over the triangular cross-sectional area of the element, as discussed in Section 8.3.2, it is more easily done numerically using the quadrature rule discussed earlier in this chapter. Expressed in terms of this rule,

$$\{F_b\} = \frac{2\pi A_e}{3} \sum_{k=1}^{3} ([N^T(r_k, \; z_k)]\{b(r_k, \; z_k)\}r_k)$$

$$= \frac{2\pi A_e}{3} \sum_{k=1}^{3} \begin{bmatrix} r_k N_1(r_k, \; z_k) b_r(r_k, \; z_k) \\ r_k N_1(r_k, \; z_k) b_z(r_k, \; z_k) \\ r_k N_2(r_k, \; z_k) b_r(r_k, \; z_k) \\ r_k N_2(r_k, \; z_k) b_z(r_k, \; z_k) \\ r_k N_3(r_k, \; z_k) b_r(r_k, \; z_k) \\ r_k N_3(r_k, \; z_k) b_z(r_k, \; z_k) \end{bmatrix} . \tag{11.37c}$$

If the stiffness matrices for triangular axisymmetric elements are being computed using generalized dof, the equivalent nodal forces for body forces should be also. Equations 6.38c and 10.33 may be adapted for this purpose:

$$\{F_b\} = 2\pi[h^{-T}] \int_A [g^T]\{b(r, \; z)\}r \, dA$$

$$= 2\pi[h^{-T}] \int_A \begin{bmatrix} 1 & 0 \\ r & 0 \\ z & 0 \\ 0 & 1 \\ 0 & r \\ 0 & z \end{bmatrix} \begin{Bmatrix} b_r(r, \; z) \\ b_z(r, \; z) \end{Bmatrix} r \, dA = 2\pi[h^{-T}] \begin{Bmatrix} \int_A r b_r(r, \; z) \, dA \\ \int_A r^2 b_r(r, \; z) \, dA \\ \int_A rz b_r(r, \; z) \, dA \\ \int_A r b_z(r, \; z) \, dA \\ \int_A r^2 b_z(r, \; z) \, dA \\ \int_A rz b_z(r, \; z) \, dA \end{Bmatrix} . \tag{11.38a}$$

The matrix h^{-T} is the inverse of the transpose of the matrix h for the triangular axisymmetric element, given by equation 11.13.

Equation 11.38a may be integrated explicitly over the area of the element's triangular cross section. However, it is usually more efficient to do it numerically using the quadrature rule discussed earlier in this chapter. Expressed in terms of this rule,

$$\{F_b\} = \frac{2\pi A_e}{3} [h^{-T}] \sum_{k=1}^{3} \begin{Bmatrix} r_k b_r(r_k, \; z_k) \\ r_k^2 b_r(r_k, \; z_k) \\ r_k z_k b_r(r_k, \; z_k) \\ r_k b_z(r_k, \; z_k) \\ r_k^2 b_z(r_k, \; z_k) \\ r_k z_k b_z(r_k, \; z_k) \end{Bmatrix} . \tag{11.38b}$$

As long as $b_r(r, z)$ and $b_z(r, z)$ are linear polynomials in r or quadratic in z, or of a lesser degree, the results will be exact using either equations 11.37c or 11.38b. If the body force varies as a higher degree polynomial or some other function, a higher order quadrature rule should be employed.

Equivalent nodal forces for body forces acting on 4-node, 8-dof quadrilateral axisymmetric elements should be computed in ζ–η space, with corresponding calculations for the stiffness matrices of these elements. Equation 6.22e adapted for these body forces is

$$\{\mathbf{F}_b\} = 2\pi \int_{-1}^{1} \int_{-1}^{1} [\mathbf{N}^{\mathrm{T}}(\zeta, \eta)]\{\mathbf{b}(\zeta, \eta)\}r(\zeta, \eta)|J(\zeta, \eta)| \, d\zeta \, d\eta$$

$$= 2\pi \int_{-1}^{1} \int_{-1}^{1} \begin{bmatrix} N_1(\zeta, \eta) & 0 \\ 0 & N_1(\zeta, \eta) \\ N_2(\zeta, \eta) & 0 \\ 0 & N_2(\zeta, \eta) \\ N_3(\zeta, \eta) & 0 \\ 0 & N_3(\zeta, \eta) \\ N_4(\zeta, \eta) & 0 \\ 0 & N_4(\zeta, \eta) \end{bmatrix} \begin{Bmatrix} b_r(\zeta, \eta) \\ b_z(\zeta, \eta) \end{Bmatrix} r(\zeta, \eta)|J(\zeta, \eta)| \, d\zeta \, d\eta$$

$$= 2\pi \begin{bmatrix} \int_{-1}^{1} \int_{-1}^{1} N_1(\zeta, \eta)b_r(\zeta, \eta)r(\zeta, \eta)|J(\zeta, \eta)| \, d\zeta \, d\eta \\ \int_{-1}^{1} \int_{-1}^{1} N_1(\zeta, \eta)b_z(\zeta, \eta)r(\zeta, \eta)|J(\zeta, \eta)| \, d\zeta \, d\eta \\ \int_{-1}^{1} \int_{-1}^{1} N_2(\zeta, \eta)b_r(\zeta, \eta)r(\zeta, \eta)|J(\zeta, \eta)| \, d\zeta \, d\eta \\ \int_{-1}^{1} \int_{-1}^{1} N_2(\zeta, \eta)b_z(\zeta, \eta)r(\zeta, \eta)|J(\zeta, \eta)| \, d\zeta \, d\eta \\ \int_{-1}^{1} \int_{-1}^{1} N_3(\zeta, \eta)b_r(\zeta, \eta)r(\zeta, \eta)|J(\zeta, \eta)| \, d\zeta \, d\eta \\ \int_{-1}^{1} \int_{-1}^{1} N_3(\zeta, \eta)b_z(\zeta, \eta)r(\zeta, \eta)|J(\zeta, \eta)| \, d\zeta \, d\eta \\ \int_{-1}^{1} \int_{-1}^{1} N_4(\zeta, \eta)b_r(\zeta, \eta)r(\zeta, \eta)|J(\zeta, \eta)| \, d\zeta \, d\eta \\ \int_{-1}^{1} \int_{-1}^{1} N_4(\zeta, \eta)b_z(\zeta, \eta)r(\zeta, \eta)|J(\zeta, \eta)| \, d\zeta \, d\eta \end{bmatrix}. \tag{11.39a}$$

The shape functions $N_i(\zeta, \eta)$ are expressed in equations 11.22, and the radial coordinate $r(\zeta, \eta)$ is obtained from the transformation equation 11.23a. The Jacobian $|J(\zeta, \eta)|$ of the transformation is obtained from equations 11.25 and 11.26.

Equation 11.39a must be evaluated numerically using a Gauss quadrature rule. When the body force is constant, the 2×2 rule is exact. For a variable body force, the 3×3 rule may be necessary. Equation 11.39a in terms of a Gauss quadrature rule is

$$
\{F_b\} = 2\pi \sum_{k=1}^{n}
\begin{Bmatrix}
N_1(\zeta_k,\ \eta_k)b_r(\zeta_k,\ \eta_k)r(\zeta_k,\ \eta_k)|J(\zeta_k,\ \eta_k)|H_k \\
N_1(\zeta_k,\ \eta_k)b_z(\zeta_k,\ \eta_k)r(\zeta_k,\ \eta_k)|J(\zeta_k,\ \eta_k)|H_k \\
N_2(\zeta_k,\ \eta_k)b_r(\zeta_k,\ \eta_k)r(\zeta_k,\ \eta_k)|J(\zeta_k,\ \eta_k)|H_k \\
N_2(\zeta_k,\ \eta_k)b_z(\zeta_k,\ \eta_k)r(\zeta_k,\ \eta_k)|J(\zeta_k,\ \eta_k)|H_k \\
N_3(\zeta_k,\ \eta_k)b_r(\zeta_k,\ \eta_k)r(\zeta_k,\ \eta_k)|J(\zeta_k,\ \eta_k)|H_k \\
N_3(\zeta_k,\ \eta_k)b_z(\zeta_k,\ \eta_k)r(\zeta_k,\ \eta_k)|J(\zeta_k,\ \eta_k)|H_k \\
N_4(\zeta_k,\ \eta_k)b_r(\zeta_k,\ \eta_k)r(\zeta_k,\ \eta_k)|J(\zeta_k,\ \eta_k)|H_k \\
N_4(\zeta_k,\ \eta_k)b_z(\zeta_k,\ \eta_k)r(\zeta_k,\ \eta_k)|J(\zeta_k,\ \eta_k)|H_k
\end{Bmatrix}.
\tag{11.39b}
$$

11.6.5. Equivalent Force Calculations

Figure 11.11 shows an axisymmetric 4-node, 8-dof quadrilateral element subjected to several forces. One is an axisymmetric uniform line force acting at node 1, as shown. Another is a linearly increasing axisymmetric surface force acting on and normal to side n_2–n_3 as shown. Body forces act throughout the element due to its weight and to the centrifugal force occurring as a result of spinning about its axis of symmetry at a constant angular velocity of 100 revolutions per minute. The unit weight is 0.282 lb/in^3. The element dof and a line variable L varying from zero at node 2 to L_{23} at node 3 are shown also. The coordinates are in inches. The element's equivalent force vectors will be computed.

From equation 11.32a, the magnitude of the uniform line force is

$$
P_1 = 2\pi(4)(10) = 80\pi = 251.33 \text{ lb.}
$$

Its r–z components are inserted directly into the element's force vector in positions 1 and 2, corresponding to edof 1 and 2:

$$
P_{1r} = 251.33 \ \sin \ 30° = 125.66 \text{ lb.} \qquad P_{1z} = 251.33 \ \cos \ 30° = 217.66 \text{ lb.}
$$

The equivalent forces for the distributed surface force are computed following the procedure indicated for equations 11.34.

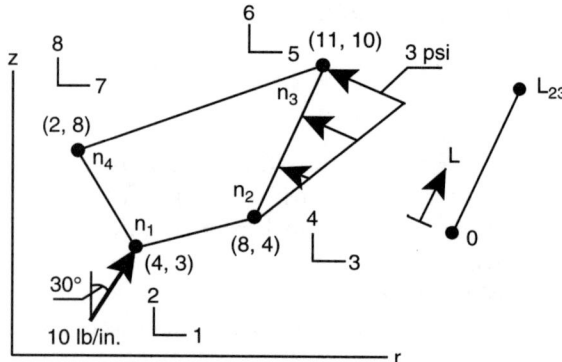

Figure 11.11. Applied forces.

The length of side n_2–n_3 is

$$L_{23} = \sqrt{(11-8)^2 + (10-4)^2} = 3\sqrt{5}\,\text{in.}$$

The shape functions along side n_2–n_3 in terms of the variable L are

$$N_2(L) = \frac{3\sqrt{5}-L}{3\sqrt{5}} \quad \text{and} \quad N_3(L) = \frac{L}{3\sqrt{5}}.$$

The distributed surface load as a function of the variable L is

$$b(L) = \frac{3L}{3\sqrt{5}}.$$

The radial coordinate along side n_2–n_3 as a function of the variable L is

$$r = 8 + \frac{11-8}{3\sqrt{5}}L = \frac{24\sqrt{5}-3L}{3\sqrt{5}}.$$

The equivalent force at node 2 normal to side n_2–n_3 is

$$F_{b2} = 2\pi \int_0^{3\sqrt{5}} \left(\frac{3\sqrt{5}-L}{3\sqrt{5}}\right)\left(\frac{3L}{3\sqrt{5}}\right)\left(\frac{24\sqrt{5}-3L}{3\sqrt{5}}\right) dL = 19.5\pi\sqrt{5}\,\text{lb} = 136.98\,\text{lb},$$

and at node 3, it is

$$F_{b3} = 2\pi \int_0^{3\sqrt{5}} \left(\frac{L}{3\sqrt{5}}\right)\left(\frac{3L}{3\sqrt{5}}\right)\left(\frac{24\sqrt{5}-3L}{3\sqrt{5}}\right) dL = 34.5\pi\sqrt{5}\,\text{lb} = 242.36\,\text{lb}.$$

The integrals may be evaluated exactly, either explicitly or numerically with the two-point Gauss quadrature rule.

Resolving these forces into their r–z components, the vector of equivalent forces for the distributed surface load is

$$\{\mathbf{F}_{bs}\} = \begin{Bmatrix} 0 \\ 0 \\ -39\pi \\ 19.5\pi \\ -69\pi \\ 34.5\pi \\ 0 \\ 0 \end{Bmatrix} \text{lb} = \begin{Bmatrix} 0 \\ 0 \\ -122.52 \\ 61.26 \\ -217.77 \\ 108.38 \\ 0 \\ 0 \end{Bmatrix} \text{lb.}$$

The equivalent forces for body forces will be computed using the procedure suggested in equation 11.39b using the 2×2 Gauss quadrature rule.

There is a component of body force in the r direction caused by the mass and constant angular velocity of the element:

$$b_r(r, \; z) = \omega^2 \rho r = \left(\frac{2\pi \cdot 100}{60}\right)^2 \left(\frac{0.282}{386.4}\right) r = 0.080033 r.$$

The component of body force in the z direction is the unit weight of the material of the element:

$$b_z(r, \; z) = -0.282 \, \text{lb/in}^3.$$

The shape functions are written from equation 11.22, the radial coordinate is written from the transformation equation 11.23b, and the Jacobian is obtained from equations 11.25 and 11.26.

The quadrilateral element shown in Figure 11.10 is identical to the one in Figures 11.6 and 9.18, used in the example calculations for element stiffness matrices. Thus the evaluations made at the four integration points for shape functions $N(\zeta_k, \; \eta_k)$, the radial coordinate $r(\zeta_k, \; \eta_k)$, and the Jacobian $|J(\zeta_k, \; \eta_k)|$ are the same as before. Repeating the relevant data listed in Sections 9.5.2 and 11.5.3 yields the following:

	Integration Point 1	Integration Point 2	Integration Point 3	Integration Point 4
N_1	0.62201	0.16667	0.16667	0.04466
N_2	0.16667	0.62201	0.04466	0.16667
N_3	0.04466	0.16667	0.16667	0.62201
N_4	0.16667	0.04466	0.62201	0.16667
r	4.64594	7.56538	4.10128	8.68739
J	6.87361	6.72928	10.77073	10.62639

Combining the data as indicated in equation 11.39b but without multiplying by 2π at this time results in the following:

F_{bb1}	7.3858	5.1375	2.4166	2.8664
F_{bb2}	−5.6015	−2.3928	−2.0762	−1.1626
F_{bb3}	1.9790	19.1732	0.6475	10.6975
F_{bb4}	−1.5009	−8.9299	−0.5563	−4.3388
F_{bb5}	0.5303	5.1375	2.4166	39.9236
F_{bb6}	−0.4022	−2.3928	−2.0762	−16.1927
F_{bb7}	1.9790	1.3766	9.0188	10.6975
F_{bb8}	−1.5009	−0.6411	−7.7484	−4.3388

When the four entries in each row for each F_{bbi} are added and the sum is multiplied by 2π, the result is the vector of equivalent forces for the body forces $\mathbf{F}_{bb}$ acting on the element. When these equivalent forces are then added to those obtained for the distributed surface force acting on the element, and the r–z components for the uniform line force are included, the result is the total equivalent force vector $\mathbf{F}_b$ for this element:

$$\{\mathbf{F}_{bb}\} = \begin{Bmatrix} 111.88 \\ -70.58 \\ 204.19 \\ -96.30 \\ 301.64 \\ -132.35 \\ 144.97 \\ -89.41 \end{Bmatrix} \text{lb.}, \qquad \{\mathbf{F}_b\} = \begin{Bmatrix} 361.21 \\ 147.08 \\ 81.67 \\ -35.04 \\ 83.87 \\ -23.97 \\ 144.97 \\ -89.41 \end{Bmatrix} \text{lb.}$$

11.7. CLOSING REMARKS

Axisymmetric triangular finite elements are the simplest formulation for a finite element analysis of an axisymmetric structure. They are the simplest to understand and require the least calculations to obtain results. They are the least accurate, however. They use linear shape functions, which results in constant strain and stress within each element. Thus achieving a desired level of accuracy may require the analyst to use very many elements with the consequent large computation effort.

Axisymmetric quadrilateral finite elements formulated with the isoparametric transformation scheme provide a more accurate way to model and analyze axisymmetric solid structures. The 4-node quadrilateral element is more accurate than the 3-node triangular element because its shape functions are higher degree, permitting more variation of strain and stress within the element. However, more calculations are required for 4-node quadrilateral elements than for 3-node triangular elements. Both elements are versatile at modeling irregularly shaped axisymmetric structures, though each models the boundaries of such structures with piecewise straight lines.

Six and 7-node triangular axisymmetric elements can be developed in the manner discussed for comparable plane stress elements in Chapter 8. They will use second-degree polynomials for shape functions, permitting more variation of stress and strain within elements, and thus can achieve more accuracy in calculating stress and strain within the structure.

Eight and 9-node quadrilateral axisymmetric elements may be developed also, following the process described for comparable plane stress elements in Chapter 9. Such elements use second-degree polynomials in ζ and η, with more polynomial terms than the 6- or 7-node triangular elements, permitting even more variation of strain and stress within the element and the consequent greater accuracy for the analysis. In addition, the isoparametric transformation used to achieve the required calculations for element stiffness matrices results in parabolic curved-sided elements in r–z space. These curved sides permit the boundaries of irregularly shaped structures to be modeled with piecewise parabolic curves.

Elements with more nodes require more calculation effort to obtain their stiffness matrices and equivalent forces than those with fewer nodes. The added effort comes about because of more shape functions, more and higher degree terms in the shape functions, more nodal coordinates to combine with these shape functions in the calculation algorithm, and often the necessity to use a higher order quadrature rule to achieve the accuracy desired. The analyst must consider whether to achieve the level of accuracy needed for the overall analysis with a large number of elements which each possess only three or four nodes or with fewer elements which each possess six, eight, or more nodes.

Upon comparing the procedures and equations of Chapters 8 and 9 with those of this chapter, it is apparent that the plane stress and axisymmetric finite elements may be formulated in a very similar manner. The axisymmetric formulation is somewhat more complicated. It must account for the interaction of the lateral circumferential stress and strain with other stresses and strains. As a result the matrix of elastic constants and the stress and strain vectors are larger. The axisymmetric formulation also includes the radial coordinate r in each integral as a result of integrating in the circumferential direction around the axes of symmetry. Thus the integrands are higher degree, occasionally needing a higher order quadrature rule.

BIBLIOGRAPHY

Cook, R. D., Malkus, D. S., and Plesha, M. E., *Concepts and Applications of Finite Element Analysis* (Chapter 10), Wiley, New York, 1989.

Logan, D. L., *A First Course in the Finite Element Method* (Chapter 10), PWS-Kent, Boston, 1992.

Weaver, W. Jr., and Johnston, P. R., *Finite Element for Structural Analysis* (Chapter 5), Prentice-Hall, Englewood Cliffs, NJ, 1984.

Yang, T. Y., *Finite Element Structural Analysis* (Chapter 10), Prentice-Hall, Englewood Cliffs, NJ, 1986.

Zienkiewcz, O. C., and Taylor, R. L., *The Finite Element Method* (Chapter 4), 4th ed., McGraw-Hill, Berkshire, England, 1989.

PROBLEMS

The use of computer software, such as MATLAB or a spreadsheet, which manipulates individual matrices and evaluates functions at a point in their space will facilitate solution of many of these problems.

11.1. The nodal coordinates for the axisymmetric triangular finite element shown are in inches.

  a. Write the element shape functions.
  b. Determine the element's matrix **B** evaluated at the first integration point.
  c. Determine the element's stiffness matrix **K** using physical dof.
  d. Determine the element's stiffness matrix **K** using generalized dof.

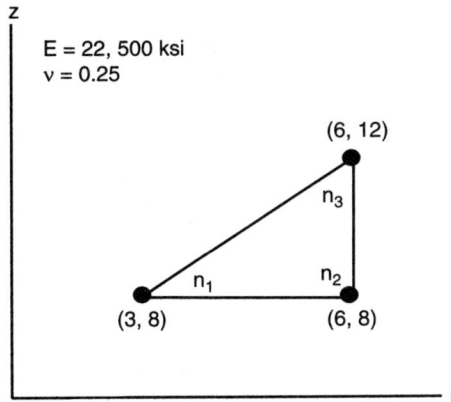

Problem 11.1

11.2. Repeat Problem 11.1 for the axisymmetric triangular finite element shown. Coordinates are in inches.

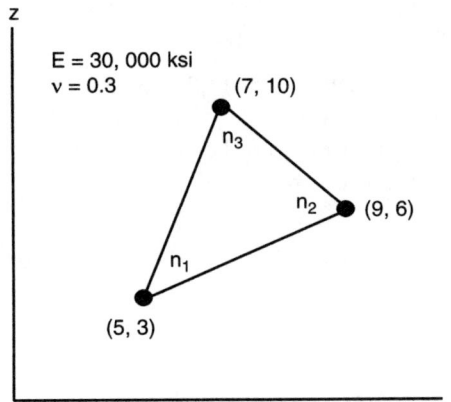

Problem 11.2

11.3. The axisymmetric triangular finite element shown has its nodal coordinates expressed in feet. Write the vector of equivalent nodal forces for:

    a. A 10-kip/ft axisymmetric line force acting at coordinates (6, 4) and sloping up and right at 3 axial on 4 radial. Use physical dof, and then repeat with generalized dof.

    b. An axisymmetric uniformly distributed surface force acting along edge n_1–n_3 with an intensity of $25 \, lb/ft^2$ of edge surface and sloping down and right at 4 axial on 3 radial.

    c. An axisymmetric distributed surface force acting radially outward with no axial component on edge n_2–n_3, increasing linearly from 0 at node 3 to $20 \, lb/ft^2$ of edge surface at node 2.

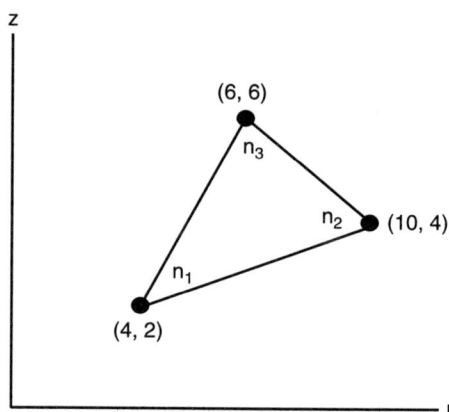

Problem 11.3

11.4. The element of Problem 11.2 has a unit weight of $485 \, lb/ft^3$ acting in the negative z direction. It is also subjected to an axisymmetric line force of $25 \, lb/in.$ acting at (7.5, 6.5) on the element's cross section. The line force slopes down and right at an angle of $30°$ from the radial direction. Determine the element's vector of equivalent nodal forces, first using physical dof and then using generalized dof.

11.5. The element described in Problem 11.4 is also subjected to an axisymmetric uniformly distributed surface force of $10 \, lb/in.^2$ of edge surface which acts normal to side n_1–n_2 and out from the element. Determine the element's total vector of equivalent nodal forces.

11.6. The axisymmetric triangular finite element shown has its nodal coordinates expressed in inches.

a. Write the shape functions $N_i(r, z)$ for the element by writing the area coordinates of a point on the element in terms of r, z, and the nodal coordinates.

b. Integrate each shape function over the area of the element.

c. Determine the vector of equivalent nodal forces for body forces acting on the element. One is the result of the element's unit weight $\gamma = 485\,\text{lb/ft}^3$ which acts in the negative z direction. The other occurs because the element rotates about its axis of symmetry at a constant angular velocity of 120 rpm.

d. An axisymmetric distributed surface force acts radially outward without an axial component along edge n_1–n_3 with a linearly increasing intensity from 0 at node 1 to $4\,\text{lb/in}^2$ of edge surface at node 3. Determine the vector of equivalent nodal forces for this surface force.

e. Determine the element's total vector of equivalent nodal forces.

f. Determine the element's stiffness matrix **K** using physical dof.

g. Determine the element's stiffness matrix **K** using generalized dof.

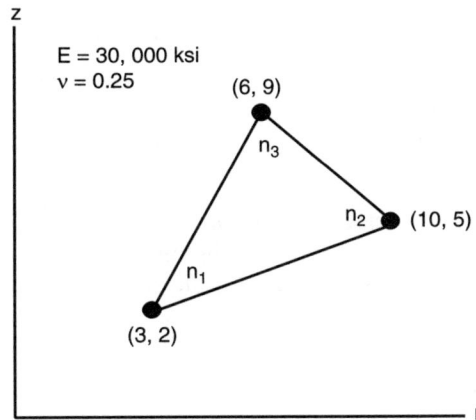

Problem 11.6

11.7. The structure $ABCD$ is an axisymmetric ring with two axisymmetric line forces at A and B and restraints at C and D which work like pin supports on its cross section $ABCD$. The axis of symmetry passes through point C and is perpendicular to lines CD and AB. The structure rotates about the axis of symmetry with a constant angular velocity of 30 rpm. Its unit weight is 485 lb/ft^3. Its modulus $E = 30,000$ ksi and Poisson's ratio $v = \frac{1}{4}$. Using the two axisymmetric 3-node, 6-dof triangular finite elements (a and b) indicated, determine the displacement of point B and the support reactions.

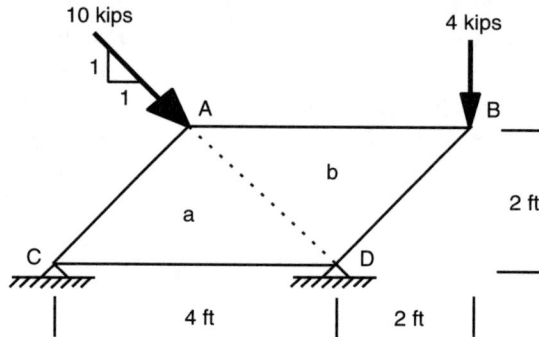

Problem 11.7

11.8. Repeat Problem 11.7, but use four 3-node, 6-dof triangular axisymmetric finite elements. The four elements should be obtained by dividing element a of Problem 11.7 into two elements with a vertical line from point A to side CD and dividing element b into two elements with a vertical line from point D to side AB.

11.9. Using the process described in Section 8.5 for a plane stress 6-node, 12-dof linear strain triangular element, extend the equations developed for the 3-node, 6-dof axisymmetric triangular finite element for use on the 6-node, 12-

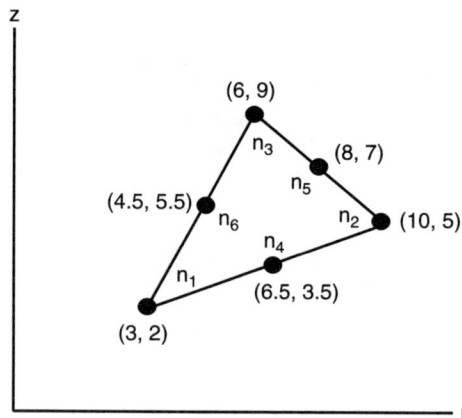

Problem 11.9

dof axisymmetric triangular finite element shown in the sketch. Its nodal coordinates are in inches.

a. Write the shape functions $N_i(r, z)$ for the element.

b. Integrate each shape function over the area of the element.

c. Determine the vector of equivalent nodal forces for body forces acting on the element. One is the result of the element's unit weight $\gamma = 485\,\text{lb/ft}^3$ which acts in the negative z direction. The other is because the element rotates about its axis of symmetry at a constant angular velocity of 120 rpm.

d. An axisymmetric distributed surface force acts radially inward and normal to edge n_1–n_6–n_3 with a linearly increasing intensity from 0 at node 1 to 4 lb/in.^{2} of edge surface at node 3. Determine the vector of equivalent nodal forces for this surface force.

e. Determine the element's total vector of equivalent nodal forces.

11.10. An axisymmetric quadrilateral element has modulus $E = 30{,}000\,\text{ksi}$ and Poisson's ratio $v = \frac{1}{4}$. The coordinates shown are in inches. Using the 2×2 Gauss quadrature rule and the numerical procedure described and illustrated in Section 11.5.3:

a. Determine the matrix $\mathbf{B}_1$ for the element, evaluated numerically at the first integration point.

b. Determine the element's matrix $\mathbf{K}_1$, evaluated numerically at the first integration point.

c. Determine the stiffness matrix $\mathbf{K}$ for the element.

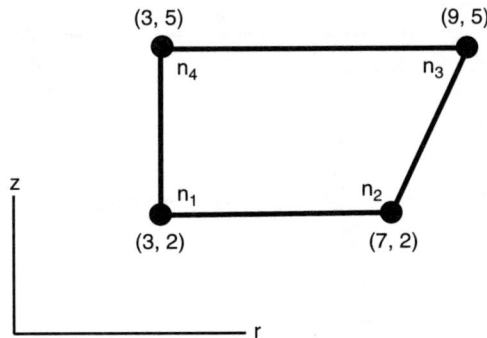

Problem 11.10

11.11. Repeat Problem 11.10, but use the axisymmetric quadrilateral element shown. Coordinates are in inches. Properties of the element are as in Problem 11.10.

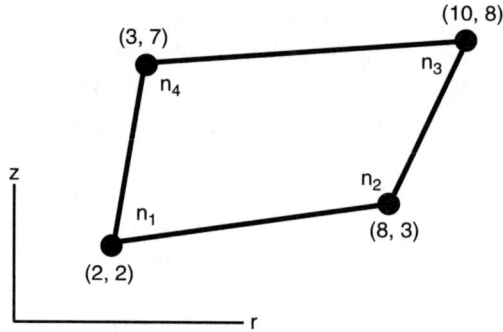

Problem 11.11

11.12. The axisymmetric line force and axisymmetric linearly distributed surface force shown on the axisymmetric quadrilateral element in the sketch each act normal to their respective element sides. The element rotates about the axis of symmetry at a constant angular velocity of 150 rpm. Its unit weight is $500\,\text{lb/ft}^3$ and it acts in the negative z direction. Coordinates are in inches. Determine the vector of equivalent nodal forces for the element.

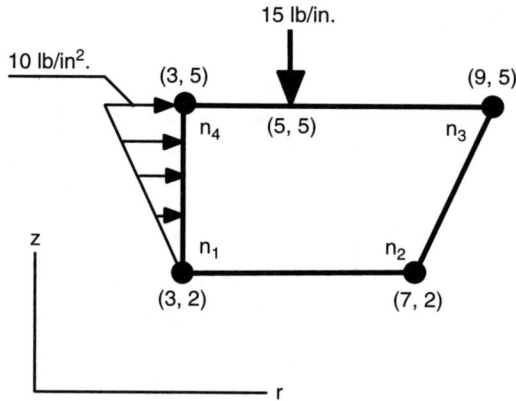

Problem 11.12

11.13. The axisymmetric uniform and linearly distributed surface forces shown on the axisymmetric quadrilateral element in the sketch each act normal to their respective element sides. The element rotates about its axes of symmetry at a constant angular velocity of 120 rpm. Its unit weight is $500\,\text{lb/ft}^3$ and it acts in the negative z direction. Coordinates are in inches. Determine the vector of equivalent nodal forces for the element.

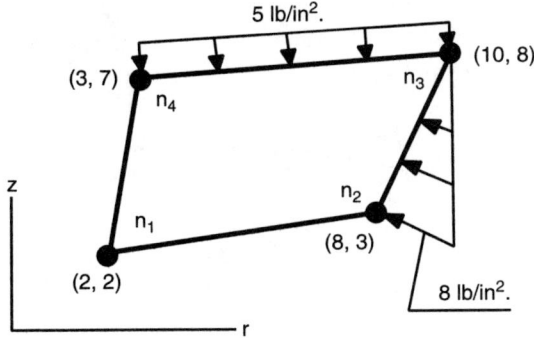

Problem 11.13

11.14. Using the process described in Section 9.6 for a plane stress 8-node, 16-dof serendipity quadrilateral element, extend the equations developed for the 4-node, 8-dof axisymmetric quadrilateral finite element for use on the 8-node, 16-dof axisymmetric quadrilateral finite element shown on page 384. Its nodal coordinates are in inches.

a. Using first the 2×2 and then the 3×3 Gauss quadrature rules:
  (1) Determine the cross-sectional area of the quadrilateral element.
  (2) Determine the volume of the quadrilateral ring element.
  (3) Evaluate the integral of the function $f(r, z) = 5r$ throughout the volume of the element.
  (4) Compare the results between the two quadrature rules. Explain the similarities and differences.

b. Find the vector of equivalent nodal forces for the element if it rotates about the axis of symmetry at a constant angular velocity of 100 rpm and if its unit weight $\gamma = 500\,\text{lb/ft}^3$ and it acts in the negative z direction.

c. Find the vector of equivalent nodal forces for the element for an axisymmetric uniformly distributed surface force which is $15\,\text{lb/in}^2$. of edge surface acting radially outward on side n_2–n_6–n_3, with no axial component.

d. Find the vector of equivalent nodal forces for the element for an axisymmetric linearly distributed surface force acting radially outward on side n_2–n_6–n_3, with no axial component. The load is $20\,\text{lb/in}^2$. of edge surface at node 2 and zero at node 3.

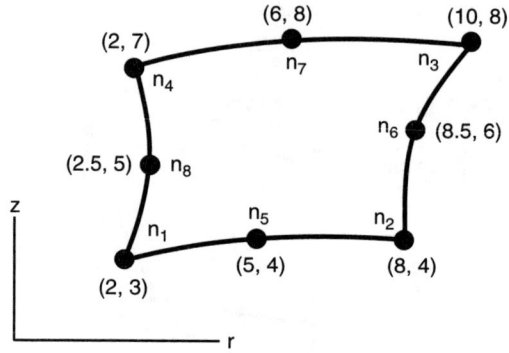

Problem 11.14

CHAPTER 12

STRUCTURAL FINITE ELEMENTS IN PERSPECTIVE

12.1. INTRODUCTION

The behavior of linear elastic structures can be modeled with great versatility by the finite element method. However, understanding the method in depth is elusive, not because of an inherent complexity of concepts, but because the method comprises an apparent bewildering assembly of many basic ideas drawn from mathematics, mechanics, and structural theory. None of these ideas is difficult to grasp individually, but all must be, as must their roles in the finite element method. This book was written to develop and explain the method to readers who do not have a firm understanding of it. To this end, considerable attention has been devoted to the basic mathematics, mechanics, and structural theory which underpin the method. Moreover, only the most basic finite elements have been discussed to make the presentation of the method in its entirety as understandable as possible. Finite elements whose geometry is characterized as 1D came first. The complexities of dealing with finite elements whose geometry is characterized by more than one dimension were presented in the treatment of flat plate 2D elements and axisymmetric elements, after mastery of the finite element method using 1D elements was achieved. More sophisticated 2D and 3D elements were not treated. Their development is a small extension from the elements discussed and is left to other books on the finite element method.

12.2. MODELING STRUCTURES WITH FINITE ELEMENTS

One-dimensional elements are 1D only in the geometric character of the element. They are very much larger in one spatial dimension than in the other two. One-dimensional elements may have nodal displacements and actions in two or three spatial dimensions, such as do the beam, plane frame, and space frame elements.

One-dimensional elements may be pieced together to model 1D columns or beams, 2D and 3D plane or space trusses, or 2D and 3D plane or space frames.

Most 1D finite elements are prismatic. A tapered structure may be successfully modeled by an appropriate number of prismatic elements, each with different transverse cross sections which approximate the geometry of the tapered structure. They may also be modeled, perhaps more accurately, with tapered elements whose geometry replicates the geometry of the tapered structure. Tapered 1D column and beam elements may be developed from the principle of virtual work following the procedures described in Chapter 5. For tapered columns, the transverse cross-sectional area should be treated as a function rather than as a constant in the applicable integral expressions. The quadratic displacement shape functions developed for the prismatic 3-node axial element should provide acceptable accuracy. For tapered beams, the centroidal moment of inertia of the transverse cross section should be treated as a function rather than as a constant in the applicable integral expressions. The cubic polynomial displacement shape functions used for prismatic beam elements can provide acceptable accuracy. However, other shape functions may be developed which are specifically adapted for the taper in question, if desired.

Flat plate finite elements are 2D in that they are very much larger in two spatial dimensions than in the third. They may have displacements and actions at nodes in two or three spatial dimensions. The flat plate finite elements treated in earlier chapters were those subjected to in-plane loads resulting in the conditions of plane stress and those subjected to transverse loads resulting in plate flexure. Rectangular, triangular, and quadrilateral flat plate elements were discussed. The elements were constant in thickness and were composed of either isotropic or orthotropic materials. They can be pieced together to model flat plates in plane stress or flexure.

The stiffness equations for a rectangular plane stress flat plate element may be combined with those for a rectangular flat plate flexure element to develop the stiffness equations for a rectangular flat plate flexure element which accounts for small in-plane displacements. Both the plane stress and flexure rectangular elements must be the same size and have the same number and locations of nodes. The combining process is similar to the one described in Chapter 3, where the stiffness equations for the 2-node axial and beam elements were combined to produce stiffness equations for a 2-node plane frame element. Such combined flat plate elements could be pieced together to model space structures constructed of flat plates. Appropriate transformation matrices would be required to transform element stiffness equations from local coordinate axes to the structure's global coordinate axes. The transformations discussed in Chapter 3 for 1D elements in 3D space may be extended for this purpose.

Axisymmetric solids are 3D structures. As long as their applied loads, material properties, and structural restraints are axisymmetric as well, the structure's distributions of displacements, internal forces, strains, and stresses will be axisymmetric. When these conditions obtain, the mathematical formulation of the axisymmetric finite element analysis is a 2D one. Triangular and quadrilateral axisymmetric elements were discussed in this way in Chapter 11. Their formulations followed closely those of the plane stress triangular and quadrilateral elements discussed in Chapters 8 and 9.

12.3. LINEAR ALGEBRA BASIS

The finite element method models its elements and structures using systems of linear algebraic equations. Consequently, a clear understanding of and manipulative skill with matrix algebra and algebraic equation systems are essential for understanding the method, its power, and its limitations. Chapter 1 initially treats those topics from matrix algebra encountered in the method and then follows with treatments of the algebraic equation systems which the method employs. Matrix manipulations studied include addition, subtraction, multiplication, transposing, inversion, rearranging, and partitioning. Equation systems were studied for solution procedures which included the submatrix, solution without rearrangement, and penalty methods. Reduction procedures were examined as a means to reduce the amount of calculation required for the solution procedures.

12.4. THE FINITE ELEMENT METHOD

The development of a structure's global stiffness equations with the system's restraints imposed on them as boundary conditions is a primary goal of the finite element analysis of the structure. These global stiffness equations are a system of linear algebraic equations which reflect the interaction of the nodal displacements, applied forces and moments, support reactions, and support displacements on one another. It is the solution of these equations which yields the structure's displacements at nodes.

Once the nodal displacements are known, the support reactions may be calculated by straightforward matrix multiplication. Moreover, the one-to-one correspondence of the structural system and element nodal displacements may be called on to identify what the nodal displacements are for each finite element of the structural system. Then the element stiffness equations for each finite element may be employed to calculate, with straightforward matrix multiplication, the internal forces and moments at the nodes of the elements.

In finite element analyses of 2D or 3D continuous structural systems (plates, solid structures, etc.), the nodal displacements are used further to calculate strains in the structure. The strains are then substituted into the constitutive relationships for the material of the structure to compute its stresses. Finally, the stresses may be transformed to obtain principal stresses and maximum shear stresses.

The essential steps in the entire process may be summarized as follows:

- Identify the finite elements and nodes which will be employed to model the structural system. The finite elements chosen must have the appropriate nodal dof identified, and reference axes for the elements and the structural system must be defined. Numbering schemes for elements, element dof, and global dof must be developed. This process is the discretizing of the structural system.

- Write the element stiffness equations for each finite element of the structural system. In this step, it may be convenient to retain the element stiffness equations in two forms. One will be the form referred to the element's own local reference axes. This is the form written from previously developed explicit equations or integral expressions evaluated numerically. The other form is referred to the structural system's global reference axes and is usually developed from mathematical transformations of the initial form. For 2D elements, the transformation to global reference axes may be interwoven with the determination of element stiffness equations.

● Assemble the global stiffness matrix from the superposition of each of the system's element stiffness matrices using the form of the element stiffness matrices referred to the structure's global axes. Write the global load vector and impose the restraints existing on the structural system in the global stiffness equations as its boundary conditions.

● Solve the global stiffness equations, which have the restraints of the structural system imposed, for the system's nodal displacements. Using the values of these nodal displacements, calculate the support reactions, internal forces and moments at nodes, and stresses in the structure.

12.5. THE FINITE ELEMENT

The heart of the application of the finite element method to a structural system is the definition of the finite elements which will be employed to model the system. When pieced together to define the structure geometrically, the assembly of elements must describe the behavior of the structure as faithfully as possible, consistent with good engineering judgment. To this end, the structural behavior of the elements must replicate the anticipated behavior of the structural system. The elements and their structural behavior are characterized by their geometric shapes, the dof they possess, the nodes where these dof are assigned, their material properties, and the character of the loads imposed on them. The geometric shapes, dof, and nodes of the elements will determine their polynomial shape functions. The element shape functions, geometry, material properties, and loads then will determine the element stiffness equations based on structural theory.

APPENDIX: MATRIX OPERATIONS FOR THE FINITE ELEMENT METHOD USING MATLAB

A.1. INTRODUCTION

MATLAB® is an interactive matrix computation environment that can be used to good advantage to complement the reader's study and use of matrices and matrix operations in the implementation of the finite element method.

MATLAB has a complete menu of matrix and array operations which can be used in a step-by-step manner, displaying intermediate results after each step. Performing the matrix operations for the finite element method in this manner is an important exercise for the reader if he or she is to grasp fully the workings of the method. All of the matrix manipulations needed for the finite element method are available in MATLAB.

MATLAB also allows the user to program the steps in an analysis so that they are accomplished automatically and rapidly and may be done repeatedly. These programmed routines are called m-files. The programming can be effected so that simplified data input is used and the results of the analysis are displayed in the format the user desires. The programming is simpler and allows more sophisticated steps to be undertaken in fewer statements than can be done by programming solely in FORTRAN, BASIC, or similar structured languages.

What follows first are illustrations of the step-by-step use of MATLAB in the finite element analysis of a plane truss and a plane frame. Next come illustrations of the use of m-files written by the author, which analyze plane trusses, plane frames and beams, plane stress thin plates, and thin flexure plates.

A.2. MATLAB MATRIX MANIPULATIONS BY MANUAL ENTRY

The MATLAB platform provides the user with a variety of matrix operations which are implemented easily by typing instructions and then striking the "enter" key. MATLAB performs the operations and displays the results of each operation immediately. When it is desired to disable the immediate display of results, it is done by typing a semicolon at the end of the instruction before striking the "enter" key. Detailed information on the syntax of these instructions is contained in the *MATLAB User's Guide*, the *MATLAB Reference Guide*, and the MATLAB on-line help facility.

Two examples on how to use MATLAB in its manual entry, or the step-by-step mode, to analyze structures with the finite element method follow as illustrations. In the examples, the typed data entry or instructions and the resulting MATLAB display are indicated by a note to that effect in pointed brackets.

A.2.1. Plane Truss Example by Manual Entry

To illustrate the use of MATLAB in the step-by-step by mode of operation, the illustrative example described in Chapter 4 as Example 4.5.1 will be reanalyzed here using MATLAB. Shown in Figure A.1 is the physical plane truss with supports and loads. The modulus of elasticity $E = 30,000$ ksi. MATLAB will be used to obtain the joint displacements, support reactions, and member forces.

The discretized representation of the truss with the nodes, gdof, and element numbers chosen is shown in Figure A.2. The nodes are designated n_1–n_3. The elements are designated 1–3 and shown in the boxes. The arrow on each box indicates the sequence from the start node to the end node for each element. For example, the start node for element 2 is n_3 and the end node is n_2. The gdof are designated 1–6 and shown in the circles. They are either horizontal or vertical in conformance with the global axes x–y chosen. The edof are shown with respect to the global axes as 1–4 for each element. Their sequence is consistent with the designation of start and end nodes for each element.

Data entry begins with the basic axial element stiffness matrix with respect to local axes without the scalar associated with each element:

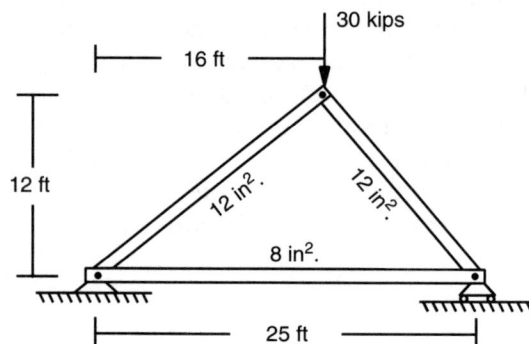

Figure A.1. Plane truss.

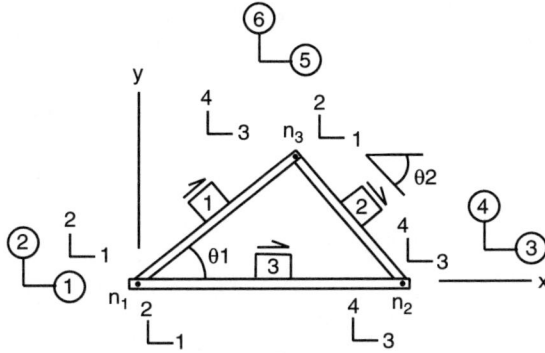

Figure A.2. Discretized truss.

≫ ke = [1 -1; -1 1]   < data entry >
ke =
        1   -1   < MATLAB display >
       -1   1

Next the element's elastic modulus (ksi) is entered:

≫ E = 30000   < data entry >
E =
        30000   < MATLAB display >

The vector of element cross-sectional areas (in.^{2}) is entered in sequence beginning with element 1:

≫ A = [12 12 8]   < data entry >
A =
       12   12   8   < MATLAB display >

The vector of element lengths (feet) is entered in sequence beginning with element 1 and multiplied by the scalar 12 to convert the unit to inches so the analysis will treat consistent units:

≫ L = [20 15 25]*12   < data entry >
L =
      240   180   300   < MATLAB display >

The area vector A, elastic modulus scalar E, and length vector L are combined to obtain a vector of local element stiffness matrix scalar multipliers (unit is kips/in.):

≫ S = A*E./L   < equation entry >
S =
     1500   2000   800   < MATLAB display >

The scalar multipliers are applied to the basic element axial stiffness matrix to obtain each element stiffness matrix with respect to local axes (unit is kips/in.):

≫ kel = S(1)*ke   <equation entry>
kel =
    1500   -1500   <MATLAB display>
   -1500   1500
≫ ke2 = S(2)*ke   <equation entry>
ke2 =
    2000   -2000   <MATLAB display>
   -2000   2000
≫ ke3 = S(3)*ke   <equation entry>
ke3 =
     800   -800   <MATLAB display>
    -800   800

The matrix of direction cosines for each element is entered so that element matrices with respect to local axes can be transformed to matrices with respect to the global axes:

≫ t1 = [.8 .6 0 0;0 0 .8 .6]   <data entry>
t1 =
  0.8000   0.6000   0   0   <MATLAB display>
       0   0   0.8000   0.6000
≫ t2 = [.6 -.8 0 0;0 0 .6 -.8]   <data entry>
t2 =
  0.6000   -0.8000   0   0   <MATLAB display>
       0   0   0.6000 -0.8000
≫ t3 = [1 0 0 0;0 0 1 0]   <data entry>
t3 =
    1   0   0   0   <MATLAB display>
    0   0   1   0

With the element transformation matrices, obtain element stiffness matrices with respect to the global (unit is kips/in.):

≫ k1 = t1'*kel*t1   <equation entry>
k1 =
   960   720   -960   -720   <MATLAB display>
   720   540   -720   -540
  -960   -720   960   720
  -720   -540   720   540
≫ k2 = t2'*ke2*t2   <equation entry>
k2 =
   720   -960   -720   960   <MATLAB display>
  -960   1280   960 -1280
  -720   960   720   -960
   960 -1280   -960   1280

$\gg$ k3 = t3'*ke3*t3   <equation entry>
k3 =

800	0	-800	0
0	0	0	0
-800	0	800	0
0	0	0	0

<MATLAB display>

Vectors of element gdof numbers are entered with the gdof numbers in sequence beginning with the gdof number corresponding to edof 1. Each vector must include all of the gdof numbers for the structure if it is to be used in the rearrangement of element stiffness matrices and the assembly of the global stiffness matrix. Thus those gdof numbers not associated with a member are inserted in sequence, beginning with the lowest gdof number not so associated and starting at position 5 of each vector:

$\gg$ u1 = [1 2 5 6 3 4]   <data entry>
u1 =

1	2	5	6	3	4

<MATLAB display>

$\gg$ u2 = [5 6 3 4 1 2]   <data entry
u2 =

5	6	3	4	1	2

<MATLAB display>

$\gg$ u3 = [1 2 3 4 5 6]   <data entry>
u3 =

1	2	3	4	5	6

<MATLAB display>

Element stiffness matrices with respect to global axes are expanded to the size of the global stiffness matrix and rearranged in accordance with their gdof numbers (unit is kips/in.):

$\gg$ k1(6,6) = 0   <equation entry>
k1 =

960	720	-960	-720	0	0
720	540	-720	-540	0	0
-960	-720	960	720	0	0
-720	-540	720	540	0	0
0	0	0	0	0	0
0	0	0	0	0	0

<MATLAB display>

$\gg$ k1 (u1,u1) = k1   <equation entry>
k1 =

960	720	0	0	-960	-720
720	540	0	0	-720	-540
0	0	0	0	0	0
0	0	0	0	0	0
-960	-720	0	0	960	720
-720	-540	0	0	720	540

<MATLAB display>

≫ k2(6,6) = 0   <equation entry>
k2 =

720	-960	-720	960	0	0	<MATLAB display>
-960	1280	960	-1280	0	0	
-720	960	720	-960	0	0	
960	-1280	-960	1280	0	0	
0	0	0	0	0	0	
0	0	0	0	0	0	

≫ k2(u2,u2) = k2   <equation entry>
k2 =

0	0	0	0	0	0	<MATLAB display>
0	0	0	0	0	0	
0	0	720	-960	-720	960	
0	0	-960	1280	960	-1280	
0	0	-720	960	720	-960	
0	0	960	-1280	-960	1280	

≫ k3(6,6) = 0   <equation entry>
k3 =

800	0	-800	0	0	0	<MATLAB display>
0	0	0	0	0	0	
-800	0	800	0	0	0	
0	0	0	0	0	0	
0	0	0	0	0	0	
0	0	0	0	0	0	

≫ k3 (u3,u3) = k3   <equation entry>
k3 =

800	0	-800	0	0	0	<MATLAB display>
0	0	0	0	0	0	
-800	0	800	0	0	0	
0	0	0	0	0	0	
0	0	0	0	0	0	
0	0	0	0	0	0	

The global stiffness matrix is assembled by adding the rearranged element stiffness matrices, those developed with respect to global axes, expanded to the size of the global stiffness matrix, and rearranged in accordance with their gdof numbers (unit is kips/in.):

≫ kg = k1 + k2 + k3   <equation entry>
kg =

1760	720	-800	0	-960	-720	<MATLAB display>
720	540	0	0	-720	-540	
-800	0	1520	-960	-720	960	
0	0	-960	1280	960	-1280	
-960	-720	-720	960	1680	-240	
-720	-540	960	-1280	-240	1820	

A vector of restraint gdof numbers is entered in a sequence beginning with the lowest gdof number:

```
>> nr = [1 2 4]                              <data entry>
nr =
     1     2     4                            <MATLAB display>
```

A vector of free gdof numbers is written by deleting the restraint gdof numbers from the complete vector of gdof numbers:

```
>> nf = [1:6]                                <data entry>
nf =
     1     2     3     4     5     6          <MATLAB display>
>> nf(nr) = [ ]                              <equation entry
nf =
     3     5     6                            <MATLAB display>
```

The submatrices needed for the submatrix method of solution of the global stiffness equations are extracted from the global stiffness matrix (unit is kips/in.):

```
>> kff = kg(nf,nf)                           <equation entry>
kff =
      1520      -720       960               <MATLAB display>
      -720      1680      -240
       960      -240      1820
>> ksf = kg(nr,nf)                           <equation entry>
ksf =
      -800      -960      -720               <MATLAB display>
         0      -720      -540
      -960       960     -1280
```

If any of the restraint values are nonzero, the submatrices kfs and kss and a vector of restraint values qo would be required also.

The vector of forces applied to the free gdof is entered in sequence beginning with the lowest numbered gdof (i.e., at gdof 3, 5, 6). There must be an entry for each free gdof (unit is kips):

```
>> ff = [0 0 -30]                            <data entry>
ff =
     0     0    -30                           <MATLAB display>
```

The displacements of the unrestrained nodes, or free gdof, are computed. They are in the sequence of the free gdof in vector nf (unit is inches):

```
>> qf = kff\ff'                              <equation entry>
qf =
      0.01800                                <MATLAB display>
      0.00408
     -0.02544
```

The truss reactions are computed. They are in the order of the restraint gdof in vector nr (unit is kips):

```
>> rs = ksf*qf                               <equation entry>
rs =
      0.0000                                 <MATLAB display>
     10.8000
     19.2000
```

The complete vector of all joint displacements qg is written from the vector of calculated displacements of free gdof (qf) and the vector restraint values (qo = [0 0 0]) and then rearranged in gdof order (unit is inches):

```
>> qg = [qf' zeros(1,3)]                     <equation entry>
qg =
      0.0180    0.0041   -0.0254    0    0    0    <MATLAB display>
>> nt = [nf nr]                              <equation entry>
nt =
      3    5    6    1    2    4               <MATLAB display>
>> qg(nt) = qg                               <equation entry>
qg =
      0    0    0.0180    0    0.0041   -0.0254    <MATLAB display>
```

The gdof, or joint displacements, with respect to global axes for each member are extracted from the rearranged vector of all gdof (unit is inches):

```
>> q1 = qg([1 2 5 6])                        <equation entry>
q1 =
      0    0    0.0041   -0.0254              <MATLAB display>
>> q2 = qg([5 6 3 4])                        <equation entry>
q2 =
      0.0041   -0.0254    0.0180    0         <MATLAB display>
>> q3 = qg([1 2 3 4])                        <equation entry>
q3 =
      0    0    0.0180    0                   <MATLAB display>
```

Member forces in each element with respect to local axes are calculated from element stiffness matrices with respect to local axes and element gdof transformed to reference to local axes (unit is kips):

```
>> f1 = kel*t1*q1'                           <equation entry>
f1 =
     18                                      <MATLAB display>
    -18
>> f2 = ke2*t2*q2'                           <equation entry>
f2 =
     24.0000                                 <MATLAB display>
    -24.0000
```

≫ f3 = ke3*t3*q3′   < equation entry >
f3 =

    -14.4000   < MATLAB display >
    14.4000

Note that by reviewing the sign convention for element local axes, the member force in element 1 is 18 kips compression, in element 2 is 24 kips compression, and in element 3 is 14.4 kips tension.

A.2.2. Plane Frame Example by Manual Entry

To illustrate the use of MATLAB in the step-by-step, or manual entry, mode of operation, a 2-element plane frame will be analyzed. Shown in Figure A.3 is the physical plane frame with supports and loads. The modulus of elasticity $E = 30,000$ ksi. MATLAB will be used to obtain nodal displacements, support reactions, and element axial forces, shears, and moments.

The discretized frame with nodes, gdof, and element numbers chosen is shown in Figure A.4. The nodes are designated n_1–n_3. The elements are designated 1–2 and shown in the boxes. The arrow on each box indicates the sequence from the start node to the end node for each element. For example, the start node for element 2 is n_2 and its end node is n_3. The gdof are designated 1–9 and shown in the circles. They are either horizontal or vertical, in conformance with the global axes chosen. The edof are shown with respect to element local axes as 1–6 for each element. Their sequence is consistent with the designation of start and end nodes for each element.

The elements' elastic modulus (ksi) is entered:

≫ E = 30000   < data entry >
E =

    30000   < MATLAB display >

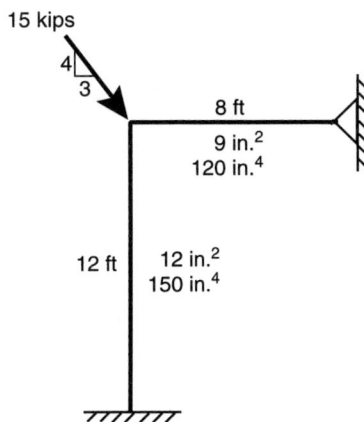

Figure A.3. Plane frame.

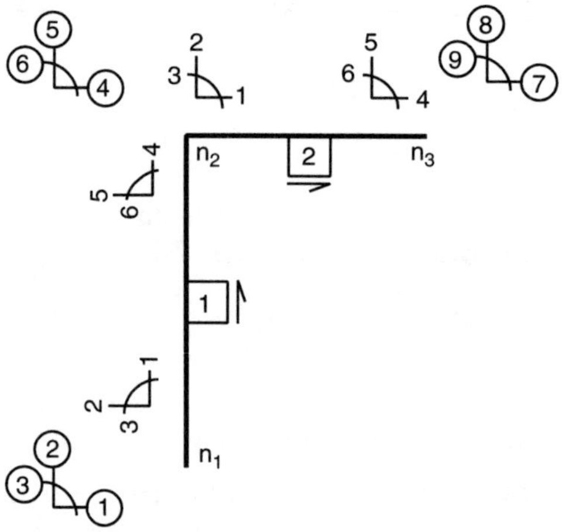

Figure A.4. Discretized frame.

The vectors of element cross-sectional areas (in.^{2}), moments of inertia (in.^{4}), and lengths (feet) are entered in sequence beginning with element 1. The length vector is multiplied by 12 to convert it to inches so that consistent units are treated:

```
≫ A = [12 9]                                  <data entry>
A =
     12     9                                 <MATLAB display>
≫ I = [150 120]                               <data entry>
I =
    150      120                              <MATLAB display>
≫ L = [12 8]*12                               <data entry>
L =
    144      96                               <MATLAB display>
```

The expressions to develop the upper left 3 × 3 submatrix of the plane frame element stiffness matrix for element 1 are entered (unit is kips/in., kips, or kip-in.):

```
≫ k111 = [A(1) 0 0;0 12*I(1)/L(1)^2  6*I(1)/L(1); 0 6*I(1)/L(1)   4I]
                                              <equation entry>
k111 =
    12.0000          0           0            <MATLAB display>
         0      0.0868      6.2500
         0      6.2500    600.0000
```

The expressions to develop the upper right 3 × 3 submatrix of the plane frame element stiffness matrix for element 1 are entered (unit is kips/in., kips, or kip-in.):

```
≫ k112 = -k111;                               <equation entry>
≫ k112(2,3) = -k112(2,3);                     <equation entry>
```

≫ k112(3,3) = -k112(3,3)/2;   <equation entry>
≫ k112
k112 =
    -12.0000   0   0   <MATLAB display>
         0   -0.0868   6.2500
         0   -6.2500   300.0000

The expressions to develop the lower left 3×3 submatrix of the plane frame element stiffness matrix for element 1 is entered (unit is kips/in., kips, or kip-in.):

≫ k121 = k112'   <equation entry>
k121 =
    -12.0000   0   0   <MATLAB display>
         0   -0.0868   -6.2500
         0   6.2500   300.0000

The expressions to develop the lower right 3×3 submatrix of the plane frame element stiffness matrix for element 1 are entered (unit is kips/in., kips, or kip-in.):

≫ k122 = k111;   <equation entry>
≫ k122(2,3) = -k122(2,3);   <equation entry>
≫ k122(3,2) = k122(2,3):   <equation entry>
≫ k122
k122 =
     12.0000   0   0   <MATLAB display>
         0   0.0860   -6.2500
         0   -6.2500   600.0000

The expression is entered to assemble the stiffness matrix for element 1 from the element 1 submatrices just developed (unit is kips/in., kips, or kip-in.):

≫ k1 = (E/L(1))*[k111 k112;k121 k122]   <equation entry>
k1 =
    1.0e + 005 *   <MATLAB display>
    0.0250   0   0   -0.0250   0   0
         0   0.0002   0.1301   0   -0.0002   0.0130
         0   0.0130   1.2500   0   -0.0130   0.6250
   -0.2500   0   0   0.0250   0   0
         0   -0.0002   -0.0130   0   0.0002   -0.0130
         0   0.0130   0.6250   0   -0.0130   1.2500

The element 1 matrix entry process is repeated for element 2:

≫ k211 = [A(2) 0 0;0 12*I(2)/L(2)^2   6*I(2)/L(2);0 6*I(2)/L(2)   4I]
                                                  <equation entry>

k211 =
     9.0000   0   0   <MATLAB display>
         0   0.1562   7.5000
         0   7.5000   480.0000

```
>> k212 = -k211                                      < equation entry >
>> k212(2,3) = -k212(2,3);                           < equation entry >
>> k212(3,3) = -k212(3,3)/2:                         < equation entry >
>> k212
k212 -
      -9.0000             0              0           < MATLAB display >
            0       -0.1562         7.5000
            0       -7.5000       240.0000
>> k221 = k212'                                      < equation entry >
k221 =
      -9.0000             0              0           < MATLAB display >
            0       -0.1562      -7.50000
            0        7.5000       240.0000
>> k222 = k211;                                      < equation entry >
>> k222(2,3) = -k222(2,3);                           < equation entry >
>> k222(3,2) = k222(2,3);                            < equation entry >
>> k222
k222 =
       9.0000             0              0           < MATLAB display >
            0        0.1562        -7.5000
            0       -7.5000       480.0000
>> k2 = (E/L(2))*[k211 k212;k221 k222]               < equation entry >
k2 =
       1.0e + 005 *                                  < MATLAB display >
       0.0281           0            0     -0.0281         0          0
            0      0.0005       0.0234          0    -0.0005     0.0234
            0      0.0234       1.5000          0    -0.0281     0.7500
      -0.0281          0            0      0.0281         0          0
            0     -0.0005      -0.0234          0     0.0005    -0.0234
            0      0.0234       0.7500          0    -0.0234     1.5000
```

The element 1 transformation submatrix is entered:

```
>> t1 = [0 1 0;-1 0 0;0 0 1]                         < data entry >
t1 =
       0       1       0                             < MATLAB display >
      -1       0       0
       0       0       1
```

Assemble the element 1 transformation matrix from its transformation submatrix:

```
>> t1 = [t1 zeros (3) ;zeros(3) t1]                  < equation entry >
t1 =
       0       1       0       0       0       0     < MATLAB display >
      -1       0       0       0       0       0
       0       0       1       0       0       0
       0       0       0       0       1       0
       0       0       0      -1       0       0
       0       0       0       0       0       1
```

The element 1 transformation matrix entry process is repeated for element 2:

≫ t2 = [1 0 0; 0 1 0;0 0 1]   < data entry >
t2 =

1	0	0
0	1	0
0	0	1

< MATLAB display >

≫ t2 = [t2 zeros (3) ;zeros(3) t2]   < equation entry >
t2 =

1	0	0	0	0	0
0	1	0	0	0	0
0	0	1	0	0	0
0	0	0	1	0	0
0	0	0	0	1	0
0	0	0	0	0	1

< MATLAB display >

Since t_2 is a unit matrix, the element 2 stiffness matrix need not be transformed.

The expressions to transform the elements' stiffness matrices from local to global references are entered (unit is kips/in., kips, or kip-in.):

≫ ke1 = t1′ *k1*t1   < equation entry >
ke1 =

1.0e + 005 *   < MATLAB display >

0.0002	0	-0.0130	-0.0002	0	-0.0130
0	0.0250	0	0	-0.0250	0
-0.0130	0	1.2500	0.0130	0	0.6250
-0.0002	0	0.0130	0.0002	0	0.0130
0	-0.0250	0	0	0.0250	0
-0.0130	0	0.6250	0.0130	0	1.2500

≫ ke2 = k2   < equation entry >
ke2 =

1.0e + 005 *   < MATLAB display >

0.0281	0	0	-0.0281	0	0
0	0.0005	0.0234	0	-0.0005	0.0234
0	0.0234	1.5000	0	-0.0234	0.7500
-0.0281	0	0	0.0281	0	0
0	-0.0005	-0.0234	0	0.0005	-0.0234
0	0.0234	0.7500	0	-0.0234	1.5000

The vectors of element gdof numbers for each element are entered in sequence beginning with each vector with the gdof number corresponding to edof 1. Each vector must include all of the gdof numbers for the structure if it is to be used in the rearrangement of the element stiffness matrices and assembly of the global stiffness matrix. Thus those gdof not associated with an element are inserted in sequence, beginning with the lowest gdof number not so associated and starting at position 7 of each vector:

```
>> u1 = [1 2 3 4 5 6 7 8 9]                         < equation entry >
u1 =
    1    2    3    4    5    6    7    8    9        < MATLAB display >
>> u2 = [4 5 6 7 8 9 1 2 3]                         < equation entry >
u2 =
    4    5    6    7    8    9    1    2    3        < MATLAB display >
```

Element stiffness matrices with respect to global axes are expanded to the size of the global stiffness matrix and rearranged in accordance with their gdof numbers (unit is kips/in., kips, or kip-in.):

```
>> ke1(9,9) = 0;                                    < equation entry >
>> ke1(u1,u1) = ke1                                 < equation entry >
ke1 =
    1.0e+005 *
    0.0002        0   -0.0130   -0.0002        0   -0.0130    0    0    0
         0   0.0250         0         0   -0.0250        0    0    0    0
   -0.0130        0    1.2500    0.0130        0    0.6250    0    0    0
   -0.0002        0    0.0130    0.0002        0    0.0130    0    0    0
         0  -0.0250         0         0    0.0250        0    0    0    0
   -0.0130        0    0.6250    0.0130        0    1.2500    0    0    0
         0        0         0         0        0        0    0    0    0
         0        0         0         0        0        0    0    0    0
         0        0         0         0        0        0    0    0    0
                                                    < MATLAB display >
>> ke2(9,9) = 0;                                    < equation entry >
>> ke2(u2,u2) = ke2                                 < equation entry >
ke2 =
    1.0e+005 *
    0    0    0        0         0         0        0         0         0
    0    0    0        0         0         0        0         0         0
    0    0    0        0         0         0        0         0         0
    0    0    0   0.0281         0         0  -0.0281         0         0
    0    0    0        0    0.0005    0.0234        0   -0.0005    0.0234
    0    0    0        0    0.0234    1.5000        0   -0.0234    0.7500
    0    0    0  -0.0281         0         0   0.0281         0         0
    0    0    0        0   -0.0005   -0.0234        0    0.0005   -0.0234
    0    0    0        0    0.0234    0.7500        0   -0.0234    1.5000
                                                    < MATLAB display >
```

The global stiffness matrix is assembled by adding the element stiffness matrices just expanded and rearranged (unit is kips/in., kips, or kip-in.):

≫ kg = ke1 + ke2   <equation entry>
kg =

     1.0e+005 *

0.0002	0	-0.0130	-0.0002	0	-0.0130	0	0	0
0	0.0250	0	0	-0.0250	0	0	0	0
-0.0130	0	1.2500	0.0130	0	0.6250	0	0	0
-0.0002	0	0.0130	0.0283	0	0.0130	-0.0281	0	0
0	-0.0250	0	0	0.0255	0.0234	0	-0.0005	0.0234
-0.0130	0	0.6250	0.0130	0.0234	2.7500	0	-0.0234	0.7500
0	0	0	-0.0281	0	0	0.0281	0	0
0	0	0	0	-0.0005	-0.0234	0	0.0005	-0.0234
0	0	0	0	0.0234	0.7500	0	-0.0234	1.5000

                                                   <MATLAB display>

A vector of restraint gdof numbers is entered in sequence beginning with the lowest restraint gdof number:

≫ nr = [1 2 3 7 8]   <equation entry>
nr =

     1   2   3   7   8   <MATLAB display>

A vector of free gdof numbers is written by deleting the restraint gdof numbers from the complete vector of gdof numbers:

≫ nf = [1:9]   <equation entry>
nf =
     1   2   3   4   5   6   7   8   9   <MATLAB display>
≫ nf(nr) = [ ]   <equation entry>
nf =
     4   5   6   9   <MATLAB display>

The submatrices needed for the submatrix method of solution of the global stiffness equations are extracted from the global stiffness matrix (unit is kips/in., kips, or kip-in.):

≫ kff = kg(nf,nf)   <equation entry>
kff =

     1.0e+005 *

| 0.0283 | 0 | 0.0130 | 0 | <MATLAB display>
0	0.0255	0.0234	0.0234
0.0130	0.0234	2.7500	0.7500
0	0.0234	0.7500	1.5000

≫ ksf = kg(nr,nf)   <equation entry>

ksf =

   1.0e + 004 *

-0.0018	0	-0.1302	0	< MATLAB display >
0	-0.2500	0	0	
0.1302	0	6.2500	0	
-0.2812	0	0	0	
0	-0.0049	-0.2344	-0.2344	

If any of the restraint values are nonzero, the submatrices kfs and kss and the vector qo would be required also.

    The vector of forces and moments applied to the free gdof is entered in sequence beginning with the lowest numbered gdof (i.e., at gdof 4, 5, 6, and 9). There must be an entry for each free gdof (unit is kips or kip-in.):

  ≫ ff = [9 -12 0 0]   < data entry >
ff =

    9   -12   0   0   < MATLAB display >

The displacements of the unrestrained nodes, or free gdof, are computed. They are in the sequence of the free gdof numbers in vector nf (unit is inches or radians):

  ≫ qf = kff\ff′   < equation entry >
qf =

     0.00317672   < MATLAB display >
    -0.00477955
     0.00000617
     0.00007160

The frame reactions are computed. They are in the order of the restraint gdof numbers in vector nr (unit is kips or kip-in.):

  ≫ rs = ksf*qf   < equation entry >
rs =

    -0.0655   < MATLAB display >
    11.9489
     4.5218
    -8.9345
     0.0511

The complete vector of all joint displacements qg is written from the vector of calculated displacements of free gdof qf and the vector restraint values (qo = [0 0 0 0 0]) and then rearranged in gdof order (unit is inches or radians):

  ≫ qg= [qf′ zeros(1,5)]   < equation entry >
qg =

    0.0032   -0.0048   0.0000   0.0001   0   0   0   0   0
                                                      < MATLAB display >
  ≫ nt = [nf nr]   < equation entry >

nt =
    4   5   6   9   1   2   3   7   8   < MATLAB display >
≫ qg(nt) = qg   < equation entry >
qg =
    0   0   0   0.0032   -0.0048   0.0000   0   0   0.0001
                 < MATLAB display >

The gdof, or joint displacements, with respect to global axes for each element are taken from the rearranged vector of all gdof (unit is inches or radians):

≫ q1 = qg(1:6)   < equation entry >
q1 =
    0   0   0   0.0032   -0.0048   0.0000   < MATLAB display >
≫ q2 = qg(4:9)   < equation entry >
q2 =
    0.0032   -0.0048   0.0000   0   0   0.0001   < MATLAB display >

Axial forces, shears, and moments in each element with respect to local axes are calculated from element stiffness matrices with respect to local axes and element gdof transformed to reference to local axes. (Units are kips and kip-inches.) The sequence is axial force, shear, and moment at the start node, and this repeats at the end node:

≫ f1 = k1*t1*q1′   < equation entry >
f1 =
      11.9489   < MATLAB display >
      0.0655
      4.5218
   -11.9489
    -0.0655
     4.9072
≫ f2 = k2*q2′   < equation entry >
f2 =
    8.9345   < MATLAB display >
   -0.0511
   -4.9072
   -8.9345
    0.0511
    0.0000

A.3. MATLAB PROGRAMMING

The MATLAB platform allows the analyst to program as m-files the sequence of steps that would otherwise be entered manually for each analysis. The m-file may be typed by any word processor and saved as a file in ASCII format with the extension "m" in its file name. The *MATLAB User's Guide* and *Reference Guide* and on-line help facility provide information on writing m-files. What follows are illustrations of eleven m-files, written by the author, for plane truss analysis (truss.m and trussi.m), plane frame or beam analysis (frame.m), plane stress thin plate analysis (pstri.m, pstriu.m, psquad.m, and psquadu.m), and thin plate flexure analysis (flxrt.m, flxrtu.m, flxtri.m, and flxtriu.m). Shown for each m-file is an example problem with the discretizing of the structure, the required data input, and the displayed results. Listings for truss.m and frame.m, suitable for direct operation from the user's working subdirectory of MATLAB v4.3 or higher are in Sections A.4.4 and A.5.4. All the m-files discussed in this appendix can be downloaded from the John Wiley & Sons FTP server. If you have Internet access and would like to download the m-files, go to:

ftp://ftp.wiley.com/public/products/subject/engineering/carroll

The files are also available on the MathWorks FTP server at:

ftp://ftp.mathworks.com/pub/books/carroll

For questions about downloading the files, call Wiley Technical Support at (212) 850-6753 or techhelp@wiley.com.

A.4. EXAMPLE PLANE TRUSS ANALYSIS BY truss.m

The m-file truss.m employs the 2-node, 2-dof straight, prismatic axial finite element discussed in Chapters 3 and 5. Data entry is by node and element, permitting a variety of plane truss configurations. Numbering schemes for global node numbers and gdof numbers is at the discretion of the analyst. The gdof numbers are related to

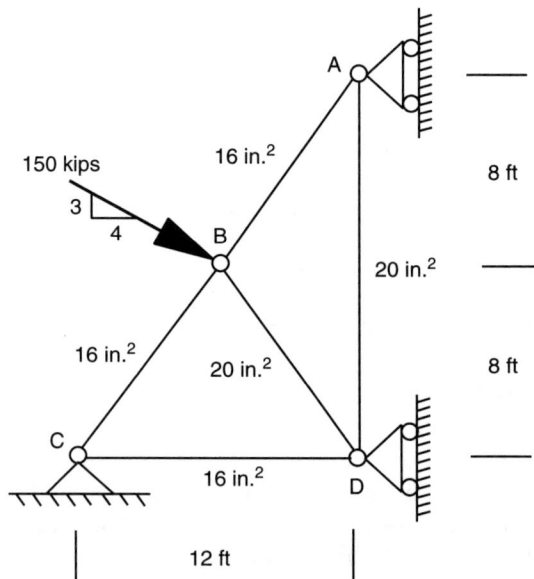

Figure A.5. Plane truss.

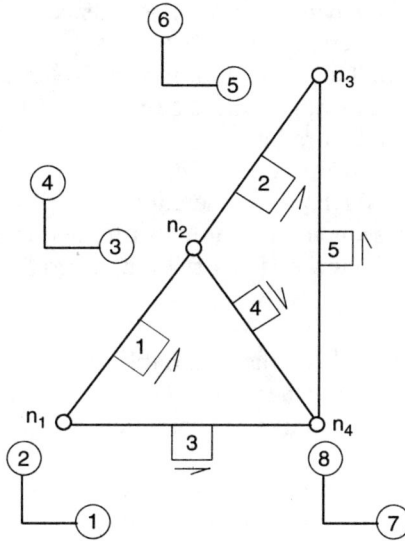

Figure A.6. Discretized truss.

the global node numbers. At global node number i, the number of the gdof parallel to the global x axis is $2i - 1$, and parallel to the global y axis is $2i$.

To illustrate the use of the m-file truss.m, consider the statically indeterminate truss shown in Figure A.5. The joint displacements, support reactions, and member forces for the truss will be determined. Modulus of elasticity $E = 30,000$ ksi.

The truss is discretized as shown in Figure A.6. The nodes are n_1–n_4. The elements are 1–5 and shown in the boxes. The arrow on each box indicates the sequence from the start node to the end node. For example, the start node for element 4 is n_2 and the end node is n_4. The gdof are 1–8 and are shown in the circles. They are either horizontal or vertical, in conformance with the global axes chosen.

A.4.1. Variables Needed for truss.m

nn   Number of nodes.

mnn   Row vector of element global node numbers in a sequence of paired numbers. Each pair is the start and end node for an element. The pairs are in sequence beginning with element 1.

rn   Row vector of restraint gdof numbers in sequence from lowest to highest.

rv   Row vector of nonzero restraint values in a sequence of paired numbers. Each pair is the restraint gdof number and its nonzero value. The pairs are in sequence beginning with the lowest gdof restraint number. Restraint values are inches for truss.m (millimeters for trussi.m). Restraints whose values are zero are not entered.

A   Row vector of the element areas in sequence beginning with element 1. Units are in.^{2} (mm$^2 \times 10^3$ for trussi.m).

E   Modulus of elasticity for the elements. Units are ksi (MPa for trussi.m).

L   Row vector of element lengths in sequence beginning with element 1.
          Units are feet (meters for trussi.m).

t   Row vector of angles of orientation of each element in sequence beginning
          with element 1. The angles must be measured from the positive global x
          axis and expressed in radians.

f   Row vector of nonzero nodal forces applied to free gdof in a sequence of
          paired numbers. Each pair is the gdof number and the value of the force.
          The pairs are in sequence beginning with the lowest gdof number. Units of
          force are kips (kilonewtons for trussi.m). Zero forces at free gdof need not
          be entered.

Program truss.m requires each of these variables exactly in the format described
above and illustrated below. Furthermore, the variables must be declared as "global" as illustrated, so that their values are communicated to truss.m. The program is
listed in Section A.4.4.

   Program trussi.m operates in exactly the same way as truss.m, using exactly the
same variables. It uses metric units for the variables rv, A, E, L, and f as indicated
above.

A.4.2. Data Entry for truss.m

Declare the global variables needed for truss.m:

≫ global nn mnn rn rv A E L t f   < data entry >
Enter the number of nodes on the truss.
≫ nn = 4   < data entry >
nn =
    4   < MATLAB display >

Enter the row vector of paired element node numbers:

≫ mnn = [1 2 2 3 1 4 2 4 4 3]   < data entry >
mnn =
    1   2   2   3   1   4   2   4   4   3
                                                   < MATLAB display >

Enter the row vector of restraint gdof numbers:

≫ rn = [1 2 5 7]   < data entry >
rn =
    1   2   5   7   < MATLAB display >

Enter the row vector of nonzero restraint pairs. Restraint values are in inches. In this
example all restraint values are zero, so this entry could be omitted:

≫ rv = [ ]   < data entry >
rv =
    [ ]   < MATLAB display >

Enter the row vector of cross-sectional areas of truss elements (unit is in^2.).

```
>> A = [16 16 16 20 20]                          <data entry>
A =
      16    16    16    20    20                  <MATLAB display>
```

Enter the modulus of elasticity (unit is ksi):

```
>> E = 30000                                      <data entry>
E =
      30000                                       <MATLAB display>
```

Enter the row vector of element lengths. Unit is feet (truss.m converts to inches for analysis and back again to feet for display purposes):

```
>> L = [10 10 12 10 16]                           <data entry>
L =
      10    10    12    10    16                   <MATLAB display>
```

Enter the row vector of element orientation angles measured from the positive global *x* axis. Unit is radians. Note the functions available in MATLAB that make entry of these angles a simple exercise:

```
>> t = [atan(16/12) atan(16/12) 0 atan((-8)/6) pi/2]    <data entry>
t =
      0.9273    0.9273    0    -0.9273    1.5708          <MATLAB display>
```

Enter the row vector of nonzero applied force pairs. Unit for forces is kips:

```
>> f = [3 120 4 -90]                              <data entry>
f =
      3    120    4    -90                         <MATLAB display>
```

Enter the command to execute the programmed steps in the m-file truss.m:

```
>> truss                                          <data entry>
```

A.4.3. Results Displayed by truss.m

PLANE TRUSS ANALYSIS
nr of nodes: 4, nr of mbrs: 5, nr of gdof: 8
elastic modulus: 30000 ksi

member properties:

mbr	length (ft)	area (in2)
1	10.00	16.00
2	10.00	16.00

3	12.00	16.00
4	10.00	20.00
5	16.00	20.00

restraints and reactions (global axes):

rstr	restraint	reaction
gdof	value (in)	(kips)
1	0.0000	67.500
2	0.0000	90.000
5	0.0000	-93.750
7	0.0000	-93.750

applied forces and displacements of free gdof (global axes):

free	applied	displacement
gdof	force (kips)	(in)
3	120.00	0.085260
4	-90.00	-0.099102
6	0.00	-0.083984
8	0.00	-0.123984

member axial forces (local member axes):

	start	end
mbr	(kips)	(kips)
1	112.500	-112.500
2	156.250	-156.250
3	0.000	0.000
4	156.250	-156.250
5	-125.000	125.000

A.4.4. Listing of truss.m

```
function y = truss
global nn A E L t mnn rn rv f
n = 2*nn;
nm = length(A);
ke = [1 -1;-1 1];
S = A./12*E./L;
% set up the vector to of sin and cos from vector t and
% the transformation matrix tt from vector to
to = zeros(1,2*nm):
for i = 1:nm
    to(2*i-1) = cos(t(i));
    to(2*i) = sin(t(i));
end
tt = zeros(2*nm,4);
for i = 1:nm
    tt([(2*i-1) : (2*i)],:) = [to(2*i-1) to(2*i) 0 0; ...
                        0 0 to (2*i-1) to(2*i)];
end
```

```
% set up the complete matrix of mbr gdof for the global k
u = zeros (nm,4);
for i = 1:nm
    u(i,1) = 2*mnn(2*i-1)-1;
    u(i,2) = 2*mnn(2*i-1);
    u(i,3) = 2*mnn(2*i)-1;
    u(i,4) = 2*mnn(2*i);
end
ux = u;
if n > 4
    ux(nm,n) = 0;
end
for i = 1:nm
    jj = 4;
    for j = 1:n
        if j~ = ux(i,1) & j~ = ux(i,2) & j~ = ux(i,3) & j~ = ux(i,4)
            jj = jj + 1;
            ux(i,jj) = j;
        end
    end
end
% compute elem stiffness matrices wrt global axes and
% assemble the global stiffness matrix
kg = zeros(n,n);
for i = 1:nm
    k = S(i)*tt([(2*i-1) : (2*i)],:)'*ke*tt([(2*i-1) : (2*i)],:);
    k(n,n) = 0;
    uu = ux(i,:);
    k(uu,uu) = k;
    kg = kg + k;
end
% construct the global load vector ff from the vector f of
% input applied point loads at nodes
vf = length(f)/2;
ff = zeros(1,n);
for i = 1:vf
    for j = 1:n
        if j = = f(2*i-1)
            ff(j) = f(2*i);
            break
        end
    end
end
rr = ff(rn);
ff(rn) = [ ];
% extract the submatrices for computation of joint displacements
% and truss reactions by the submatrix method of solution.
nt = [1:n];
```

```
nf = nt;
nf(rn) = [ ];
kff = kg(nf,nf);
kfs = kg(nf,rn);
ksf = kg(rn,nf);
kss = kg(rn,rn);
vr = length(rn);
qo = zeros(1,vr);
vv = length(rv)/2;
for i = 1:vv
    for j = 1:vr
        if rv(2*i-1) = = rn(j)
            qo(j) = rr(2*i);
        end
    end
end
qf = kff\(ff'-kfs*qo');
rs = ksf*qf + kss*qo';
% construct the total final displacement global vector and
% the individual element displacement vectors, and
% compute the member bar forces
qg = [qf' qo]';
nt = [nf rn];
qg(nt) = qg;
fk = zeros(2,nm);
for i = 1:nm
    fk(1:2,i) = S(i)*ke*tt([2*i-1,2*i],:)*qg(u(i,:));
end
% print out controlling data input, joint displacements,
% reactions, and members forces
fprintf(1,'\n')
fprintf(1,'          PLANE TRUSS ANALYSIS\n')
fprintf(1,'\n')
fprintf(1,'    nr of nodes:%3.0f,  nr of mbrs:%3.0f,  nr of gdof:%3.0f\n',nn,nm,n)
fprintf(1,'\n')
fprintf(1,'          elastic modulus: %6.0f ksi\n',E)
fprintf(1,'\n')
fprintf(1,'          member properties:\n')
fprintf(1,'\n')
fprintf(1,'          mbr      length        area\n')
fprintf(1,'                   (ft)          (in2)\n')
fprintf(1,'\n')
nmi = [1:nm];
qnm = [nmi' L' A']';
fprintf(1,'          %2.0f      %8.2f        %8.2f\n',qnm)
fprint(1,'\n')
fprintf(1,'          restraints and reactions (global axes):\n')
fprintf(1,'\n')
```

```
fprintf(1,'              rstr        restraint       reaction\n')
fprintf(1,'              gdof        value (in)      (kips)\n')
fprintf(1,'\n')
vrsi = [rn' qo' rs]';
fprintf(1,'              %2.0f        %10.4f         %10.3f\n',vrsi)
fprintf(1,'\n')
fprintf(1,'         applied forces and displacements of free gdof (global axes):\n')
fprintf(1,'\n')
fprintf(1,'              free        applied         displacement\n')
fprintf(1,'              gdof        force (kips)        (in)\n')
fprint(1,'\n')
nfqi = [nf' ff' qf]';
fprintf(1,'              %2.0f        %8.2f          %12.6f\n',nfqi)
fprintf(1,'\n')
fprintf(1,'         member axial forces (local member axes):\n')
fprintf(1,'\n')
fprintf(1,'                          start       end\n')
fprintf(1,'              mbr         (kips)      (kips)\n')
fprintf(1,'\n')
nmfk = [nmi' fk']';
fprintf(1,'              %2.0f        %10.3f         %10.3f\n',nmfk)
fprintf(1,'\n')
```

A.5. EXAMPLE PLANE FRAME ANALYSIS BY frame.m

The m-file frame-m. employs the 2-node, 6-dof straight, prismatic plane frame element discussed in Chapter 3. Data entry is by node and element, permitting a variety of plane frame configurations. Numbering schemes for global node numbers and gdof numbers are at the discretion of the analyst. The gdof numbers are related to the global node numbers. At global node number i, the number of the axial translation gdof is $3i - 2$, of the transverse translation is $3i - 1$, and of the in-plane rotation is $3i$.

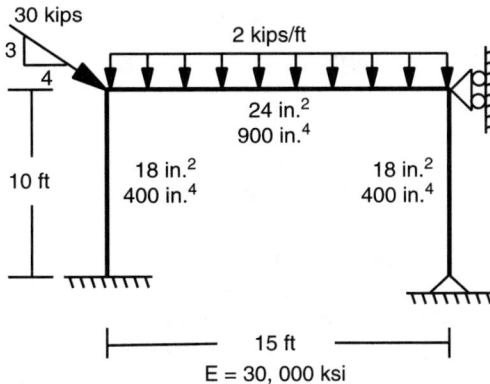

Figure A.7. Plane frame.

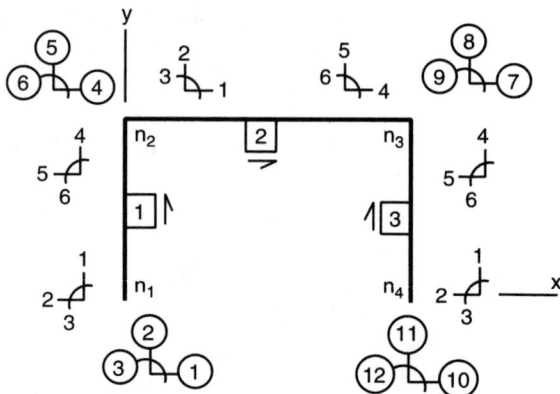

Figure A.8. Discretized frame.

To illustrate the use of the m-file frame.m, consider the statically indeterminate three-member frame shown in Figure A.7. The joint displacements, support reactions, and member axial forces, shears, and moments for the frame will be determined.

The frame has a fixed support on its lower left corner, a pinned support on its lower right corner, and a roller support on its upper right corner. The fixed support is fully restrained in all three gdof. The pinned support displaces 0.100 in. vertically downward but is fully restrained for horizontal translation. The roller support does not provide complete horizontal restraint; instead it displaces 0.25 in. to the left under the action of the loads. There is an inclined concentrated load at the upper left corner and a vertical, downward acting uniformly distributed load across the top of the frame.

The frame is discretized as shown in Figure A.8. The nodes are n_1–n_4. The elements are 1–3 and shown in the boxes. The arrow on each box indicates the sequence from the start node to the end node for each element. For example, the start node for element 3 is n_4 and its end node is n_3. The gdof are 1–12 and shown in the circles. They are horizontal or vertical, in conformance with the global axes.

A.5.1. Variables Needed for frame.m

nn Number of nodes.

mnn Row vector of element global node numbers in a sequence of paired numbers. Each pair is the start and end node for an element. The pairs are in sequence beginning with element 1.

rn Row vector of restraint gdof numbers in sequence from lowest to highest.

rv Row vector of nonzero restraint values in a sequence of paired numbers. Each pair is the restraint gdof number and its nonzero value. The pairs are in sequence beginning with the lowest gdof restraint number. Restraint values are in inches or radians. Restraints whose values are zero are not entered.

A   Row vector of element cross-sectional areas in sequence beginning with element 1. Unit is in.^{2}

I   Row vector of element cross-sectional moments of inertia in sequence beginning with element 1. Unit is in.^{4}

E   Modulus of elasticity. Unit is ksi.

L   Row vector of element lengths in sequence beginning with element 1. Unit is feet.

t   Row vector of angles of orientation of each element in sequence beginning with element 1. The angles must be measured from the positive global x axis and are expressed in radians.

f   Row vector of nonzero nodal forces and couples applied to free gdof in a sequence of paired numbers. Each pair is the gdof number and the value of the force or couple. The pairs are in sequence beginning with the lowest gdof number. Unit of force is kips and that of moments is kip-feet. Zero forces or couples at free gdof need not be entered.

w   Row vector of paired numbers, in sequence, reflecting the nonzero uniformly distributed loads applied transversely to the element. Each pair is the element number and the value of the distributed load in kips per feet with a sign consistent with the element local reference axes. The pairs are in sequence beginning with the lowest element number.

Program frame.m requires each of these variables in the format described and illustrated below. They must be declared as "global."

A.5.2. Data Entry for frame.m

Declare the global variables needed for frame.m:

≫ global nn mnn rn rv A E I L t f w   < data entry >

Enter the number of nodes on the frame.

≫ nn = 4   < data entry >
nn =
      4   < MATLAB display >

Enter the row vector of paired element node numbers:.

≫ mnn = [1 2 2 3 4 3]   < data entry >
mnn =
      1   2   2   3   4   3   < MATLAB display >

Enter the row vector of restraint gdof numbers:

≫ rn = [1 2 3 7 10 11]   < data entry >
rn =
      1   2   3   7   10   11   < MATLAB display >

Enter the row vector of nonzero restraint pairs. Restraint values are in inches or radians, as appropriate:

```
>> rv = [7 0.25 11 -0.1000]                        <data entry>
rv =
      7.0000      0.2500     11.0000     -0.1000    <MATLAB display>
```

Enter the row vector of cross-sectional areas of frame elements (unit is in.^{2}):

```
>> A = [18 24 18]                                  <data entry>
A =
     18     24     18                               <MATLAB display>
```

Enter the row vector of cross-sectional moments of inertia of frame elements (unit is in.^{4}):

```
>> I = [400 900 400]                               <data entry>
I =
    400    900    400                               <MATLAB display>
```

Enter the modulus of elasticity (unit is ksi):

```
>> E = 30000                                       <data entry>
E =
   30000                                            <MATLAB display>
```

Enter the row vector of element lengths. Unit is feet (frame.m converts them to inches for analysis and back again to feet for display purposes):

```
>> L = [10 15 10]                                  <data entry>
L =
     10     15     10                               <MATLAB display>
```

Enter the row vector of element orientation angles measured from the positive global x axis. Unit is radians. Note the functions available in MATLAB that make entry of these angles a simple exercise:

```
>> t = [pi/2 0 pi/2]                               <data entry>
t =
    1.5708        0    1.5708                       <MATLAB display>
```

Enter the row vector of nonzero applied force or couple pairs. Unit for force is kips and that for couples is kip-feet. In this example there are no applied couples:

```
>> f = [4 24 5 -18]                                <data entry>
f=
      4     24      5     -18                       <MATLAB display>
```

Enter the row vector of nonzero uniformly distributed loads applied to members (unit is kips per feet):

≫ w = [2 -2]   <data entry>
w =
    2   -2   <MATLAB display>

Enter the command to execute the programmed steps in the m-file frame.m:

≫ frame   <data entry>

A.5.3. Results Displayed by frame.m

PLANE FRAME ANALYSIS
nr of nodes: 4, nr of mbrs: 3, nr of gdof: 12
elastic modulus: 30000 ksi

member properties and uniform distributed loads (local axes):

mbr	length (ft)	area (in2)	mom inertia (in4)	distr load (kip/ft)
1	10.00	18.00	400.00	0.000
2	15.00	24.00	900.00	-2.000
3	10.00	18.00	400.00	0.000

restraints and reactions (global axes):

rstr gdof	restraint value	reaction
1	0.0000 in	-10.082 kips
2	0.0000 in	27.403 kips
3	0.0000 rad	68.814 ft-kips
7	0.2500 in	-8.723 kips
10	0.0000 in	-5.194 kips
11	-0.1000 in	20.597 kips

applied forces/moms and displacements of free gdof (global axes):

free gdof	applied force/moment	displacement
4	24.00 kips	0.2534794 in
5	-18.00 kips	-0.0060895 in
6	0.00 ft-kips	-0.0022082 rad
8	0.00 kips	-0.1045771 in
9	0.00 ft-kips	-0.0000055 rad
12	0.00 ft-kips	-0.0031222 rad

member axial forces, shears, and moments (local members axes):

mbr	start axial (kips)	start shear (kips)	start moment (ft-kips)	end axial (kips)	end shear (kips)	end moment (ft-kips)
1	27.403	10.082	68.814	-27.403	-10.082	32.011
2	13.918	9.403	-32.011	-13.918	20.597	-51.945
3	20.597	5.194	0.000	-20.597	-5.194	51.945

A.5.4. Listing of frame.m

```
function y = frame
global nn A E I L t mnn rn rv f w
n = 3*nn;
nm = length(A);
LL = 12*L;
% compute element stiffness matrices wrt local axes
ke = zeros(nm*6,6);
for i = 1:nm
    ke1 = [A(i)              0                      0
           0         12*I(i)/LL(i)^2           6*I(i)/LL(i)
           0          6*I(i)/LL(i)             4*I(i)];
    ke2 = -ke1;
    ke2(2,3) = -ke2(2,3);
    ke2(3,3) = -ke2(3,3)/2;
    ke3 = ke2';
    ke4 = ke1;
    ke4(2,3) = -ke4(2,3);
    ke4(3,2) = ke4(2,3);
    ke([(6*i-5):6*i],:) = E/LL(i)*[ke1 ke2;ke3 ke4];
end
% set of transformation matrix tt from vector t
tt = zeros(nm*6,6);
for i = i:nm
    c = cos(t(i));
    s = sin(t(i));
    cs = [c s 0;-s c 0;0 0 1];
    tt([(6*i-5):6*i],:) = [cs zeros(3,3);zeros(3,3) cs];
end
% set up complete matrix of mbr gdof nrs for the global k
u = zeros(nm,6);
for i = 1:nm
    u(i,1) = 3*mnn(2*i-1)-2;
    u(i,2) = 3*mnn(2*i-1)-1;
    u(i,3) = 3*mnn(2*i-1);
    u(i,4) = 3*mnn(2*i)-2;
    u(i,5) = 3*mnn(2*i)-1;
    u(i,6) = 3*mnn(2*i);
end
ux = u;
if n > 6
    ux(nm,n) = 0;
end
for i = 1:nm
    jj = 6;
    for j = 1:n
        if j~ = ux(i,1) & j~ = ux(i,2) & j~ = ux(i,3) & ...
```

```
            j~ = ux(i,4) & j~ = ux(i,5) & j~ = ux(i,6)
            jj = jj + 1;
            ux(i,jj) = j;
        end
    end
end
% compute elem stiffness matrices wrt to global axes and
% assemble the global stiffness matrix
kg = zeros(n,n);
for i = 1:nm
    k = tt([(6*i-5):6*i],:)'*ke([(6*i-5):6*i],:)*tt([(6*i-5):6*i],:);
    k(n,n) = 0;
    uu = ux(i,:);
    k(uu,uu) = k;
    kg = kg + k;
end
% construct the global load vector from input vectors of point loads
% on nodes and uniformly distributed transverse loads on members
vf = length(f)/2;
ff = zeros(1,n);
for i = 1:vf
    for j = 1:n
        if j = = f(2*i-1)
            ff(j) = f(2*i);
            break
        end
    end
end
fff = ff;
fff(rn) = [ ];
for i = 1:n/3
    ff(3*i) = ff(3*i)*12;
end
vw = length(w)/2;
fw = zeros(1,nm);
for i = 1:vw
    for j = i:nm
        if j = = w(2*i-1)
            fw(j) = w(2*i);
            break
        end
    end
end
fwl = zeros(nm,6);
for i = 1:nm
    fwl(i,2) = fw(i)*L(i)/2;
    fwl(i,3) = fw(i)*L(i)*LL(i)/12;
    fwl(i,5) = fw(i)*L(i)/2;
```

```
        fwl(i,6) = -fw(i)*L(i)*LL(i)/12;
end
fwg = zeros(nm,6);
for i = 1:nm
    fwg(i,:) = (tt([(6*i-5):6*i],:)'*fwl(i,:)')')';
end
fwg(nm,n) = 0;
for i = 1:nm
    uu = ux(i,:);
    fwg(i,uu) = fwg(i,:);
    ff = ff + fwg(i,:);
end
rr = ff(rn);
ff(rn) = [ ];
% extract the submatrices for the computation of joint
% displacements and frame reactions by the submatrix
% method of solution
nt = [1:n];
nf = nt;
nf(rn) = [ ];
kff = kg(nf,nf);
kfs = kg(nf,rn);
ksf = kg(rn,nf);
kss = kg(rn,rn);
vr = length(rn);
qo = zeros(1,vr);
vv = length(rv)/2;
for i = 1:vv
    for j = 1:vr
        if rv(2*i-1) = = rn(j)
            qo(j) = rv(2*i);
        end
    end
end
qf = kff\(ff'-kfs*qo');
rs = ksf*qf + kss*qo'-rr';
% construct the total final displacement vector and
% the individual member displacement vectors, and
% compute member axial and shear forces and moments
qg = [qf' qo]';
nt = [nf rn];
qg(nt) = qg;
fk = zeros(6,nm);
for i = 1:nm
    fk(1:6,i) = ke([(6*i-5):6*i],:)*tt([(6*i-5):6*i],:)*qg u(i,:)) ...
        -fwl(i,:)';
end
% print out controlling data input, joint displacements,
```

```
% reactions, and members axial and shear forces and moments
fprintf(1,'\n')
fprintf(1,'          PLANE FRAME ANALYSIS\n')
fprintf(1,'\n')
fprintf(1,'     nr of nodes:%3.0f, nr of mbrs:%3.0f, nr of gdof:%3.0f\n',n/3,nm,n)
fprintf(1,'\n')
fprintf(1,'          elastic modulus: %6.0f ksi\n',E)
fprintf(1,'\n')
fprintf(1,'member properties and uniform distributed loads (local member axes):\n')
fprintf(1,'\n')
fprintf(1,'          mbr     length     area     mom inertia     distr load\n')
fprintf(1,'                  (ft)       (in2)     (in4)          (kip/ft)\n')
fprintf(1,'\n')
nmi = [1:nm];
qnm = [nmi' L' A' I' fw']';
fprintf(1,'          %2.0f      %8.2f     %7.2f     %8.2f      %9.3f\n',qnm)
fprintf(1,'\n')
fprintf(1,'          restraints and reactions (global axes):\n')
fprintf(1,'\n')
fprintf(1,'          rstr     restraint\n')
fprintf(1,'          gdof     value          reaction\n')
fprintf(1,'\n')
for i = 1:vr
   if rem(rn(i),3) = = 0
      rs(i) = rs(i)/12;
   end
end
vrsi = [rn' qo' rs]';
for i = 1:vr
   if rem(rn(i),3)~ = 0
      fprintf(1,'          %2.0f      %10.4f in      %9.3f      kips\n',vrsi(:,i))
   else
      fprintf(1,'          %2.0f      %10.4f rad      %9.3f      ft-kips\n',vrsi(:,i))
   end
end
fprintf(1,'\n')
fprintf(1,' applied forces/moments and displacements of free gdof (global axes):\n')
fprintf(1,'\n')
fprintf(1,'          free     applied\n')
fprintf(1,'          gdof     force/moment      displacement\n')
fprintf(1,'\n')
nfqi = [nf' fff' qf]';
for i = 1:n-vr
   if rem(nf(i),3)~ = 0
      fprintf(1,'          %2.0f      %8.2f kips          %12.7f      in\n',nfqi(:,i))
   else
      fprintf(1,'          %2.0f      %8.2f ft-kips       %12.7f      rad\n',nfqi(:,i))
   end
```

```
end
fprintf(1,'\n')
fprintf(1,'            member axial forces, shears, and moments (local member axes):\n')
fprintf(1,'\n')
fprintf(1,'                    start    start    start     end     end    end\n')
fprintf(1,'            mbr    axial    shear   moment    axial   shear  moment\n')
fprintf(1,'                   (kips)   (kips)  (ft-kips) (kips)  (kips)  (ft-kips)\n')
fprintf(1,'\n')
fk(3,:) = fk(3,:)/12;
fk(6,:) = fk(6,:)/12;
nmfk = [nmi' fk']';
fprintf(1,'%2.0f   %10.3f   %10.3f   %10.3f   %10.3f   %10.3f   %10.3f\n',nmfk)
fprintf(1,'\n')
```

A.6. EXAMPLE PLANE STRESS THIN PLATE ANALYSIS BY pstri.m AND pstriu.m

The m-files pstri.m and pstriu.m employ the 3-node, 6 dof triangular plane stress finite element discussed in Chapter 8. The file pstri.m requires data entry for individual nodes and elements and thereby permits irregularity in the shape of the plate and the variation in the sizes and shapes of the triangular elements. Numbering schemes for global nodes, gdof, and elements are at the discretion of the user. Node numbering within an element must be counterclockwise around the element for the element's three nodes.

The file pstriu.m establishes a relatively uniform finite element mesh from simplified data entry. It requires entry of the coordinates for just the four corner nodes of a quadrilateral plate, which it will use to establish a pattern of nodes which define a mesh of quadrilaterals across the plate. To do so, it also requires the numbers of rows and columns of these quadrilaterals which discretized plate. Each quadrilateral is divided by pstriu.m into two triangular elements. They will be on either side of each quadrilateral's diagonal, which runs from its lower left corner to its upper right.

Element numbering by pstriu.m begins in the left column of triangles with the bottom triangle and proceeds upward to the top edge of the plate. Element numbering continues with the triangle at the bottom of the second column of triangles, proceeding upward again to the top edge of the plate. It continues in like manner until all of the triangular elements are numbered.

Global node numbering by pstriu.m begins with the lower left corner of the quadrilateral plate structure and proceeds upward along its left edge to its upper left corner. It continues from the bottom of the second column of nodes and proceeds upward again to the top of the plate. It continues in like manner until all of the nodes are numbered. Local node numbering within an element starts at its lowest numbered global node and proceeds counterclockwise around the element for the element's three nodes.

Numbering for gdof by pstriu.m is tied to the global node number. For node i, the gdof in the global x axis is number $2i - 1$; the gdof in the global y axis is number $2i$.

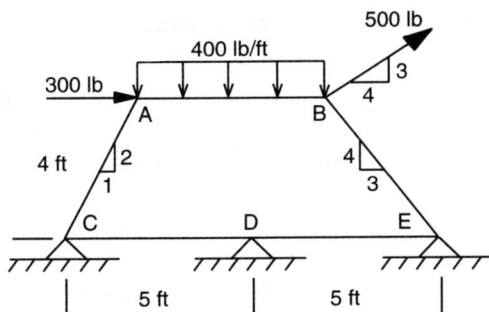

Figure A.9. Plane stress plate.

The triangular elements established by pstriu.m will be fairly uniform in size and shape and the plate must be a quadrilateral. The data entry, however, is greatly simplified for a fine mesh with a large number of elements.

When using pstriu.m, if it is required that the plate have a node in addition to those at its four corners and that it be specifically located, such as at a support which is not at a corner, care must be exercised to choose the correct numbers of rows and columns of quadrilaterals so that pstriu.m puts a node where it is required.

To illustrate the use of pstri.m and pstriu.m, consider the statically indeterminate thin plate shown in Figure A.9. There are in-plane concentrated loads at the upper left and upper right corners of the plate and an in-plane uniformly distributed load across its upper horizontal edge. The three-pin supports provide complete restraint in the horizontal and vertical directions. The plate thickness is 1.00 in., elastic modulus is 30,000 ksi, and Poisson's ratio is 0.25. The nodal displacements, support reactions, and element stresses will be determined.

The plate is discretized into four 3-node, 6-dof triangular plane stress elements, as shown in Figure A.10. The nodes are n_1–n_6. The elements are 1–4 with the element numbers shown in the boxes. Each element's local node numbers are shown by the numerals 1–3 within each element. The gdof are 1–12 and shown in the circles. They are horizontal or vertical, in conformance with the global axes chosen.

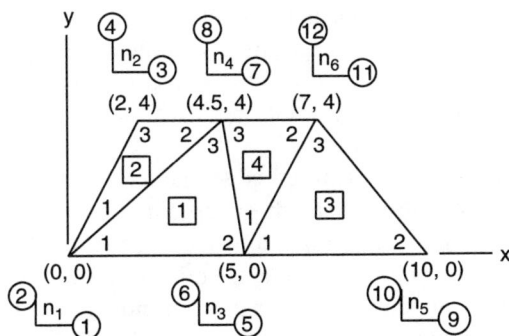

Figure A.10. Discretized plate.

A.6.1. Variables Needed for pstri.m and pstriu.m

For pstri.m Only

nc   Matrix of nodal coordinates in a sequence of paired numbers beginning
       with node 1. The matrix has two columns and as many rows as there are
       nodes. Column 1 contains the x coordinates and column 2 the y coordi-
       nates. The unit is feet, which pstri.m converts to inches for the analysis.

mnn   Matrix of element global node numbers in groups of three numbers. Each
       group is the global node number for the start node, center node, and end
       node for an element. The groups are in sequence beginning with element 1.
       The matrix has three columns and as many rows as there are elements.

For pstriu.m Only

kp   Matrix of coordinates for the nodes at the four corners of the quadrilat-
       eral plate. It is entered in the same format as described for matrix nc
       above. It has four rows and two columns. The unit is feet, which pstriu.m
       converts to inches for the analysis.

rce   Row vector of two numbers. The first is the number of rows of quadri-
       lateral figures described by the nodes on the plate. The second is the
       number of columns of these quadrilaterals. Each quadrilateral consists
       of two 3-node, 6-dof triangular elements on either side of the quadrilater-
       al's diagonal, which runs from its lower left corner to its upper right.

For both pstri.m and pstriu.m

rdof   Row vector of restraint gdof numbers in sequence from lowest to highest.
rval   Row vector of nonzero restraint values in a sequence of paired numbers.
       Each pair is the restraint gdof number and its nonzero value. The pairs are
       in sequence beginning with the lowest gdof restraint number. Restraint
       values are in inches. Restraints whose values are zero are not entered.
E   Modulus of elasticity for the elements (unit is ksi).
nu   Poisson's ratio for the elements.
th   Plate thickness (unit is inches).
f   Row vector of nonzero nodal forces applied to free gdof in a sequence of
       paired numbers. Each pair is the gdof number and the value of the force.
       The pairs are in sequence beginning with the lowest gdof number. Unit of
       force is kips. Note that zero forces at free gdof need not be entered.
bf   Row vector of the x and y components of the structure's body force. The
       vector has just two entries (unit is lb/ft^3).
w   Matrix of uniformly distributed edge line loads on elements in a sequence
       of four grouped numbers. Each group contains the element number, the
       element side number, and the x and y components of the line load. Side 1
       is taken as the side between element nodes 1 and 2, side 2 between element
       nodes 2 and 3, and side 3 between element nodes 3 and 1. Line load unit is
       kips/ft. The matrix has four columns and as many rows as there are
       applied edge line loads.

Programs pstri.m and pstriu.m require each of these variables exactly in the format described above and illustrated below. Furthermore, the variables must be declared as "global" as illustrated, so that their values are communicated to them.

A.6.2. Data Entry for pstri.m and pstriu.m

For pstri.m Only.   Declare the global variables needed for pstri.m:

$\gg$ global nc mnn rdof rval E nu th f bf w   < data entry >

Enter the matrix of nodal coordinates on the plate (unit is feet):

$\gg$ nc = [0 0;2 4;5 0;4.5 4;10 0;7 4]   < data entry >
nc =
0	0
2.0000	4.0000
5.0000	0
4.5000	4.0000
10.0000	0
7.0000	4.0000

< MATLAB display >

Enter the matrix of element global node numbers.

$\gg$ mnn = [1 3 4;1 4 2;3 5 6;3 6 4]   < data entry >
mnn =
1	3	4
1	4	2
3	5	6
3	6	4

< MATLAB display >

For pstriu.m Only.   Declare the global variables needed for pstriu.m:

$\gg$ global kp rce rdof rval E nu th f bf w   < data entry >

Enter the matrix of coordinates of nodes at the four corners of the quadrilateral plate (unit is feet):

$\gg$ kp = [0 0;2 4;10 0;7 4]   < data entry >
kp =
0	0
2	4
10	0
7	4

< MATLAB display >

Enter the row vector of numbers of rows and columns of quadrilateral figures on the plate:

```
≫ rce = [1 2]                                          <data entry>
rce =
     1     2                                           <MATLAB display>
```

For both pstri.m and pstriu.m. Enter the row vector of restraint gdof numbers:

```
≫ rdof = [1 2 5 6 9 10]                                <data entry>
rdof =
     1     2     5     6     9     10                  <MATLAB display>
```

Enter the row vector of nonzero restraint pairs. Restraint values are in inches. For this example all restraint values are zero so this entry could be omitted:

```
≫ rval = [ ]                                           <data entry>
rval =
     [ ]                                               <MATLAB display>
```

Enter the modulus of elasticity (unit is ksi):

```
≫ E = 30000                                            <data entry>
E =
     30000                                             <MATLAB display>
```

Enter the Poisson's ratio:

```
≫ nu = 0.25                                            <data entry>
nu =
     0.2500                                            <MATLAB display>
```

Enter the plate thickness (unit is inches):

```
≫ th = 1                                               <data entry>
th =
     1                                                 <MATLAB display>
```

Enter the row vector of applied nonzero nodal forces (unit is kips):

```
≫ f = [3 .3 11 .4 12 .3]                               <data entry>
f =
     3.0000    0.3000    11.0000    0.4000    12.0000    0.3000
                                                       <MATLAB display>
```

Enter the row vector of x and y components of the structure's body force (unit is lb/ft^3):

$\gg$ bf = [0 -500]	< data entry >
bf =	
0 -500	< MATLAB display >

Enter the matrix of nonzero uniformly distributed line loads applied to edges of elements (unit is kips/ft):

$\gg$ w = [2 2 0 -.4;4 2 0 -.4]	< data entry >
w =	
2.0000 2.0000 0 -.4000	< MATLAB display >
4.0000 2.0000 0 -.4000	

Enter the command to execute the programmed steps in the m-file pstri.m or pstriu.m:

$\gg$ pstri	< data entry >
or	
$\gg$ pstriu	< data entry >

A.6.3. Results Displayed by pstri.m or pstriu.m

PLANE STRESS THIN PLATE ANALYSIS
(3 node, 6 dof triangular elements)
nr of nodes: 6, nr of mbrs: 4, nr of gdof: 12
elastic modulus: 30000 ksi, Poissons ratio: 0.250
body force: bx = 0 lb/ft3, by = -500 lb/ft3

nodal coordinates: plate thickness: 1.000 inches

node	x (ft)	y (ft)
1	0.00	0.00
2	2.00	4.00
3	5.00	0.00
4	4.50	4.00
5	10.00	0.00
6	7.00	4.00

element properties:

elem	nodes			area (ft2)	side	distributed edge loads wx(kip/ft)	wy(kip/ft)
1	1	3	4	10.00	0	0.000	0.000
2	1	4	2	5.00	2	0.000	-0.400
3	3	5	6	10.00	0	0.000	0.000
4	3	6	4	5.00	2	0.000	-0.400

restraints and reactions (global axes):

rstr gdof	restraint value (in)	reaction (kips)
1	0.0000	0.1201
2	0.0000	0.4240
5	0.0000	-0.5853
6	0.0000	1.9175
9	0.0000	-0.2348
10	0.0000	0.6085

applied forces and displacements of free gdof (global axes):

free gdof	applied force (kips)	displacement (in/1000)
3	0.300	0.029326
4	0.000	-0.035171
7	0.000	0.025113
8	0.000	-0.047260
11	0.400	0.035676
12	0.300	-0.031940

element constant stress:

elem	normal x (ksi)	normal y (ksi)	shear xy (ksi)
1	-0.003938	-0.015753	0.003139
2	-0.004372	-0.009061	0.001669
3	-0.002662	-0.010647	0.004459
4	0.001951	-0.013324	0.006467

A.7. EXAMPLE PLANE STRESS THIN PLATE ANALYSIS BY psquad.m AND psquadu.m

The m-files psquad.m and psquadu.m employ the 4-node, 8-dof isoparametric quadrilateral finite element and 2 × 2 Gauss quadrature both discussed in Chapter 9. The file psquad.m requires data entry for individual nodes and elements and thereby permits irregularity in the shape of the plate and variation in the sizes and shapes of the quadrilateral elements. Numbering schemes for global nodes, gdof, and elements are at the discretion of the user. Node numbering within an element must be counterclockwise around the element for its four nodes.

The file psquadu.m establishes a relatively uniform finite element mesh from simplified data entry. It requires entry of the coordinates for just the four corner nodes of a quadrilateral plate. It will use these to establish a pattern of nodes which define a mesh of quadrilateral elements across the plate. To do so, it also requires the numbers of rows and columns of these elements which are desired in the discretized plate. The elements will be fairly uniform in size and shape and the plate must be a quadrilateral, but the data entry is greatly simplified for a fine mesh with a large number of elements. Numbering of elements, global nodes, local element nodes, and gdof done by psquadu.m is done in the manner that the file pstriu.m numbers them as described in Section A.6. It is illustrated and explained below.

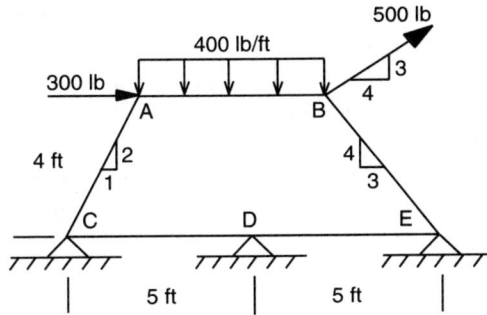

Figure A.11. Plane stress plate.

To illustrate the use of psquad.m and psquadu.m, consider the statically indeterminate thin plate shown in Figure A.11. There are in-plane concentrated loads at the upper right and upper left corners of the plate and an in-plane uniformly distributed load across its horizontal edge. The three-pin supports provide complete restraint in the horizontal and vertical directions. The plate thickness is 1.00 in., elastic modulus is 30,000 ksi, and Poisson's ratio is 0.25. The nodal displacements, support reactions, and element stresses will be determined.

The plate is discretized into two 4-node, 8-dof quadrilateral plane stress finite elements, as shown in Figure A.12. The nodes are n_1–n_6. The two elements' numbers are shown in the boxes. Each element's node numbering sequence is shown by the numerals 1–4 within each element. The gdof are 1–12 and are shown in the circles. They are horizontal or vertical, in conformance with the global axes chosen.

When using psquad.m, discretizing is partially at the discretion of the user. The global axes shown must be used and the orientation and sequence of gdof shown at nodes must be used since they are what is used for edof. The element node numbering sequence 1–4 must be as shown. Nodes may be located as desired, except that their coordinates must reflect quadrilateral elements. The assembly of quadrilateral elements need not be a quadrilateral, however.

When using psquadu.m, the program discretizes as shown. Nodes and elements are numbered in sequence approximately in the direction of the y axis to the upper

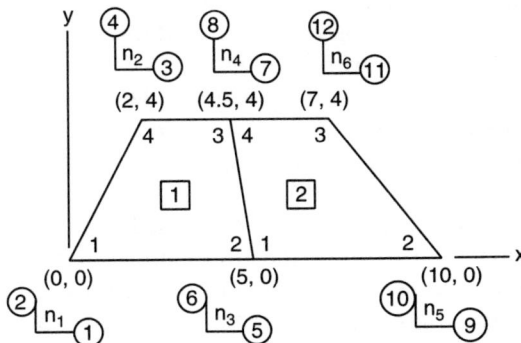

Figure A.12. Discretized plate.

edge of the plate and then continued an element width to the right in sequence along a line approximately parallel to the y axis. The sequence of gdof developed by the program is as shown. The elements will be fairly uniform quadrilaterals, and the assembly of elements will be a quadrilateral roughly similar in shape to the elements. If it is required that the plate have a node in addition to those at its four corners and specifically located, such as at the support D in Figures A.11 and A.12, care must be exercised to choose the correct number of elements so that psquadu.m puts a node at that location.

A.7.1. Variables Needed for psquad.m and psquadu.m

For psquad.m Only

nc	Matrix of nodal coordinates in a sequence of paired numbers beginning with node 1. The matrix has two columns and as many rows as there are nodes. Column 1 contains the x coordinates and column 2 the y coordinates. The unit is feet, which psquad.m converts to inches for the analysis.
mnn	Matrix of element global node numbers in a sequence of four grouped numbers. Each group is the global node number for the first, second, third, and fourth nodes for an element. The groups are in sequence beginning with element 1. The matrix has four columns and as many rows as there are elements.

For psquadu.m Only

kp	Matrix of coordinates for the nodes at the four corners of the quadrilateral plate. It is entered in the format described for matrix nc above. It has four rows and two columns. The unit is feet, which psquadu.m converts to inches for the analysis.
rce	Row vector of two numbers. The first is the number of rows of elements used to discretize the plate; the second is the number of columns of elements.

For both psquad.m and psquadu.m

rdof	Row vector of restraint gdof numbers in sequence from lowest to highest.
rval	Row vector of nonzero restraint values in a sequence of paired numbers. Each pair is the restraint gdof number and its nonzero value. The pairs are in sequence beginning with the lowest gdof restraint number. Restraint values are in inches. Restraints whose values are zero are not entered.
E	Modulus of elasticity for the elements (unit is ksi).
nu	Poisson's ratio for the elements.
th	Plate thickness (unit is inches).
f	Row vector of nonzero nodal forces applied to free gdof in a sequence of paired numbers. Each pair is the gdof number and the value of the force. The pairs are in sequence beginning with the lowest gdof number. The unit of force is kips. Note that zero forces at free gdof need not be entered.
bf	Row vector of the x and y components of the structure's body force. The vector has just two entries (unit is lb/ft^3).

w   Matrix of uniformly distributed edge line loads on elements in a sequence of four grouped numbers. Each group contains the element number, the element side number, and the x and y components of the line load. Side 1 is taken as the side between element nodes 1 and 2, side 2 between element nodes 2 and 3, side 3 between element nodes 3 and 4, and side 4 between element nodes 4 and 1. The line load unit is lb/ft. The matrix has four columns and as many rows as there are applied edge line loads.

Programs psquad.m and psquadu.m require each of these variables exactly in the format described and illustrated below. Furthermore, the variables must be declared as "global" as illustrated, so that their values are communicated to them.

A.7.2. Data Entry for psquad.m and psquadu.m

For psquad.m Only.   Declare the global variables needed for psquad.m:

≫ global nc mnn rdof rval E nu th f bf w   < data entry >

Enter the matrix of nodal coordinates on the plate (unit is feet):

≫ nc = [0 0;2 4;5 0;4.5 4;10 0;7 4]   < data entry >
nc =
           0   0   < MATLAB display >
    2.0000   4.0000
    5.0000   0
    4.5000   4.0000
  10.0000   0
    7.0000   4.0000

Enter the matrix of element global node numbers:

≫ mnn = [1 3 4 2;3 5 6 4]   < data entry >
mnn =
    1   3   4   2   < MATLAB display >
    3   5   6   4

For psquadu.m Only.   Declare the global variables needed for psquadu.m:

≫ global kp rce rdof rval E nu th f bf w   < data entry >

Enter the matrix of coordinates of nodes at the four corners of the quadrilateral (unit is feet):

≫ kp = [0 0;2 4;10 0;7 4]   < data entry >

kp =

0	0
2	4
10	0
7	4

< MATLAB display >

Enter the row vector of numbers of rows and columns of elements:

≫ rce = [1 2]   < data entry >
rce =
      1   2   < MATLAB display >

For both psquad.m and psquadu.m.   Enter the row vector of restraint gdof numbers:

≫ rdof = [1 2 5 6 9 10]   < data entry >
rdof =
      1   2   5   6   9   10   < MATLAB display >

Enter the row vector of nonzero restraint pairs. Restraint values are in inches. For this example all restraint values are zero so this entry could be omitted:

≫ rval = [ ]   < data entry >
rval =
      [ ]   < MATLAB display >

Enter the modulus of elasticity (unit is ksi):

≫ E = 30000   < data entry >
E =
      30000   < MATLAB display >

Enter the Poisson's ratio:

≫ nu = 0.25   < data entry >
nu =
      0.2500   < MATLAB display >

Enter the plate thickness (unit is inches):

≫ th = 1   < data entry >
th =
      1   < MATLAB display >

Enter the row vector of nonzero applied nodal forces (unit is kips):

≫ f = [3 .3 11 .4 12 .3]   < data entry >

f =

| 3.0000 | 0.3000 | 11.0000 | 0.4000 | 12.0000 | 0.3000 |

<MATLAB display>

Enter the row vector x and y components of the structure's body force (unit is lb/ft^3):

≫ bf = [0 -500]   <data entry>
bf =
      0   -500   <MATLAB display>

Enter the matrix of nonzero uniformly distributed line loads applied to edges of elements (unit is kips/ft):

≫ w = [1 3 0 -.4;2 3 0 -.4]   <data entry>
w =
| 1.0000 | 3.0000 | 0 | -.4000 |
| 2.0000 | 3.0000 | 0 | -.4000 |

<MATLAB display>

Enter the command to execute the programmed steps in the m-file psquad.m or psquadu.m:

≫ psquad   <data entry>
or
≫ psquadu   <data entry>

A.7.3. Results Displayed by psquad.m or psquadu.m

PLANE STRESS THIN PLATE ANALYSIS
(4 node 8 dof isoparametric quadrilateral elements)
nr of nodes: 6, nr of mbrs: 2, nr of gdof: 12
elastic modulus: 30000 ksi, Poissons ratio: 0.250
body force: bx = 0 lb/ft3, by = -500 lb/ft3

nodal coordinates: plate thickness: 1.000 inches

node	x (ft)	y (ft)
1	0.00	0.00
2	2.00	4.00
3	5.00	0.00
4	4.50	4.00
5	10.00	0.00
6	7.00	4.00

element properties:

elem	nodes				area (ft2)	side	distributed edge loads wx(kip/ft)	wy(kip/ft)
1	1	3	4	2	15.00	3	0.000	-0.400
2	3	5	6	4	15.00	3	0.000	-0.400

restraints and reactions (global axes):

rstr gdof	restraint value (in)	reaction (kips)
1	0.0000	0.1042
2	0.0000	0.4625
5	0.0000	-0.5447
6	0.0000	1.8405
9	0.0000	-0.2595
10	0.0000	0.6470

applied forces and displacements of free gdof (global axes):

free gdof	applied force (kips)	displacement (in/1000)
3	0.300	0.030765
4	0.000	-0.028899
7	0.000	0.027598
8	0.000	-0.053394
11	0.400	0.036379
12	0.300	-0.031361

element average Gauss point stress:

elem	normal x (ksi)	normal y (ksi)	shear xy (ksi)
1	-0.007686	-0.025872	0.003618
2	-0.002472	-0.023397	0.011978

A.8. EXAMPLE THIN PLATE FLEXURAL ANALYSIS BY flxrt.m AND flxrtu.m

The m-files flxrt.m and flxrtu.m employ the 4-node, 12-dof rectangular thin plate finite element in terms of generalized dof as discussed in Chapter 10. The file flxrt.m requires data entry for individual nodes and elements and thereby permits irregularity in the shape of the plate and variation in the sizes and shapes of the rectangular elements desired by the user. Numbering schemes for global nodes, gdof, and elements are at the discretion of the user. Node numbering within an element must be counterclockwise around the element for the element's four nodes.

The file flxrtu.m establishes a uniform finite element mesh from simplified data entry. It requires entry of the coordinates for just the four corner nodes of a rectangular plate. It will use these to establish a pattern of nodes which define a mesh of rectangular elements across the plate. To do so, it also requires the numbers of rows and columns of these elements which are desired in the discretized plate. The elements will be uniform in size and shape and the plate must be rectangular, but the data entry is greatly simplified for a fine mesh with a larger number of elements. Numbering of elements, global nodes, and local elements done by flxrtu.m is done in the manner that the file pstriu.m numbers them, as described in Section A.6. It is also illustrated and explained below.

Numbering for gdof by flxrtu.m is tied to the global node number. For node i, the gdof for the transverse translation along the global z axis is number $3i - 2$; the gdof

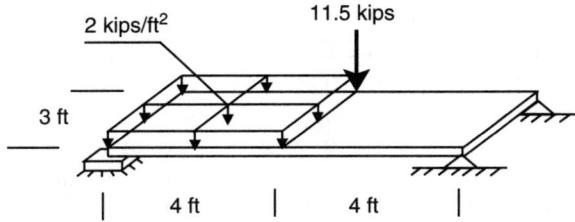

Figure A.13. Thin flexure plate.

for the rotation about the global x axis is number $3i - 1$; the gdof for the rotation about the global y axis is number $3i$.

To illustrate the use of flxrt.m and flxrtu.m, consider the statically indeterminate thin plate shown in Figure A.13. There are a transverse concentrated load at the center of one edge of the plate and a transverse uniformly distributed load across the left half of the plate. The two supports on the left corners of the plate are clamped to provide complete restraint for transverse displacement and for rotation about horizontal axes. The two supports on the right corners of the plate are pins which provide only complete restraint for transverse displacement.

The plate thickness is 2.50 in, its modulus of elasticity is 30,000 ksi, and its Poisson ratio is 0.25. The nodal displacements, support reactions, and element distributed moments will be determined.

The plate is discretized into two 4-node, 12-dof rectangular thin plate finite elements, as shown in Figure A.14. The nodes are n_1-n_6. The two element numbers are shown in the boxes. Each element's node numbering sequence is shown by the numerals 1–4 within each element. The gdof are 1–18 and are shown in the circles. They are horizontal or vertical, in conformance with the global axes chosen.

When using flxrt.m, discretizing is partially at the discretion of the user. The global axes shown must be used and the orientation and sequence of gdof shown at nodes must be used since they are what has been used for edof. The element local node numbering sequence 1–4 must be as shown. Nodes may be located on the plate as desired, except that their coordinates must reflect rectangular elements. The assembly of rectangular elements need not be a rectangle, however.

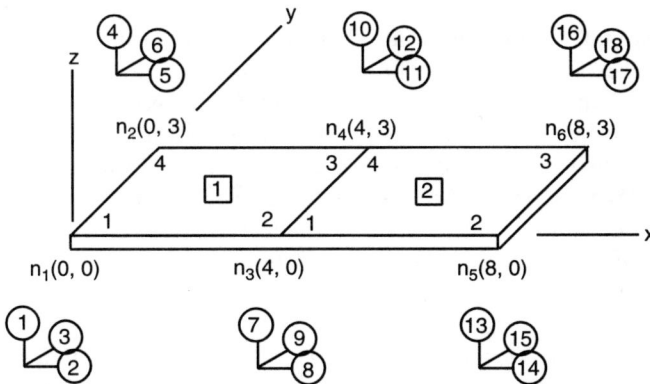

Figure A.14. Discretized plate.

When using flxrtu.m, the program discretizes as shown above. Nodes and elements are numbered in sequence along the y axis to the upper edge of the plate and then continued an element width to the right in sequence along a line parallel to the y axis. The sequence of gdof is as shown. The elements will be uniform, identical rectangles, and the assembly of elements will be a rectangle geometrically similar to the elements. If it is required that the plate have nodes in addition to those at its four corners and that they be specifically located, such as at the edge of the distributed load in the illustration above, care must be exercised to choose the correct number of elements so that flxrtu.m puts nodes at those locations.

A.8.1. Variables Needed for flxrt.m and flxrtu.m

For flxrt.m Only

nc	Matrix of nodal coordinates in a sequence of paired numbers beginning with node 1. The matrix has two columns and as many rows as there are nodes. Column 1 contains the x coordinates and column 2 the y coordinates. The unit is feet, which flxrt.m converts to inches for the analysis.
mnn	Matrix of element global node numbers in a sequence of four grouped numbers. Each group is the global node number for the first, second, third, and fourth nodes for an element. The groups are in sequence beginning with element 1. The matrix has four columns and as many rows as there are elements.

For flxrtu.m Only

kp	Matrix of coordinates for the nodes at the four corners of the rectangular plate. It is entered in the same format as described for matrix nc above. It has four rows and two columns. The unit is feet, which flxrtu.m converts to inches for the analysis.
rce	Row vector of two numbers. The first is the number of rows of elements used to discretize the plate; the second is the number of columns of elements.

For both flxrt.m and flxrtu.m

rdof	Row vector of restraint gdof numbers in sequence from lowest to highest.
rval	Row vector of nonzero restraint values in a sequence of paired numbers. Each pair is the restraint gdof number and its nonzero value. The pairs are in sequence beginning with the lowest gdof restraint number. Restraint values are in inches for transverse translation restraint and radians for rotation restraint about the global x or y axes. Restraints whose values are zero are not entered.
E	Modulus of elasticity for the elements (unit is ksi).
nu	Poisson's ratio for the elements.
th	Plate thickness (unit is inches).
f	Row vector of nonzero transverse forces applied to nodes in a sequence of paired numbers. Each pair is the node number and the value of the force. The pairs are in sequence beginning with the lowest node number. The

nodes cannot have their transverse gdof restrained. The unit of force is kips. Note that zero forces at such unrestrained nodes need not be entered.

bf   Value of the structure's transverse body force (unit is lb/ft^3).

w   Matrix of uniformly distributed transverse loads on elements in a sequence of pairs of numbers. Each pair contains the element number and its transverse distributed load. The unit is lb/ft^2. The matrix has two columns and as many rows as there are applied uniformly distributed loads.

Programs flxrt.m and flxrtu.m require each of these variables exactly in the format described and illustrated below. Furthermore, the variables must be declared as "global" as illustrated, so that their values are communicated to them.

A.8.2. Data Entry for flxrt.m and flxrtu.m

For flxrt.m Only.   Declare the global variables needed for flxrt.m:

≫ global nc mnn rdof rval E nu th f bf w   < data entry >

Enter the matrix of nodal coordinates on the plate (unit is feet):

≫ nc = [0 0;0 3;4 0;4 3;8 0;8 3]   < data entry >
nc =
```
    0    0                                          < MATLAB display >
    0    3
    4    0
    4    3
    8    0
    8    3
```

Enter the matrix of element global node numbers:

≫ mnn = [1 3 4 2;3 5 6 4]   < data entry >
mnn =
```
    1    3    4    2                                 < MATLAB display >
    3    5    6    4
```

For flxrtu.m Only.   Declare the global variables needed for flxrtu.m:

≫ global kp rce rdof rval E nu th f bf w   < data entry >

Enter the matrix of coordinates of nodes at the four corners of the rectangle (unit is feet):

≫ kp = [0 0;0 3;8 0;8 3]   < data entry >

```
kp =
    0   0
    0   3
    8   0
    8   3
```
<MATLAB display>

Enter the row vector of numbers of rows and columns of elements:

```
>> rce = [1 2]                                      <data entry>
rce =
    1    2
```
<MATLAB display>

For both flxrt.m and flxrtu.m: Enter the row vector of restraint gdof numbers:

```
>> rdof = [1 2 3 4 5 6 13 16]                       <data entry>
rdof =
    1    2    3    4    5    6   13   16
```
<MATLAB display>

Enter the row vector of nonzero restraint pairs. Restraint values are in inches for transverse restraint and radians for rotational restraint. For this example all restraint values are zero so this entry could be omitted:

```
>> rval = [ ]                                       <data entry>
rval =
    [ ]
```
<MATLAB display>

Enter the modulus of elasticity (unit is ksi):

```
>> E = 30000                                        <data entry>
E =
    30000
```
<MATLAB display>

Enter Poisson's ratio:

```
>> nu = 0.25                                        <data entry>
nu =
    0.2500
```
<MATLAB display>

Enter the plate thickness (unit is inches):

```
>> th = 2.5                                         <data entry>
th =
    2.5000
```
<MATLAB display>

Enter the row vector of nonzero applied nodal forces (unit is kips):

```
>> f = [4 -11.5]                                    <data entry>
```

f =
      4.0000   -11.5000   < MATLAB display >

Enter the structure's transverse body force (unit is lb/ft^3):

≫ bf = -500   < data entry >
bf =
     -500   < MATLAB display >

Enter the matrix of nonzero uniformly distributed loads applied to elements (unit is kips/ft^2):

≫ w = [1 -2]   < data entry >
w =
     1.0000   -2.0000   < MATLAB display >

Enter the command to execute the programmed steps in the m-file flxrt.m or flxrtu.m:

≫ flxrt   < data entry >

or

≫ flxrtu   < data entry >

A.8.3. Results Displayed by flxrt.m or flxrtu.m

FLEXURAL THIN PLATE ANALYSIS
(4 node 12 dof rectangular elements)
nr of nodes: 6, nr of elems: 2, nr of gdof: 18
elastic modulus: 30000 ksi, Poissons ratio: 0.250
plate thickness: 2.500 inches

nodal coordinates:

node	x (ft)	y(ft)
1	0.00	0.00
2	0.00	3.00
3	4.00	0.00
4	4.00	3.00
5	8.00	0.00
6	8.00	3.00

element properties:

elem	nodes				area (ft2)
1	1	3	4	2	12.00
2	3	5	6	4	12.00

applied transverse concentrated loads:

node	Load
4	-11.500 kips

applied transverse uniformly distributed loads:

elem	Load
1	-2.000 kips/ft2

transverse body force: bf = -500 lb/ft3

restraints and reactions (global axes):

rstr node	rstr gdof	restraint value	reaction
1	1	0.0000 in	11.770 kips
1	2	0.0000 rad	4.899 ft-kips
1	3	0.0000 rad	-21.176 ft-kips
2	4	0.0000 in	19.107 kips
2	5	0.0000 rad	-6.404 ft-kips
2	6	0.0000 rad	-25.842 ft-kips
5	13	0.0000 in	0.978 kips
6	16	0.0000 in	6.145 kips

applied forces and displacements of free gdof (global axes):

free node	free gdof	displacement
3	7	-0.1024526 in
3	8	-0.0014778 rad
3	9	0.0007191 rad
4	10	-0.1589053 in
4	11	-0.0017727 rad
4	12	0.0007929 rad
5	14	0.0003689 rad
5	15	-0.0039060 rad
6	17	-0.0003306 rad
6	18	-0.0052140 rad

element distributed moments at z = 0.00, e = 0.00

elem	Mxx (kip-in/in)	Myy (kip-in/in)	Mxy (kip-in/in)
1	-0.699	-0.335	-1.497
2	4.471	0.578	1.239

A.9. EXAMPLE THIN PLATE FLEXURAL ANALYSIS BY flxtri.m AND flxtriu.m

The m-files flxtri.m and flxtriu.m employ the 3-node, 9-dof triangular thin plate finite element in terms of generalized dof as discussed in Chapter 10. The file flxtri.m requires data entry for individual nodes and elements and thereby permits irregularity in the shape of the plate and variation in the sizes and shapes of the triangular elements desired by the user. Numbering schemes for global nodes, gdof, and elements are at the discretion of the user. Node numbering within an element must be counterclockwise around the element for the element's three nodes.

The file flxtriu.m establishes a relatively uniform finite element mesh from simplified data entry. It requires entry of the coordinates for just the four corner nodes of a quadrilateral plate. It will use these to establish a pattern of nodes which define a mesh of quadrilaterals across the plate. To do so, it also requires the numbers of rows and columns of these quadrilaterals which are desired in the discretized plate. Each quadrilateral is divided by flxtriu.m into two triangular elements. They will be on either side of each quadrilateral's diagonal, which runs from its lower left corner to its upper right. The triangular elements will be fairly uniform in size and shape and the plate must be a quadrilateral, but the data entry is greatly simplified for a fine mesh with a large number of elements. Numbering of elements, global nodes, and local element nodes by flxtriu.m is done in the manner that the file pstriu.m numbers them as described in Section A.6. It is illustrated and explained below.

Numbering for gdof by flxtriu.m is tied to the global node number. For node i, the gdof for the transverse translation along the global z axis is number $3i - 2$; the gdof for the rotation about the global x axis is number $3i - 1$; the gdof for the rotation about the global y axis is number $3i$.

When using flxtriu.m, if it is required that the plate have a node in addition to those at its four corners and that it be specifically located, such as at a support which is not at a corner or where a concentrated transverse load is applied, care must be exercised to choose the correct numbers of rows and columns of quadrilaterals so that flxtriu.m puts these nodes where required.

To illustrate the use of flxtri.m and flxtriu.m, consider the statically indeterminate thin plate shown in Figure A.15. There are a transverse concentrated load at the center of one edge of the plate and a transverse uniformly distributed load across the left half of the plate. The two supports on the left corners of the plate are clamped to provide complete restraint for transverse displacement and for rotation about horizontal axes. The two supports on the right corners of the plate are pins which provide only complete restraint for transverse displacement.

The plate thickness is 2.50 in., its modulus of elasticity is 30,000 ksi, and its Poisson ratio is 0.25. The nodal displacements and support reactions will be determined.

The plate is discretized into four 3-node, 9-dof triangular thin plate elements, as shown in Figure A.16. The global nodes are n_1–n_6. The four elements' numbers are shown in the boxes. Each element's node numbering sequence is shown by the

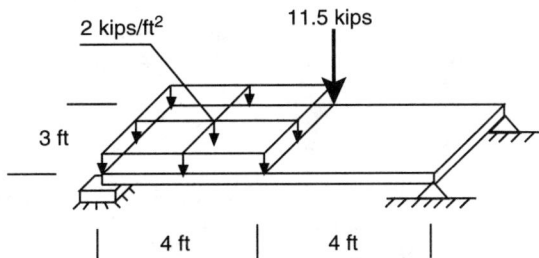

Figure A.15. Thin flexure plate.

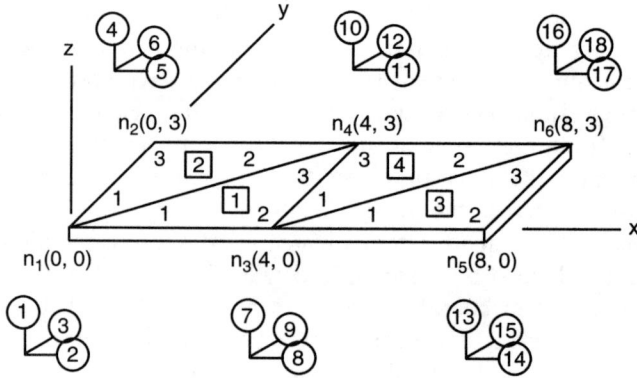

Figure A.16. Discretized plate.

numerals 1–3 within each element. The gdof are 1–18 and are shown in the circles. They are horizontal or vertical, in conformance with the global axes chosen.

When using flxtri.m, discretizing is partially at the discretion of the user. The global axes shown must be used and the orientation and sequence of gdof shown at nodes must be used for consistency with the sequences of edof. The element local node numbering sequence 1–3 must be as shown. Nodes may be located on the plate as desired, except that their coordinates must reflect triangular elements. The assembly of triangular elements may be an irregularly shaped flat plate.

When using flxtriu.m, the program discretizes as shown above. Nodes and elements are numbered in sequence along the y axis to the upper edge of the plate and then continued an element width to the right in sequence along a line parallel to the y axis. The sequence of gdof is as shown. The elements will be relatively uniform triangles and the assembly of elements must be a quadrilateral.

A.9.1. Variables Needed for flxtri.m and flxtriu.m

For flxtri.m Only

nc	Matrix of nodal coordinates in a sequence of paired numbers beginning with node 1. The matrix has two columns and as many rows as there are nodes. Column 1 contains the x coordinates and column 2 the y coordinates. The unit is feet, which flxtri.m converts to inches for the analysis.
mnn	Matrix of element global node numbers in a sequence of three grouped numbers. Each group is the global node number for the first, second, and third nodes for an element. The groups are in sequence beginning with element 1. The matrix has three columns and as many rows as there are elements.

For flxtriu.m Only

kp	Matrix of coordinates for the nodes at the four corners of the rectangular plate. It is entered in the same format as described for matrix nc above. It has four rows and two columns. The unit is feet, which flxtriu.m converts to inches for the analysis.

rce   Row vector of two numbers. The first is the number of rows of quadri-laterals, each comprising two triangular elements, used to discretize the plate; the second is the number of columns of quadrilaterals.

For both flxtri.m and flxtriu.m

rdof   Row vector of restraint gdof numbers in sequence from lowest to highest.

rval   Row vector of nonzero restraint values in a sequence of paired numbers. Each pair is the restraint gdof number and its nonzero value. The pairs are in sequence beginning with the lowest gdof restraint number. Restraint values are in inches for transverse translation restraint and radians for rotation restraint about the global x or y axis. Restraints whose values are zero are not entered.

E   Modulus of elasticity for the elements (unit is ksi).

nu   Poisson's ratio for the elements.

th   Plate thickness (unit is inches).

f   Row vector of nonzero transverse forces applied to nodes in a sequence of paired numbers. Each pair is the node number and the value of the force. The pairs are in sequence beginning with the lowest node number. The nodes cannot have their transverse gdof restrained. The unit of force is kips. Note that zero forces at such unrestrained nodes need not be entered.

bf   Value of the structure's transverse body force (unit is lb/ft^3).

w   Matrix of uniformly distributed transverse loads on elements in a sequence of pairs of number. Each pair contains the element number and its transverse distributed load. The unit is lb/ft^2. The matrix has two columns and as many rows as there are applied uniformly distributed loads.

Programs flxtri.m and flxtriu.m require each of these variables exactly in the format described and illustrated below. Furthermore, the variables must be declared as "global" as illustrated, so that their values are communicated to them.

A.9.2. Data Entry for flxtri.m and flxtriu.m

For flxtri.m Only.   Declare the global variables needed for flxtri.m:

≫ global nc mnn rdof rval E nu th f bf w   < data entry >

Enter the matrix of nodal coordinates on the plate (unit is feet):

≫ nc = [0 0;0 3;4 0;4 3;8 0;8 3]   < data entry >

nc =

                        < MATLAB display >

     0   0

     0   3

     4   0

     4   3

     8   0

     8   3

Enter the matrix of element global node numbers:

```
≫ mnn = [1 3 4;1 4 2;3 5 6;3 6 4]          <data entry>
mnn =
     1    3    4                            <MATLAB display>
     1    4    2
     3    5    6
     3    6    4
```

For flxtriu.m Only.   Declare the global variables needed for flxtriu.m:

```
≫ global kp rce rdof rval E nu th f bf w          <data entry>
```

Enter the matrix of coordinates of nodes at the four corners of the rectangle (unit is feet):

```
≫ kp = [0 0;0 3;8 0;8 3]                    <data entry>
kp =
     0    0                                 <MATLAB display>
     0    3
     8    0
     8    3
```

Enter the row vector of numbers of rows and columns of elements:

```
≫ rce = [1 2]                              <data entry>
rce =
     1    2                                 <MATLAB display>
```

For both flxrt.m and flxrtu.m.   Enter the row vector of restraint gdof numbers:

```
≫ rdof = [1 2 3 4 5 6 13 16]               <data entry>
rdof =
     1    2    3    4    5    6   13   16    <MATLAB display>
```

Enter the row vector of nonzero restraint pairs. Restraint values are in inches for transverse restraint and radians for rotational restraint. For this example all restraint values are zero so this entry could be omitted:

```
≫ rval = [ ]                               <data entry>
rval =
     [ ]                                    <MATLAB display>
```

Enter the modulus of elasticity (unit is ksi):

```
≫ E = 30000                                <data entry>
E =
     30000                                  <MATLAB display>
```

Enter Poisson's ratio:

$\gg$ nu = 0.25	< data entry >
nu =	
0.2500	< MATLAB display >

Enter the plate thickness (unit is inches):

$\gg$ th = 2.5	< data entry >
th =	
2.5000	< MATLAB display >

Enter the row vector of nonzero applied nodal forces (unit is kips):

$\gg$ f = [4 -11.5]	< data entry >
f =	
4.0000 -11.5000	< MATLAB display >

Enter the structure's transverse body force (unit is lb/ft^3):

$\gg$ bf = -500	< data entry >
bf =	
-500	< MATLAB display >

Enter the matrix of nonzero uniformly distributed loads applied to elements (unit is $kips/ft^2$):

$\gg$ w = [1 -2 2 -2]	< data entry >
w =	
1.0000 -2.0000 1.0000 -2.0000	< MATLAB display >

Enter the command to execute the programmed steps in the m-file flxtri.m or flxtriu.m:

$\gg$ flxtri	< data entry >
or	
$\gg$ flxtriu	< data entry >

A.9.3. Results Displayed by flxtri.m or flxtriu.m

FLEXURAL THIN PLATE ANALYSIS
(3 node 9 dof triangular elements)
nr of nodes: 6, nr of elems: 4, nr of gdof: 18
elastic modulus: 30000 ksi, Poissons ratio: 0.250
plate thickness: 2.500 inches

nodal coordinates:

node	x (ft)	y (ft)
1	0.00	0.00
2	0.00	3.00
3	4.00	0.00
4	4.00	3.00
5	8.00	0.00
6	8.00	3.00

element properties:

elem	nodes			area (ft2)
1	1	3	4	6.00
2	1	4	2	6.00
3	3	5	6	6.00
4	3	6	4	6.00

applied transverse concentrated loads:

node	load
4	-11.500 kips

applied transverse uniformly distributed loads:

elem	load
1	-2.000 kips/ft2
2	-2.000 kips/ft2

transverse body force: bf = -500 lb/ft3

restraints and reactions (global axes):

rstr node	rstr gdof	restraint value	reaction
1	1	0.0000 in	12.891 kips
1	2	0.0000 rad	9.479 ft-kips
1	3	0.0000 rad	-19.409 ft-kips
2	4	0.0000 in	18.407 kips
2	5	0.0000 rad	-9.301 ft-kips
2	6	0.0000 rad	-30.975 ft-kips
5	13	0.0000 in	0.418 kips
6	16	0.0000 in	6.284 kips

applied forces and displacements of free gdof (global axes):

free node	free gdof	displacement
3	7	-0.1095440 in
3	8	-0.0011874 rad
3	9	0.0011166 rad
4	10	-0.1579414 in
4	11	-0.0017941 rad
4	12	0.0001803 rad
5	14	-0.0000173 rad
5	15	-0.0037469 rad
6	17	0.0002220 rad
6	18	-0.0054832 rad

ANSWERS TO SELECTED PROBLEMS

CHAPTER 1

1.1.

b.
$$[\mathbf{A}^{-1}] = \begin{bmatrix} -2.6 & -1.8 & 3 & 0.2 \\ 8.12 & 5.56 & -9.4 & -0.24 \\ -6.6 & -4.8 & 8 & 0.2 \\ 6.12 & 4.56 & -7.4 & -0.24 \end{bmatrix}$$

c.
$$[\mathbf{AB}] = \begin{bmatrix} 48 & 35 & 49 & 71 \\ 92 & 114 & 106 & 103 \\ 94 & 96 & 104 & 119 \\ 57 & 61 & 36 & 90 \end{bmatrix}$$

d. det $\mathbf{A} = 25$

e. det $\mathbf{B} = 30$

f. det $\mathbf{A}^{\mathrm{T}} = 25$

g. det $\mathbf{A}^{-1} = 0.04$

i. det $\mathbf{AB} = 750$

j. det $\mathbf{A}^{\mathrm{mod1}} = 25$

k. det $\mathbf{A}^{\mathrm{mod2}} = -25$

1.2.

a.
$$[\mathbf{U}] = \begin{bmatrix} 1 & 3 & 4 \\ 0 & 1 & 2 \\ 0 & 0 & 1 \end{bmatrix}, \quad [\mathbf{L}] = \begin{bmatrix} 1 & 0 & 0 \\ 2 & 2 & 0 \\ 3 & -6 & 4 \end{bmatrix}$$

b. det $\mathbf{U} = 1$, det $\mathbf{L} = 8$

c. det $\mathbf{A} = 8$

1.3. $x_1 = 12$, $x_2 = -8$, $x_3 = 1$

1.4. $x = 2$, $y = 4$, $z = 3$

1.5. $x_1 = 2$, $x_2 = 3$, $x_3 = 5$

1.6.

$$[\mathbf{x}] = \begin{bmatrix} x_{11} & x_{12} & x_{13} \\ x_{21} & x_{22} & x_{23} \\ x_{31} & x_{32} & x_{33} \end{bmatrix} = \begin{bmatrix} 2 & 2 & 1 \\ 4 & 3 & 4 \\ 3 & 5 & 2 \end{bmatrix}$$

1.8.

$$[\mathbf{C}] = \begin{bmatrix} 23 & 39 & 20 & 84 \\ 118 & 93 & 37 & -58 \\ 91 & 31 & 31 & -17 \\ 1 & -82 & 8 & 31 \end{bmatrix}, \quad [\mathbf{C}^{-1}] = \begin{bmatrix} -0.014183 & -0.092595 & 0.135307 & -0.060609 \\ 0.005179 & 0.008754 & -0.012610 & -0.004571 \\ 0.038737 & 0.243002 & -0.326766 & 0.170490 \\ 0.004161 & -0.036569 & 0.046607 & -0.021875 \end{bmatrix}$$

1.9. $x_1 = 5$, $x_2 = -12$, $x_3 = 1$, $x_4 = 7$, $x_5 = -5$

1.10. $x_1 = 8$, $x_2 = 4$, $x_3 = 12$, $x_4 = 8$

1.11. $x_1 = 8$, $x_2 = 4$, $x_3 = 12$, $x_4 = 8$

1.12. $x_1 = 2$, $x_2 = -1$, $x_3 = 4$, $x_4 = -3$

1.13. $x_1 = 3$, $x_2 = 5$, $x_3 = 4$, $b_4 = 68$, $b_5 = -41$

1.14. $x_1 = 44$, $x_3 = 40$, $x_5 = -2$, $b_2 = -2$, $b_4 = 251$

1.15. $x_1 = -9$, $x_2 = -2$, $x_3 = 6$, $x_4 = -8$, $b_5 = 2$, $b_6 = -91$

1.16. $\lambda_1 = 8$, $\lambda_2 = 9$, $\lambda_3 = 13$

CHAPTER 3

3.1. c. $\delta_{AB} = 0.003440$ in. elongation

$\delta_{BC} = 0.004920$ in. shortening

d. $F_{AB} = 8.600$ kips tension

$F_{BC} = 9.225$ kips compression

e. $P = 12.61$ kips at $6.12°$ CCW from positive x axis

3.2. c. $\delta_{AB} = 0.003440$ in. elongation

$\delta_{BC} = 0.004920$ in. shortening

$\delta_{DB} = 0.006000$ in. elongation

d. $F_{AB} = 6.880$ kips tension

$F_{BC} = 7.380$ kips compression

$F_{DB} = 6.000$ kips tension

e. $P = 16.07$ kips at $3.84°$ CCW from positive x axis

3.3. c. $\delta_{AB} = 6.364$ mm elongation

$\delta_{BC} = 5.470$ mm shortening

$\delta_{BD} = 3.000$ mm elongation

d. $F_{AB} = 2800$ kN tension

$F_{BC} = 1030$ kN compression

$F_{BD} = 1080$ kN tension

e. $P = 3933$ kN at $41.7°$ CW from positive x axis

3.4. d. $u_B = 0.004853$ in.

$v_B = -0.0006400$ in.

e. $\delta_{AB} = 0.002400$ in. elongation

$\delta_{BC} = 0.004267$ in. shortening

f. $F_{AB} = 6.000$ kips tension

$F_{BC} = 8.000$ kips compression

3.5. c. Applied $M_A = 187.5$ in.-kips CW

Applied $M_B = 156.3$ in.-kips CCW

Applied $M_C = 62.50$ in.-kips CW

Reaction $R_A = 1.465$ kips $\downarrow$

Reaction $R_B = 1.953$ kips $\uparrow$

Reaction $R_C = 0.488$ kips $\downarrow$

3.6. c. Applied $M_A = 180.0$ in.-kips CW

Applied $M_B = 154.4$ in.-kips CCW

Applied $M_c = 62.50$ in.-kips CW

Reaction $R_A = 1.125$ kips $\downarrow$

Reaction $R_B = 1.613$ kips $\uparrow$

Reaction $R_C = 0.488$ kips $\downarrow$

3.7. c. Applied $M_A = 75.00$ in.-kips CW

Applied $M_B = 307.0$ in.-kips CCW

Applied $M_C = 660.2$ in.-kips CCW

Reaction $R_A = 1.875$ kips $\downarrow$

Reaction $R_B = 13.54$ kips ↑

Reaction $R_C = 11.64$ kips ↓

3.8. c. $\theta_B = 0.00009600$ rad CCW

e. Reaction $M_A = 150.0$ in.-kips CCW

Reaction $M_C = 30.00$ in.-kips CCW

Reaction $R_A = 5.938$ kips ↑

Reaction $R_B = 5.000$ kips ↑

Reaction $R_c = 0.938$ kips ↓

3.9.

a.
$$\{\mathbf{F}_E\} = \begin{Bmatrix} -36.80 \text{ kips} \\ -118.4 \text{ ft-kips} \\ -27.20 \text{ kips} \\ 105.6 \text{ ft-kips} \end{Bmatrix}$$

b.
$$\{\mathbf{F}_E\} = \begin{Bmatrix} -39.62 \text{ kips} \\ -91.72 \text{ ft-kips} \\ -32.38 \text{ kips} \\ 72.27 \text{ ft-kips} \end{Bmatrix}$$

c.
$$\{\mathbf{F}_E\} = \begin{Bmatrix} -10.50 \text{ kips} \\ -16.33 \text{ ft-kips} \\ 3.50 \text{ kips} \\ 0.00 \text{ ft-kips} \end{Bmatrix}$$

3.10.

a.
$$\{\mathbf{F}'_E\}_{AB} = \begin{Bmatrix} 0.00 \\ -20.00 \text{ kips} \\ -66.67 \text{ ft-kips} \\ 0.00 \\ -20.00 \text{ kips} \\ 66.67 \text{ ft-kips} \end{Bmatrix}, \quad \{\mathbf{F}_E\}_{AB} = \begin{Bmatrix} 12.00 \text{ kips} \\ -16.00 \text{ kips} \\ -66.67 \text{ ft-kips} \\ 12.00 \text{ kips} \\ -16.00 \text{ kips} \\ 66.67 \text{ ft-kips} \end{Bmatrix}$$

$$\{\mathbf{F}'_E\}_{BC} = \{\mathbf{F}_E\}_{BC} = \begin{Bmatrix} 0.00 \\ -12.50 \text{ kips} \\ -50.00 \text{ ft-kips} \\ 0.00 \\ -12.50 \text{ kips} \\ 50.00 \text{ ft-kips} \end{Bmatrix}$$

b. $\{\mathbf{F}_E\}_{AB} = \left\{\begin{array}{c} 4.05\,\text{kips} \\ -5.40\,\text{kips} \\ -22.50\,\text{ft-kips} \\ 9.45\,\text{kips} \\ -12.60\,\text{kips} \\ 33.75\,\text{ft-kips} \end{array}\right\}, \quad \{\mathbf{F}_E\}_{BC} = \left\{\begin{array}{c} 0.00\,\text{kips} \\ -15.75\,\text{kips} \\ -33.75\,\text{ft-kips} \\ 0.00\,\text{kips} \\ -6.75\,\text{kips} \\ 22.50\,\text{ft-kips} \end{array}\right\}$

c. $\{\mathbf{F}_E\}_{AB} = \left\{\begin{array}{c} 15.00\,\text{kips} \\ 0.00\,\text{kips} \\ -37.50\,\text{ft-kips} \\ 15.00\,\text{kips} \\ 0.00\,\text{kips} \\ 37.50\,\text{ft-kips} \end{array}\right\}, \quad \{\mathbf{F}_E\}_{BC} = \left\{\begin{array}{c} -5.00\,\text{kips} \\ -8.66\,\text{kips} \\ -38.97\,\text{ft-kips} \\ -5.00\,\text{kips} \\ -8.66\,\text{kips} \\ 38.97\,\text{ft-kips} \end{array}\right\}$

CHAPTER 4

4.1. c, d. $u_B = 0.0099419\,\text{in.} \rightarrow, \; v_B = 0.027685\,\text{in.} \downarrow$

$A_x = 19.42\,\text{kips} \rightarrow, \; A_y = 25.89\,\text{kips} \uparrow$

$C_x = 29.48\,\text{kips} \leftarrow, \; C_y = 22.11\,\text{kips} \uparrow$

$D_x = 9.942\,\text{kips} \leftarrow, \; D_y = 0.00$

e. $F_{AB} = 32.36\,\text{kips (C)}, \; F_{CB} = 36.85\,\text{kips (C)}, \; F_{DB} = 9.942\,\text{kips (T)}$

4.2. c, d. $u_B = 3.3610\,\text{mm} \rightarrow, \; v_B = 1.2723\,\text{mm} \downarrow$

$A_x = 1048\,\text{kN} \leftarrow, \; A_y = 1048\,\text{kN} \uparrow$

$C_x = 551.6\,\text{kN} \leftarrow, \; C_y = 306.4\,\text{kN} \downarrow$

$D_x = 0.00, \; D_y = 458.0\,\text{kN} \uparrow$

e. $F_{AB} = 1483\,\text{kN (T)}, \; F_{CB} = 458\,\text{kips (T)}, \; F_{DB} = 631.0\,\text{kips (C)}$

4.3. b. $A_y = 5.938\,\text{kips} \uparrow, \; M_A = 150.0\,\text{in.-kips CCW},$

$B_y = 5.000\,\text{kips} \uparrow,$

$C_y = 0.938\,\text{kips} \downarrow, \; M_C = 30.00\,\text{in.-kips CCW}$

$V_{AB} = 5.938\,\text{kips} \uparrow, \; M_{AB} = 150.00\,\text{in.-kips CCW},$

$V_{BA} = 4.063\,\text{kips} \uparrow, \; M_{BA} = 60.00\,\text{in.-kips CW}$

$V_{BC} = 0.938$ kips ↑, $M_{BC} = 60.00$ in.-kips CCW

$V_{CB} = 0.938$ kips ↓, $M_{CB} = 30.00$ in.-kips CCW

4.4. b. $A_y = 13.68$ kips ↑, $M_A = 307.4$ in.-kips CCW

$B_y = 11.41$ kips ↑, $C_y = 1.096$ kips ↓

$V_{AB} = 13.68$ kips ↑, $M_{AB} = 307.4$ in.-kips CCW

$V_{BA} = 10.32$ kips ↑, $M_{BA} = 105.3$ in.-kips CW

$V_{BC} = 1.096$ kips ↑, $M_{BC} = 105.3$ in.-kips CCW

$V_{CB} = 1.096$ kips ↓

4.5. b. $A_y = 4.996$ kips ↑, $B_y = 21.76$ kips ↑, $C_y = 9.245$ kips ↑

$V_{AB} = 4.996$ kips ↑,

$V_{BA} = 7.004$ kips ↑, $M_{BA} = 264.5$ in.-kips CW

$V_{BC} = 14.76$ kips ↑, $M_{BC} = 264.5$ in.-kips CCW

$V_{CB} = 9.245$ kips ↑

4.6.

$$\{\mathbf{F}_G\} = \begin{Bmatrix} R_1 + 12.305\,\text{kN} \\ R_2 \\ M_2 - 21.094\,\text{kN-m} \\ 11.095\,\text{kN} \\ -7.2000\,\text{kN} \\ -2.3444\,\text{kN-m} \\ 0.90000\,\text{kN} \\ -16.190\,\text{kN} \\ 23.333\,\text{kN-m} \\ 16.000\,\text{kN} \\ 4.0000\,\text{kN} \\ -8.3333\,\text{kN-m} \\ R_{13} \\ R_{14} \\ M_{15} \end{Bmatrix}$$

4.7.

$$\{\mathbf{F}_G\} = \begin{Bmatrix} A_x + 9.504 \text{ kips} \\ A_y \\ M_A - 38.88 \text{ ft-kips} \\ 47.50 \text{ kips} \\ -3.600 \text{ kips} \\ -31.68 \text{ ft-kips} \\ 18.00 \text{ kips} \\ -24.00 \text{ kips} \\ 30.00 \text{ ft-kips} \\ D_x \\ D_y \\ M_D \\ 0.00 \\ -36.00 \text{ kips} \\ 72.00 \text{ ft-kips} \\ 0.00 \\ -24.00 \text{ kips} \\ -48.00 \text{ ft-kips} \end{Bmatrix}$$

4.8. Invert two 32×32 submatrices. Number of arithmetic operations: Gauss–Jordan inversion, 258,024; direct inversion, 4.47769×10^{37}.

4.9. $A_x = 93.75 \text{ kips} \leftarrow$

$C_x = 67.50 \text{ kips} \rightarrow$, $C_y = 90.00 \text{ kips} \uparrow$

$D_x = 93.75 \text{ kips} \leftarrow$

$F_{AB} = 156.25 \text{ kips (C)}$, $F_{BC} = 112.50 \text{ kips (C)}$

$F_{AD} = 125.00 \text{ kips (T)}$, $F_{BD} = 156.25 \text{ kips (C)}$

$F_{CD} = 0.00$

4.10. $A_x = 80.00 \text{ kips} \leftarrow$, $A_y = 28.89 \text{ kips} \downarrow$

$B_y = 66.03 \text{ kips} \uparrow$

$C_y = 22.86 \text{ kips} \uparrow$

$F_{AB} = 80.00 \text{ kips (T)}$, $F_{BC} = 30.48 \text{ kips (T)}$

$F_{AD} = 28.89 \text{ kips (T)}$, $F_{BD} = 82.55 \text{ kips (C)}$

$F_{CD} = 38.10 \text{ kips (C)}$

4.11. $A_x = 18.71 \text{ kN} \rightarrow$, $A_y = 27.21 \text{ kN} \uparrow$, $M_A = 57.66 \text{ kN-m CCW}$

$B_x = 18.71 \text{ kN} \leftarrow$, $B_y = 17.79 \text{ kN} \uparrow$, $M_B = 10.07 \text{ kN-m CCW}$

4.12. $A_x = 1.678 \text{ kips} \leftarrow$, $A_y = 7.215 \text{ kips} \uparrow$, $M_A = 260.2 \text{ in.-kip CCW}$

$D_x = 8.322 \text{ kips} \leftarrow$, $D_y = 10.79 \text{ kips} \uparrow$, $M_D = 578.2 \text{ in.-kip CCW}$

4.13. $A_x = 3.543 \text{ kips} \leftarrow$, $A_y = 2.896 \text{ kips} \uparrow$, $M_A = 306.5 \text{ in.-kip CCW}$

$D_x = 20.46 \text{ kips} \leftarrow$, $D_y = 17.10 \text{ kips} \uparrow$, $M_D = 824.2 \text{ in.-kip CCW}$

CHAPTER 5

5.1. $N_1(\zeta) = 1 - \frac{11}{2}\zeta + 9\zeta^2 - \frac{9}{2}\zeta^3$, $N_2(\zeta) = 9\zeta - \frac{45}{2}\zeta^2 + \frac{27}{2}\zeta^3$

$N_3(\zeta) = -\frac{9}{2}\zeta + 18\zeta^2 - \frac{27}{2}\zeta^3$, $N_4(\zeta) = \zeta - \frac{9}{2}\zeta^2 + \frac{9}{2}\zeta^3$

5.2. See equations 5.2.

5.3. $N_1(\zeta) = 1 - 11\zeta^2 + 18\zeta^3 - 8\zeta^4$, $N_2(\zeta) = L(\zeta - 4\zeta^2 + 5\zeta^3 - 2\zeta^4)$

$N_3(\zeta) = 16\zeta^2 - 32\zeta^3 + 16\zeta^4$, $N_4(\zeta) = -5\zeta^2 + 14\zeta^3 - 8\zeta^4$,
$N_5(\zeta) = L(\zeta^2 - 3\zeta^3 + 2\zeta^4)$

5.4. See equation 5.38.

5.5.

$$[\mathbf{K}] = \frac{AE}{600L} \begin{bmatrix} 2220 & -2835 & 810 & -195 \\ -2835 & 6480 & -4455 & 810 \\ 810 & -4455 & 6480 & -2835 \\ -195 & 810 & -2835 & 2220 \end{bmatrix}$$

5.6. See equation 3.5.

5.7.

$$[\mathbf{K}] = \frac{EI}{15L} \begin{bmatrix} \frac{948}{L^2} & \frac{282}{L} & \frac{-1536}{L^2} & \frac{588}{L^2} & \frac{-102}{L} \\ \frac{282}{L} & 108 & \frac{-384}{L} & \frac{102}{L} & -18 \\ \frac{-1536}{L^2} & \frac{-384}{L} & \frac{3072}{L^2} & \frac{-1536}{L^2} & \frac{384}{L} \\ \frac{588}{L^2} & \frac{102}{L} & \frac{-1536}{L^2} & \frac{948}{L^2} & \frac{-282}{L} \\ \frac{-102}{L} & -18 & \frac{384}{L} & \frac{-282}{L} & 108 \end{bmatrix}$$

5.8. See Section 5.2.3.

5.9. See Figure 3.7.

5.10.

$$\{\mathbf{F}'_E\} = \frac{w_o L}{60} \begin{Bmatrix} 14 \\ L \\ 32 \\ 14 \\ -L \end{Bmatrix}$$

5.11.

$$\{\mathbf{F}'_E\} = \frac{-w_o L}{(n+3)(n+4)} \begin{Bmatrix} \frac{6}{n+1} \\ \frac{2L}{n+2} \\ n+6 \\ -L \end{Bmatrix}$$

5.12.

$$\{\mathbf{F}'_E\} = \begin{Bmatrix} 18\,\text{kips} \\ 24\,\text{kips} \\ 21\,\text{ft-kips} \\ 18\,\text{kips} \\ 24\,\text{kips} \\ -21\,\text{ft-kips} \end{Bmatrix}$$

5.13.

$$\{\mathbf{F}'_E\} = \begin{Bmatrix} \frac{2}{3}P\,\cos\,\theta \\ -\frac{20}{27}P\,\sin\,\theta \\ -\frac{4}{27}PL\,\sin\,\theta \\ \frac{1}{3}P\,\cos\,\theta \\ -\frac{7}{27}P\,\sin\,\theta \\ \frac{2}{27}PL\,\sin\,\theta \end{Bmatrix}$$

5.14. Using the three elements *AB*, *BC*, and *CD*:

$$\{\mathbf{F}_G\} = \begin{Bmatrix} A_y - 12.276\,\text{kips} \\ M_A - 12.730\,\text{ft-kips} \\ -60.867\,\text{kips} \\ 28.643\,\text{ft-kips} \\ -75\,\text{kips} \\ 0 \\ D_y\,\text{kips} \\ 0 \end{Bmatrix}, \quad \{\mathbf{q}_G\} = \begin{Bmatrix} 0 \\ 0 \\ -0.13261\,\text{in.} \\ -0.0032225\,\text{rad} \\ -0.18949\,\text{in.} \\ 0.0015968\,\text{rad} \\ 0 \\ 0.0051322\,\text{rad} \end{Bmatrix}$$

$A_y = 102.2\,\text{kips} \uparrow$, $M_A = 276.6\,\text{ft-kips}$ CCW, $D_y = 45.92\,\text{kips} \uparrow$

5.15. Using the three elements AB, BC, and CD:

$$\{\mathbf{q}_G\} = \begin{Bmatrix} -0.22923 \text{ in.} \\ -0.0009080 \text{ rad} \\ -0.26039 \text{ in.} \\ -0.0000363 \text{ rad} \\ -0.23216 \text{ in.} \\ 0.0014252 \text{ rad} \\ -0.13165 \text{ in.} \\ 0.0023067 \text{ rad} \end{Bmatrix} \quad \{\mathbf{F}_{AB}\} = \begin{Bmatrix} 0 \\ 0 \\ 19.51 \text{ kips} \\ 39.88 \text{ ft-kips} \end{Bmatrix},$$

$$\{\mathbf{F}_{BC}\} = \begin{Bmatrix} -19.51 \text{ kips} \\ -39.88 \text{ ft-kips} \\ -34.95 \text{ kips} \\ 72.71 \text{ ft-kips} \end{Bmatrix}, \quad \{\mathbf{F}_{CD}\} = \begin{Bmatrix} -40.05 \text{ kips} \\ -72.71 \text{ ft-kips} \\ 0 \\ 0 \end{Bmatrix}$$

CHAPTER 6

6.3. $\lfloor \mathbf{B(x)} \rfloor = -\dfrac{2y}{L^2} \left\lfloor \dfrac{3(2x-L)}{L} \quad (3x-2L) \quad \dfrac{3(L-2x)}{L} \quad (3x-L) \right\rfloor$

6.4. See equation 3.5.

6.5. See equation 3.5.

6.7. a. See Figure 3.7.

b.
$$\{\mathbf{F}'_E\} = \frac{w_o L^2}{20} \begin{Bmatrix} 3 \\ \frac{2}{3}L \\ 7 \\ -L \end{Bmatrix}$$

c.
$$\{\mathbf{F}'_E\} = \frac{w_o L^{n+1}}{(n+3)(n+4)} \begin{Bmatrix} \frac{6}{n+1} \\ \frac{2L}{n+2} \\ n+6 \\ -L \end{Bmatrix}$$

6.8.

$$\left\{ \begin{array}{c} u(x, \ y) \\ v(x, \ y) \end{array} \right\} = \left[\begin{array}{cccccccc} N_1 & 0 & N_2 & 0 & N_3 & 0 & N_4 & 0 \\ 0 & N_1 & 0 & N_2 & 0 & N_3 & 0 & N_4 \end{array} \right] \left\{ \begin{array}{c} u_1 \\ v_1 \\ u_2 \\ v_2 \\ u_3 \\ v_3 \\ u_4 \\ v_4 \end{array} \right\}$$

6.9.

$$[\mathbf{B(x, \ y)}] = \left[\begin{array}{cccccccc} \frac{\partial N_1}{\partial x} & 0 & \frac{\partial N_2}{\partial x} & 0 & \frac{\partial N_3}{\partial x} & 0 & \frac{\partial N_4}{\partial x} & 0 \\ 0 & \frac{\partial N_1}{\partial y} & 0 & \frac{\partial N_2}{\partial y} & 0 & \frac{\partial N_3}{\partial y} & 0 & \frac{\partial N_4}{\partial y} \\ \frac{\partial N_1}{\partial y} & \frac{\partial N_1}{\partial x} & \frac{\partial N_2}{\partial y} & \frac{\partial N_2}{\partial x} & \frac{\partial N_3}{\partial y} & \frac{\partial N_3}{\partial x} & \frac{\partial N_4}{\partial y} & \frac{\partial N_4}{\partial x} \end{array} \right]$$

6.10.

$$[\mathbf{B}_i(\mathbf{x, \ y})] = \left[\begin{array}{cc} \frac{\partial N_i}{\partial x} & 0 \\ 0 & \frac{\partial N_i}{\partial y} \\ \frac{\partial N_i}{\partial y} & \frac{\partial N_i}{\partial x} \end{array} \right], \quad i = 1, 2, 3, 4$$

CHAPTER 7

7.5. a. $9.00 \, \text{ksi}$, b. $8.00 \, \text{ksi}$, c. $4.00 \, \text{ksi}$, d. $4.00 \, \text{ksi}$

7.6. a. $9.00 \, \text{ksi}$, b. $6.75 \, \text{ksi}$, c. $3.00 \, \text{ksi}$, d. $2.00 \, \text{ksi}$

7.7.

a.
$$\left\{ \begin{array}{c} \epsilon_x \\ \epsilon_y \\ \gamma_{xy} \end{array} \right\} = \left\{ \begin{array}{c} 6.3333 \\ -0.3333 \\ 5.0000 \end{array} \right\} \times 10^{-4}$$

b.
$$\left\{ \begin{array}{c} \sigma_{p1} \\ \sigma_{p2} \\ \tau_p \end{array} \right\} = \left\{ \begin{array}{c} 22.000 \\ 2.000 \\ 0.000 \end{array} \right\} \text{ksi}, \quad \left\{ \begin{array}{c} \epsilon_{p1} \\ \epsilon_{p2} \\ \gamma_p \end{array} \right\} = \left\{ \begin{array}{c} 7.1667 \\ -1.1667 \\ 0.0000 \end{array} \right\} \times 10^{-4}$$

c.
$$\left\{ \begin{array}{c} \sigma_a \\ \sigma_a \\ \tau_m \end{array} \right\} = \left\{ \begin{array}{c} 12.000 \\ 12.000 \\ 10.000 \end{array} \right\} \text{ksi}, \quad \left\{ \begin{array}{c} \epsilon_a \\ \epsilon_a \\ \gamma_m \end{array} \right\} = \left\{ \begin{array}{c} 3.0000 \\ 3.0000 \\ 8.3333 \end{array} \right\} \times 10^{-4}$$

d. $\tau_m = 11.000 \, \text{ksi}$

e. $\sigma_z = 0$, $\epsilon_z = 2.0000 \times 10^{-4}$

7.8.

a.
$$\begin{Bmatrix} \epsilon_x \\ \epsilon_y \\ \gamma_{xy} \end{Bmatrix} = \begin{Bmatrix} 5.8333 \\ -0.8333 \\ 5.0000 \end{Bmatrix} \times 10^{-4}$$

b.
$$\begin{Bmatrix} \sigma_{p1} \\ \sigma_{p2} \\ \tau_p \end{Bmatrix} = \begin{Bmatrix} 22.000 \\ 2.000 \\ 0.000 \end{Bmatrix} \text{ksi}, \quad \begin{Bmatrix} \epsilon_{p1} \\ \epsilon_{p2} \\ \gamma_p \end{Bmatrix} = \begin{Bmatrix} 6.6667 \\ -1.6667 \\ 0.0000 \end{Bmatrix} \times 10^{-4}$$

c.
$$\begin{Bmatrix} \sigma_a \\ \sigma_a \\ \tau_m \end{Bmatrix} = \begin{Bmatrix} 12.000 \\ 12.000 \\ 10.000 \end{Bmatrix} \text{ksi}, \quad \begin{Bmatrix} \epsilon_a \\ \epsilon_a \\ \gamma_m \end{Bmatrix} = \begin{Bmatrix} 2.5000 \\ 2.5000 \\ 8.3333 \end{Bmatrix} \times 10^{-4}$$

d. $\tau_m = 10.000\,\text{ksi}$

e. $\sigma_z = 6.000\,\text{ksi (T)}, \quad \epsilon_z = 0$

7.9.

a.
$$\begin{Bmatrix} \epsilon_x \\ \epsilon_y \\ \gamma_{xy} \end{Bmatrix} = \begin{Bmatrix} 3.5000 \\ -1.5000 \\ 6.6667 \end{Bmatrix} \times 10^{-4}$$

b.
$$\begin{Bmatrix} \sigma_{p1} \\ \sigma_{p2} \\ \tau_p \end{Bmatrix} = \begin{Bmatrix} 14.000 \\ -6.000 \\ 0.000 \end{Bmatrix} \text{ksi}, \quad \begin{Bmatrix} \epsilon_{p1} \\ \epsilon_{p2} \\ \gamma_p \end{Bmatrix} = \begin{Bmatrix} 5.1667 \\ -3.1667 \\ 0.0000 \end{Bmatrix} \times 10^{-4}$$

c.
$$\begin{Bmatrix} \sigma_a \\ \sigma_a \\ \tau_m \end{Bmatrix} = \begin{Bmatrix} 4.000 \\ 4.000 \\ 10.000 \end{Bmatrix} \text{ksi}, \quad \begin{Bmatrix} \epsilon_a \\ \epsilon_a \\ \gamma_m \end{Bmatrix} = \begin{Bmatrix} 1.0000 \\ 1.0000 \\ 8.3333 \end{Bmatrix} \times 10^{-4}$$

d. $\tau_m = 10.000\,\text{ksi}$

e. $\sigma_z = 0, \quad \epsilon_z = 0.6667 \times 10^{-4}$

7.10.

a.
$$\begin{Bmatrix} \epsilon_x \\ \epsilon_y \\ \gamma_{xy} \end{Bmatrix} = \begin{Bmatrix} 3.3333 \\ -1.6667 \\ 6.6667 \end{Bmatrix} \times 10^{-4}$$

b. $\begin{Bmatrix} \sigma_{p1} \\ \sigma_{p2} \\ \tau_p \end{Bmatrix} = \begin{Bmatrix} 14.000 \\ -6.000 \\ 0.000 \end{Bmatrix}$ ksi, $\begin{Bmatrix} \epsilon_{p1} \\ \epsilon_{p2} \\ \gamma_p \end{Bmatrix} = \begin{Bmatrix} 5.0000 \\ -3.3333 \\ 0.0000 \end{Bmatrix} \times 10^{-4}$

c. $\begin{Bmatrix} \sigma_a \\ \sigma_a \\ \tau_m \end{Bmatrix} = \begin{Bmatrix} 4.000 \\ 4.000 \\ 10.000 \end{Bmatrix}$ ksi, $\begin{Bmatrix} \epsilon_a \\ \epsilon_a \\ \gamma_m \end{Bmatrix} = \begin{Bmatrix} 0.8333 \\ 0.8333 \\ 8.3333 \end{Bmatrix} \times 10^{-4}$

d. $\tau_m = 10.000$ ksi

e. $\sigma_z = 2.000$ ksi (T), $\epsilon_z = 0$

7.11.

$[\mathbf{E}_m] = \begin{bmatrix} 32{,}349 & 7{,}117 & 0 \\ 7{,}117 & 21{,}566 & 0 \\ 0 & 0 & 9{,}494 \end{bmatrix}$ ksi, $\begin{Bmatrix} \epsilon_x \\ \epsilon_y \\ \gamma_{yx} \end{Bmatrix} = \begin{Bmatrix} 2.3807 \\ 1.1040 \\ 1.8040 \end{Bmatrix} \times 10^{-3}$

7.12.

$[\mathbf{E}_m] = \begin{bmatrix} 35{,}744 & 9{,}007 & 0 \\ 9{,}007 & 23{,}075 & 0 \\ 0 & 0 & 10{,}000 \end{bmatrix}$ ksi, $\begin{Bmatrix} \epsilon_x \\ \epsilon_y \\ \gamma_{xy} \end{Bmatrix} = \begin{Bmatrix} 2.3157 \\ 0.9177 \\ 1.7329 \end{Bmatrix} \times 10^{-3}$

CHAPTER 8

8.1. a. $N_1 = 2 - \frac{1}{3}x,$ $N_2 = 1 + \frac{1}{3}x - \frac{1}{4}y,$ $N_3 = -2 + \frac{1}{4}y$

b. $[\mathbf{K}] = \begin{bmatrix} 16 & 0 & -16 & 3 & 0 & -3 \\ 0 & 6 & 4.5 & -6 & -4.5 & 0 \\ -16 & 4.5 & 19.375 & -7.5 & -3.375 & 3 \\ 3 & -6 & -7.5 & 15 & 4.5 & -9 \\ 0 & -4.5 & -3.375 & 4.5 & 3.375 & 0 \\ -3 & 0 & 3 & -9 & 0 & 9 \end{bmatrix}$ kip/in.$\times 1000$

8.2.

a. $\{\mathbf{F}_b\} = \begin{Bmatrix} 160 \\ 120 \\ 80 \\ 60 \\ 100 \\ 120 \end{Bmatrix}$ kips

b.

$$\{F_b\} = 5\sqrt{5}\begin{Bmatrix} 3 \\ -4 \\ 0 \\ 0 \\ 3 \\ -4 \end{Bmatrix} \text{ lb}$$

c.

$$\{F_b\} = \frac{20\sqrt{5}}{3}\begin{Bmatrix} 1 \\ 0 \\ 0 \\ 0 \\ 2 \\ 0 \end{Bmatrix} \text{ lb}$$

8.3.

$$[K] = \begin{bmatrix} 195.55 & 58.44 & -298.95 & -28.10 & 103.40 & -30.34 \\ 58.44 & 107.89 & -15.73 & -65.18 & -42.71 & -42.71 \\ -298.95 & -15.73 & 566.43 & -102.27 & -267.48 & 118.01 \\ -28.10 & -65.18 & -102.27 & 237.70 & 130.27 & -172.51 \\ 103.40 & -42.71 & -267.48 & 130.37 & 164.09 & -87.66 \\ -30.34 & -42.71 & 118.01 & -172.51 & -87.66 & 215.22 \end{bmatrix} \text{ kips/ft} \times 1000$$

8.4.

$$\{F_E\} = \begin{Bmatrix} 205.98 \\ -281.29 \\ 432.56 \\ -386.95 \\ 267.77 \\ -310.11 \end{Bmatrix} \text{ lb}$$

8.5.

$$\{\mathbf{F}_E\} = \begin{Bmatrix} 505.98 \\ -681.29 \\ 732.56 \\ -786.95 \\ 267.77 \\ -310.11 \end{Bmatrix} \text{ lb}$$

8.6. $C_x = 0$, $C_y = 267.99 \text{ lb} \uparrow$

$D_x = 393.29 \text{ lb} \leftarrow$, $D_y = 1857.97 \text{ lb} \uparrow$

$E_x = 306.71 \text{ lb} \leftarrow$, $E_y = 824.03 \text{ lb} \uparrow$

8.7. $C_x = 201.11 \text{ lb} \leftarrow$, $C_y = 271.16 \text{ lb} \uparrow$

$D_x = 0$, $D_y = 1852.70 \text{ lb} \uparrow$

$E_x = 498.89 \text{ lb} \leftarrow$, $E_y = 826.14 \text{ lb} \uparrow$

8.8. $C_x = 5.548 \text{ kips} \leftarrow$, $C_y = 4.000 \text{ kips} \downarrow$

$D_x = 12.129 \text{ kips} \leftarrow$, $D_y = 29.678 \text{ kips} \uparrow$

$q_{Bx} = 1.143 \times 10^{-3} \text{ in.} \rightarrow$, $q_{By} = 1.389 \times 10^{-3} \text{ in.} \downarrow$

8.9. $C_x = 6.528 \text{ kips} \leftarrow$, $C_y = 4.000 \text{ kips} \downarrow$

$D_x = 11.50 \text{ kips} \leftarrow$, $D_y = 29.678 \text{ kips} \uparrow$

$q_{Bx} = 2.067 \times 10^{-3} \text{ in.} \rightarrow$, $q_{By} = 4.111 \times 10^{-3} \text{ in.} \downarrow$

8.10. $A_x = 0$, $A_y = 6.000 \text{ kips} \uparrow$

$C_x = 0$, $C_y = 6.000 \text{ kips} \uparrow$

$q_{Bx} = 1.600 \times 10^{-3} \text{ in.} \rightarrow$, $q_{By} = 10.867 \times 10^{-3} \text{ in.} \downarrow$

8.11. $A_x = 6.876 \text{ kips} \rightarrow$, $A_y = 6.000 \text{ kips} \uparrow$

$C_x = 6.876 \text{ kips} \leftarrow$, $C_y = 6.000 \text{ kips} \uparrow$

$q_{Bx} = 0.0809 \times 10^{-3} \text{ in.} \leftarrow$, $q_{By} = 8.995 \times 10^{-3} \text{ in.} \downarrow$

8.12. $A_x = 0$, $A_y = 6.000 \text{ kips} \uparrow$

$C_x = 0$, $C_y = 6.000 \text{ kips} \uparrow$

$q_{Bx} = 1.800 \times 10^{-3} \text{ in.} \rightarrow$, $q_{By} = 11.333 \times 10^{-3} \text{ in.} \downarrow$

8.13. $A_x = 7.200 \text{ kips} \rightarrow$, $A_y = 6.000 \text{ kips} \uparrow$

$C_x = 7.200 \text{ kips} \leftarrow$, $C_y = 6.000 \text{ kips} \uparrow$

$q_{Bx} = 0$, $q_{By} = 9.333 \times 10^{-3} \text{ in.} \downarrow$

8.14. a. $N_1 = \frac{1}{10}(18 - x - 2y)$, $N_2 = \frac{1}{10}(-6 + 2x - y)$, $N_3 = \frac{1}{10}(-2 - x + 3y)$

b. $\int_A N_1 \, dA = \int_A N_2 \, dA = \int_A N_3 \, dA = \frac{10}{3}$

c.
$$\{F_b\} = -\frac{1}{3} \begin{Bmatrix} 0 \\ 485 \\ 0 \\ 485 \\ 0 \\ 485 \end{Bmatrix} \text{ lb}$$

d.
$$\{F_b\} = \frac{1}{3} \begin{Bmatrix} 0 \\ 3.5 \\ 0 \\ 4.0 \\ 0 \\ 4.5 \end{Bmatrix} \text{ lb}$$

e.
$$\{F_b\} = \frac{20\sqrt{5}}{3} \begin{Bmatrix} 1 \\ 0 \\ 0 \\ 0 \\ 2 \\ 0 \end{Bmatrix} \text{ lb}$$

8.15. $D_x = 201.9 \, \text{lb} \leftarrow,\ D_y = 1630.2 \, \text{lb} \uparrow$

$F_x = 293.1 \, \text{lb} \leftarrow,\ F_y = 3735.5 \, \text{lb} \uparrow$

8.16. a. $N_1 = 4.68 - 0.62x - 1.24y + 0.02x^2 + 0.08xy + 0.08y^2$

$N_2 = 1.32 - 0.68x + 0.34y + 0.08x^2 - 0.08xy + 0.02y^2$

$N_3 = 0.28 + 0.18x - 0.54y + 0.02x^2 - 0.12xy + 0.18y^2$

$N_4 = -4.32 + 1.68x - 0.24y - 0.08x^2 - 0.12xy + 0.08y^2$

$N_5 = 0.48 + 0.08x - 0.64y - 0.08x^2 + 0.28xy - 0.12y^2$

$N_6 = -1.44 - 0.64x + 2.32y + 0.04x^2 - 0.04xy - 0.24y^2$

b. $\int_A N_1(x,\ y) \, dA = \int_A N_2(x,\ y) \, dA = \int_A N_3(x,\ y) \, dA = 0$

$\int_A N_4(x,\ y) \, dA = \int_A N_5(x,\ y) \, dA = \int_A N_6(x,\ y) \, dA = \frac{10}{3} \text{ ft}^2$

c.

$$\{\mathbf{F}_b\} = -\frac{1}{3}\begin{Bmatrix} 0 \\ 0 \\ 0 \\ 0 \\ 0 \\ 0 \\ 0 \\ 485 \\ 0 \\ 485 \\ 0 \\ 485 \end{Bmatrix} \text{lb}$$

d.

$$\{\mathbf{F}_b\} = \frac{1}{3}\begin{Bmatrix} 0 \\ 0 \\ 0 \\ 0 \\ 0 \\ 0 \\ 0 \\ 3 \\ 0 \\ 5 \\ 0 \\ 4 \end{Bmatrix} \text{lb}$$

e.

$$\{\mathbf{F}_b\} = \frac{20\sqrt{5}}{3}\begin{Bmatrix} 0 \\ 0 \\ 0 \\ 0 \\ 1 \\ 0 \\ 0 \\ 0 \\ 0 \\ 0 \\ 2 \\ 0 \end{Bmatrix} \text{lb}$$

8.17. $C_x = 8.854$ kips $\leftarrow$, $C_y = 4.000$ kips $\downarrow$

$D_x = 8.824$ kips $\leftarrow$, $D_y = 29.678$ kips $\uparrow$

$q_{Bx} = 4.015 \times 10^{-3}$ in. $\rightarrow$, $q_{By} = 7.737 \times 10^{-3}$ in. $\downarrow$

CHAPTER 9

9.1. a. $N_1 = \frac{1}{4}(1 - \frac{x}{10})(1 - \frac{y}{5})$,

$N_2 = \frac{1}{4}(1 + \frac{x}{10})(1 - \frac{y}{5})$,

$N_3 = \frac{1}{4}(1 + \frac{x}{10})(1 + \frac{y}{5})$,

$N_4 = \frac{1}{4}(1 - \frac{x}{10})(1 + \frac{y}{5})$.

b.

$$[\mathbf{B}] = \frac{1}{200} \begin{bmatrix} -5(1-\frac{y}{5}) & 0 & 5(1-\frac{y}{5}) & 0 & 5(1+\frac{y}{5}) & 0 & -5(1+\frac{y}{5}) & 0 \\ 0 & -10(1-\frac{x}{10}) & 0 & -10(1+\frac{x}{10}) & 0 & 10(1+\frac{x}{10}) & 0 & 10(1-\frac{x}{10}) \\ -10(1-\frac{x}{10}) & -5(1-\frac{y}{5}) & -10(1+\frac{x}{10}) & 5(1-\frac{y}{5}) & 10(1+\frac{x}{10}) & 5(1+\frac{y}{5}) & 10(1-\frac{x}{10}) & -5(1+\frac{y}{5}) \end{bmatrix}$$

(units are in.$^{-1}$ if coordinates x and y are in inches)

c. $k_{42} = 4333.33$ kips/in.

d. $k_{55} = 6666.67$ kips/in.

9.2. b.

$$[\mathbf{B}] = \frac{1}{32} \begin{bmatrix} -2(1-\frac{y}{2}) & 0 & 2(1-\frac{y}{2}) & 0 & 2(1+\frac{y}{2}) & 0 & -2(1+\frac{y}{2}) & 0 \\ 0 & -4(1-\frac{x}{4}) & 0 & -4(1+\frac{x}{4}) & 0 & 4(1+\frac{x}{4}) & 0 & 4(1-\frac{x}{4}) \\ -4(1-\frac{x}{4}) & -2(1-\frac{y}{2}) & -4(1+\frac{x}{4}) & 2(1-\frac{y}{2}) & 4(1+\frac{x}{4}) & 2(1+\frac{y}{2}) & 4(1-\frac{x}{4}) & -2(1+\frac{y}{2}) \end{bmatrix}$$

(units are ft^{-1} if coordinates x and y are in feet)

c. $k_{14} = -1250.00$ kips/in.

9.3. a. $I = 18 + \ln(4) = 19.386294$

b,c. $I = 19.363636$

d. $I = 19.383673$

9.4. a. $I = 402/7 - 2 \ln(7) = 53.536751$

b,c. $I = 53.357066$

d. $I = 53.005917$

9.5. a. $I = 8 + 2 \ln(5) = 11.218876$

b,c. $I = 11.130435$

d,e. $I = 11.205387$

9.6. a,b. $k_{42} = 4333.33$ kips/in.

9.7. a,b. $k_{55} = 6666.67$ kips/in.

9.8.

$$[K] = \begin{bmatrix} 6,666.67 & 2,500.00 & -666.67 & -500.00 & -3,333.33 & -2,500.00 & -2,666.67 & 500.00 \\ 2,500.00 & 11,666.67 & 500.00 & 4,333.33 & -2,500.00 & -5,833.33 & -500.00 & -10,166.67 \\ -666.67 & 500.00 & 6,666.67 & -2,500.00 & -2,666.67 & -500.00 & -3,333.33 & 2,500.00 \\ -500.00 & 4,333.33 & -2,500.00 & 11,666.67 & 500.00 & -10,166.67 & 2,500.00 & -5,833.33 \\ 3,333.33 & -2,500.00 & -2,666.67 & 500.00 & 6,666.67 & 2,500.00 & -666.67 & -500.00 \\ -2,500.00 & -5,833.33 & -500.00 & -10,166.67 & 2,500.00 & 11,666.67 & 500.00 & 4,333.33 \\ -2,666.67 & -500.00 & -3,333.33 & 2,500.00 & -666.67 & 500.00 & 6,666.67 & -2,500.00 \\ 500.00 & -10,166.67 & 2,500.00 & -5,833.33 & -500.00 & 4,333.33 & -2500 & 11,666.67 \end{bmatrix} \text{ kips/in}$$

9.9. b. $A = 21.00000$

c. $I = 1743.58321$

9.10. a,b. $A = 31.00000 \text{ ft}^2$

c. $I = 6211.61007$

9.11. b. $I = 20\pi + 22.5 = 85.331853$

d. $I = 85.215851$ (2 × 2 Gauss rule), $I = 85.332955$ (3 × 3 Gauss rule)

9.12. a.

$$[B] = \begin{bmatrix} -0.178326 & 0 & 0.178326 & 0 & 0.047782 & 0 & -0.047782 & 0 \\ 0 & -0.237768 & 0 & -0.095565 & 0 & 0.063710 & 0 & 0.269623 \\ -0.237768 & -0.178326 & -0.095565 & 0.178326 & 0.063710 & 0.047782 & 0.269623 & -0.047782 \end{bmatrix} \text{ ft}^{-1}$$

b. $k_{42} = 2533.79$ kips/in.

9.13.

$$[K] = \begin{bmatrix} 9,648.65 \\ 2,635.14 & 13,466.21 \\ -3,648.65 & 1,864.86 & 12,648.65 \\ 364.86 & 2,533.79 & -4,864.86 & 21,466.21 \\ -3,567.57 & -3,242.24 & -2,432.43 & 243.24 & 7,621.62 \\ -3,242.24 & -3,939.19 & -1,256.76 & -12,060.81 & 2,837.84 & 10,290.54 \\ -2,432.43 & -1,256.76 & -6,567.57 & 4,256.76 & -1,621.62 & 1,662.16 & 10,621.62 \\ 243.24 & -12,060.81 & 4,256.76 & -11,939.19 & 162.16 & 5,709.46 & -4,662.16 & 18,290.54 \end{bmatrix} \text{ kips/in.}$$

(symmetric)

9.14. a.

$$[B] = \begin{bmatrix} -0.105702 & 0 & 0.139208 & 0 & 0.028323 & 0 & -0.061829 & 0 \\ 0 & -0.132127 & 0 & -0.075990 & 0 & 0.035403 & 0 & 0.172714 \\ -0.132127 & -0.105702 & -0.075990 & 0.139208 & 0.035403 & 0.028323 & 0.172714 & -0.061829 \end{bmatrix} \text{ ft}^{-1}$$

b. $k_{42} = 1902.87$ kips/in.

9.15.

$$[K] = \begin{bmatrix}
8,040.30 \\
1,776.85 & 9,976.62 & & & & \text{(symmetric)} \\
-4,965.76 & 1,532.95 & 14,934.42 \\
32.94 & 1,902.87 & -6,488.49 & 19,221.28 \\
-2,161.56 & -2,983.98 & -1,091.12 & 1,210.77 & 7,327.72 \\
-2,983.98 & -2,505.39 & -289.23 & -9,600.51 & 1,923.17 & 9,071.02 \\
-912.98 & -325.81 & -8,877.53 & 5,244.78 & -4,075.04 & 1,350.04 & 13,865.56 \\
1,174.19 & -9,374.10 & 5,244.78 & -11,523.65 & -149.96 & 3,034.88 & -6,269.01 & 17,862.88
\end{bmatrix} \quad \text{kips/in.}$$

9.16.

$$\{F_E\} = \begin{Bmatrix}
600 \\
-109.375 \\
3000 \\
-2109.375 \\
1500 \\
-1625 \\
1200 \\
-1125
\end{Bmatrix} \quad \text{lb}$$

9.17.

$$\{F_E\} = \begin{Bmatrix}
0 \\
-236.98 \\
-3333.33 \\
1098.96 \\
-1416.67 \\
-1330.73 \\
250 \\
-2000
\end{Bmatrix} \quad \text{lb}$$

9.18. b.

$$\{F_E\} = \begin{Bmatrix}
-700 \\
-1410.42 \\
74.92 \\
-1314.58 \\
849.84 \\
-1677.08 \\
0 \\
-2072.92
\end{Bmatrix} \quad \text{lb}, \qquad k_{63} = 12,745.65 \text{ kips/in.}$$

$$[\mathbf{K}] = \begin{bmatrix} 21,043.74 \\ 3,575.52 & 39,588.10 & & & \text{(symmetric)} \\ -16,198.67 & -7,572.49 & 87,444.01 \\ -13,572.49 & 3,287.54 & -18,805.30 & 53,528.12 \\ -1,667.90 & -6,321.77 & -20,300.01 & 18,745.65 & 35,462.97 \\ -6,321.77 & -23,044.02 & 12,745.65 & -34,451.96 & 1,531.80 & 51,899.96 \\ -3,177.17 & 10,318.74 & -50,945.33 & 13,632.15 & -13,495.07 & -7,955.69 & 67,617.57 \\ 16,318.74 & -19,831.62 & 13,632.15 & -22,363.70 & -13,955.69 & 5,596.02 & -15,995.15 & 36,599.30 \end{bmatrix} \text{ kips/in.}$$

c.
$$\{\mathbf{F}_E\} = \begin{Bmatrix} 0 \\ -1531.25 \\ 1067.19 \\ -1458.33 \\ 533.59 \\ -1218.75 \\ 0 \\ -1291.67 \end{Bmatrix} \text{ lb,} \qquad k_{44} = 29,787.45 \text{ kips/in.}$$

$$[\mathbf{K}] = \begin{bmatrix} 29,995.83 \\ 12,398.46 & 52,625.96 & & & \text{(symmetric)} \\ -15,925.85 & -6,701.54 & 45,425.20 \\ -12,701.54 & 6,671.22 & -10,715.33 & 29,787.45 \\ -15,732.63 & -12,016.92 & -1,780.18 & 12,227.90 & 44,750.85 \\ -12,016.92 & -36,933.38 & 6,227.90 & -26,151.66 & 12,793.77 & 68,859.67 \\ 1,662.66 & 6,320.00 & -27,719.16 & 11,188.97 & -27.238.04 & -7,004.15 & 53,294.54 \\ 12,320.00 & -22,363.80 & 11,188.97 & -10,307.01 & -13,004.15 & -5,774.62 & -10,504.82 & 38,445.43 \end{bmatrix} \text{ kips/in.}$$

9.19. $A_x = 0$, $A_y = 6.000$ kips $\uparrow$

$C_x = 0$, $C_y = 6.000$ kips $\uparrow$

$q_{Bx} = 4.867 \times 10^{-3}$ in. $\rightarrow$, $q_{By} = 26.667 \times 10^{-3}$ in. $\downarrow$

9.20. $A_x = 12.882$ kips $\rightarrow$, $A_y = 6.000$ kips $\uparrow$

$C_x = 12.882$ kips $\leftarrow$, $C_y = 6.000$ kips $\uparrow$

$q_{Bx} = 0.0000 \times 10^{-3}$ in. $\leftarrow$, $q_{By} = 16.361 \times 10^{-3}$ in. $\downarrow$

9.21. $A_x = 0$, $A_y = 6.000$ kips $\uparrow$

$C_x = 0$, $C_y = 6.000$ kips $\uparrow$

$q_{Bx} = 2.586 \times 10^{-3}$ in. $\rightarrow$, $q_{By} = 17.244 \times 10^{-3}$ in. $\downarrow$

9.22. $A_x = 9.108$ kips $\rightarrow$, $A_y = 6.000$ kips $\uparrow$

$C_x = 9.108$ kips $\leftarrow$, $C_y = 6.000$ kips $\uparrow$

$q_{Bx} = 0.010 \times 10^{-3}$ in. $\rightarrow$, $q_{By} = 13.456 \times 10^{-3}$ in. $\downarrow$

9.23. $A_x = 0$, $A_y = 6.000$ kips $\uparrow$

$C_x = 0$, $C_y = 6.000$ kips ↑

$q_{Bx} = 2.583 \times 10^{-3}$ in. →, $q_{By} = 17.612 \times 10^{-3}$ in. ↓

9.24. $A_x = 9.140$ kips →, $A_y = 6.000$ kips ↑

$C_x = 9.140$ kips ←, $C_y = 6.000$ kips ↑

$q_{Bx} = 0$, $q_{By} = 13.689 \times 10^{-3}$ in. ↓

9.25. a. $A = 22.66667$ ft^2 (2 × 2 or 3 × 3 Gauss rule)

b. $I = 3574.61728$ ft^5 (2 × 2 Gauss rule)

$I = 3573.65333$ ft^5 (3 × 3 Gauss rule)

9.26. a. $A = 25.00000$ ft^2 (2 × 2 or 3 × 3 Gauss rule)

b. $I = 5408.68313$ ft^5 (2 × 2 Gauss rule)

$I = 5423.00267$ ft^5 (3 × 3 Gauss rule)

9.27. Gauss three-point rule:

$$\{\mathbf{F}_E\} = \begin{Bmatrix} 0 \\ 0 \\ 1.57762 \\ 0 \\ 1.57762 \\ 0 \\ 0 \\ 0 \\ 0 \\ 0 \\ 6.07873 \\ 0 \\ 0 \\ 0 \\ 0 \\ 0 \end{Bmatrix} \text{ kips}$$

9.28 Gauss three-point rule:

$$\{\mathbf{F}_E\} = \begin{Bmatrix} 0 \\ 0 \\ 2.36643 \\ 0 \\ 0 \\ 0 \\ 0 \\ 0 \\ 0 \\ 0 \\ 4.55905 \\ 0 \\ 0 \\ 0 \\ 0 \\ 0 \end{Bmatrix} \text{ kips}$$

9.29. (Gauss 2×2 rule)

$$\{\mathbf{F}_E\} = \begin{Bmatrix} 0 \\ 144.676 \\ 0 \\ 134.259 \\ 0 \\ 122.685 \\ 0 \\ 119.213 \\ 0 \\ -483.796 \\ 0 \\ -527.778 \\ 0 \\ -557.870 \\ 0 \\ -513.889 \end{Bmatrix} \text{ lb,}$$

(Gauss 3×3 rule)

$$
\{\mathbf{F}_E\} = \begin{Bmatrix} 0 \\ 138.194 \\ 0 \\ 127.778 \\ 0 \\ 112.500 \\ 0 \\ 109.028 \\ 0 \\ -481.944 \\ 0 \\ -516.667 \\ 0 \\ -548.611 \\ 0 \\ -502.778 \end{Bmatrix} \text{ lb}
$$

CHAPTER 10

10.1. a.

$$
\begin{Bmatrix} \sigma_x \\ \sigma_y \\ \tau_{xy} \end{Bmatrix} = \begin{Bmatrix} -8.400 \\ -6.600 \\ 6.300 \end{Bmatrix} \text{ ksi}
$$

b.

$$
\begin{Bmatrix} \sigma_{p1} \\ \sigma_{p2} \\ \tau_p \end{Bmatrix} = \begin{Bmatrix} -13.864 \\ -1.136 \\ 0.000 \end{Bmatrix} \text{ ksi}, \quad \begin{Bmatrix} \sigma_m \\ \sigma_m \\ \tau_{max} \end{Bmatrix} = \begin{Bmatrix} -7.500 \\ -7.500 \\ 6.364 \end{Bmatrix} \text{ ksi}
$$

c.

$$
\begin{Bmatrix} M_{xx} \\ M_{yy} \\ M_{xy} \end{Bmatrix} = \begin{Bmatrix} 0.7875 \\ 0.6188 \\ -0.5906 \end{Bmatrix} \text{ kip-in.}
$$

d. $\begin{Bmatrix} M_{p1} \\ M_{p2} \\ M_p \end{Bmatrix} = \begin{Bmatrix} 1.2998 \\ 0.1065 \\ 0.0000 \end{Bmatrix}$ kip-in., $\begin{Bmatrix} M_m \\ M_m \\ M_{max} \end{Bmatrix} = \begin{Bmatrix} 0.7031 \\ 0.7031 \\ -0.5966 \end{Bmatrix}$ kip-in.

10.2.

a. $\begin{Bmatrix} \sigma_x \\ \sigma_y \\ \tau_{xy} \end{Bmatrix} = \begin{Bmatrix} 5.280 \\ 6.720 \\ 2.400 \end{Bmatrix}$ ksi

b. $\begin{Bmatrix} \sigma_{p1} \\ \sigma_{p2} \\ \tau_p \end{Bmatrix} = \begin{Bmatrix} 3.494 \\ 8.506 \\ 0.000 \end{Bmatrix}$ ksi, $\begin{Bmatrix} \sigma_m \\ \sigma_m \\ \tau_{max} \end{Bmatrix} = \begin{Bmatrix} 6.000 \\ 6.000 \\ 2.506 \end{Bmatrix}$ ksi

c. $\begin{Bmatrix} M_{xx} \\ M_{yy} \\ M_{xy} \end{Bmatrix} = \begin{Bmatrix} -0.7425 \\ -0.9450 \\ -0.3375 \end{Bmatrix}$ kip-in.

d. $\begin{Bmatrix} M_{p1} \\ M_{p2} \\ M_p \end{Bmatrix} = \begin{Bmatrix} -0.4914 \\ -1.1961 \\ 0.0000 \end{Bmatrix}$ kip-in., $\begin{Bmatrix} M_m \\ M_m \\ M_{max} \end{Bmatrix} = \begin{Bmatrix} -0.8438 \\ -0.8438 \\ -0.3524 \end{Bmatrix}$ kip-in.

10.3. $N_{11} = \frac{1}{8}(2 - 3\zeta - 3\eta + 4\zeta\eta + \zeta^3 + \eta^3 - \zeta^3\eta - \zeta\eta^3)$

10.4. $N_{21} = \frac{1}{8}(2 + 3\zeta - 3\eta - 4\zeta\eta - \zeta^3 + \eta^3 + \zeta^3\eta + \zeta\eta^3)$

10.5. $N_{31} = \frac{1}{8}(2 + 3\zeta + 3\eta + 4\zeta\eta - \zeta^3 - \eta^3 - \zeta^3\eta - \zeta\eta^3)$

10.6. $N_{12} = \frac{1}{4}(1 - \zeta - \eta + \zeta\eta - \eta^2 + \zeta\eta^3 + \eta^3 - \zeta\eta^3)$

10.7. $N_{13} = \frac{3}{8}(-1 + \zeta + \eta + \zeta^2 - \zeta\eta - \zeta^3 - \zeta^2\eta + \zeta^3\eta)$.

10.8. a. $k_{74} = \int_{-1}^{1}\int_{-1}^{1} I_{k74}\, d\zeta\, d\eta$

$$I_{74} = 0.476837\zeta^2(1 - \eta^2) - 0.423855\zeta\eta^2(1 + \zeta)^2$$
$$- 1.507041\eta^2(1 + \zeta)^2 - 0.0353213(4 - 3\zeta^2 - 3\eta^2)^2$$

b. $k_{74} = -3.0088$ kips/in.

c,d. $k_{74} = -3.2349 \text{ kips/in.}$

10.9. a. $k_{42} = \int_{-1}^{1} \int_{-1}^{1} I_{k42} \, d\zeta \, d\eta$

$$I_{42} = 1.271566\zeta(1 - \zeta)(1 - \eta)(1 - 3\eta) - 9.042245\eta(1 - \zeta^2)(1 - 3\eta)$$
$$- 0.635783(1 + 3\eta)(1 - \eta)(4 - 3\zeta^2 - 3\eta^2)$$

b. $k_{42} = 20.7218 \text{ kips/in.}$

c,d. $k_{42} = 18.6873 \text{ kips/in.}$

10.10.

$$[\mathbf{K}] = \begin{bmatrix}
4.506 & 53.651 & -27.579 & -0.488 & 18.687 & -23.058 & -0.784 & 22.078 & -7.460 & -3.235 & 50.260 & -2.939 \\
53.651 & 1255.064 & -162.760 & 18.687 & 481.047 & 0.000 & -22.078 & 313.766 & 0.000 & -50.260 & 554.290 & 0.000 \\
-27.579 & -162.760 & 824.653 & 23.058 & 0.000 & 282.118 & 7.460 & 0.000 & 206.163 & -2.939 & 0.000 & 151.910 \\
-0.488 & 18.687 & 23.058 & 4.506 & 53.651 & 27.579 & -3.235 & 50.260 & 2.939 & -0.784 & 22.078 & 7.460 \\
18.687 & 481.047 & 0.000 & 53.651 & 1255.064 & 162.760 & -50.260 & 554.290 & 0.000 & -22.078 & 313.766 & 0.000 \\
-23.058 & 0.000 & 282.118 & 27.579 & 162.760 & 824.653 & 2.939 & 0.000 & 151.910 & -7.460 & 0.000 & 206.163 \\
-0.784 & -22.078 & 7.460 & -3.235 & -50.260 & 2.939 & 4.506 & -53.651 & 27.579 & -0.488 & -18.687 & 23.058 \\
22.078 & 313.766 & 0.000 & 50.260 & 554.290 & 0.000 & -53.651 & 1255.064 & -162.760 & -18.687 & 481.047 & 0.000 \\
-7.460 & 0.000 & 206.163 & 2.939 & 0.000 & 151.910 & 27.579 & -162.760 & 824.653 & -23.058 & 0.000 & 282.118 \\
-3.235 & -50.620 & -2.939 & -0.784 & -22.078 & -7.460 & -0.488 & -18.687 & -23.058 & 4.506 & -53.651 & -27.579 \\
50.260 & 554.290 & 0.000 & 22.078 & 313.766 & 0.000 & -18.687 & 481.047 & 0.000 & -53.651 & 1255.064 & 162.760 \\
-2.939 & 0.000 & 151.910 & 7.460 & 0.000 & 206.163 & 23.058 & 0.000 & 282.118 & -27.579 & 162.760 & 824.653
\end{bmatrix}$$

$$[\mathbf{K}_{ij}] = \begin{bmatrix}
\text{kips/in.} & \text{kip} & \text{kip} \\
\text{kip} & \text{kip-in.} & \text{kip-in.} \\
\text{kip} & \text{kip-in.} & \text{kip-in.}
\end{bmatrix}$$

10.11.

$$[\mathbf{K}] = \begin{bmatrix}
52.296 & 573.333 & -80.000 & 17.148 & 260.000 & -52.222 & -18.926 & 271.111 & -1.111 & -50.519 & 562.222 & 26.667 \\
573.333 & 9102.222 & -666.667 & 260.000 & 4231.111 & 0.000 & -271.111 & 2275.556 & 0.000 & -562.222 & 4391.111 & 0.000 \\
-80.000 & -666.667 & 2755.556 & 52.222 & 0.000 & 377.778 & 1.111 & 0.000 & 688.889 & 26.667 & 0.000 & -622.222 \\
17.148 & 260.000 & 52.222 & 52.296 & 573.333 & 80.000 & -50.519 & 562.222 & -26.667 & -18.926 & 271.111 & 1.111 \\
260.000 & 4231.111 & 0.000 & 573.333 & 9102.222 & 666.667 & -562.222 & 4391.111 & 0.000 & -271.111 & 2275.556 & 0.000 \\
-52.222 & 0.000 & 377.778 & 80.000 & 666.667 & 2755.556 & -26.667 & 0.000 & -622.222 & -1.111 & 0.000 & 688.889 \\
-18.926 & -271.111 & 1.111 & -50.519 & -562.222 & -26.667 & 52.296 & -573.333 & 80.000 & 17.1458 & -260.000 & 52.222 \\
271.111 & 2275.556 & 0.000 & 562.222 & 4391.111 & 0.000 & -573.333 & 9102.222 & -666.667 & -260.000 & 4231.111 & 0.000 \\
-1.111 & 0.000 & 688.889 & -26.667 & 0.000 & -622.222 & 80.000 & -666.667 & 2755.556 & -52.222 & 0.000 & 377.778 \\
-50.519 & -562.222 & 26.667 & -18.926 & -271.111 & -1.111 & 17.148 & -260.000 & -52.222 & 52.296 & -573.333 & -80.000 \\
562.222 & 4391.111 & 0.000 & 271.111 & 2275.556 & 0.000 & -260.000 & 4231.111 & 0.000 & -573.333 & 9102.222 & 666.667 \\
26.667 & 0.000 & -622.222 & 1.111 & 0.000 & 688.889 & 52.222 & 0.000 & 377.778 & -80.000 & 666.667 & 2755.556
\end{bmatrix}$$

$$[\mathbf{K}_{ij}] = \begin{bmatrix}
\text{kips/in.} & \text{kip} & \text{kip} \\
\text{kip} & \text{kip-in.} & \text{kip-in.} \\
\text{kip} & \text{kip-in.} & \text{kip-in.}
\end{bmatrix}$$

10.17.

$$\{\mathbf{F}_k\} = \begin{Bmatrix} -12.361 \\ -80.000 \\ 90.000 \\ -2.454 \\ -26.667 \\ -30.000 \\ -0.671 \\ 13.333 \\ -15.000 \\ -4.514 \\ 40.000 \\ 45.000 \end{Bmatrix}, \quad \{\mathbf{F}_{ki}\} = \begin{Bmatrix} \text{kips} \\ \text{kip-in.} \\ \text{kip-in.} \end{Bmatrix}$$

10.18.

$$\{\mathbf{F}_k\} = \begin{Bmatrix} -5 \\ -45 \\ 60 \\ -5 \\ -45 \\ -60 \\ -5 \\ 45 \\ -60 \\ -5 \\ 45 \\ 60 \end{Bmatrix}, \quad \{\mathbf{F}_{ki}\} = \begin{Bmatrix} \text{kips} \\ \text{kip-in.} \\ \text{kip-in.} \end{Bmatrix}$$

10.19.

$$\{\mathbf{F}_k\} = \begin{Bmatrix} -4.32 \\ -9.72 \\ 12.96 \\ -4.32 \\ -9.72 \\ -12.96 \\ -4.32 \\ 9.72 \\ 12.96 \\ -4.32 \\ 9.72 \\ -12.96 \end{Bmatrix}, \quad \{\mathbf{F}_{ki}\} = \begin{Bmatrix} \text{kips} \\ \text{kip-in.} \\ \text{kip-in.} \end{Bmatrix}$$

10.20.

$$\mathbf{F}_b\} = \begin{Bmatrix} -2.592 \\ -17.280 \\ 27.648 \\ -6.048 \\ -34.560 \\ -41.472 \\ -6.048 \\ 34.560 \\ -41.472 \\ -2.592 \\ 17.280 \\ 27.648 \end{Bmatrix}, \quad \{\mathbf{F}_{ki}\} = \begin{Bmatrix} \text{kips} \\ \text{kip-in.} \\ \text{kip-in.} \end{Bmatrix}$$

10.21.

$$\{\mathbf{F}_b\} = \begin{Bmatrix} -4.860 \\ -32.400 \\ 64.800 \\ -11.340 \\ -64.800 \\ -97.200 \\ -11.340 \\ 64.800 \\ -97.200 \\ -4.860 \\ 32.400 \\ 64.800 \end{Bmatrix}, \quad \{\mathbf{F}_{ki}\} = \begin{Bmatrix} \text{kips} \\ \text{kip-in.} \\ \text{kip-in.} \end{Bmatrix}$$

10.22.

$$[\mathbf{K}_E] = \begin{bmatrix}
8.770 & 74.135 & -54.891 & 1.026 & 27.530 & -6.824 & -0.298 & 16.924 & -0.587 & -1.722 & 38.085 & -14.729 \\
74.135 & 1383.034 & -297.130 & 27.530 & 545.033 & 89.580 & -16.924 & 265.777 & 59.720 & -38.085 & 458.312 & -89.580 \\
-54.891 & -297.130 & 1052.156 & 6.824 & -89.580 & 111.490 & 0.587 & 59.720 & 120.849 & -14.729 & 89.580 & 265.661 \\
1.026 & 27.530 & 6.824 & 8.770 & 74.135 & 54.891 & -1.722 & 38.085 & 14.729 & -0.298 & 16.924 & 0.587 \\
27.530 & 545.033 & -89.580 & 74.135 & 1383.034 & 297.130 & -38.085 & 458.312 & 89.580 & -16.924 & 265.777 & -59.720 \\
-6.824 & 89.580 & 111.490 & 54.891 & 297.130 & 1052.156 & 14.729 & -89.580 & 265.661 & -0.587 & -59.720 & 120.849 \\
-0.298 & -16.924 & 0.587 & -1.722 & -38.085 & 14.729 & 8.770 & -74.135 & 54.891 & 1.026 & -27.530 & 6.824 \\
16.924 & 265.777 & 59.720 & 38.085 & 458.312 & -89.580 & -74.135 & 1383.034 & -297.130 & -27.530 & 545.033 & 89.580 \\
-0.587 & 59.720 & 120.849 & 14.729 & 89.580 & 265.661 & 54.891 & -297.130 & 1052.156 & -6.824 & -89.580 & 111.490 \\
-1.722 & -38.085 & -14.729 & -0.298 & -16.924 & -0.587 & 1.026 & -27.530 & -6.824 & 8.770 & -74.135 & -54.891 \\
38.085 & 458.312 & 89.580 & 16.924 & 265.777 & -59.720 & -27.530 & 545.033 & -89.580 & -74.135 & 1383.034 & 297.130 \\
-14.729 & -89.580 & 265.661 & 0.587 & -59.720 & 120.849 & 6.824 & 89.580 & 111.490 & -54.891 & 297.130 & 1052.156
\end{bmatrix}$$

$$[\mathbf{K}_{ij}] = \begin{bmatrix} \text{kips/in.} & \text{kip} & \text{kip} \\ \text{kip} & \text{kip-in.} & \text{kip-in.} \\ \text{kip} & \text{kip-in.} & \text{kip-in.} \end{bmatrix}$$

10.23. a. 58.50000, b. 412.20833, c. 534.08333, d. 794.08333

10.26.

$$[\mathbf{K}] = \begin{bmatrix} 55.466 & 25.388 & -68.933 & -19.372 & -15.508 & -38.440 & -36.094 & 42.938 & -97.770 \\ 25.388 & 82.272 & -33.432 & 37.309 & 55.659 & 34.939 & -62.696 & 24.770 & -3.053 \\ -68.933 & -33.432 & 132.089 & 15.784 & -9.649 & 23.083 & 53.148 & -47.432 & 83.195 \\ -19.372 & 37.309 & 15.784 & 107.124 & 167.006 & 121.502 & -87.753 & 78.315 & 135.078 \\ -15.508 & 55.659 & -9.649 & 167.006 & 341.371 & 186.405 & -151.497 & 72.971 & 203.781 \\ -38.440 & 34.939 & 23.083 & 121.502 & 186.405 & 187.782 & -83.061 & 66.280 & 147.458 \\ -36.094 & -62.696 & 53.148 & -87.753 & -151.497 & -83.061 & 123.847 & -121.253 & -37.309 \\ 42.938 & 24.770 & -47.432 & 78.315 & 72.971 & 66.280 & -121.253 & 223.081 & 8.967 \\ -97.770 & -3.053 & 83.195 & 135.078 & 203.781 & 147.458 & -37.309 & 8.967 & 332.813 \end{bmatrix}$$

$$[\mathbf{K}_{ij}] = \begin{bmatrix} \text{kips/in.} & \text{kip} & \text{kip} \\ \text{kip} & \text{kip-in.} & \text{kip-in.} \\ \text{kip} & \text{kip-in.} & \text{kip-in.} \end{bmatrix}$$

10.30. $F_{Az} = +21.74$ kips, $M_{Ax} = 0$, $M_{Ay} = 0$

$F_{Bz} = +4.250$ kips, $M_{Bx} = 0$, $M_{By} = 0$

$F_{Cz} = +29.250$ kips, $M_{Cx} = 0$, $M_{Cy} = 0$

$F_{Dz} = +11.750$ kips, $M_{Dx} = 0$, $M_{Dy} = 0$

$\delta_{Ez} = -1.5731$ in.

10.31. $F_{Az} = +22.150$ kips, $M_{Ax} = +5.177$ kip-ft, $M_{Ay} = -30.926$ kip-ft

$F_{Bz} = +1.975$ kips, $M_{Bx} = -1.165$ kip-ft, $M_{By} = -18.103$ kip-ft

$F_{Cz} = +31.525$ kips, $M_{Cx} = +8.927$ kip-ft, $M_{Cy} = +38.426$ kip-ft

$F_{Dz} = +11.350$ kips, $M_{Dx} = -4.915$ kip-ft, $M_{Dy} = +25.603$ kip-ft

$\delta_{Ez} = -0.4577$ in.

10.32. $F_{Az} = +13.000$ kips, $M_{Ax} = 0$, $M_{Ay} = 0$

$F_{Bz} = +13.000$ kips, $M_{Bx} = 0$, $M_{By} = 0$

$F_{Cz} = +20.500$ kips, $M_{Cx} = 0$, $M_{Cy} = 0$

$F_{Dz} = +20.500$ kips, $M_{Dx} = 0$, $M_{Dy} = 0$

$\delta_{Ez} = -1.3686$ in.

10.33. $F_{Az} = +12.429$ kips, $M_{Ax} = 0$, $M_{Ay} = -20.928$ kip-ft

$F_{Bz} = +12.429$ kips, $M_{Bx} = 0$, $M_{By} = -20.928$ kip-ft

$F_{Cz} = +21.071$ kips, $M_{Cx} = 0$, $M_{Cy} = +25.493$ kip-ft

$F_{Dz} = +21.071$ kips, $M_{Dx} = 0$, $M_{Dy} = +25.493$ kip-ft

$\delta_{Ez} = -0.4868$ in.

10.34 $F_{Az} = +27.090$ kips, $M_{Ax} = 0$, $M_{Ay} = 0$

$F_{Bz} = -1.090$ kips, $M_{Bx} = 0$, $M_{By} = 0$

$F_{Cz} = +23.910$ kips, $M_{Cx} = 0$, $M_{Cy} = 0$

$F_{Dz} = +17.090$ kips, $M_{Dx} = 0$, $M_{Dy} = 0$

$\delta_{Ez} = -1.9014$ in.

10.35 $F_{Az} = +18.651$ kips, $M_{Ax} = +0.963$ kip-ft, $M_{Ay} = -25.817$ kip-ft

$F_{Bz} = +2.321$ kips, $M_{Bx} = -1.180$ kip-ft, $M_{By} = -12.110$ kip-ft

$F_{Cz} = +32.219$ kips, $M_{Cx} = +15.297$ kip-ft, $M_{Cy} = +54.035$ kip-ft

$F_{Dz} = +13.809$ kips, $M_{Dx} = -15.469$ kip-ft, $M_{Dy} = +24.119$ kip-ft

$\delta_{Ez} = -0.5200$ in.

10.36. $F_{Az} = +15.688$ kips, $M_{Ax} = 0$, $M_{Ay} = 0$

$F_{Bz} = +10.312$ kips, $M_{Bx} = 0$, $M_{By} = 0$

$F_{Cz} = +17.812$ kips, $M_{Cx} = 0$, $M_{Cy} = 0$

$F_{Dz} = +23.188$ kips, $M_{Dx} = 0$, $M_{Dy} = 0$

$\delta_{Ez} = -1.5050$ in.

10.37. $F_{Az} = +12.285$ kips, $M_{Ax} = 0$, $M_{Ay} = -20.111$ kip-ft

$F_{Bz} = +12.283$ kips, $M_{Bx} = 0$, $M_{By} = -24.573$ kip-ft

$F_{Cz} = +21.215$ kips, $M_{Cx} = 0$, $M_{Cy} = +31.326$ kip-ft

$F_{Dz} = +21.217$ kips, $M_{Dx} = 0$, $M_{Dy} = +24.814$ kip-ft

$\delta_{Ez} = -0.4752$ in.

CHAPTER 11

11.1. a. $N_1 = 2 - \frac{r}{3}$, $N_2 = 1 + \frac{r}{3} - \frac{z}{4}$, $N_3 = -2 + \frac{z}{4}$

b.

$$[\mathbf{B}_1] = \frac{1}{24}\begin{bmatrix} -8 & 0 & 8 & 0 & 0 & 0 \\ 4 & 0 & 1 & 0 & 1 & 0 \\ 0 & 0 & 0 & -6 & 0 & 6 \\ 0 & -8 & -6 & 8 & 6 & 0 \end{bmatrix} \text{ in.}^{-1}$$

c,d.

$$[\mathbf{K}] = \begin{bmatrix} 531.215 & 0.000 & -547.494 & 113.097 & -19.706 & -113.097 \\ 0.000 & 188.496 & 141.372 & -188.496 & -141.372 & 0.000 \\ -547.494 & 141.372 & 778.401 & -311.018 & -52.265 & 169.646 \\ 113.097 & -188.496 & -311.018 & 506.582 & 113.097 & -318.086 \\ -19.706 & -141.372 & -52.265 & 113.097 & 137.516 & 28.274 \\ -113.097 & 0.000 & 169.646 & -318.086 & 28.274 & 318.086 \end{bmatrix} \text{ kip/in.} \times 1000$$

11.2. a. $N_1 = \frac{1}{11}(24 - 2r - z), \quad N_2 = \frac{1}{22}(-29 + 7r - 2z), \quad N_3 = \frac{1}{22}(3 - 3r + 4z)$

b.

$$[\mathbf{B}_1] = \frac{1}{396}\begin{bmatrix} -72 & 0 & 126 & 0 & -54 & 0 \\ 44 & 0 & 11 & 0 & 11 & 0 \\ 0 & -36 & 0 & -36 & 0 & 72 \\ -36 & -72 & -36 & 126 & 72 & -54 \end{bmatrix} \text{ in.}^{-1}$$

c,d.

$$[\mathbf{K}] = \begin{bmatrix} 622.873 & 194.427 & -995.962 & -59.317 & 300.506 & -135.110 \\ 194.427 & 346.015 & -186.189 & -161.473 & -116.986 & -184.541 \\ -995.962 & -186.189 & 2337.617 & -439.933 & -835.510 & 626.122 \\ -59.317 & -161.473 & -439.933 & 726.631 & 390.502 & -565.157 \\ 300.506 & -116.986 & -835.510 & 390.502 & 505.720 & -273.516 \\ -135.110 & -184.541 & 626.122 & -565.157 & -273.516 & 749.698 \end{bmatrix} \text{ kip/in.} \times 1000$$

11.3.

a.

$$\{\mathbf{F}_k\} = 4.8\pi \begin{Bmatrix} 8 \\ 6 \\ 4 \\ 3 \\ 8 \\ 6 \end{Bmatrix} \text{ kips}$$

b.

$$\mathbf{F}_{b13}\} = \frac{20\pi\sqrt{5}}{3} \begin{Bmatrix} 21 \\ -28 \\ 0 \\ 0 \\ 24 \\ -32 \end{Bmatrix} \text{lb}$$

c.

$$\{\mathbf{F}_{b23}\} = \frac{80\pi\sqrt{5}}{3} \begin{Bmatrix} 0 \\ 0 \\ 9 \\ 0 \\ 4 \\ 0 \end{Bmatrix} \text{lb}$$

11.4.

$$\{\mathbf{F}_b\} = \begin{Bmatrix} 231.878 \\ -133.875 \\ 486.943 \\ -281.137 \\ 301.441 \\ -174.037 \end{Bmatrix} + \begin{Bmatrix} 0 \\ -42.030 \\ 0 \\ -48.497 \\ 0 \\ -45.263 \end{Bmatrix} = \begin{Bmatrix} 231.878 \\ -175.905 \\ 486.943 \\ -329.634 \\ 301.441 \\ -219.300 \end{Bmatrix} \text{lb}$$

11.5.

$$\{\mathbf{F}_b\} = \frac{\pi}{3} \begin{Bmatrix} 570 \\ -760 \\ 690 \\ -920 \\ 0 \\ 0 \end{Bmatrix} + \begin{Bmatrix} 231.878 \\ -175.905 \\ 486.943 \\ -329.634 \\ 301.441 \\ -219.300 \end{Bmatrix} = \begin{Bmatrix} 828.781 \\ -971.775 \\ 1209.509 \\ -1293.056 \\ 301.441 \\ -219.300 \end{Bmatrix} \text{lb}$$

11.6. a. $N_1 = \frac{1}{10}(15 - r - z)$, $\quad N_2 = \frac{1}{40}(-15 + 7r - 3z)$, $\quad N_3 = \frac{1}{40}(-5 - 3r + 7z)$

b. $\int_A N_1 \, dA = \int_A N_2 \, dA = \int_A N_3 \, dA = 6.66667 \text{ in.}^2$

c.
$$\{\mathbf{F}_b\} = \begin{Bmatrix} 153.618 \\ -64.662 \\ 261.524 \\ -85.236 \\ 192.656 \\ -73.480 \end{Bmatrix} \text{ lb}$$

d.
$$\{\mathbf{F}_b\} = 2\pi\sqrt{58} \begin{Bmatrix} 3 \\ 0 \\ 0 \\ 0 \\ 7 \\ 0 \end{Bmatrix} \text{ lb}$$

e.
$$\{\mathbf{F}_b\} = \begin{Bmatrix} 297.171 \\ -64.662 \\ 261.524 \\ -85.236 \\ 527.615 \\ -73.480 \end{Bmatrix} \text{ lb}$$

f,g.

$$[\mathbf{K}] = \begin{bmatrix} 437.024 & 140.743 & -328.858 & -133.204 & 27.991 & -7.540 \\ 140.743 & 382.018 & -145.770 & 47.752 & -145.770 & -429.770 \\ -328.858 & -145.770 & 1204.932 & -288.398 & -394.470 & 434.168 \\ -133.204 & 47.752 & -288.398 & 453.646 & 308.504 & -501.398 \\ 27.991 & -145.770 & -394.470 & 308.504 & 501.035 & -162.734 \\ -7.540 & -429.770 & 434.168 & -501.398 & -162.734 & 931.168 \end{bmatrix} \text{ kip/in.} \times 1000$$

11.7. $\delta_{Br} = 0.018832$ in. $\times 10^{-3} \rightarrow,$ $\delta_{Bz} = 0.016879$ in. $\times 10^{-3} \downarrow$

$C_r = 11.364$ kips $\leftarrow,$ $C_z = 4.078$ kips $\downarrow$

$D_r = 70.106$ kips $\leftarrow,$ $D_z = 125.644$ kips $\uparrow$

11.8. $\delta_{Br} = 0.021721$ in. $\times 10^{-3} \rightarrow$, $\qquad \delta_{Bz} = 0.034577$ in. $\times 10^{-3} \downarrow$

$\qquad C_r = 2.219$ kips $\leftarrow$, $\qquad C_z = 0.035$ kips $\downarrow$

$\qquad D_r = 61.484$ kips $\leftarrow$, $\qquad D_z = 121.530$ kips $\uparrow$

11.9. a. $N_1(r,\ z) = \frac{1}{50}(150 - 25r - 25z + r^2 + 2rz + z^2)$

$\qquad N_2(r,\ z) = \frac{1}{800}(525 - 350r + 150z + 49r^2 - 42rz + 9z^2)$

$\qquad N_3(r,\ z) = \frac{1}{800}(125 + 90r - 210z + 9r^2 - 42rz + 49z^2)$

$\qquad N_4(r,\ z) = \frac{1}{100}(-225 + 120r - 30z - 7r^2 - 4rz + 3z^2)$

$\qquad N_5(r,\ z) = \frac{1}{400}(75 + 10r - 90z - 21r^2 + 58rz - 21z^2)$

$\qquad N_6(r,\ z) = \frac{1}{100}(-75 - 40r + 110z + 3r^2 - 4rz - 7z^2)$

b. $\int_A N_1\, dA = \int_A N_2\, dA = \int_A N_3\, dA = 0$

$\qquad \int_A N_4\, dA = \int_A N_5\, dA = \int_A N_6\, dA = 6.66667\, \text{in}^2.$

c.
$$\{F_b\} = \begin{Bmatrix} -32.654 \\ 6.532 \\ 39.283 \\ -7.185 \\ -6.629 \\ 0.653 \\ 209.228 \\ -75.112 \\ 235.254 \\ -80.991 \\ 163.316 \\ -67.275 \end{Bmatrix} \text{lb}$$

d.

$$\{\mathbf{F}_{b163}\} = \frac{2\pi}{5} \begin{Bmatrix} 7 \\ -3 \\ 0 \\ 0 \\ -133 \\ 57 \\ 0 \\ 0 \\ 0 \\ 0 \\ -224 \\ 96 \end{Bmatrix} \text{ lb}$$

e.

$$\{\mathbf{F}_{b\text{tot}}\} = \begin{Bmatrix} -23.858 \\ 2.762 \\ 39.283 \\ -7.185 \\ -173.761 \\ 72.281 \\ 209.228 \\ -75.112 \\ 235.254 \\ -80.991 \\ -118.170 \\ 53.363 \end{Bmatrix} \text{ lb}$$

11.10. a.

$$[\mathbf{B}_1] = \begin{bmatrix} -0.17833 & 0 & 0.17833 & 0 & 0.04778 & 0 & -0.04778 & 0 \\ 0.15809 & 0 & 0.04236 & 0 & 0.01135 & 0 & 0.04236 & 0 \\ 0 & -0.23777 & 0 & -0.09556 & 0 & 0.06371 & 0 & 0.26962 \\ -0.23777 & -0.17833 & -0.09556 & 0.17833 & 0.06371 & 0.04778 & 0.26962 & -0.04778 \end{bmatrix} \text{ in}^{-1}.$$

b.

$$[\mathbf{K}_1] = \begin{bmatrix} 167.803 & 46.459 & -31.442 & -39.820 & -29.322 & -12.449 & -33.028 & 5.810 \\ 46.459 & 198.185 & -34.864 & 35.786 & -25.015 & -53.104 & -46.044 & -180.867 \\ -31.442 & -34.864 & 123.027 & -37.522 & 24.566 & 9.342 & -39.771 & 63.045 \\ -39.820 & 35.786 & -37.522 & 58.253 & 5.619 & -9.589 & 47.823 & -84.450 \\ -29.322 & -25.015 & 24.566 & 5.619 & 12.182 & 6.703 & 13.041 & 12.693 \\ -12.449 & -53.104 & 9.342 & -9.589 & 6.703 & 14.229 & 12.337 & 48.463 \\ -33.028 & -46.044 & -39.771 & 47.823 & 13.041 & 12.337 & 79.589 & -14.116 \\ 5.810 & -180.867 & 63.045 & -84.450 & 12.693 & 48.463 & -14.116 & 216.854 \end{bmatrix} \quad \text{kip/in.} \times 1000$$

c.

$$[\mathbf{K}] = \begin{bmatrix} 422.429 & 54.341 & -162.083 & 58.756 & -152.257 & -148.079 & -39.853 & 34.982 \\ 54.341 & 537.807 & 83.889 & 115.645 & -198.345 & -177.628 & -90.682 & -475.824 \\ -162.083 & 83.889 & 803.439 & -372.915 & -81.690 & 9.849 & -311.431 & 279.177 \\ 58.756 & 115.645 & -372.915 & 1316.921 & -115.814 & -777.416 & 203.779 & -655.150 \\ -152.257 & -198.345 & -81.690 & -115.814 & 553.694 & 236.384 & -61.576 & 77.776 \\ -148.079 & -177.628 & 9.849 & -777.416 & 236.384 & 625.092 & 52.643 & 329.952 \\ -39.853 & -90.682 & -311.431 & 203.779 & -61.576 & 52.643 & 498.865 & -165.740 \\ 34.982 & -475.824 & 279.177 & -655.150 & 77.776 & 329.952 & -165.740 & 801.021 \end{bmatrix} \quad \text{kip/in.} \times 1000$$

11.11. a.

$$[\mathbf{B}_1] = \begin{bmatrix} -0.10570 & 0 & 0.13921 & 0 & 0.02832 & 0 & -0.06183 & 0 \\ 0.17651 & 0 & 0.04730 & 0 & 0.01267 & 0 & 0.04730 & 0 \\ 0 & -0.13213 & 0 & -0.07599 & 0 & 0.03540 & 0 & 0.17271 \\ -0.13213 & -0.10570 & -0.07599 & 0.13921 & 0.03540 & 0.02832 & 0.17271 & -0.06183 \end{bmatrix} \quad \text{in}^{-1}.$$

b.

$$[\mathbf{K}_1] = \begin{bmatrix} 212.378 & 9.140 & 20.843 & -47.131 & -6.520 & -2.449 & 11.732 & 40.440 \\ 9.140 & 125.977 & -32.928 & 30.543 & -18.157 & -33.755 & -32.385 & -122.764 \\ 20.843 & -32.928 & 166.109 & -49.068 & 27.833 & 8.823 & -56.650 & 73.173 \\ -47.131 & 30.543 & -49.068 & 72.761 & 3.595 & -8.184 & 49.854 & -95.120 \\ -6.520 & -18.157 & 27.833 & 3.595 & 9.634 & 4.865 & 6.374 & 9.697 \\ -2.449 & -33.755 & 8.823 & -8.184 & 4.865 & 9.045 & 8.678 & 32.895 \\ 11.732 & -32.385 & -56.650 & 49.854 & 6.374 & 8.678 & 83.582 & -26.146 \\ 40.440 & -122.764 & 73.173 & -95.120 & 9.697 & 32.895 & -26.146 & 184.990 \end{bmatrix} \quad \text{kip/in.} \times 1000$$

c.

$$[\mathbf{K}] = \begin{bmatrix} 475.584 & 4.362 & -196.718 & 32.063 & -71.954 & -138.967 & 62.003 & 102.542 \\ 4.362 & 399.093 & 69.762 & 82.519 & -201.798 & -124.900 & -60.821 & -356.713 \\ -196.718 & 69.762 & 1102.116 & -527.311 & 29.886 & 99.386 & -427.688 & 358.164 \\ 32.063 & 82.519 & -527.311 & 1233.687 & -63.977 & -684.512 & 257.633 & -631.695 \\ -71.954 & -201.798 & 29.886 & -63.977 & 627.272 & 205.365 & -162.281 & 60.410 \\ -138.967 & -124.900 & 99.386 & -684.512 & 205.365 & 595.860 & 22.711 & 213.551 \\ 62.003 & -60.821 & -427.688 & 257.633 & -162.281 & 22.711 & 701.815 & -219.523 \\ 102.542 & -356.713 & 358.164 & -631.695 & 60.410 & 213.551 & -219.523 & 774.856 \end{bmatrix} \quad \text{kip/in.} \times 1000$$

11.12.

$$\{F_b\} = 50\pi \begin{Bmatrix} 0 \\ 0 \\ 0 \\ 0 \\ 0 \\ -1 \\ 0 \\ -2 \end{Bmatrix} + 30\pi \begin{Bmatrix} 1 \\ 0 \\ 0 \\ 0 \\ 0 \\ 0 \\ 2 \\ 0 \end{Bmatrix} + \begin{Bmatrix} 90.102 \\ -29.089 \\ 158.855 \\ -39.088 \\ 209.162 \\ -47.875 \\ 114.223 \\ -34.847 \end{Bmatrix} = \begin{Bmatrix} 184.349 \\ -29.089 \\ 158.855 \\ -39.088 \\ 209.162 \\ -204.955 \\ 302.719 \\ -349.005 \end{Bmatrix} \text{lb}$$

11.13.

$$\{F_b\} = \pi \begin{Bmatrix} 0 \\ 0 \\ 0 \\ 0 \\ 23 \\ -161 \\ 44 \\ -308 \end{Bmatrix} + \frac{8\pi}{3} \begin{Bmatrix} 0 \\ 0 \\ -85 \\ 34 \\ -45 \\ 18 \\ 0 \\ 0 \end{Bmatrix} + \begin{Bmatrix} 124.385 \\ -61.283 \\ 253.267 \\ -89.438 \\ 313.279 \\ -102.392 \\ 157.203 \\ -71.106 \end{Bmatrix} = \begin{Bmatrix} 124.385 \\ -61.283 \\ -458.828 \\ 195.400 \\ 8.545 \\ -457.392 \\ 295.433 \\ -1038.717 \end{Bmatrix} \text{lb}$$

11.14. a. (1) $A = 25.00 \text{ in.}^2$ (2×2 and 3×3 rules)

(2) $V = 875.46 \text{ in.}^3$ (2×2 and 3×3 rules)

(3) Int $= 27,143 \text{ in.}^4$ (2×2 rule)
Int $= 27,235 \text{ in.}^4$ (3×3 rule)

b.

$$\{\mathbf{F}_b\} = \begin{Bmatrix} -50.055 \\ 26.474 \\ -33.508 \\ 19.774 \\ -14.426 \\ 13.837 \\ -50.608 \\ 24.353 \\ 130.069 \\ -76.381 \\ 201.327 \\ -101.654 \\ 167.127 \\ -92.496 \\ 95.869 \\ -67.223 \end{Bmatrix} \text{lb} (2 \times 2 \text{ rule}), \quad \{\mathbf{F}_b\} = \begin{Bmatrix} -45.558 \\ 25.358 \\ -27.394 \\ 18.229 \\ -4.490 \\ 11.284 \\ -45.686 \\ 23.255 \\ 123.171 \\ -75.574 \\ 196.135 \\ -99.346 \\ 158.329 \\ -90.959 \\ 92.792 \\ -65.571 \end{Bmatrix} \text{lb} (3 \times 3 \text{ rule})$$

c.	Two-Point Rule	Three-Point Rule	Four-Point Rule
	$F_{b3} = 438.883\,\text{lb}$	$F_{b3} = 464.257\,\text{lb}$	$F_{b3} = 464.347\,\text{lb}$
	$F_{b5} = 820.329\,\text{lb}$	$F_{b5} = 847.451\,\text{lb}$	$F_{b5} = 847.568\,\text{lb}$
	$F_{b11} = 2518.423\,\text{lb}$	$F_{b11} = 2466.763\,\text{lb}$	$F_{b11} = 2466.541\,\text{lb}$

d.	Two-Point Rule	Three-Point Rule	Four-Point Rule
	$F_{b3} = 669.943\,\text{lb}$	$F_{b3} = 670.101\,\text{lb}$	$F_{b3} = 670.076\,\text{lb}$
	$F_{b5} = 84.766\,\text{lb}$	$F_{b5} = 51.093\,\text{lb}$	$F_{b5} = 50.947\,\text{lb}$
	$F_{b11} = 1509.417\,\text{lb}$	$F_{b11} = 1542.324\,\text{lb}$	$F_{b11} = 1542.466\,\text{lb}$

INDEX